Grzimek's ANIMAL LIFE ENCYCLOPEDIA

Grzimek's
ANIMAL LIFE
ENCYCLOPEDIA

Editor-in-Chief

Dr. Dr. h.c. Bernhard Grzimek

Professor, Justus Liebig University of Giessen
Director, Frankfurt Zoological Garden, Germany
Trustee, Tanzania and Uganda National Parks, East Africa

 VAN NOSTRAND REINHOLD COMPANY

New York   Cincinnati   Toronto   London   Melbourne

First published in paperback in 1984

Copyright © 1968 Kindler Verlag A.G. Zurich

Library of Congress Catalog Card Number 79-183178

ISBN 0-442-23048-6

Printed in Federal Republic of Germany

Van Nostrand Reinhold Company Inc.
135 West 50th Street
New York, New York 10020

Van Nostrand Reinhold Company Limited
Molly Millars Lane
Wokingham, Berkshire RG11 2PY, England

Van Nostrand Reinhold
480 Latrobe Street
Melbourne, Victoria 3000, Australia

Macmillan of Canada
Division of Gage Publishing Limited
164 Commander Boulevard
Agincourt, Ontario M1S 3C7 Canada

16 15 14 13 12 11 10 9 8 7 6 5 4 3 2 1

EDITORS AND CONTRIBUTORS

Editor-in-Chief
DR. DR. H.C. BERNHARD GRZIMEK
Professor, Justus Liebig University of Giessen, Germany
Director, Frankfurt Zoological Garden, Germany
Trustee, Tanzania and Uganda National Parks, East Africa

DR. JACQUES GERY — ST. GENIES, FRANCE

DR. WOLFGANG GEWALT
Director, Animal Park — DUISBURG, GERMANY

DR. DR. H.C. DR. H.C. VIKTOR GOERTTLER
Professor Emeritus, University of Jena — JENA, GERMANY

DR. FRIEDRICH GOETHE
Director, Institute of Ornithology, Heligoland Ornithological Station — WILHELMSHAVEN, GERMANY

DR. ULRICH F. GRUBER
Herpetological Section, Zoological Research Institute and A. Koenig Museum — BONN, GERMANY

DR. H. R. HAEFELFINGER
Museum of Natural History — BASEL, SWITZERLAND

DR. THEODOR HALTENORTH
Director, Mammalogy, Zoological Collection of the State of Bavaria — MUNICH, GERMANY

BARBARA HARRISSON
Sarawak Museum, Kuching, Borneo — ITHACA, NEW YORK, U.S.A.

DR. FRANCOIS HAVERSCHMIDT
President, High Court (retired) — PARAMARIBO, SURINAM

DR. HEINZ HECK
Director, Catskill Game Farm — CATSKILL, NEW YORK, U.S.A.

DR. LUTZ HECK
Professor (retired), and Director, Zoological Garden, Berlin — WIESBADEN, GERMANY

DR. DR. H.C. HEINI HEDIGER
Director, Zoological Garden — ZURICH, SWITZERLAND

DR. DIETRICH HEINEMANN
Director, Zoological Garden, Münster — DÖRNIGHEIM, GERMANY

DR. HELMUT HEMMER
Institute for Physiological Zoology, University of Mainz — MAINZ, GERMANY

DR. W. G. HEPTNER
Professor, Zoological Museum, University of Moscow — MOSCOW, U.S.S.R.

DR. KONRAD HERTER
Professor Emeritus and Director (retired), Zoological Institute, Free University of Berlin — BERLIN, GERMANY

DR. HANS RUDOLF HEUSSER
Zoological Museum, University of Zurich — ZURICH, SWITZERLAND

DR. EMIL OTTO HÖHN
Associate Professor of Physiology, University of Alberta — EDMONTON, CANADA

DR. W. HOHORST
Professor and Director, Parasitological Institute, Farbwerke Hoechst A.G. — FRANKFURT-HÖCHST, GERMANY

DR. FOLKHART HÜCKINGHAUS
Director, Senckenbergische Anatomy, University of Frankfurt a.M. — FRANKFURT A.M., GERMANY

FRANCOIS HÜE
National Museum of Natural History — PARIS, FRANCE

DR. K. IMMELMANN
Professor, Zoological Institute, Technical University of Braunschweig — BRAUNSCHWEIG, GERMANY

DR. JUNICHIRO ITANI
Kyoto University — KYOTO, JAPAN

DR. RICHARD F. JOHNSTON
Professor of Zoology, University of Kansas — LAWRENCE, KANSAS, U.S.A.

OTTO JOST
Oberstudienrat, Freiherr-vom-Stein Gymnasium — FULDA, GERMANY

DR. PAUL KÄHSBAUER
Curator, Fishes, Museum of Natural History — VIENNA, AUSTRIA

DR. LUDWIG KARBE
Zoological State Institute and Museum — HAMBURG, GERMANY

DR. N. N. KARTASCHEW
Docent, Department of Biology, Lomonossow State University — MOSCOW, U.S.S.R.

DR. WERNER KÄSTLE
Oberstudienrat, Gisela Gymnasium — MUNICH, GERMANY

DR. REINHARD KAUFMANN
Field Station of the Tropical Institute, Justus Liebig University, Giessen, Germany — SANTA MARTA, COLOMBIA

DR. HELMUT O. WAGNER
Director (retired), Overseas Museum, Bremen   MEXICO CITY, MEXICO

DR. FRITZ WALTHER
Professor, Texas A & M University   COLLEGE STATION, TEXAS, U.S.A.

JOHN WARHAM
Zoology Department, Canterbury University   CHRISTCHURCH, NEW ZEALAND

DR. SHERWOOD L. WASHBURN
University of California at Berkeley   BERKELEY, CALIFORNIA, U.S.A.

EBERHARD WAWRA
First Zoological Institute, University of Vienna   VIENNA, AUSTRIA

DR. INGRID WEIGEL
Zoological Collection of the State of Bavaria   MUNICH, GERMANY

DR. B. WEISCHER
Institute of Nematode Research, Federal   MÜNSTER/WESTFALEN,
Biological Institute   GERMANY

HERBERT WENDT
Author, Natural History   BADEN-BADEN, GERMANY

DR. HEINZ WERMUTH
Chief Curator, State Nature Museum, Stuttgart   LUDWIGSBURG, GERMANY

DR. WOLFGANG VON WESTERNHAGEN   PREETZ/HOLSTEIN, GERMANY

DR. ALEXANDER WETMORE
United States National Museum, Smithsonian Institution   WASHINGTON, D.C., U.S.A.

DR. DIETRICH E. WILCKE   RÖTTGEN, GERMANY

DR. HELMUT WILKENS
Professor and Director, Institute of Anatomy, School of Veterinary Medicine   HANNOVER, GERMANY

DR. MICHAEL L. WOLFE
Utah State University   UTAH, U.S.A.

HANS EDMUND WOLTERS
Zoological Research Institute and A. Koenig Museum   BONN, GERMANY

DR. ARNFRID WÜNSCHMANN
Research Associate, Zoological Garden   BERLIN, GERMANY

DR. WALTER WÜST
Instructor, Wilhelms Gymnasium   MUNICH, GERMANY

DR. HEINZ WUNDT
Zoological Collection of the State of Bavaria   MUNICH, GERMANY

DR. CLAUS-DIETER ZANDER
Zoological Institute and Museum, University of Hamburg   HAMBURG, GERMANY

DR. DR. FRITZ ZUMPT
Director, Entomology and Parasitology, South African Institute for Medical Research   JOHANNESBURG, SOUTH AFRICA

DR. RICHARD L. ZUSI
Curator of Birds, United States National Museum, Smithsonian Institution   WASHINGTON, D.C., U.S.A.

Volume 13

MAMMALS IV

Edited by:

A. G. BANNIKOW
BERNHARD GRZIMEK
LUTZ HECK
DIETRICH HEINEMANN
W. G. HEPTNER
HEINZ-GEORG KLÖS
ERNST M. LANG
ERICH THENIUS
FRITZ WALTHER

ENGLISH EDITION

GENERAL EDITOR:
George M. Narita

PRODUCTION DIRECTOR:
James V. Leone

ART DIRECTOR:
Lorraine K. Hohman

SPECIAL CONSULTANT:
Marvin L. Jones

SCIENTIFIC EDITOR:
Erich Klinghammer

TRANSLATORS:
Lisel Walther
Erich Klinghammer

ASSISTANT EDITOR:
Frances A. Wilke

EDITORIAL ASSISTANTS:
Detlef K. Onderka
Karen Boikess

INDEX:
Suzanne C. Klinghammer

CONTENTS

For a more complete listing
of animal names, see systematic classification or the index.

1 Tapirs

Suborder:
Ceratomorpha—
rhinoceros relatives

Among the odd-toed ungulates (Perissodactyla), the tapirs and rhinoceros form a suborder of the Ceratomorpha, which are related to the rhinoceros. The other suborder of the genus *Perissodactyla,* the Hippomorpha, which are related to horses, are discussed in Volume XII.

Long ago, in the Tertiary Period, the rhinoceros-related species were a large group with many diverging forms. Today we find only a few survivors: the family of tapirs (Tapiridae) and the rhinoceros (Rhinocerotidae, Ch. 2). The various prehistorical tapir forms cannot be taken into one family and, therefore, are considered to be the members of the superfamily of the Tapiroidea. Today there exists only one family, consisting of one genus and four species.

Family: tapirs

Distinguishing
characteristics

TAPIRS (family Tapiridae, genus *Tapirus*): HRL 180–250 cm, TL 5–10 cm, BH 75–120 cm, weight 225 to more than 300 kg. The ♂♂ in most cases are somewhat smaller than ♀♀. The head, on a medium sized neck, is relatively short, with rather flat sides, and it is slightly arched upward, with a short, movable trunk. The eyes are small, and the ears are short and round. The rump is bulky, slightly flat at the sides, with a higher back arching toward the shoulder. The legs are short and graceful. The first digit is missing; the third digit is extremely strong, mainly carrying the weight; the second and fourth digits are weaker, and the fifth digit is very short, touching the ground only on a soft surface. The hind feet are three-toed, with the third digit the strongest and longest. The hoofs, with callous sole balls, are relatively stronger than in rhinoceros. There is a bald callosity at the forearm that corresponds to horses' "chestnuts." The small, fat tail is not very conspicuous from a distance. The 42–44 teeth are arranged as follows: $\frac{3 \cdot 1 \cdot 4 \cdot 3}{3 \cdot 1 \cdot 3\text{-}4 \cdot 3}$. The outer upper incisor is rather long, the canine tooth is conical, and the molars are flat with ridges and protuberances across. The gestation period is between 390 and 400 days; only one young is born.

With good reason, the tapirs have been called "living fossils," survivors from past eras. Not only is their physical structure rather primitive, but their presently widely separated areas of distribution indicate that they originated from a group of animals from an ancient geological era. In the past, the Tapiridae were much more widespread, with a remarkable variety of species, than today. Due to numerous discoveries of fossils, one could not only trace their history, but also one could prove the earlier existence of several tribes which have long since become extinct. The Isectolophidae belong to the geologically oldest tapirs from the Eocene of North America and Asia. These very primitive odd-toed ungulates, who lived approximately fifty million years ago, are scarcely different from the common primitive forms of all the rhinoceros-relatives. In addition, the Lophiodontidae, some of which were the size of a rhinoceros of the European old Tertiary Age, and the widely distributed Helaletidae of the same era of North America and Asia, are not too different from the original type odd-toed ungulates. All those early Tapiridae lack the movable extension of the upper lip and the nose which is so significant for the actual tapirs. This can be demonstrated by looking at the skull; all Proboscideans have an especially large nasal cavity which reaches far to the back. The significance of the nasal cavity in the skull is indicated only in the oldest forms of genuine Tapiridae of the genus *Protapirus*. They lived in Europe in the Oligocene, approximately forty to twenty-five million years ago. Similar forms, which may well belong to the same genus, have been found in North America. These genuine Tapiridae probably were derived either from the Helaletidae or perhaps from the *Heptodon* in the older Eocene of North America.

The modifications in the phylogeny of the tapir affected mainly the skull. The set of teeth and the limbs have been transformed only very insignificantly. As early as the Oligocene or—as G. H. R. von Koenigswald suggests—in the Miocene, the genuine tapirs must have split into two branches whose derivatives are still living currently in America and Asia. The tapirs reached South America not sooner than the extensive migration in the beginning of the Pleistocene, approximately a million years ago, when the Central American landbridge began to form. At the same time they became extinct in Europe, giant forms like the megatapirs developed temporarily in Asia. Actually, there are only three species of tapirs left in tropical America and another one in the area of East India and Malaya.

1. LOWLAND TAPIR (*Tapirus terrestris*; Color plate, p. 21 and p. 22): this rather small species is the most common. The HRL is approximately 200 cm, and the BH is about 100 cm. The back of the neck is distinct, starting in front of the ears. The short, slick coat is much softer and longer in young animals than in adults. There are two subspecies.

2. CENTRAL AMERICAN TAPIR (⚥ *Tapirus bairdi*; Color plate, p. 22): This is the largest species of New World tapir; the BH is up to 120 cm, and

Phylogeny
by Erich Thenius

Fig. 1-1.
1. Lowland Tapir (*Tapirus terrestris*). 2. Central American Tapir (*Tapirus bairdi*).

Fig. 1-2.
Mountain Tapir (*Tapirus pinchaque*).

Present-day tapirs
by H. Frädrich

the weight is over 300 kg. The neck ridge is barely visible and has a small brush mane.

3. MOUNTAIN TAPIR (◊ *Tapirus pinchaque*; Color plate, p. 22): this is the smallest and most graceful species. The HRL is approximately 180 cm; the BH, 75–80 cm; and the weight, approximately 225–250 kg. The coat, in contrast to other tapirs, is not short, smooth, and slick, but strikingly soft and woolly, especially at the underside. The individual hairs are sometimes several centimeters long. On the croup there is a nearly bald patch, divided by a ridge of skin. This mountain dweller lives in altitudes of 2000–4000 meters.

4. MALAYAN TAPIR (*Tapirus indicus*; Color plate, p. 22): this, the only Old World species, is larger than the American tapirs. The HRL is 250 cm. The trunk is longer and stronger; the feet are more robust. There is no distinct neck ridge or mane. The short, smooth and slick coat is strikingly colored with the rear half above the legs being white. Although the Malayan tapir of Asia had already been mentioned in old Chinese text and school books, the Europeans, strangely enough, learned about him much later than about his New World relatives. The Central American tapir, already known to the Mayans, probably served them as a model for a human figure with a trunk as is found on several of their temples. The first discoverers of America, Columbus, Pinzón, and Cabral, probably had not yet encountered the tapirs. However, toward the end of the year 1500, Pietro Martyr described an animal "of the size of an ox," of the color of cattle, which has "an elephant's trunk and hooves like a horse," but which, after all, is neither cattle, nor elephant, nor horse. The later explorers of America soon became familiar with the lowland tapir, although initially they did not know to which group of animals he actually belonged. However, in regard to the Malayan tapir, even 300 years later, there were only rumors that a "two-colored" rhinoceros lived on the island of Sumatra. At the beginning of the nineteenth century, travellers brought back to Europe accurate information on the Asian tapir, precisely at the time when the great French zoologist Georges Cuvier had declared emphatically that there was little hope in the future for the discovery of any more species of large mammals.

It was even longer before the zoologists recognized that tapirs belonged to the odd-toed ungulates. Explorers of earlier centuries had classified them partly as relatives of the hippopotamus. More often they were compared to giant pigs, probably because of their barrel shape and the trunk. "I think that this is a rather unfortunate comparison," writes Hans Krieg, a zoologist with a good knowledge of South America, "because, whenever I saw a tapir, it reminded me of an animal similar to a horse or a donkey. The movements as well as the shape of the animal, especially the high neck with the small brush mane, even the expression on the face is much more like a horse's than

Distinguishing characteristics

The discovery of the tapirs

Fig. 1-3.
Malayan Tapir (*Tapirus indicus*).

a pig's. When watching a tapir on the alert, turning the white-rimmed ears and the moveable trunk, as he picks himself up when recognizing danger, taking off in a gallop, almost nothing remains of the similarity to a pig." Today it is common knowledge that the tapir, the rhinoceros, and the horse are related. The tapir is the geologically earlier form, much closer to the general basic pattern of mammals than the rhinoceros and the horses. "Many details prove this," says Adolf Portmann, "for example, the teeth are less specialized in tapirs, and the limbs have several digits and hooves, while in horses only one digit with the hoof has evolved. The height of the horse is another such characteristic; the tapir is lower and more plump. The latest findings in brain research confirm these differences: The cortical index (which measures how many times heavier the cortex is when compared to the lowest remnant of the cerebellum known in animals of this size) in the tapir is 12.6; the zebra is 29.5, which is about the highest in ungulates. This also corresponds with the fact that tapirs are olfactory animals whose noses are extended into their trunks, much more so than in horses. These contrasts become apparent also in the shape and coloring. In the tapir, the head is not emphasized, hardly distinguished in color. In the horse the neck is raised; the head is carried above the back and is further enhanced by the mane on the neck. In the zebra species this is achieved by special markings, which especially emphasize the structures of the head and the sense organs."

At first glance, the tapirs' movements also are not similar to those of their relatives, the rhinoceros and the horses. In a slow walk, they usually keep the head lowered. In a trot, they lift their heads and move their legs in an elastic manner. The amazingly fast gallop is seen only when the animals are in flight, playing, or when they are extremely excited. The tapirs can also climb quite well, even though one would not expect this because of their bulky figure. Even steep slopes do not present obstacles. They jump vertical fences or walls, rising on their hindlegs and leaping up. Some zoo people have found that they are able to squeeze themselves through unbelievably narrow gaps or between bars, or that they "sneak" out under lower bars with their backs arched. These abilities certainly are of advantage in the wild when they wander through jungles of bamboo and reed.

Getting up and lying down for the tapir is similar to the procedure for the rhinoceros. These two movements are phylogenetically very instructive. According to E. Zannier-Tanner, the tapirs do not support themselves on their wrists, as do horses, but in the process of lying down they briefly assume a sitting position, which is characteristic of less advanced ungulates. Here, then, we see that they are the most primitive contemporary of the odd-toed ungulates.

The tapirs prefer to stay in the vicinity of water. They are excellent swimmers and cross even wide streams without great effort. In order

Like this young lowland tapir (*Tapirus terrestris*), all young tapirs have a horizontally banded and spotted first coat.

to feed on aquatic plants or to escape when pursued, they also are able to dive quite well. The Malayan tapir—similar to the hippopotamus—is said to be able to walk on the bottom of streams.

All species of tapir, except for the mountain tapir, have a rather coarse leathery skin. For this reason they are frequently hunted in their native countries. The people tan the skins and cut them into long straps for reins and whips. The outer layer of the skin is rather delicate. Since in flight the lowland tapirs dash recklessly through the brush, they are usually covered with scars, cuts, and bruises. Hans Krieg believes that "this does not bother them because the fibrous tissue layer which yields the leather after tanning is very thick. There is no doubt that their skin is able to quickly repair slight damages with new growth of skin." In order to remove dandruff, hair, or insects, the tapirs scrub themselves on objects, or scratch their chest and front legs with their hind feet. While lying down and resting, they sometimes lick their front legs. They also protect the surface of their body against insect bites by rolling in the mud; the dry mud remains as a thin protective film on the skin.

Except for the mountain tapir, the tapirs are short-haired; only on the back of the neck is longer hair found, frequently in the form of a short mane. Newborn tapirs always have horizontal stripes and dots of a yellow-white color (Color plate, p. 21). According to the American zoologist Frederic A. Ulmer they look "like banded watermelons with legs." Young lowland tapirs, for example, have on each side of their body four rows of bright stripes and dots; the legs are striped horizontally. In the other species, the young are marked with similar patterns of stripes and dots. This striking coat of the young, which strongly reminds one of the color pattern in young wild boar piglets, does not occur in any other group of odd-toed ungulates and may well be considered further evidence for the fact that the tapirs are phylogenetically a less evolved family. After the first year the stripes have usually disappeared.

The tapir's food

The tapirs feed mainly on leaves, fresh sprouts, and small branches, tearing them from bushes and low trees. In addition, they eat all varieties of fruits, grasses, and swamp and aquatic plants. Wolfgang von Richter points out that those odd-toed ungulates which predominantly feed on leaves have the largest and most moveable upper lips. In tapirs, the extension of the upper lip is especially well developed, having become a trunk with a button-shaped ridge underneath the tip. Not only is the trunk easily moved in all directions, it also may be stretched out and contracted with special muscles. When tapirs are not able to reach a favorite food because it is too high, they will rise on their hind legs, planting their front feet firmly against fences or walls. Studying the Malayan tapirs of the Nürnberg Zoo, Wolfgang von Richter discussed how the trunk is used in feeding: "When a tapir finds a branch

Tapirs

1. Central American Tapir (Tapirus bairdi)
2. Lowland Tapir (Tapirus terrestris)
3. Mountain Tapir (Tapirus pinchaque)
4. Malayan Tapir (Tapirus indicus).

or leaves within his reach, he extends his trunk until it meets the branch. With the aid of this ridge, the branch is pressed against the lower side of the trunk, thus being surrounded by the trunk from all sides; then the trunk is contracted. Thin branches or leaves then will come off. Thick branches are brought to the mouth in the same manner and then are torn off. In grazing, the tapirs bite off grass with their incisors, like horses do, keeping their trunk contracted. However, the trunk is not only a plucking tool; the animals frequently smell and touch their food with it before they eat. Even though the tapirs in the wild probably are genuine vegetarians, they may well take other food in captivity." Professor H. Hediger, the director of the Zurich Zoo, reports that the tapirs there occasionally liked to eat fish.

Being genuine forest animals, the tapirs have a relatively good sense of smell; their eyesight, on the other hand, is very poor. The trunk, whose tip is equipped with hair-like tactile bristles, plays an important role as a tactile organ in exploring a new area. "When moving about in the pen or in the barn, the tapir constantly extends and contracts his trunk," according to Wolfgang von Richter. "In this manner, he not only tests the ground, the walls, and all strange objects, but also his pen fellows and other animals. When brought to a strange environment, the activity of the trunk will increase immediately."

Sensory performance and habits

All species of tapir are similar in their habits. They are unsocial, cautious creatures of the forest, who avoid open territory and depend on the vicinity of water. In denser populated areas of human settlements, the South American lowland tapirs are considered to be strictly nocturnal animals, according to Hans Krieg. One hardly ever sees them unless they are routed out of their hiding places by dogs. "In places with few people," Krieg continues, "it is possible to see tapirs at any time, except for the hottest time around noon; most likely, however, one can see them in the morning and in the evening. But these encounters were not at all commonplace, even though, with some knowledge of the area, one could expect to see them at certain places at a certain time."

Tapirs remain fairly close to one locality in their habitat and they adhere to regular paths. These may eventually become tunnels in the vegetation. At the river banks, according to Krieg, "these paths run like small tunnels out of the brush at the edge of the forest. At the steep slopes of river banks between the brush and the water, they form chutes and stairs which are readily visible. It is not recommended that a tapir's path be used, for one frequently ends up with hundreds of ticks all over one's body." As long as the animals are not disturbed, they hardly ever leave their paths. However, when they are pursued by a predator or suddenly confronted by people, they dash through the jungle with considerable speed with their heads lowered, creating a path through the densest jungle in spite of the entangled vines and

Fig. 1-4.
Tapirs like to sit on their hindquarters.

thorny brush. In this way, they may well brush off a predator clinging on their back. According to A. Cabrera and J. Yepes, they also manage to brush off leeches during such wild chases.

Tapirs like to bathe

Water plays such an important part in the habitat of all tapirs that they really should have a pool in their zoo pens. A substitute in zoos may be an occasional shower with a hose. The animals frequently extend their daily bath during the hottest time of the day; not only does this cool off the animals, but it also protects them from biting insects and other parasites of the skin. It is easy to recognize a tapir's wallow, not only by the characteristic three-toed footprints but also by the feces which are frequently found in the surroundings. Hans Krieg often saw manure scattered on the ground of the forest and floating on the water. Whenever possible, the tapirs defecate into the water or in the immediate vicinity of a waterhole or water container. Wolfgang von Richter assumes that the vagus nerve (which also affects the digestive system) is being stimulated as soon as the animals come into contact with water. A female Malayan tapir of the Nürnberg Zoo regularly defecated when sprayed with water.

Behavior while defecating

Tapirs who live in the same enclosure usually use a regular "restroom area," a region limited strictly to defecation. In this case their behavior is rather strange: They briefly smell at the place and paw the ground by alternating the left and right hind leg. They drop the feces, and then paw again without ever touching the excrements. While some rhinoceros scatter their feces with pawing movements in a rather striking manner which makes the excrement visible from a distance, the tapirs' "restroom areas" in captivity probably do not represent marking points of their habitat. However, the males do leave striking "calling cards" with their urine. In many zoo enclosures these spots are easily recognized; as a result of continuous use they are covered with whitish crystals. After a tapir has smelled such a place, he turns around and sprays it with urine with the penis arched backward and—he never misses. Afterwards or sometimes before, he paws the ground with his hind legs. The meaning of these scent marks was unknown until now. One might imagine, however, that they repel strange male tapirs, and thus prevent potential rivals from entering the habitat. When highly excited, tapirs will spray their urine about diffusely. For example, in the Frankfurt Zoo a male tapir was forced out of his warm stable into the cold winter air. At first he resisted all attempts to make him leave his quarters. Eventually, he stood on the threshold and smelled the cold air, and then suddenly he sprayed backward. The keeper, who was gently trying to push him outside, was sprayed full force with urine.

They are unsociable loners

Except in the mating season, tapirs in the wild are usually unsociable loners. One hardly ever sees more than three animals together. Keeping them in pairs or even in family groups the year round, as is usually done in our zoological gardens, is basically unbiological. In spite of this fact, they usually get along well with one another, even in relatively small enclosures, or, better, they coexist beside each other. They pay hardly any attention to their pen mates. Serious squabbles occur as rarely as does playing together. There also seems to be no rank order within the group. Since they are olfactory rather than visually oriented animals, their means of visible expressions are few. Neither do they show display gestures which might impress their partner, nor do they show postures which might indicate submission to a higher ranking individual. Only with the ears and the mouth region can they transmit "signals." When two tapirs meet, their white-rimmed ears point forward. When enraged or sniffed at, the ears point backward. Some kind of facial expression is seen only in "flehmen" (lip-curling) after they have sniffed or licked the urine of a conspecific. The tapir then rolls his tongue inside the mouth and around the trunk several times in a manner like a gourmet would enjoy his wine. According to Wolfgang von Richter, the tapir then points his head upward in a slanting position so that the lower jaw and throat form a straight line; the lower and upper jaws are then completely visible. In addition, the mouth is opened and the tongue is moved inside the mouth.

"Flehmen" (lip-curl) and threatening gestures

Only occasionally do tapirs show a threatening gesture. Then their only and rarely used weapons are exhibited. They will pull their lips apart until all their teeth are visible. The tapir's teeth do not really look frightening. However, if a threatening tapir does attack with his teeth, he may well cause serious injuries to his opponent. Instances are known where seemingly harmless tapirs suddenly went wild and attacked everything within their reach with their teeth. They hardly ever use these weapons against their natural enemies, the large cats. In most cases they try to escape by flight.

In zoological gardens, female tapirs become pubescent at the age of three to four years. There is no correct information as to the onset of the age of reproduction. Tapirs in captivity do not have a mating season; therefore, young are born rather regularly throughout the year. Generally, female tapirs seem to come in heat every fifty to eighty days. Normally heat will last for two days. However, it may well last considerably longer. Before copulation, the animals are highly excited, frequently uttering short, wheezing sounds or shrill, piercing whistles, and spraying large quantities of urine. Usually a female ready for mating will be pursued by the male; but in the Malayan tapir, the female may, in the beginning, occasionally pursue the male. The animals walk towards each other and stand parallel to each other, but facing in opposite directions so that one can smell the other's anal region. From this position, frequent circling or carousel-movements may develop. The male tries to push with his head under the female's underside or to snap at her hind legs. The female in turn snaps at the

Mating, birth, raising of the young

male's hind legs. Simultaneously each animal tries to get its hind legs out of the other's reach, while the partner follows with the head. These "carousel" movements are a behavior which in the tapir is found only during the early phases of courtship; phylogenetically it may have been derived from fighting positions which still occur in the horses. After copulation, which may be repeated several times in short succession, the female may become aggressive and ward the male off by biting.

Before giving birth, the female starts walking to and fro restlessly. Her vagina is swollen and the udder is filled with milk. As soon as parturition starts, she lays down and stretches out on her side, rolls, gets up again, and may then go to her "restroom area," assuming a perching posture. She gives birth in this posture, and the young appears head first. Since the young's body is oval without a long neck or long legs it usually takes only about three to six minutes until it is out of the mother's body. When first lying on the ground, the young does not move, but the mother immediately turns towards it, licking and nibbling it, whereupon the young starts to struggle, to shake its head, to move the ears, and then tries to get up on its feet.

As soon as the newborn has risen, it starts to search for the udder. Rosl Kirchshofer reports on the lowland tapirs in the Frankfurt Zoo: "The first suckling movements of the young at the mother's body are at her head and neck. However, right from the beginning the young attempts to circle the mother, thus reaching her underside from the rear. The circling movements around the mother are repeated frequently and seem to be purposeful, but undirected. The female, meanwhile, is standing still or resting in the typical posture for tapirs with legs pulled under the body, pushes the young several times with her head towards the rear, thus clearly aiding him. In addition, she rolls over to the side whenever the young starts to search at her hind legs or her flank and she offers her udder with her hind legs stretched away. If the young's search is unsuccessful, she will change her position again."

When the young tapir has become surer in his movements, he will let his mother know when he is hungry. He pushes the mother's flank, thus causing her to lie down and offer the udder. Tapirs apparently enjoy being patted on their flanks or the underside. They then will lie down like a pig, close their eyes and remain in this position "in expectation" for a while. In order to find out whether a tapir mother has sufficient milk, one can usually make her roll on the side in this manner and then actually milk her.

Like young pigs, the young tapir lies down for suckling. However, the suckling is much more quiet than in similar circumstances in a group of piglets. A young tapir does not need to massage the mother's udder in order to obtain milk. As long as the young is small, the mother stays in his immediate vicinity. When the mother and young lie down, she always puts her head beside the young. In this position, she licks him persistently until finally they both fall asleep. When the young tapir walks off, the mother follows him and calls him back with short calls. This close bond becomes less so with increasing age of the young. Compared with the behavior of zebra or rhinoceros mothers, it is most surprising how little the tapir's mother is disturbed when her young is touched or handled in her presence. Except for a slight uneasiness, she does not show any reactions. I do not know of any case where a tapir mother has attacked an "intruder" in order to protect or to defend her young.

With increasing age, the young tapir becomes more and more venturesome and does not need as much sleep as he did during the earlier part of his life. Although he now gallops around his mother, gamboling and throwing his head, she still is the focal point of his activities. Occasionally he will try to entice her to take part in his playing by pushing her lightly with his head, but she rarely does so. Exhausted from such wild chases, the young tapir lies down for a short while, only to begin the game all over again. Young tapirs who have lost their mother easily become tame with their keeper and they soon learn how to drink their milk from a bottle or a bowl. At a few weeks of age, they also eat mashed food and cooked vegetables, eventually accepting raw vegetables. They grow up relatively quickly: a male lowland tapir born in the Berlin Zoo in 1966 increased his weight during the first six months by 3 kg each week, thus reaching a weight eight times his birth weight, which had been 10 kg.

Under favorable conditions, female tapirs may give birth to quite a number of young during their lifetime. A female tapir who had come to the Frankfurt Zoo as a fully grown animal gave birth to a total of ten young from 1953 until 1967; and, doubtless, there would have been more if it had not been necessary to replace the male by a subadult male after the second young. If it is possible to return the male to the female and young shortly after birth, it is easily possible to have a young born every fifteen months. Usually tapir fathers are very friendly with their newborn offspring. However, a few cases have been observed where they attacked and bit the young. There is little information concerning the beginning of puberty, the birth rate, and the life span of tapirs in the wild. Up to now all the available data have come from observations in zoological gardens.

Breeding of tapirs in zoological gardens

The best known species, and the one most frequently displayed in zoos, is the South American LOWLAND TAPIR (*Tapirus terrestris*). As the name indicates, his preferred habitat is the lowlands. According to Hans Krieg, he is definitely not a nocturnal animal. However, in areas where he is disturbed by man, he will hide during day time in the jungle. At night he may well visit the settlements, invading the planta-

The four species of tapirs

tions and eating cane, mangoes, melons, cocoa plants, and vegetables. Other than in settled areas, he feeds mainly on leaves from trees, buds, and several kinds of fruit. This information was obtained by analyzing the contents of their stomachs. Like all his relatives, the lowland tapir has a special craving for salt. The lowland tapirs will walk long distances in order to reach a salt lick.

In all places in their South American habitat where the land is being cultivated, the number of tapirs decreases steadily. The South American Indians kill them for their skin, their meat, or both. They use poisoned arrows and occasionally chase them with dogs. When pursued, the animals plunge into the water. Then they will be killed from a boat with spears and knives. However, the tapir population is not really endangered by the hunting Indians. Furthermore, some Indian tribes prohibit the killing of tapirs for religious reasons. Their main enemies are the white or half-white settlers who in most cases kill these harmless vegetarians "just for the fun of it." In the villages, one often finds young orphan tapirs whose mothers have been killed. They become as tame as dogs within a few days. They like to be petted and even let the children ride on their backs. In spite of these characteristics, which are suitable for domestication, there have been few attempts to actually domesticate tapirs. According to several reports, only in the last century have the German-Brazilian settlers in Santa Caterina occasionally tamed tapirs. On remote farms, they have even used them to pull their ploughs.

Tapirs have been kept in European zoos for a long time, and they require very little care. They usually reproduce without difficulty and may reach a considerable age on a simple diet of vegetables, fruit, herbs, hay, and some grain. A tapir in the Frankfurt Zoo reached the age of almost thirty and one half years. In American zoos, lowland tapirs have reached an age of more than thirty years.

Chances for the preservation of the three other species are remote. Especially endangered is the CENTRAL AMERICAN TAPIR (Tapirus bairdi) which is the largest mammal of the American tropics. According to I. T. Sanderson, this tapir's narrow winding paths lead along even the steepest slopes and down to water. Humans like to use these tapirs' paths because of their convenient locations, initially as a path for pedestrians and later on as a street. Even though the meat of these giant tapirs is considered to be extremely fatty and not very tasty, they are being pursued everywhere in their habitat. In spite of all regulations for the protection of the Central American tapir, he already has become so rare in Mexico that his complete extinction can soon be expected. In the Central American Republics, too, the tapir population is decreasing constantly because of the destruction of forests, plantations, and hunting. Therefore, the International Union for the Conservation of Nature (I.U.C.N.) has placed the Central American tapir on the list of the endangered species. So far, only very few Central American tapirs have been kept in zoological gardens. By chance, an occasional animal may come into one of the smaller Central American zoos. This large tapir probably does as well in captivity as the lowland tapir and may, according to L. S. Crandall, reach a similarly old age. In the New York zoo a male lived for fourteen years; in Chicago a female reached approximately twenty-seven years. In 1967 each of the zoos in Philadelphia and San Francisco had one female Central American tapir. So far, reproduction occured only in a few individual cases. As N. Alvarez del Toro reports, the zoo of Tuxtla Gutierrez (Mexico), in 1954 received a male which had been seriously injured by a jaguar. He was approximately half a year old. In 1955 a female, four weeks old, was put together with him. The first young of this couple was born in 1960; however, it was not raised. In 1964 another young was born which, interestingly enough, was part albino. In the San Francisco Zoo a lowland tapir male and a Central American tapir female had several offspring. The young were raised and some of these interesting hybrids are now displayed in the Los Angeles Zoo.

The graceful MOUNTAIN TAPIR *(Tapirus pinchaque)* who is adapted to his alpine habitat is much more delicate and more difficult to keep. His tracks may be found up to the snow line of the Andes. His dense, woolly fleece offers effective protection against the very low night temperatures at these elevations. The mountain tapir, also, seems to be an endangered species. Hardly anything is known about his behavior. So far the mountain tapirs are extremely rare in zoological gardens and the attempts at keeping them have been rather unsuccessful. They turn out to be extraordinarily susceptible in captivity. Apparently, they are not able to adjust to the climatic conditions of the lowland. In New York a mountain tapir lived for two and a quarter years which to date comprises the rather poor record of keeping them.

The MALAYAN TAPIR *(Tapirus indicus)* is different from the American species by the extraordinary contrasting pattern of his coat. One would imagine that such a striking coloring does not contribute to hiding the animals from their enemies; but apparently just the opposite is true. In their natural habitat, the brush of the forest which has changing light and shade patterns, this "tripartition" practically breaks up the outline of the body. Lying down during the daytime, a Malayan tapir is said to resemble a pile of rocks, and therefore is hard to detect along the rock-strewn creeks of his habitat. As in all the other species of tapir, the young Malayan tapir also has horizontal stripes and dots on his coat. The white part begins to show at the age of about sixty-eight days, according to G. H. Pournelle's observations in the San Diego Zoo. After this time the light coloration becomes more and more pronounced, while the stripes fade out rapidly. Finally, 155 days from the date of birth, the stripes disappear almost completely.

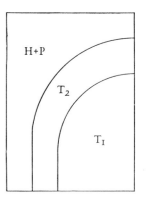

The Phylogeny of the Odd-toed Ungulates
A. Horse-related animals (Hippomorpha, described in Vol. XII)
1. Brontotheria
2. Paleotheria
3. Horses: with a detailed genealogical tree of the horses in Vol. XIII
B. Rhinoceros-related animals (Ceratomorpha)
I. Rhinocerotidae
4. Hyrachyidae
5. Hyracodontidae
6. Amynodontidae
7. Rhinoceros
II. Tapiridae
8. Tapirs
C. Ankylopoda
9. Chalikotheria
Present-day species of animals are shown in color and extinct ones are shown in gray. Extinct phyla are marked with a cross.

Geological ages (see figure above):
(T₁) Early Tertiary, approximately seventy to twenty-five million years ago.
(T₂) Late Tertiary, approximately twenty-five to two million years ago.
(P+H) Pleistocene (Glacial Period) and Holocene (Present) from approximately two million years ago onward.

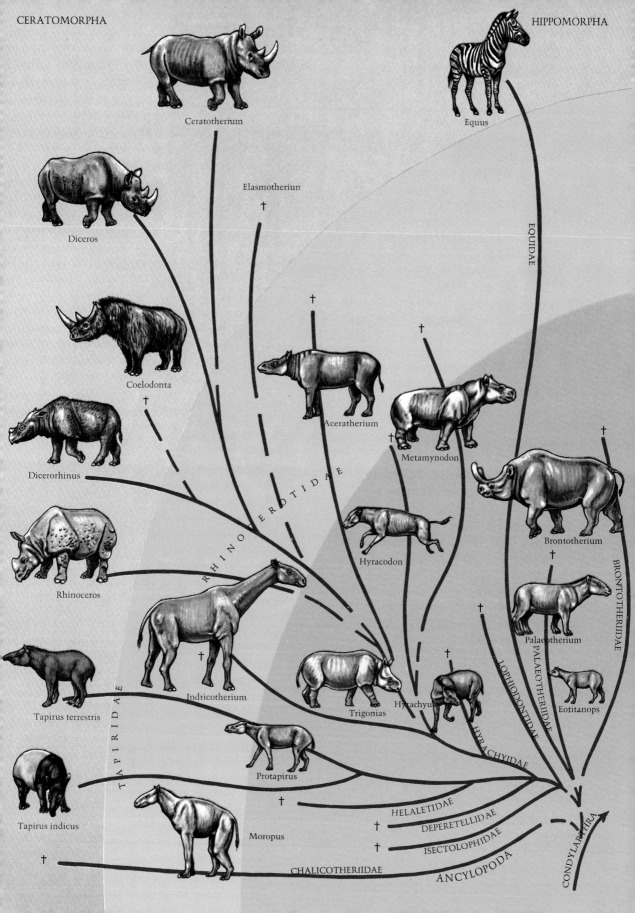

CERATOMORPHA

HIPPOMORPHA

Ceratotherium

Equus

Elasmotherium †

Diceros

EQUIDAE

Coelodonta

Dicerorhinus †

Aceratherium †

Metamynodon †

Brontotherium †

RHINOCEROTIDAE

Rhinoceros

Hyracodon

BRONTOTHERIIDAE

Indricotherium †

Palaeotherium

PALAEOTHERIIDAE

Tapirus terrestris

Trigonias

Hyrachyus †

Eotitanops

LOPHIODONTIDAE

TAPIRIDAE

Protapirus

HYRACHYIDAE

Tapirus indicus

HELALETIDAE

Moropus

DEPERETELLIDAE †

ISECTOLOPHIDAE †

CONDYLARTHRA

CHALICOTHERIIDAE

ANCYLOPODA

†

Muntjacus

Cervus

Giraffa

Antilocapra

Odocoileus

Moschus

Bos

Sivatherium

CERVIDAE

TRAGULINA

Tragulus

BOVIDAE

GIRAFFIDAE

ANTILOCAPRIDAE

Camelus

TRAGULIDAE

Mery:codus

PECORA

CAENOTHERIIDAE

Alticamelus

Syndyoceras

Lama

Procamelus

TYLOPODA

Proto ceras

HYPERTRAGULIDAE

Caeno therium

Caenotherium

PROTOCERATIDAE

GELOCIDAE

LEPTOMERYCIDAE

Poëbro therium

Hippopotamus

CAMELIDAE

OREODONTIDAE

Meryco choerus

Protylo pus

Sus

ENTELODONTIDAE

Daeodon

Archaeo meryx

Potamochoerus

NONRUMINANTIA

ANTHRACO- THERIIDAE

HIPPOPOTAMIDAE

SUIDAE

LEPTOCHOERIDAE

?

Bothriodon

ANOPLOTHERIIDAE

Tayassu

Hyotherium

DICHOBUN

TAYASSUIDAE

CEBOCHOERIDAE

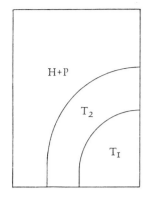

The Phylogeny of the
Even-toed Ungulates
A. Nonruminantia
1. Peccaries
2. Pigs
3. Hippopotamuses
4. Anthracotheriidae
5. Entelodontidae
6. Oreodontidae (lead in
to Typlopodes)
B. Tylopodes
7. Camels
C. Ruminants (Rumin-
antia)
It is not known whether
the earliest ancestors of
the ruminants did
ruminate.
8. Hypertrapulidae
9. Protoceratidae
10. Mouse Deer
11. Deer
12. Giraffes
13. Pronghorns
14. Pecoras: a detailed
genealogical tree of the
Pecoras is on p. 264.
D. Caenotheria
Geological Ages, see
p. 30.

The number of Malayan tapirs is decreasing because of the increased clearing of the forests in his habitat. Since the Muslims of Malaysia and Sumatra consider the tapir a pig, they scarcely hunt him, as pork is considered impure. However, in other areas of Southeast Asian habitat, the Malayan tapir is severely persecuted, especially because he likes to feed on the young, green corn plants of the plantations. According to statements of the Nürnberg Zoo director, Alfred Seitz, he has no protection at all in Thailand; his meat is even sold on the markets under the name of "mu-nam." The population of the Malayan tapir, whose only enemy, other than man, is the tiger, seems not to be endangered in most areas of his habitat. However, with the increasing civilization of these countries, this may soon change.

Malayan tapirs are not often kept in zoological gardens, but they do make striking, remarkable, and much admired objects for display. Nice groups of these large Asian tapirs currently live in the zoos of Berlin, Basel, Nürnberg, and in some American zoos, and they will occasionally reproduce. To date the longevity record for keeping them in captivity is over twenty-nine years. During the summer, the Malayan tapirs of the Nürnberg Zoo may feed on a fenced-in meadow with trees and bushes. There it is possible to observe from close by their method of grazing and tearing off leafy branches with their extremely movable trunks. Alfred Seitz reports: "At first they ate all the leafy bushes of hornbeam and oak bushes as far as they could reach them, as well as some of the bark. The animals were not interested in the leaves of poplars and small willow trees. They then ate the tender branches of the multiflora rose. They liked best the freshly cut green corn plants which seem to be very appropriate for their crushing method of chewing. Twice a day our tapirs ate a large bowl of cracked rice, together with fresh apples and carrots, and some calcium. Good quality hay proved to be indispensable for their well being the whole year round."

The Nürnberg pair had two female young, one in 1964 and one in 1966, this being the first reproduction of Malayan tapir in a German zoo since 1936. Such events are especially important in our days; they may, according to Seitz, "eventually become, already in the near future, the way to save a species from extinction."

Hans Frädrich
Erich Thenius

2 Rhinoceros

The presently living rhinoceros are a well defined group of animals whose members closely resemble each other, in spite of the fact that two of the species live on the African continent and three in Asia. Similar to the history of the tapirs, the superfamily Rhinocerotoidea consisted of many more species during the Tertiary Period and were classified into several rather different families. Among these families were light-footed runners with long, slender limbs, the *Hyrachyidae* and the Hydracodontidae, which appeared in North America during the Eocene age, approximately fifty million years ago. Even though they resembled the contemporary members of the horse family, they were originally like hornless rhinoceros. The Hydracondontidae became extinct during the Oligocene (about 40–45 million years ago), leaving no descendents; we probably would have to search among the original Hyrachyidae in order to find the original ancestors of all the other Rhinocerotoidea (Color plate, p. 31). However, during the early Tertiary Period, there were rhinoceros-like animals which looked quite different, for example the Amynodontidae of Eurasia and North America whose appearance resembled that of a hippopotamus, and was rather plump.

The true rhinoceros of the Rhinocerotidae family during the Tertiary were also a widely distributed group of many species. During the Eocene, a hornless, small form with slender feet, probably not too different from the other odd-toed ungulates of this era, first appeared. The skull was low and flat, without any indications of horns. The molar teeth consisted of premolars and molars with low crowns and ridges across and on the sides. This basic structure, in spite of some variations, is the same as found in the later rhinoceros. Fossils of members of the subfamily Caenopodinae (*Eotrigonias, Caenopus*, etc.), which belong to the most primitive, oldest rhinoceros, and fossils of several such forms have been found in the Early Tertiary stratum of North America and Europe. These slender-footed, hornless, primitive rhinoceros still had a complete set of front teeth and molars.

Phylogeny
by Erich Thenius

Rhinoceros once in
Europe

Of the contemporary rhinoceros, the ASIATIC TWO-HORNED RHINOC-
EROS *(Dicerorhiniae)* may be traced back approximately forty million
years into the Oligocene. At first they occurred as small animals, *Dicerorhinus tagicus,* which were less than the size of a tapir and soon split
up into different lines. One line led to the well known, early glacial
WOOLLY RHINOCEROS *(Coelodonta antiquitatis).* He was a cold-resistant
species with a long-haired, thick coat. Knowledge about him is gained
not only from bones, but also from complete bodies with skin and fur
which were discovered in the Siberian permafrost soil. In addition, the
people of the Early Stone Age have portrayed him on their cave drawings. He was extinct by the end of the glacial period. The woolly rhinoceros had a long drawn out skull and high crowned molars which
were especially adapted for the crushing of the hard grass of the
steppe. His front teeth had completely disappeared, as in the contemporary SQUARE-LIPPED RHINOCEROS (see p. 64), which also is a grass-eating
animal of the steppe. But this similarity in adaptation does not indicate
any closer relationship. We find a similar evolution in the somewhat
older relative of the woolly rhinoceros, the STEPPE-RHINOCEROS of the
earlier and middle glacial period of Europe *(Dicerorhinus hemitoechus).*
The larger MERCK RHINOCEROS *(Dicerorhinus kirchbergensis)* from the same
glacial periods, however, was rather a forest type. The only contemporary species of this group, the SUMATRAN RHINOCEROS *(Dicerorhinus
sumatrensis)* is much closer to the phylogenetically older forms than his
glacial relatives, a fact which is frequently found in the inhabitants of
the tropical prime forests. Since he still has front teeth and molars with
low crowns, which are not suitable for the crushing of hard steppe
grass, we have to consider him as a slightly modified survivor from
the Tertiary Period.

The GREAT INDIAN RHINOCEROS, which live in South Asia today, can
also be traced back to the Tertiary (Miocene, approximately twenty-
five to ten million years ago). *Gaindatherium browni* from the lower and
middle Siwalik strata of India can easily be traced from the Early Tertiary genus *Caenopus,* and thus represents the original form of the glacial species of *Rhinoceros sivalensis* and *Rhinoceros sinensis* as well as of
the present-day Great Indian and JAVAN RHINOCEROS *(Rhinoceros unicornis*
and *Rhinoceros sondaicus).* The Javan rhinoceros is the older of the two
species, remaining almost unchanged since the late Pliocene of more
than a million years ago.

The African rhinoceros form a separate branch (subfamily Dicero-
tinae) which includes the present-day BLACK RHINOCEROS *(Diceros bicornis),* which originally fed on foliage, and the SQUARE-LIPPED RHINOC-
EROS *(Ceratotherium simum),* which is a more highly evolved grass-eater.

From the Eocene until the Miocene (approximately sixty to ten million years ago), the Paraceratheriae or Baluchitheriae (subfamily Para-
ceratheriinae) lived in Eurasia. They were hornless, long-necked

rhinoceros with huge, column-shaped legs. The largest terrestrial mammals of all times belonged to this group, the genera *Paracerathrium*, *Indricotherium*, and *Benaratherium*. The *Indricotherium asiaticum* was five meters in height and seven meters long. The bones of this giant animal, which were approximately thirty-five million years old, were found in Kazahhstan on the banks of the Tschulka River. Those giant rhinoceros became extinct during the Miocene without leaving any descendants.

However, this list nowhere near exhausts the multitude of prehistoric forms of rhinoceros. There were slender-footed, long-legged rhinoceros, for example, the predominantly hornless *Aceratherium* which had long tusks in the lower jaw; furthermore, there were short-footed savannah types like the genus *Teleoceras* from North America and the genus *Brachypotherium* from Europe; and finally, there was the North American-Eurasian genus *Diceratherium*, which had two horns side by side on the nose. Another extinct line of the rhinoceros are the Elasmotheria from the glacial period of Eurasia. *Elasmotherium* was a giant form with a skull almost one meter long. This skull bore on its forehead a huge bony pad on which a correspondingly large horn must have sat. The dental enamel of the molars was ruffled, which is unknown in any other rhinoceros.

Compared to this multitude of forms in the tertiary and glacial rhinoceros, the surviving four genera appear rather stunted in spite of their size. They all live in remote habitats, seemingly because they have not been able to compete any longer with the other ungulates, especially the ruminants. Above all, however, human influence has basically changed wide areas of Africa and Asia, thus making them uninhabitable for rhinoceros. Since man first pursued animals, the rhinoceros have been hunted. The pictures in the Early Stone Age caves of Pech-Merle, Rouffignac, Colombière, and Les Trois Frères tell an obvious story. But they also show that these animals already had mystical significance in earlier times.

The present-day rhinoceros (family Rhinocerotidae) are either hairless or barely villous. HRL 200–400 cm, TL 60–76 cm, BH (shoulder) 100–200 cm, weight 1000–3600 kg. The surface of skin is distinctly sectioned, especially in the Asiatic species. On the nasal bone are one to two horns. There are 24–34 teeth arranged as follows: $\frac{0\text{-}1 \cdot 0 \cdot 3\text{-}4 \cdot 3}{0\text{-}1 \cdot 0\text{-}1 \cdot 3\text{-}4 \cdot 3}$. The gestation period is 419–550 days. One young is born.

It is commonly held that rhinoceros horns consist of matted hair. This is not quite correct. The horns consist throughout of ceratin, and they do not have a bony pith like the horns of cattle. Under a microscope, however, one can see that the individual rods are not coated with an individual protective layer as is real hair. They adhere densely together in layers, thus they resemble neither the hair nor the horn of a ruminant, but rather the material of the hoof. This construction

▷

Rhinoceros

1. Great Indian Rhinoceros (*Rhinoceros unicornis*)
2. Javan Rhinoceros (*Rhinoceros sondaicus*)
3. Sumatran Rhinoceros (*Dicerorhinus sumatrensis*)
4. Square-lipped Rhinoceros (*Ceratotherium simum*)
5. Black Rhinoceros (*Diceros bicornis*)

▷▷

A black rhinoceros mother with her sub-adult young. The African black rhinoceros (*Diceros bicornis*) is the only species of rhinoceros which still occurs rather frequently in many areas of its distribution.

Present-day rhinoceros
by E. M. Lang

Distinguishing characteristics

Of what does the horn consist?

gives the nose horn a stiffness and quality similar to a ruminant's horn with a pith. The nose horn sits on a bony dome formed by the nasal bone; it may unravel in places, causing it to look like a growth of hair. If it is torn off by accident, only a lightly bleeding area remains on the nose. Soon a new horn begins to grow. In young animals a horn may be replaced completely.

Extinction because of superstition

Except for the elephants, we find the largest terrestrial mammals among the rhinoceros. However, these handsome mammals provide a classical example of the extent to which man is responsible for the decrease and extinction of large mammals. Superstition played a dominant and especially destructive part in the disappearance of many rhinoceros species. The Chinese, as well as other Asiatic peoples, believe that powdered rhinoceros horns make an aphrodisiac. Many centuries ago the powder made from these horns was sold in East Asiatic pharmacies at a high price. Since the rhinoceros are easy to kill, they have been poached ever since; now, after the almost total extinction of the Asiatic species, those in Africa are poached. Years ago, on the black market, people in Africa paid approximately fifteen dollars for a kilogram of nose horn. The medium-sized horn of a great Indian rhinoceros literally is worth its weight in gold, as well-informed people have confirmed. In 1965 the price for an Asiatic horn was no less than $1125 for one kilogram. The strong faith in the healing powers of these "remedies" increases the prices constantly and stimulates natives and agents to kill even the very last rhinoceros without regard for the laws protecting the animals.

John A. Hunter, who may claim the sad record for having killed the highest number of rhinoceros, shredded rhinoceros horn and made it into a dark brown tea. "Even though I drank several portions of the brew," he writes, "I am sorry to say that I did not feel any reaction whatsoever, perhaps because I did not believe in it, or maybe because I was not in the right company." The possible medicinal effect of the horn has recently been carefully tested, thanks to the initiative of A. Schaurte. There, too, not the slightest effect could be demonstrated. Perhaps the Asiatic superstition is based on the fact that the great Indian rhinoceros do copulate for about one hour during which the bull ejaculates approximately every three minutes. To become capable of such sexual prowess seems to be desirable to many Asiatic people.

Black rhinoceros (*Diceros bicornis*) like to take dust baths. These gray giants are often accompanied by cattle egrets (*Bubulcus ibis*) which do not, as formerly presumed, collect the parasites from the rhinos, but rather eat the insects the rhinos stir up.

Skillfully carved cups of rhinoceros horn used by Indian and Far Eastern potentates to test beverages for the possibility of containing poison, may indicate a similar belief. Today these rhinoceros horn cups are rare collector's items. An example from the Calcutta Zoo shows how far superstition can go: All great Indian rhinoceros who die there are immediately removed by the keepers, cut up, and sold at an immense price to fanciers. In Assam, when one of the few surviving great Indian rhinoceros was shot with official approval, as happened a few years ago when the English Queen visited the country, not only the nose horn but also the skin, the skeleton, the muscles, and even each hair went to the black market trade. As was confirmed not only by Professor Ullrich, the director of the Dresden Zoo, but also by other visitors of the Kazirange Game Reserve in Assam, poachers even in the nineteen sixties still dug many traps there for the rhino.

Therefore, all species of rhinoceros are threatened by extinction and urgently need all possible protection. In Africa, only in the National Parks and in the protected areas will one find a good rhinoceros population. The situation in Asia is much more critical. Of the once abundant great Indian rhinoceros, there are presently only a few hundred animals left, whose further existence is not at all assured. The closely related Javan rhinoceros' extinction is imminent; it is only in a tiny area, the Udjung-Kulon Reserve in Java, that 25 to 40 animals are found. The number of surviving Sumatran rhinoceros on the Malayan continent is unknown; according to official statements, there are 170 to 600 animals left. If the World's Nature Conservation efforts do not succeed in establishing effective measures for protection, our descendants will not see a living Javan or Sumatran rhinoceros. Unfortunately, not even the most general data of the life and behavior of these animals are known. The few surviving ones have to lead such a secretive life that any close survey or research is technically impossible.

The original, yet also the smallest, living species of rhinoceros is the SUMATRAN RHINOCEROS (⊹ *Dicerorhinus sumatrensis*, Color plate, p. 37). HRL 250–280 cm, BH (shoulder) 110–150 cm. It is the sole villous rhinoceros. There are two nose horns; the maximal length of the anterior one is 25 cm; the second one is, in most cases, only a blunt protuberance (or hump). The skin is only slightly sectioned (semi-plated); the ears are fringed with hair; and the coat, while dense, thins out in older animals. Formerly the distribution was over all of East India and Indonesia; presently there are only infrequent sightings. These animals are very rare.

The great Marco Polo (1254–1324), on his travels through East Asia, had seen the Sumatran rhinoceros in the Malayan Archipelago and described it. However, there is hardly anything known about the life in the wild of this animal which will soon become extinct. Earlier zoologists distinguished between the original form on the island of Sumatra *(Dicerorhinus sumatrensis)* and a continental form *(Dicerorhinus sumatrensis lasiotis)*, which was also called the rough-eared rhinoceros. But in comparison with specimens in museums and pictures from the wild, this opinion is open to question. According to cautious present-day estimates, there are only a few hundred of these animals on the island of Sumatra, some others on the island of Borneo, in Burma, in Siam, and in the Malaysian preserve of Sungei Dusun (Selangor). But the timber industry, the establishment of rubber plan-

Asiatic rhinoceros

Fig. 2-1.
Former and present distribution of the Sumatran rhinoceros *(Dicerorhinus sumatrensis)*. This species now exists only in those few places which are marked by triangles on the map.

Sumatran Rhinoceros

Great Indian Rhinoceros

Javan Rhinoceros
Fig. 2-2.
The skin folds at the shoulder and base of the tail are differently arranged in the various species of Asiatic rhinoceros.

tations, and other similar interferences with the natural landscape destroy the original habitat of this animal to such an extent that it is uncertain whether the species may be preserved.

Lately only a very few Europeans have encountered Sumatran rhinoceros. On March 14, 1957, a rhinoceros was seen on a coconut plantation at the Slim River near Perak in Malaysia. A farmer was able to take a photograph of it. At first sight it seemed to be a Javan rhinoceros (see p. 44). Some people, therefore, presumed that there might be some surviving specimen of the Javan rhinoceros on the South East Asian continent. Therefore, the photographs were widely distributed. But the lack of a horizontal skin fold on the buttocks indicated clearly that it was actually a Sumatran rhinoceros. It had a fairly dense coat and no ear tufts. From behind such a small, hairy rhinoceros almost looks like a cape buffalo.

In 1959 two female Sumatran rhinoceros were captured in the area of the Siak River on the island of Sumatra and brought to Europe. One of them arrived at the Basel Zoo in very poor health; her body height at the shoulder was 112 cm and she weighed 386 kg. After almost uninterrupted medical treatment for two years, she died of total deterioration of the kidneys. The other animal remained in good health and was to this date (1967) in the Copenhagen Zoo. Presently, it is the only Sumatran rhinoceros on earth in man's care. Unfortunately, all efforts to find a mate for this animal have been unsuccessful.

After the success achieved in keeping the great Indian rhinoceros in the Basel Zoo, it should be as easy to keep and breed the Sumatran rhinoceros. The first rhinoceros ever to be born in captivity was a Sumatran rhinoceros, born on January 30, 1889, in the Calcutta Zoo in India. Then, these small rhinoceros were not nearly as rare as they are now. However, it would be possible to preserve the species only if there were enough pairs available for the zoological gardens. But by now they have become too rare. It would be absolutely necessary to place the few remaining specimens in their original habitat under rigorous protection. However, the situation in Malaya is rather discouraging. Because many Chinese live in the already limited remaining habitat of the Sumatran rhinoceros, effective protection there seems impossible. Wherever Chinese poachers are at work or buy rhinoceros' horns from the hunting natives, the rhinoceros disappear.

The earliest species of rhinoceros to become known in Europe was the GREAT INDIAN RHINOCEROS (genus *Rhinoceros*). HRL 210-420 cm, TL 60-75 cm, BH (shoulder) 110-200 cm. The weight is 1500-2000 kg; the ♀♀ are somewhat smaller and lighter; footprints of adult ♂♂ measure 28-29 cm in diameter in the front, those of adult ♀♀ measure 26-27 cm in the front, in the rear 23.5-24.5 cm. There is only one horn. The bare skin is not very thick, is well supplied with blood vessels, and is divided into sections by large folds. On individual sections there are flat bumps which look like rivets on a hull of a ship. Hair occurs only in a few places: the tail tassel, tufts on the tips of the ears, and, in neonates, also a light hair brush at the base of the ear's outer rim. The three toes on each foot are covered by rather large nail plates, and are buffered by massive tissue pads which bulge out when the foot is lifted. The upper lip ends in a strong "finger." The two incisors in the lower jaw grind against the tooth plates in the upper jaw, and since they are razor sharp they are effectively used as weapons.

There are two species: 1. GREAT INDIAN RHINOCEROS (⚥ *Rhinoceros unicornis*; Color plate, p. 37), whose shoulder skin fold arches over the shoulder blade. 2. JAVAN RHINOCEROS (⚥*Rhinoceros sondaicus*; Color plate, p. 37), is rather similar, but smaller and lighter; ♂♂ have only a weak horn, ♀♀ are often hornless. Their shoulder fold comes up from both sides and meets above the shoulder.

The great Indian rhinoceros *(Rhinoceros unicornis)* is an impressive sight. It does not really have a hunchback as does the African square-lipped rhinoceros but the bull has a bulky, wide neck. The withers and pelvis in most cases are of the same height, but once in a while one may see "overendowed" females. Within the same population, there are both long-legged, slender animals and shorter, heavier types. The pace of the great Indian rhinoceros is a deliberate walk but may also be faster. The trot appears surprisingly elegant, and the gallop is extremely fast. When galloping on a good surface, a great Indian rhinoceros may well reach speeds of up to 35–40 km/hr.

As is true of all the rhinos, the great Indian rhinoceros is a vegetarian. It feeds on grasses and twigs, pushing them into its mouth with the finger-like extension of the upper lip. In the Basel Zoo, the great Indian rhinoceros are fed, besides the basic diet of good quality hay, which at times may be mixed with alfalfa, a special compound of highly concentrated food containing oil cake and several cereals, the necessary vitamins and minerals, and about 18 percent pure protein. An adult great Indian rhinoceros eats about 15 kg of hay per day and 4–6 kg of the Basel special compound; it drinks 80–100 liters of water.

Great Indian rhinoceros like to rest in the water or in a clay wallow which helps to keep their skin in good condition. In the Basel Zoo, which is perhaps the most experienced in the keeping and breeding of the great Indian rhinoceros, the animals have a pool, heated in the winter, which they use the whole year round. Great Indian rhinoceros are well adapted to life in the water. They are skillful swimmers and divers; even as wide a river as the Brahmaputra is frequently crossed by great Indian rhinos.

E. P. Gee, the protector and warden of the Kazirange Reserve in Assam, has found that the rhinos there defecate on certain "rhino-dung heaps." Gee is of the opinion that each great Indian rhinoceros that passes such a dung heap is attracted by the scent—whether its own

Fig. 2-3.
Former and present distribution of the great Indian rhinoceros *(Rhinoceros unicornis)*. Presently it is found only in a few protected areas (marked on the map with triangles).

Fig. 2-4.
Former and present distribution of the Javan rhinoceros *(Rhinoceros sondaicus)*. Today only a few animals live in the Udjong-Kulon Preserve in Java (see arrow).

or that of another rhino—and thus is almost stimulated to defecate as if by some compulsion. The director of the Dresden Zoo, Wolfgang Ullrich, saw dung heaps forming mounds up to 70 cm high. He writes: "How strong a stimulus for defecation such a mound represents is demonstrated by the fact that even rhinos in flight will stop there for a few seconds in order to deposit a dropping. These dung heaps are found especially often near wallows, bathing places, and grazing grounds, which are on the edge of open spaces. With their odor they mark the entrances to the tunnel-like paths in the dense elephant grass, thus enabling the rhinos to find the path by olfaction."

According to Gee, the great Indian rhinoceros are not strictly bound to a specific home range; usually a weaker bull will leave the territory after he has lost a fight with a stronger rival. Gee sometimes followed single rhinos when they were roaming through wide areas; he also found gatherings of four to six animals using the same wallow. As Gee reports, the mating season in Assam extends from the end of February to the end of April. Wolfgang Ullrich spent several weeks in the Kazirange Reserve. There he observed the great Indian rhinoceros feeding on the young sprouts of high grass and bamboo shoots, on several herbs in the swamp, and on the water hyacinths which cover the lakes like a carpet. All other animals generally avoid the rhinos, according to Ullrich's observations. The great Indian rhinoceros flee only, at least in most of the cases, from mounted elephants, and then, in contrast to the African black rhinoceros, they do not carry their tails erect but closely pressed to the body. Occasionally, they may attack a riding elephant or at least threaten him; but in most cases, the rhino will veer off just before reaching the elephant. Many of the great Indian rhinoceros have serious injuries and large scars, which in Ullrich's opinion are the result of fights over a territory. There are also some cases known where female rhinos have been injured by a bull.

In the journal "Der Zoologische Garten" Ullrich wrote on the home ranges and paths of the great Indian rhinoceros: "The grassy jungle is crossed by many paths, which are separated into 'public' and 'private' pathways. The 'public' paths are used by several rhinoceros. They connect wallows, bathing places, and grazing grounds. Wallows and bathing places belong to all rhinos and are not defended. We frequently observed several rhinos resting peacefully beside each other in wallows and ponds. In a small lake covered with water hyacinths, we found nine rhinoceros; several of them, including a calf about four months old and a subadult, were laying close together. Two of them had even put their heads on another's back. At most, only the noses, the eyes, and the ears were visible above the water. Two other rhinos were also resting closely together, about 20 meters away from the group. When three of these rhinos began to feed on the water hyacinths, there was no conflict.

Social life of the great Indian rhinoceros

Public and private rhino paths

"When a rhinoceros came to the edge of a lake where two others were already resting, those two arose uttering threatening sounds. Then the rhino on the shore would make a snorting sound. It sounded as if someone was blowing air through a hose into water. The two rhinos immediately answered with the same sound, then went back to rest and allowed the newcomer to rest with them.

"Branching off from the 'public' paths near the resting places and the grazing ground are the 'private' paths. Daily we visited a large swamp meadow which was divided into seven grazing territories which belonged to three females with three calves, another adult female, and three adult males. Except for the territory of one of the bulls, all the grazing areas, which were situated at the edge of the pasture, were about 4000 square meters in size. A short 'private' path connected each grazing area with the 'public' paths which led through the grassy jungle around the pasture. The rhinos approached the area on the 'public' paths and then entered to their 'private' paths in order to reach their part of the pasture which they defended against conspecifics. When they were disturbed while grazing, they always fled to their 'private' paths. In the same manner, 'private' paths branched off from the 'public' paths near their sleeping places which are defended in the same way as the grazing territories. The sleeping places are situated in the tall elephant grass, where the rhinos rest from midnight until sunrise and during the hot part of the day at noon."

Except for the observations of Gee, Ripley, and Ullrich about the reproductive behavior of the great Indian rhinoceros, the only other information comes from zoological gardens. Even though this species had already come to Europe during the Middle Ages, they were first bred in a zoo in 1956. A female in heat sprays urine, while the vagina opens up and "flashes." At the same time, she utters rhythmical whistling sounds that are produced by forcing the air in and out during breathing. She comes into heat every 46–48 days and remains so for approximately 24 hours; however, the intervals between heats may vary from 38 to 58 days. The bull reacts immediately to her condition. Shortly after heat begins, the animals will drive one another intensely; we often saw them galloping around the large rhinoceros pen in the Basel Zoo two dozen times. Usually a rest period of several hours follows this driving. Then the animals will stand beside each other; the bull may lie down, and sometimes the female holds her head between his hind legs. After many hours, the first attempt to mount will take place, but only after several such attempts is the bull's penis erect enough to achieve intromission. Both animals remain in the copulation posture for an average of sixty minutes. During one such copulation, we counted up to fifty-six ejaculations. The longest copulation we recorded was eighty-three minutes. After the bull dismounts, the animals pay no further attention to each other.

Reproductive behavior

Birth and the raising of young in the zoo

While the females reach sexual maturity at three years, the bulls do not become sexually mature until they are seven to nine years old. The average gestation period for twelve females in captivity was between 462 and 489 days. The birth started with episodes of labor lasting about one hour; the actual birth, however, took only 15–30 minutes. A neonate great Indian rhinoceros has an average weight of 65 kg; it has folded skin like an adult with all the "rivets" and protuberances. On both sides of the head is a light spot which the English zoologist Cave interprets as a relic of tactile hair. The plum-shaped head is especially conspicuous in the newborn great Indian rhinoceros. There is a flat, smooth, oval plate where the nose horn will grow later. At the age of five weeks, this plate begins to rise. During the development of the sub-adult, the forehead becomes depressed while the area around the ears and the horn arch upward.

The young great Indian rhinoceros grow much faster than has been thought earlier. At the Basel Zoo, we found weight gains of 2–3 kg per day; thus, the weight at birth is multiplied tenfold within one year. Shortly after birth, the BH at the shoulder is 62–64 cm. After one year, it is approximately 120 cm; at the age of two years, about 145 cm. At the age of three and a half to four and a half years, the female is fully grown, while the bulls may keep growing for up to five years. The rhino mother has to produce twenty to twenty-five liters of milk in order to increase the weight of her young to such an extent.

Can the Javan rhinoceros still be saved?

It is most regrettable that we know so much less about the exact appearance or even the behavior of the closely related Javan rhinoceros *(Rhinoceros sondaicus).* There are only a few specimens in museums and only a few photographs of this moribund species. In former times, the Javan rhinoceros were widely distributed all over East India, Sumatra, and Java. Reliable observers, like Eugen Schuhmacher, a photographer and student of animals, doubt whether the last survivors in the Udjung-Kulon Reserve in Java will be able to sustain the species. Only in 1967 did the World Wildlife Fund manage to send an ethologist, Professor Rudolf Schenkel from Basel, to Java to study the habitat and the behavior of this species, which is literally at the point of extinction. In his letters to me from the Udjung-Kulon Reserve, he reported that the Javan rhinoceros lives in the dense jungle. Its feeding grounds are at the edge of the forests of the coastal area and in the sparse mountain forest, where there are many young trees and openings caused by fallen trees. There it feeds predominantly on young trees whose trunks are no thicker than 10 cm, on the foliage of low-hanging branches, and on bushes. The rhinoceros bends these trees with the upper part of its trunk until they break; then it feeds on some of the leaves from the crown—the rest it does not touch.

Compared with other rhinoceros, the population density is rather low, which can probably be linked with the scarcity of suitable food plants. The animals are solitary; the young become independent at a relatively early age. In order to study the behavior of individual animals, which are hardly ever seen in the dense jungle, Schenkel measured their footprints. In adult animals, these are 27–39 cm wide in the front, and 25–27 cm in the rear. In the young who were on their own, Schenkel measured 21.5–22 cm for the front feet and 20–20.5 cm for the hind feet. Presumably, the females, with or without young, remain in a rather fixed home range but they take long trips from there daily. This roaming is even more pronounced in the males. The paths of the Javan rhinoceros are found predominantly on passes which cross mountain ranges and parallel to them, but the paths are most distinct near wallows. These wallows and bathing places, as well as the resting places, are different depending on the season. During the rainy season, the rhinos wallow in creeks, and less frequently in wet places in the brush. However, during the dry season most of these wallows dry out. The Javan rhinoceros also bathes where larger creeks flow into the ocean and occasionally even in the ocean itself.

The bulls spray their urine backwards and upwards at bushes. The fresh urine, which is orange to red in color, smells like horse urine. Until now these red splashes were thought to be nasal secretions. The Javan rhinoceros defecate either in creeks or on regularly visited "manure fields" of five to ten meters in diameter. Often they may leave droppings on their way. Rudolf Schenkel thinks that defecation in Javan rhinoceros is of no special significance in intraspecific communication.

The only species of rhinoceros of which there still is a good population in the wild is the AFRICAN BLACK RHINOCEROS (♢ *Diceros bicornis;* Color plate, p. 37 and pp. 38–40). HRL 300–375 cm, TL approximately 70 cm, BH (shoulder) 150–160 cm. The weight is up to two tons. There are two horns, of which the anterior one is longer (usually about 50 cm, sometimes up to 138 cm). Occasionally there is even a disposition for a third horn. The body is hairless, except for the tips of the tail and ears. Rib-like folds are on the sides of the rump. The upper lip is extended and the tip is suitable for grasping. There are no incisors or canine teeth; there are seven premolars and molars on each side of the jaw. The gestation period is fifteen to sixteen months.

A person on foot who encounters a black rhinoceros really feels rather small and insignificant. One immediately recalls the angry attacks and even fatal accidents one has read about in books on Africa. After all, the black rhinoceros is one of the largest terrestrial mammals, next to the elephant and the square-lipped rhinoceros. The most impressive attributes are the two nose horns. A visitor to a zoo, seeing the animal for the first time, may already imagine them between his ribs. But then zoo rhinos almost never have the remarkable length of horns as do rhinos in the wild. The world record probably is held by

The African black rhinoceros by Bernhard Grzimek

Fig. 2-5.
Former and present distribution of the black rhinoceros *(Diceros bicornis).* This is the only species of rhinoceros which still occurs quite frequently (black triangles) in some areas.

Gertie, one of the two adult female rhinoceros of the Amboseli Game Reserve; her front horn is bent in an unusual way forward horizontally and upward to a length of 138 cm. For many years she was the most photographed wild game animal on earth. The adult female Gladys, who lives in the same area, had a similarly monstrous horn formation. In 1965 she broke 45 cm off of it. With photos, one was able to show that the frontal horns of these animals had grown 45 cm in six to seven years.

In some areas of Africa, where the rhinoceros are now extinct, there have allegedly been groups in which the two nose horns were of the same length. The three-horned rhinoceros were found frequently in Northern Rhodesia (in the vicinity of Lake Young). There are reports even about a five-horned rhinoceros and of others with horns growing out of their bodies. The great Indian rhinoceros on the famous drawing by Albrecht Dürer, which has a small horn on the shoulder, may well have had a living model. Occasionally, rhinos without ears are born. Gertie of the Amboseli Game Reserve of Kenya, who has perfectly shaped ears herself, in 1953 gave birth to Prixie, who did not have ears. I received the impression that Prixie, whom I have observed and filmed from a very close distance, may, in spite of this defect, decrease or even close the openings of the auditory canal.

In contrast to the square-lipped rhinoceros—the "white rhinoceros" of the literature on Africa—the "black rhinoceros" is not really black, just as the white rhino is not white. Depending on the soil of the habitat where the rhino likes to roll in the mud and the dust, the originally slate-colored skin may be covered with substances that may make it look white or reddish, or, in areas with lava, black. Since it is hairless and without sweat glands, it has a special liking for mud baths. Hence, in rare cases it may happen that a rhino becomes trapped in the mud without being able to get out; then it may be attacked by hyenas.

Black rhinoceros are remarkably nearsighted. Apparently they cannot distinguish between a man and a tree at a distance of only 40 or even 20 meters. This nearsightedness explains some of their behavior, for example, their alleged "aggressiveness." Their sense of hearing is much better; their cone-shaped ears react quickly to unusual noises. The best developed of their senses is the sense of smell, which is probably comparable to the quality of a dog's. They follow conspecifics by the scent of their tracks. When mother and young have lost one another, they may be within range of clear sight. However, they do not move towards each other but rather they sniff at the ground until they encounter the other's track, which they then follow.

Animals with poor eyesight may well approach people or other objects slowly and curiously until they catch their scent. Once a rhino approached Cherry Kearton, an explorer of Africa and animal photographer, who was filming it. The animal circled closer and closer until, at a distance of 10 meters, it finally ran off. The black rhinoceros have another habit which can become fatal for them; they will attack an object which they cannot identify, approach it snorting furiously, and then veer off or just pass within a few meters. A camera man, Martin Johnson, together with his wife, jumped down a deep cliff from attacking rhinos, but then they saw that the animals had stopped five meters from the place where the two of them had just stood. In two other instances, when the Johnsons could not flee, the rhinos also turned off shortly before reaching them.

However, in most cases, travellers in Africa do not have the nerve to wait and see whether it is the exploring approach of the nearsighted animals or actually an attack. Therefore, in most cases, the hunters will shoot the rhinoceros before they know for sure. Rhinoceros sometimes attack tree trunks or termite heaps in the same manner and then simply walk on. John Owen, the director of the Tanzania National Parks, had an especially exciting experience. One day when he was climbing up hill in the Ngurdoto Crater with a woman who was a famous horsewoman, the two suddenly saw a rhino coming towards them. Owen quickly escaped into the bushes; the lady pulled herself up on a branch. The branch broke and the woman landed on the rhino's back. Both mount and rider were terrified; the lady fell off and the rhino hastily ran away.

Of course, one cannot always rely on the harmlessness of rhinoceros. This the Swiss zoologist Rudolf Schenkel learned when he observed rhinoceros and lions in the Tsavo National Park in Kenya while on foot. Many of his encounters with black rhinos indeed where harmless–but one evening a bull attacked him when he moved along at a distance of about 50 meters, silhouetted against the horizon. Schenkel ran towards the bull, roaring in order to drive him away. Since the bull was approaching at full speed, Schenkel had to dash aside in order to avoid him. He ran towards a small tree whose top half was broken off and was hanging down. There was no time left to climb into the intact part of the tree top. So he just ran around the trunk and over its broken part, while the rhino had to run around the part with the dead top. But soon the bull changed his method. While Schenkel remained on the side of the tree with the broken part of the trunk, the bull waited on the other side in order to suddenly dash forward. Schenkel tried to get into the remaining part of the tree top, but the bull caught him and threw him into the air. He came down at first on the animal's shoulder and then to the ground where he immediately crawled under the broken tree top. The bull pushed aside the broken trunk and top part of the tree. Schenkel decided to remain motionless, lifting one foot to the level of the rhino's mouth so he could push himself off in case the worst happened. At first the bull was puzzled; then he came closer until his nose touched the naked foot—the

Is the African black rhinoceros dangerous?

shoe had fallen off. When the bull no longer saw the moving object, he responded to the human scent. He suddenly turned around and trotted away with his tail erect.

So we can see that the behavior of black rhinoceros may differ substantially; it depends on the behavior of the people who share their habitat. The Wakamba in Kenya pursue these animals with poisoned arrows or sling traps for the legs. The poor rhinos may drag around such a sling of wire with a heavy piece of wood attached to it for days and weeks, while all the time the wire cuts deeper into muscles and bone. Therefore, the rhinoceros in the Wakamba territory are said to be aggressive and mean. However, in the territory of the Masai who do not hunt and leave the rhinos in peace, they are known to be rather peaceful.

Sometimes it turns out that rhinoceros who suddenly attack have been wounded before. Once in Tanzania, Oscar Koenig had shot the hindquarters of a rhino which was blocking his way. During the following nights this animal turned over three limousines and two trucks and finally had to be shot. Kearton reported that a woman hunter, who had shot at a rhino which was generally known to be peaceful with too small a caliber bullet, was killed by the animal. The following day a farmer and his wife from the area came along the road with their car. The rhino immediately attacked when he saw the car. The man quickly pulled his wife out of the car and helped her to climb a tree; he could not reach it himself in time and was killed. In 1964, a game warden in the Hluhluwe Game Reserve in Natal was thrown into the air twice by a black rhinoceros; he was seriously injured in the thighs and buttocks. When the rhino started the third attack, the game warden grasped the frontal horn and desperately clung to it. The rhino shook his head vigorously from one side to the other, trying to dislodge the man. He finally succeeded with an extremely vehement jerk. When the man flew into the bush, the rhino departed.

Rhino attacks on cars

I have experienced several attacks on cars, all of which I had provoked myself. In most cases the animals stopped just short of the car without touching it; only in one case did I get a dent in the metal. One day when I wanted to take a closer look at the ear openings of the earless Prixie, the game warden's son drove me close to the sleeping animal. Prixie suddenly jumped up on all four legs and attacked immediately, making a dent in the side of the open car right next to my buttocks. In Amboseli, too, in 1965 a rhinoceros pierced his horn through the open window of a fully occupied limousine into the metal of the roof, and then made dents all over the car. It injured the passengers with the shaft of a spear which was still stuck in his throat—so again this aggressive rhino actually was a wounded animal. Often rhinoceros work over a passing car just out of mere curiosity, sticking their heads under the fender and shaking the vehicle. A game warden of the Hluhluwe Reserve, on one such occasion, bravely got out of his car and hit the animal over the head with his belt. On the railroad from Moshi to Same a rhino once chased off all the workers and damaged their lorries.

I do not know of anyone who has even seen a black rhinoceros cross a lake or a river by swimming, although these animals love to wallow or go into shallow water to graze on the reeds. However, they are able to swim. During the damming up of the artificial Lake Kariba in Sambia, attempts were made to save the game animals from the slowly disappearing islands. It happened that a rhinoceros, while attacking the boat, went into surprisingly deep waters where it could no longer stand. However, it did not disappear completely in the floods—just the nose, ears, and eyes were slightly above the water. A few waves would have been sufficient to eventually drown the animal.

They are poor swimmers

In spite of their apparent awkwardness, black rhinoceros climb rather high up into the mountains. In East Africa, they were found at elevations from 900 to 2700 meters. They live in dense bush, in scattered forest, on open grass plains, and even in semi-desert. They do not like hot and humid areas; therefore, they have never penetrated the rain forest of the Congo Basin or the woodlands of West Africa. Thus, even in earlier times, they were never found throughout Africa. From the time the Europeans first entered Africa, the rhinos have become exterminated in wide areas of their habitat. In South Africa, south of the Zambesi River, only a few are left in protected areas. In Rhodesia and Malawi, too, they have become rare; they are somewhat more numerous in Zambia, especially in the area of the Luangwa River. The estimate for the Portuguese area of Mozambique is approximately five hundred head; for Angola it is one hundred and fifty; and for Southwest Africa, two hundred and eighty. In the French colonies of Africa, they were nearly extinct by 1930. Only then were strict laws for their protection introduced, saving some. The few rhinoceros in the Southern Sudan may have disappeared in the last years due to the civil wars going on there and the ready availability of firearms. If it had not been for the National Parks, especially in East Africa, and other protected areas, the black rhinoceros would probably be extinct by now. The total number now (as of 1967) of black rhinoceros left are only 11,000 to 13,500, three to four thousand of them in Tanzania.

How many black rhinoceros are left?

The white hunters especially have wreaked havoc among the black rhinoceros. No less than 800 rhinoceros horns were exported from the sultantate of Fort Archambault in the area of Lake Chad in 1927. The professional big game hunter Cannon has killed about 350 rhinoceros in less than four years. He and a butcher by the name of Tiran "worked" mainly in the Cameroons, in Ubangi, and Chad. At times they switched from ivory hunting to rhinoceros because killing the rhinos was easier and their horns had increased in price. These people

What white big game hunters did

supplied modern firearms to the natives who eagerly participated in the shooting. The British big game hunter John A. Hunter brags about having killed more than one thousand and six hundred rhinoceros and more than one thousand elephants, partly of his own volition but also by order of the government who wished to prepare the land, for example the Wakamba, for settlements. In 1947 he killed three hundred rhinoceros there and, in the following year, another five hundred. Later it was found that this area was hardly suitable for settlements. The most difficult to understand are the so-called "sport-hunters" who, just for the fun of it, without any economic gain, have travelled in Africa and killed as many of these unsuspecting animals as possible. There are reports about a Dr. Kolb who has killed one hundred and fifty rhinoceros in East Africa.

It might be of special interest for psychologists to analyse the mentality of such wholesale killers from their letters and reports. These "Big Game Hunters" obviously are an entirely different type of man from those hunters in Europe who care for the game and spend large sums in order to preserve or to improve the game population. Since traditionally big game hunting in Africa has been described as something worthy of heroes, one may presume that a personal feeling of inferiority, destructive tendencies, and a certain addiction for fame have led to such slaughter. However, the rhinoceros hunt especially has never been a dangerous, heroic deed. During the many years he lived in Africa, the English explorer Frederick Selous (1851–1917) had not heard of a single instance where a European rhinoceros hunter had been killed by a rhino.

Unlike the roaming elephants, rhinoceros rarely return to areas where they once were exterminated. There is only one way to reintroduce them: They must be caught in other places, transported in boxes, and set free in that area. This was done in the 1950's in the Garamba National Park in Rwanda. During the last few years, we have caught sixteen rhinoceros, many of them seriously wounded, in the hunting areas of Tansania and brought them to the island of Rubondo in Lake Victoria. Meanwhile, they have reproduced there. Following habit and instinct, rhinoceros will stay in their home range after it becomes settled by man and disturbances increase.

Ethological studies in national parks

Since we now obtain our information on these gray giants no longer from big game hunters but from patient scientists and game wardens, we have learned more about their life. Studies on their behavior really began only in 1960. In contrast to many other species of animals, black rhinoceros do not have territories from which they chase conspecifics. However, at certain times of the year as well as the day, one may find the same animal in the same place engaged in the same activity. Once a day a rhino takes a specific, well-trodden, wide path to get a drink. The distance between the pasture and the waterhole may be eight to ten kilometers. Usually the rhino begins to graze only in the afternoon, spending the rest of the day in the shadow of a tree or in a wallow. At night at the waterhole, the animals may play, chasing each other, hissing and snorting. Where they are not pursued by man, as for example in the Ngorongoro Crater or in Amboseli, they are in the open all day long.

From the more intensive observations in the Ngorongoro Crater, we learned that individual black rhinoceros do exclusively remain in a specific home range, as was originally thought. This open protected area in Tanzania measures two hundred and sixty square kilometers; it is possible to count from an airplane how many rhinoceros are there at a given time. In January, 1958, my son and I counted nineteen rhinoceros there. Molly's count in March, 1959, revealed forty-two animals. Hans Klingel, between June, 1963 and May, 1965, found a population of sixty-one rhinoceros in the Ngorongoro area; thirty four of which seemed to be more or less permanent residents of the bottom land crater. J. Goddard, a biologist who lived in the Crater for three years until 1966, knew each animal individually, and regularly took photographs of them. During this period, he saw one hundred and nine rhinoceros in the crater. These varying figures are due to the fact, as Goddard presumes, that the great majority of the rhinoceros live the whole year round in the area above the rim of the Crater. Most of the permanent residents, especially the bulls, were found regularly in distinct areas, according to Klingel; but it may also happen that single animals of both sexes lose their home range, moving permanently to another area.

Black rhinoceros like especially to eat branches, which they grip with their upper lip as with a finger of a hand. When grazing on a pasture, in many cases, they pull out tiny little bushes only. According to the observations of Fraser-Darling, a rhinoceros daily pulled out two hundred and fifty little whistling acacias. In Natal (South Africa), two black rhinoceros were seen breaking off a rather large Mtomboti tree (*Spirostachya africanus*). One of the animals held the trunk of the tree between the two horns and then pushed, slowly shifting the weight of his body with a circling movement. When the tree broke off and lay on the ground, the two animals ate the shoots from the tops of the branches. Rhinoceros also eat the very prickly branches of thornbush and do not mind the sticky white juice of the Euphorbias. Klingel repeatedly observed a group of four animals who ate Wildebeest droppings. There they did not consume any plants at all but went straight from one pile to the next. They probably satisfied a need for minerals and trace elements.

The diet of the African black rhinoceros

In some areas these gray giants dig up the mineral soil with their horns. It is said that they tear up their own dung heaps in the same manner. The usual method is just to use their hind legs such as when

Behavior while defecating

a dog covers his fresh feces by scratching soil over them. In contrast to elephants, rhinoceros do not urinate and defecate at the same time. However, different individuals, bulls as well as females, may defecate on the same heap. Only in rare cases do they pause briefly to deposit their droppings right on their paths. The dung heaps probably do not mark a specific area as an individual's territory. Rudolf Schenkel, while doing research on the black rhinoceros in the Tsavo National Park in Kenya in 1964 and 1965, thought that the animals in an area maintain olfactory contact with each other. For similar reasons, the female rhinoceros may spray urine on their paths when walking. Bulls sometimes attack bushes first with their horns and then with their feet until they finally spray urine over them.

How rhinoceros sleep

Herbert Gebbing studied the sleeping habits of rhinoceros in 1957 at the Frankfurt Zoo. Usually, the animals lie on their bellies, slightly to one side, with the front legs pulled in an angle under the body and the hind legs stretched out forward. The head rests forward on the ground. Only in rare cases does the animal lie completely on one side, stretching out all four legs. The rhinoceros seem to rest in this position in especially deep sleep. Their sleep lasts quite long, an average of eight to nine hours a night. Usually they rest without interruption for two, three, or even five hours, and they are not disturbed by familiar noises. Two or three times during the night they get up to defecate. According to Gerda Schütt, the rhinoceros of the Hanover Zoo slept for nine and a half hours; within this period they were up for almost three hours in which they ate almost without interruption. As soon as one of them got up, the other one usually would wake up too. If it did not, the first rhino would push the other with its head until it too stood up.

Rhinoceros among themselves

Except in wallows, one will find black rhinoceros always singly or in small groups of up to five animals at most. If there are two of them, in most cases, it will be a mother with her more or less grown up young or a bull and a female; rarely will two bulls be together. Rhinoceros who stand together may caress each other with their lips or rub their chin on the other animal. In 1958, game warden Ellis saw a group of adult rhinoceros females one evening in the Nairobi National Park coming out of the woods; three of the animals walked side by side, while the fourth walked behind them. The animal in the middle was obviously in labor. When the animals became aware of being observed, they stopped, but one of the females kept rubbing the flank of the mother-to-be with the side of her head and horn. Finally, they retreated into the bushes. Three days later, a newborn calf was seen there.

When rhinoceros encounter each other, the meeting sometimes may seem antagonistic, but, in most cases, it is peaceful. For example, there may be a mother standing with her child. Suddenly a big bull appears from behind a bush. All heads go up, the female snorts, and the bull snorts too; both of these huge creatures raise their tails straight up. The bull paws the ground several times with the hind legs and snorts. Then, almost simultaneously, both animals lower their heads and dash towards each other. One is prepared to hear the terrible clash of two heavyweights crashing into each other. Then, suddenly, at a distance of six meters, both stand still and look at each other with their heads erect. The ears are turned towards the other. Then the bull turns aside and walks to the water, and shortly after that, the female, too, turns around. However, a short while later all three stand together.

Elephants clearly are recognized as superior by rhinoceros, although the two species hardly ever have reason to fight. One day in Uganda, on a narrow path, an elephant and a rhinoceros were slowly walking towards each other. They did not become aware of the other's presence until they were fifteen meters apart. The elephant spread his ears widely and walked straight towards the rhinoceros, who stopped and lifted his head. When the elephant attacked, the rhino moved backwards, shaking his head from one side to the other, and snorting loudly. The next short forward movement of the elephant drove away the rhinoceros, which disappeared in a gallop in the direction from which it had come. Later, the two animals were seen grazing not far from each other without seeming to notice the other's presence. Mrs. Trappe once found a rhinoceros in the area of the Ngurdoto National Park which had apparently been pierced by elephant tusks, since the surrounding ground was covered with elephant footprints. There are several reports of similar cases. In 1960, game warden Koos observed a bitter fight between a rhinoceros bull and an elephant in the Kruger National Park. Obviously, the elephant was unwilling to let the rhinoceros drink water, but the rhino insisted. During the following fight, both animals fell three meters down the steep slope of the river bank and continued fighting in the water. Large pools of blood led to the place where the rhinoceros finally lay dead. He had four holes made by the elephant's tusks in his body as well as other injuries. It has been observed repeatedly that elephants covered rhinoceros, which they had killed, completely with branches and twigs.

The relationships between rhinoceros and other large animals are not at all as clear. A game warden in the Murchison Falls National Park once saw a black rhinoceros chase a group of twelve waterbucks over a distance of about one hundred meters. This was all the antelopes would put up with. Turning around, they attacked the grey giant who retreated quickly into the brush and did not show up any more. On another occasion, a rhinoceros attacked a herd of about three hundred and fifty Cape buffalo, who were grazing in a line about four hundred meters wide. The rhinoceros ran almost playfully along the line of the unsuspecting buffalos, chasing them in all directions, and then he

Behavior towards other animals

walked on. In the Nairobi National Park, Guggisberg also saw a group of zebras playfully attacking a rhinoceros who finally retreated. However, mutual toleration is far more frequent, sometimes even leading to a kind of friendship between rhinoceros and other species of animals. A. Ritchie reports on two rhinoceros who were seen over a long period together with a large herd of Cape buffalo. They even slept regularly in the midst of the buffalo in a clearing in the forest, lying right next to them.

In other cases, animals of a different species may help the rhinoceros to rid itself of parasites. In Natal, a female rhinoceros was rolling in a creek and two turtles were seen tugging hard at her fissured skin. This was obviously painful for the rhino because she repeatedly jumped to her feet. However, she made no attempt to attack the turtles. On another occasion, again in Natal, at least six turtles approached a rhinoceros who lay in a puddle and started to pull the ticks out of his skin. They would rise up to seventeen centimeters above the water level in order to reach the parasites. In order to pull off the ticks, the turtles would push their forefeet against the rhino's body, take the ticks in their mouths, and then pull until the parasites came out. When the turtles worked in the more sensitive parts of the rhino's skin, he quivered several times, but the turtles did not pay any attention to this.

It is also said that the cattle egret picks the parasites from rhinoceros. Indeed, the cattle egrets follow the rhinos all day long, even sitting on their backs. But they seem only interested in catching the insects stirred up by the large animals. They do not pick ticks from the rhinos; this was confirmed by an analysis of their stomach contents.

Occasionally, rhinoceros calves are killed by lions. In 1966, in the Manyara National Park (Tanzania), several lions attacked a rhinoceros mother with her calf and drove them towards the entrance gate of the park. About fifty meters from the administration building, they caught the calf while the mother called loudly as if for help. Two passing cars were chased back by the rhinoceros mother, but the furious animal was driven off by shouts and rocks. The lions left the remains of the calf and walked off. In the Ngorongoro Crater a subadult rhinoceros was found killed by lions; he had severe injuries at the throat. Since there were no signs of a fight, one may presume that the lions broke the animal's neck. Although the lions stayed for one day with the dead rhinoceros, they made no attempt to eat it. The next day they moved on.

Usually, however, rhinoceros do not pay any attention to lions, even if the large cats walk closely past them. Once in a while, especially at waterholes, they are killed by other large animals. Guggisberg once watched a rhinoceros about to drink from a clear spring-fed pond in the Tsavo National Park. There a hippopotamus surfaced, grabbed the rhino's right front leg, pulled him down, and tore him to pieces with his huge tusks. Selous even photographed an adult female rhinoceros that was pulled under water by a crocodile and drowned.

When two black rhinoceros fight with each other, which occurs rarely enough, it is quite a spectacular sight. As a rule, the two opponents are not—as is the case with deer and horned ungulates—two males that are jealous or may be fighting for a territory, but two quarreling females or a female fighting a bull. Sometimes, however, what appears to be a "fight" is actually just play. Our pair of black rhinoceros in the Frankfurt Zoo would often play for hours with horns pressed together. More often the calf will play in this manner with its mother or father. But even in serious fights rhinoceros rarely injure each other. The many wounds at their shoulders and flanks have other causes. There are many hypotheses, and for a while it was believed that they had been caused or enlarged by the pecks from the beaks of the ox peckers.

This is a distinct possibility. Because the rhinoceros are plagued by many parasites which are pulled out by these birds, such crescent-shaped wounds may easily result. J. G. Schillings found extremely thin worms in these wounds which are transferred by a mosquito. The black rhinoceros are also plagued by several other animal parasites. In their stomachs live the larvae of a species of fly which attach themselves by their mandibles to the stomach walls and live off the tissue fluids and blood. As soon as they have metamorphosed, they pass through the anus and pupate in the soil. The large-headed flies which result do not take any food, but stay close to the rhinoceros and deposit their eggs usually near their heads and horns. It is unknown how the larva get from there into the stomach. Besides these parasites, twenty-six different species of ticks were found on the black and square-lipped rhinoceros, in addition to one species of leech and several species of tapeworms. All these rhinoceros parasites are not dangerous to man or domesticated animals. Rhinos in a zoo are usually free of these parasites because there are no intermediate hosts to transfer the parasites in the new environment.

In recent times, procedures were developed that permit the immobilization of large mammals by shooting them with darts containing narcotic or paralyzing drugs (see p. 69). Since that time it has become much easier to capture rhinoceros, to move them to other areas, or to treat injured ones in the wild. Thanks to this method, it was possible in the Amboseli National Park in 1962 to remove the badly injured eye of the famous "Gertie." She recovered tolerably after twenty-four hours.

When a female rhino is in heat, the bull stands opposite to her. The animals sniff at each other's mouths, frequently making gargling sounds. Almost regularly, the female then attacks the bull and butts hard into his flank. The bull tolerates this, even though the butts are

Fig. 2–6.
The fights of rhinoceros are fair duels which are performed according to specific rules. Serious injuries rarely occur, and often these fights are mere play. (Black rhinoceros, *Diceros bicornis*.)

The mating behavior of the African black rhinoceros

sometimes so hard that he has to burp. If a second bull appears, he may dance around her in a circle. But in spite of this, the two males do not fight; the female chooses her favorite. During this part of the courtship, the animals snort, sniff, grunt, or occasionally squeak. I have never heard a loud, penetrating whistle in the wild, such as I heard from our bull in the zoo. Perhaps it expresses surprise. It is possible to attract rhinoceros by imitating the snorting and sniffing.

Black rhinoceros mate throughout the year and can have offspring at all times of the year. Martin Johnson once watched the courtship of a rhinoceros pair from his car at a very close distance. Both animals circled each other with short, stiff-legged steps. After half an hour, the bull smelled the car, snorted with surprise, and stormed into the bushes with his tail erect.

"Of course, we expected the female to do the same," Johnson's report continues. "But this did not happen. It almost seemed as if she had not even seen her mate disappear. Apparently she was quite surprised that his amorous efforts had suddenly ceased. But then she became aware of us and she started her mating behavior all over again, obviously treating our car like a rhinoceros. That a car may suddenly become the object of a rhinoceros female's admiration was quite an unusual experience for us. This new amorous adventure was not restricted to a moment only. The creature tried for fifteen minutes or so to attract the attention of our silent, motionless car. Then she retreated virtuously, and when nothing happened, she stopped, and pranced clumsily. She seductively took a tuft of grass and threw it into the air. She gracefully approached us in a stilted walk, and came even closer than before. Then she smelled our scent. With an angry snort, the animal stopped her flirting. Down came her head, up came her tail, and she headed straight for us; in the next moment she dashed against our fender. Because the clatter of the metal and our yelling were new sounds to the rhino's ears, she snorted once more angrily and then took off in the direction of the salt lick."

The actual copulation we have repeatedly observed here in the Frankfurt Zoo. This event has rarely been observed in the wild. Frank Poppleton describes a bull who stood with the soles of his front feet on the female's back and remained in this position for thirty-five minutes. The animals held their heads parallel and moved slowly forward in a circle. When the bull had dismounted, the female turned to him and the two looked at each other for some minutes. My associate, Dr. Scherpner, observed in Tsavo Park a copulation which lasted 21 to 22 minutes. In 1964 and 1965, John Goddard observed copulating rhinoceros six times in the Ngorongoro Crater. In one instance, the male and the female stayed together for four months after the mating. Two other pairs split up soon after mating, but they were observed courting one month later, and again separated after that.

South African Savannah Three Hundred Years Ago

When European settlers took over South Africa, the country was covered with vast herds of game. As he did in many places on earth, the white man destroyed what nature had entrusted to him for profit, trophies, or just for the pleasure of killing. Blue antelope, quagga, and Burchell's zebra became extinct. Bontebok, white-tailed gnu, and mountain zebra were destroyed except for a few survivors which live on several farms. These remaining animals have reproduced well during the last decades under strict laws which protect them so that these forms of life may perhaps be saved.

Even-toed ungulates: 1. Bontebok *(Damaliscus dorcas dorcas)*, Ch. 14. 2. Blue antelope *(Hippotragus equinus leucophaeus)*, Ch. 14. 3. White-tailed gnu *(Connochaetes gnou)*, Ch. 14. 4. Eland *(Taurotragus oryx)*, Ch. 12. ☐ Odd-toed ungulates; 5. Burchell's zebra *(Equus quagga burchelli)*. 6. Quagga *(Equus quagga quagga)*. 7. Cape mountain zebra *(Equus zebra zebra)*. ☐ Ostriches (see Vol. VII): 8. South African ostrich *(Struthio camelus australis)*.

Mervyn Cowie, the former director of the Kenya National parks, was present when a bull mounted two females in succession within a short time. In the time between he was attacked—as is usual in rhinoceros—by the first female.

The first successful breeding in a zoo

Since 1941 the black rhinoceros have reproduced in zoological gardens. It took place for the first time at the Brookfield Zoo in Chicago. The second "black" zoo-rhinoceros was born in Rio de Janeiro. The first one in Europe was born in 1950 in the Frankfurt Zoo. At the Frankfurt birth, 17 liters of amniotic fluid came out first. Our female rhinoceros, "Katharina the Great," was so tame that she could be milked before giving birth. The first distinct signs of labor appeared only one and a half hours before the actual birth. The cow permitted the veterinarian to pull out the baby, which weighed twenty-five kilos. After a few seconds, the ears of the newborn moved. Two minutes later the mother belatedly attacked the assistants present in her stall. Then she smelled at the baby, but she did not lick it.

The development of young

The newborn rose to its feet ten minutes later; one hour after birth it was walking around in a lively manner. After four hours, it found the mother's udder and drank. Not until nine and a half hours later did it lay down for one hour. At birth the frontal horn of the young was only a stump one centimeter thick, and the second one was just a white spot. Rhinoceros born in other zoological gardens weighed only twenty kilograms at birth, but one in Hannover was thirty-eight kilograms. So far, twin births have never been observed in rhinos. To the best of my knowledge, until now all black rhinoceros born in zoological gardens could be raised. So far we raised two at the Frankfurt Zoo. In Rio, as well as in Frankfurt, the females were mounted regularly during the gestation period, since they were always kept together with the bulls. Eight days after the birth, our rhinoceros female was again completely tame with the keeper and all persons with whom she was familiar. We could go into her stall, ride on her back, and play with the young.

As early as 1911, the Hungarian explorer Kalman Kittenberger mistakenly killed a furious black rhinoceros which was in the process of giving birth. He opened the dead animal's abdomen and managed to get the young out alive, although it died after eight days. It was only in 1963 that game wardens Malinda and Edy of the Manyara National Park observed the birth of a black rhinoceros in the wild for the first time. They found a female rhinoceros lying on the ground. Taking it for dead, they started to throw rocks at her. When they came closer, they saw the soil around the animal was soaking wet. Within the next few minutes, the rhinoceros suddenly arose; the baby emerged, apparently without causing any difficulty to the mother. After another ten minutes, the calf dropped to the ground. The mother turned around and began to remove the foetal membranes with her lips. Ten minutes later, the baby stood on its feet and shook its ears.

The young nurse for about two years at the mother's two nipples and usually remain with her for three and a half years. If one captures a nursing young rhinoceros, it will become as tame as a domestic animal. In most cases, eight to ten months will pass until the female becomes pregnant again. In Amboseli Park, the first calf stayed with its mother for two and three quarters of a year, the following one three years, and after five years she gave birth to the third one. Black rhinoceros are sexually mature at the age of approximately seven years.

Previously, when a zoological garden kept rhinoceros, they were mostly great Indian rhinoceros. Unfortunately, the great Indian rhinos are now almost extinct, and only a very few may be given to qualified zoos. The first black rhinoceros came to Germany in 1903 to the Berlin Zoo. Now black rhinoceros are the most frequently kept species of rhinoceros in zoos. In 1966, thirty-two of them were kept in zoos in the United States.

The animals usually become very tame in captivity; it is even possible to ride on some adult females' backs. They like being caressed with the palm of the hand over their closed eyes. Probably due to a lack of anything to do, they often rub their horns against concrete walls and iron fences, which reduces them to short stubs. Therefore, a trunk of soft pine wood should be placed in a rhinoceros pen where the animals may rub and polish their horns. They cannot cross a ditch of 1.75 meters in diameter at the upper rim and 1.20 meters of height at the outer wall, even though the inner wall of the ditch is inclined upwards towards the animals. How long they live we only know from zoological gardens. In Brookfield Zoo of Chicago, the breeding pair which came there on May 19, 1935, is still alive (1967). The two animals, who must be by now approximately thirty-three or thirty-four years old, do not show any symptoms of old age. Presumably, rhinoceros may reach an age of about fifty years.

A most impressive animal, which today is found only in a few savannah regions in Africa, is the SQUARE-LIPPED RHINOCEROS (*Ceratotherium simum*; Color plate, p. 37). It is the largest species of rhinoceros. The HRL is 3.6–4 m, the BH (shoulder) is 1.6–2 m, and the weight is approximately 3 tons (in one case approx. 5 t.). There are two horns. A strong shoulder lump, which consists of muscles and epidermal tissues, is not supported by the skeleton. It has wide, almost square-shaped lips that characterize it as a herbivore. Incisors are present only in the embryonic stage; later there are only high crowned premolars and molars: $\frac{(1) \cdot 0 \cdot 3 \cdot 3}{(1) \cdot 0 \cdot 3 \cdot 3}$. The gestation period is seventeen to eighteen months; one young is born. There are two subspecies: 1. SOUTHERN SQUARE-LIPPED RHINOCEROS (*Ceratotherium simum simum*). 2. NORTHERN SQUARE-LIPPED RHINOCEROS (◊ *Ceratotherium simum cottoni*).

In place of the lacking incisors, the square-lipped rhinoceros has a

The square-lipped rhinoceros by H. G. Klös

Fig. 2-7.
Former and present dis-
tribution of the two sub-
species of the square-
lipped rhinoceros:
1. Northern square-lipped
rhinoceros (Ceratotherium
simum cottoni); 2. Southern
square-lipped rhinoceros
(Ceratotherium simum
simum). Only in the areas
marked with black
triangles do square-lipped
rhinoceros still exist.

hard, horny edge on the lower lip that facilitates grazing. A too rapid abrasion of the molars by the hard grass which contains silicic acid is prevented, or at least slowed down, by a specific course of development of the teeth. The ridges on the surface of the teeth are especially high and the intermediate space is filled with dentine. Due to this construction, the tooth has achieved extraordinary durability. According to Player and Feely, the square-lipped rhinoceros in Zululand prefer specific grasses like *Urochloa, Panicum,* and *Digitaria.*

As in the black rhinoceros, the best developed of the senses is the olfactory. Hearing and vision are rather poor. According to Dieter Backhaus, square-lipped rhinoceros are able to recognize a person, approaching slowly in a favorable wind direction, only at a distance of approximately thirty to thirty-five meters. I found the same to be true in the Umfolozi Preserve. Since square-lipped rhinoceros are much more placid and less aggressive than the black rhinoceros, I could frequently observe them quite easily from a very close distance before they would raise their huge heads, uneasily turn their large ears in all directions, and finally trot away with their tails curled upward. Only once when they were greatly terrified did I see them pull their tails between their hind legs. The light trot of the square-lipped rhinoceros does increase to a considerable speed in situations of danger. Player and Feely report as follows: "The normal type of fast locomotion is an extremely rapid and graceful trot which was measured at a speed of 29 km/hr from a vehicle. When galloping short distances they may attain a speed of 40 km/hr."

The bare skin of the square-lipped rhinoceros, which has only a few bristles at the ear rims and a tail of a dull slate color, is only slightly different from the black rhinoceros. The common name "white rhinoceros" for the square-lipped rhinoceros, therefore, is as confusing as the name "black rhinoceros" is for the other African species. The name probably came from an error in translation; the Boer word *wijde* (wide) was changed, due to a misunderstanding, into the English word white. Furthermore, after the animals have wallowed in the mud, the hot African sun dries it quickly into a crust which covers the body like another skin. Depending upon the soil, these mud crusts are different colors. Therefore it is possible that the name "white rhinoceros" originally was given to animals which had wallowed in a light clay and therefore appeared to be "white."

Like in the black rhinoceros, the anterior horn of the square-lipped rhinoceros may reach a considerable length, while the second horn always remains shorter and bulkier. So far, the record measurement of an anterior horn in the Southern race was, according to Maberly, 1.58 meters. The females' anterior horns often are longer and thinner than those of the males. H. Lang thinks that the frontal horns serve, so to speak, as a bumper: "The horns are carried closely to the ground to clear the way for the short, column-shaped front legs and the barrel-shaped body. When the animal moves slowly or is grazing, he is constantly nodding his head. It is this abrasion which is the natural cause for the smooth surface of the horns, the flattening of the frontal parts, the wearing out immediately above the base and rear edge of the anterior horn, as well as the frequently found spatula-shape of the second horn. It is not due to the friction during digging and honing against stones." The nose horns, which are rather loosely attached to the skin, represent the weak point in the construction of rhinoceros. The horns, especially long ones, are easily torn off if force is applied. The female square-lipped rhinoceros Kuababa of the Berlin Zoo lost her anterior horn in May, 1963, when she was captured in the Umfolozi Conservation Area. By July, a small elevation became visible on the nose, and in December the length of the horn had reached thirteen centimeters. The new anterior horn kept growing at an average rate of half a centimeter per month. In March, 1967, it had attained a length of 34.5 cm. The behavior of the Northern subspecies has been described in 1959 by Backhaus, and that of the Southern subspecies by Player and Feely in 1960.

In contrast to the black rhinoceros, the square-lipped rhinos are quite sociable animals. One encounters them in smaller groups of sometimes up to eighteen animals, and often an adult bull is with the group. Frequently there are several mothers, each with a small and a subadult calf among them. However, the females with calves will tolerate a bull's presence only until he attempts to mount. As soon as he does this, they will reject him in bitter fights which can result in the death of the bull. The bull, however, will not tolerate a calf very close to him. Therefore, a young accompanied by a female in heat is continuously in danger of being killed by the furious bull. If such a group is alarmed, all animals stand in a circle with their behinds together so that the heads, armed with horns, point outward in all directions.

The square-lipped rhinoceros live gregariously

The individual home ranges are marked with urine by the bulls. The animal sprays the urine backwards with explosive force in two or three jets so that the bushes or grass in the area are covered with small white drops. A strange crescent-shaped drag-spoor described by Hediger in 1915 and by Backhaus in 1959 may also be a part of the marking behavior. Furthermore, the sometimes very high dung heaps, which are frequently found along the various paths of the square-lipped rhinoceros, are quite conspicuous. Apparently, the sight of such a dung heap irresistibly stimulates the square-lipped rhinoceros to defecate. After defecation, the animal makes scratching movements with the hind legs.

The daily activities of the square-lipped rhinoceros seem to depend to a great extent on the weather. In the hot sun these savannah animals retreat into the shade, reappearing only at dawn in the open pastures. They also seek shelter in the bushes from rain or when it is cool. A long period of the day is spent grazing by the rhinos; frequently they graze the whole night long. Like all species of rhinoceros they need wallows for their well-being. Frequently, they first drink at the wallows and then they spend a long time, sometimes even during the night, resting in the mud. Player describes the frequently visited wallows in the Umfolozi Preserve where turtles await the arrival of the rhinoceros, eager to take off their ticks; this agrees with the observations of the black rhinoceros. During the winter, the square-lipped rhinoceros likes to take a sand bath, which takes the place of the summer mud wallow.

Reproduction

So far there is very little known about the reproduction of the square-lipped rhinoceros. There are hardly any observations from the wild. In captivity, only one birth is known. In June 1967, a female square-lipped rhinoceros, which was pregnant when captured, gave birth to a calf in the zoo at Pretoria (South Africa). According to Owen Smith, these large rhinos are fully grown at the age of seven to ten years. While the mating season is usually from July until September, females in heat have been observed at other times of the year. Then the males have bitter fights which may end with the death of one of the opponents. Foster observed a pregnant female rhinoceros over an extended period of time in the Umfolozi Preserve. She could be distinguished by her conspicuous horn, and he estimated her gestation period to be eighteen months (547 days). Usually one young is born, although according to Maberly, there have also been twins. Twenty-four hours after birth the young is able to follow the mother around. It seems to maintain no fixed position with respect to her as it follows its mother, except when it is in danger. Then it is always ahead of the mother, apparently guided by her horn and mouth. At the age of one week, it begins to eat grass, although it usually continues to nurse for another year. Probably an adult square-lipped rhinoceros female may give birth to a calf every two and a half to three and a half years.

Former distribution and extermination

Formerly, square-lipped rhinoceros occurred in many parts of Africa. Their former range can be reconstructed only with difficulty from the cave drawings and reports of the first European settlers, hunters, and explorers. The most Northern subspecies lived in the area from Southwestern Sudan through Uganda of the Central African Republic. The habitat of the Southern subspecies reached from the Orange River in the South up to the Zambesi in the North, and from the coast of the Indian Ocean in the East to the Damara country and the Kalahari Desert in the West. In 1785 the great French naturalist Buffon thought that the rhinoceros feared neither "the steel nor the fire of the hunter." But in the nineteenth century, when the days of the European hunters in Africa began, it was shown how inappropriate

Buffon's remarks were. It is shocking to read the contemporary reports on the deaths of the southern square-lipped rhinoceros. For example, Charles J. Anderson wrote in 1858: "In South Africa a large number of rhinoceros are being killed every year. One may get a good idea of the quantity when one hears that Oswell and Vardon killed no less than eighty-nine of them during one year. During my sojourn, I myself killed almost one third of this number."

By 1892, only seventy-five years after its first discovery by the explorer Burchell, the southern square-lipped rhinoceros was considered to be extinct. Fortunately, this was incorrect since a small number of the gray giants had survived in Natal in the valley of the Umfolozi River. It is to the great merit of the South African government that it made this last refuge of the square-lipped rhinoceros into a preserve as early as 1897.

At about the same time, in 1900, Major Gibbons discovered near Lado at the Upper Nile that, besides the moribund southern square-lipped rhinoceros, there existed another northern subspecies. In the Umfolozi Preserve, the population increased steadily owing to the excellent protection it received. In 1930, according to official estimates, there had been only thirty animals, but by 1966 their number had increased to nine hundred and fifty. In contrast, the population of the northern subspecies did not show such an even increase because their habitat covers several African nations. The population in the Central African Republic consists, at most, of ten animals. In the Congo in 1963 about one thousand square-lipped rhinoceros still lived; according to Curry Lindahl, only about one hundred have survived the revolution and the following civil war. In 1928 in Uganda, there were about one hundred thirty square-lipped rhinoceros. Their number increased to three hundred animals in 1951, but in 1962 it was down to only eighty head. The estimates for the Sudan diverge greatly; some informants speak of only a few hundred, while others talk about two thousand. In this case, the second number seems more probable because of the strict laws for their protection.

The Umfolozi Preserve in Natal is two hundred an eighty-eight square kilometers in size, with hilly savannahs between the White and the Black Umfolozi Rivers. During the last few years, the number of square-lipped rhinoceros has increased to such an extent that this preserve has become overpopulated. This increased the danger of epidemics, and the pasture became more and more scarce. Therefore, it was decided to give some of the animals to other preserves and national parks as well as to zoological gardens under scientific supervision. With "Action White Rhinoceros," another interesting chapter began in the exciting history of the square-lipped rhinoceros.

With the development of the capture gun, the catching of large mammals has become easier and less cruel to the animals. Pre-

The discovery of the northern subspecies

In the Umfolozi Preserve

Rhinoceros being resettled

viously, large mammals were captured at great cost in pits or with ropes, causing many cauulties; today many of them are shot with a specially constructed gun loaded with an injection-cartridge. The cartridge, when it hits the animal's body, discharges the drug and thus anesthetizes the animal. Of course, the anesthetic in use has to meet certain standards and must be safe within a wide limit. Often it is difficult to estimate the correct weight of an animal in the wild. Therefore, a drug is needed which is equally effective with animals of different weights. The effect should be immediate, before the animal can retreat into the dense brush after being shot with the dart. Furthermore, one must be able to give an antidote afterwards which neutralizes the drug's effect. However, the animal should also be able to recover without an antidote, in case it cannot be found. The answer to all these questions is a compound developed by A. M. Harthoorn which has been tested with good results on the square-lipped rhinoceros of the Umfolozi Preserve.

In spite of all these techniques and precautions, the capturing of one of these collossi is still an adventure full of breath-taking suspense. Landrovers, trucks with boxes, and riding horses take off early in the morning for the capture. As soon as a suitable rhinoceros is tracked down in the yellow glow of the savannahs, the marksman approaches the unsuspecting animal step by step against the wind. Every cover is taken advantage of until the distance between the rhinoceros and the marksman is only a few meters. Then he fires the shot, and the animal immediately jumps onto its feet and takes off with incredible speed. Now the men must follow on their agile horses. There is no time left for reflection and cautious looking for trails. They break through brush, jump over ditches, and cause clouds of ants to pour down from the whistling acacias. Even though the horses eventually learn the hard way to avoid the holes of wart hogs, herds of cape buffalo, and thorn bush, often horse and rider still lose shreds of skin in this wild chase.

Finally, after eight or ten minutes, the anesthesia takes effect. The rhinoceros slows down, stands still, staggers, and then lies down. The riders inform their fellows with the truck. The truck comes and the transport box is unloaded. The motionless rhinoceros is given a small injection of the antidote into the vein of the ear. It arises as if on command and can then be led into the box. Ian Player, the supervisor of the Natal Game Reserves, reports: "In areas which the truck could not reach, a small dosage of the antidote was enough to make the animal move. Then it could be led to the truck. Once an adult rhinoceros was led in this manner over a distance of two miles, much to the surprise of the tourists and some native women." The procedures of capture were later modified. Now the marksman drives with the landrover right up to the rhinoceros. In Uganda even a helicopter was used for the capture. Later, in the camp, the animals are carefully familiarized with the closeness of people and substitute foods. With this procedure, it has been possible since 1962 to send forty-two "white" rhinoceros to twenty-five different zoological gardens throughout the world. (The first square-lipped rhinoceros who were captured without capture guns were in the zoos of Pretoria, Antwerp, London, Washington, and St. Louis.) Furthermore, it was possible to transfer about one hundred and fifty square-lipped rhinoceros to several other protected areas in the South African Republic, Rhodesia, and Uganda. After this success, the square-lipped rhinoceros was removed from the list of species threatened with extinction in 1966.

With "Action White Rhinoceros," it became possible to bridge the time when, in C. Harris' words, "out of each bush looked the ugly head of such a creature" and the present where the square-lipped rhinoceros, after the most dire threat to its survival, has at last regained its foothold in Africa. We hope sincerely that the cruel decades of its extermination belong irrevocably to the past.

Bernhard Grzimek
Heinz-Georg Klös
Ernst M. Lang
Erich Thenius

3   The Even-toed Ungulates

Those extremeties of the mammals which originally had five digits and five toes have undergone many and varied changes during phylogeny in the different orders. In the large group called "ungulates," the hoofed animals, such changes led to a better adaptation for walking in almost all cases, while the ability to climb and to grasp gradually disappeared in these animals. For running purposes, the greatest possible simplification of the "component parts" is of utmost advantage (see Vol. XII, "Odd-toed Ungulates"). With the increasing adaptation to walking and running, the significance of the inner and outer digits and toes decreases. The weight of the body is increasingly shifted to the center digits and toes, while superfluous parts of the skeleton finally disappear completely.

However, while in the odd-toed ungulates the center digit and the center toe mainly carry the weight of the body, in the even-toed ungulates this weight is carried by the third and fourth digits and toes. The second and the fifth ones are reduced, but they still are well developed in the hippopotamus. In most of the other even-toed ungulates, they are more or less well developed as "pseudo-toes" which no longer touch the ground or do so only on very soft ground. In camels and giraffes, they have completely disappeared. The first digit and the first toe are lacking in all contemporary even-toed ungulates.

Distinguishing characteristics

EVEN-TOED UNGULATES (order Artiodactyla): They range in size from small to very large. The HRL is 40 cm (mouse deer) to 400 cm (hippopotamus, giraffe). The BH is from 20 cm in the mouse deer to 330 cm in the giraffe. The weight is 2 kg in the mouse deer to 3200 kg for the hippopotamus. The shape of the body greatly varies. The third and fourth digit and toe are enlarged, carrying the hoof. The second and fifth digits in the hippopotamuses are weaker (carrying the hoof), while in pigs, deer, pronghorns, and horned ungulates they are rudimentary (with or without pseudo-toes). In camels and giraffes, they are completely lacking; the first digit is always lacking.

All have skin glands which play an important role in intraspecific relations of many species, for example the marking of territory, cohesion of groups of conspecifics, and attracting mates. They have usually two to four nipples; only in pigs (family Suidae) are there six to twelve nipples. Their weapons are predominantly or exclusively used in intraspecific conflicts. Some have elongated canine teeth, for example, musk deer, camels, and pigs; others have weapons on the forehead such as antlers or horns. Only in exceptional cases do some have both prolonged canines and antlers (muntjacs). The hairless nose region may be very large (as in cattle), medium sized, small, or lacking altogether (e.g. camels).

The skull and teeth vary greatly. The front teeth, connections with a predominantly or exclusivley herbivore way of feeding, are more or less reduced in most even-toed ungulates. Only in the wild boar (genus *Sus*) are they complete. The crowns of the premolars and molars are, in present-day nonruminants, hinged, while in all the others, they are selenodont, which means that each consists of four crescent-shaped bumps.

Even-toed ungulates walk on the tips of their toes. With the increasing reduction of the lateral digits, the metacarpus and metatarsus of the two center (main) digits coalesce to form the "cannon bone." There is no clavicle. The cerebrum is profusely convoluted and trenched, partly overlapping the cerebellum. The eyes are often large and especially adapted to perceive movement. Many species evidently have color vision. The large ears are readily moveable, and the sense of hearing ranges from good to very good. The olfactory organs are well developed. There is a tactile sense, at least in lips and tongue. Furthermore, there are tactile bristles around the mouth, the eyes, and on the cheeks. For a description of the digestive organs, see the suborders. The uterus is two-horned. The gestation period, which lasts from four to fifteen and a half months, yields one to twelve young. The nursing period is from a few weeks to nine months.

Of the two large main branches of the ungulates, the odd-toed ungulates belong almost exclusively to the past. In the animal kingdom, they are represented by only six genera today. The even-toed ungulates, however, have reached their prime only in recent times. With eighty-one genera, they represent the majority of present-day ungulates and in fact of the large mammals per se. Not only do they represent the most important animals of prey for the primitive hunters as well as for modern-day trophy hunters, to them belong also most of the largest, most important domesticated animals.

The even-toed ungulates are an order of mammals characterized by their uniform structure. One glance at the feet identifies an animal as a member of this group of ungulates. Some zoologists doubt a closer relationship between the even-toed ungulates and the rest of the un-

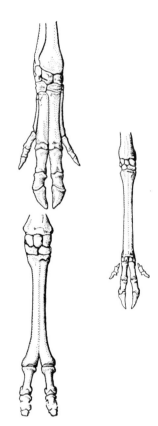

Carpal Bones of Even-toed Ungulates
Even-toed ungulates step on the enlarged center digits (3rd and 4th finger resp. toe). In pigs (top) the 2nd and 5th digit are still well developed as pseudo-claws, and the metacarpal bones are not fused. In deer (center) the lateral digits are weak, their metacarpal bone is rudimentary, and the metacarpals of the main digits are fused into the cannon bone. The hind feet are similarly structured. The camel (bottom) no longer possesses lateral digits. Here too the metacarpals have become fused into the cannon bone. The hind legs are similarly joined.

gulates because the difference in the structure of the foot compared to the odd-toed ungulates is rather large. However, the American zoologist and paleontologist G. G. Simpson was able to prove that a derivation of the even-toed ungulates from the primitive extinct ungulates (Condylarthra) is definitely a possibility. Therefore, the Condylarthrae may be considered as the common group of ancestors for all ungulates.

From fossil evidence the history of the even-toed ungulates can be traced down from the earliest Eocene (approximately sixty million years ago) to the present (see Color plate, p. 32). For the evaluation and classification of these fossils, which is the object of Paleontology, the details about the structure of the molars are of importance. The earliest even-toed ungulates are known from the early Eocene of North America. This is the genus *Diacodexis* which belongs to the Dichobunidae, a small, primitive even-toed ungulate with a complete set of teeth and that still possessed five digits. Many other fossils prove that the even-toed ungulates were a group of animals comprising an abundance of forms by the Tertiary, and that many of these former families are now extinct. But all these even-toed ungulates, extinct and living today, may be classified into three large suborders: the nonruminants (Nonruminantia), of which remain the pigs and the hippopotamuses, the Tylopoda containing the camels, and the remainder of the ruminants (Ruminantia) with the large numbers of our contemporary even-toed ungulates. The Tylopoda and the ruminants are described in subsequent chapters (see Ch. 6 and Ch. 7). In the present chapter, we shall be concerned in some detail with the nonruminants and their history.

NONRUMINANTS (suborder Nonruminantia) range in size between rabbits and hippopotamuses. The eye cavity opens toward the temporal cavity. All have the first upper incisor, and the molars are bunodont. The stomach is undivided or three-chambered without the esophageal groove characteristic of the ruminants. They are four-toed; only the peccary has a vestigial fifth hind toe. There is no metacarpus; the metatarsus occurs only in the peccary fused with the cannon bone. The contemporary forms are classified into two superfamilies with three subfamilies.

A. PIG-LIKE MAMMALS *(Suoidea)*: 1. PIGS (Family Suidae), 2. PECCARIES (Family Tayassuidae). B. HIPPOPOTAMUS-LIKE MAMMALS (Anthracotherioidea): 3. HIPPOPOTAMUSES (Family Hippopotamidae).

Even though there are some special adaptations in the nonruminants, they are the most primitive of all even-toed ungulates. The structure of their limbs and skulls, their still complete sets of teeth, and the more simple structure of their stomachs are primitive characteristics which show that the present-day nonruminants are more closely related to the oldest known even-toed ungulates from the Eocene than to the camels and ruminants. Today's pigs, peccaries, and hippopotamuses are the last survivors of this suborder, which formerly contained an abundance of widely branched forms. The oldest and most primitive nonruminants were the Dichobunidae, which are considered the most primitive group of all remaining even-toed ungulates. Closely related to the Dichobunidae are the Entelodontidae; these two and some other families comprise the Paleodonta. The Entelodontidae and their relatives were more advanced than the Dichobunidae in many respects. Their limbs had two or three digits and their metacarpus and metatarsus were not fused. Pictures of a large specimen of this highly evolved group of ungulates, the *Daeodon*, are shown on p. 32 and p. 92. They all became extinct during the Miocene, approximately ten to twenty-five million years ago.

The pig-like mammals (Suoidea) do have rather primitive teeth with bunodont molars. However, in the ancestors of the pigs (Suidae) the humps on the molars were not round but were similar to the crescent shape of the ruminants. Such teeth are called "bunoselenodont." At the beginning of the Oligocene, about forty million years ago, the pig-like mammals had split into two different lines, the true Old World pigs (family Suidae) and the peccaries (family Tayassuidae), which today occur only in the Americas.

The Anoplotheriae (family Anoplotheriidae), who appeared suddenly in the early Eocene, becoming quite numerous only to disappear again in the Oligocene, belonged to the strangest even-toed ungulates of the European Old Tertiary Period. Besides the two main digits, whose metacarpals and metatarsals were not fused, they had only the second digit. The name *Anoplotherium* means "unarmed animal."

We find large, clumsy forms among the Anthracotheriae (family Anthracotheriidae) whose fossils were found mainly in the Tertiary lignite stratum of Europe. They evolved in the Eocene presumably out of the Dichobunidae and reached their prime in the Oligocene. Then, about forty to twenty-five million years ago, there were giant forms such as the *Anthracotherium magnum*. Some genera extended as far as North America. These even-toed ungulates were especially numerous in the Old Tertiary of South Asia and Africa, becoming extinct only in the Pleistocene (glacial period). On p. 32 is a figure of the *Bothriodon*, a medium sized Holarctic form. Many paleontologists think that the hippopotamuses (family Hippopotamidae) are derived from the *Anthracotheriae*. Dehm places them with the Anoplotheriae in the infraorder Ancodonta. According to other opinions (Matthew a.o.), however, the hippopotamuses were derived from the primitive pigs. This question can not yet be finally answered with certainty in light of the presently available fossils.

The Caenotheriae (family Caenotheriidae), a rather strange hare-like even-toed ungulate, occurred in the Tertiary in the Old World. They not only walked on the tips of their toes like all other even-toed un-

gulates, but also on the entire toe and on the metacarpus and metatarsus. Their hooves were more claw-shaped. The Caenotheriae became
extinct by the Miocene. They are not immediately related to the other
even-toed ungulates.

The Oreodontae (family Agriochoeridae and Merycoidodontidae) of
the North American Tertiary were small to medium sized herbivores with
a selenodont set of teeth without gaps between them and with originally five-digit front feet and four-toed hind feet. On p. 32 is a figure
of *Merycochoerus*, a pig-like, short-legged form with a proboscis, which
presumably lived near and in water. Their systematic relationship
could not be determined for a long time because they have certain
similarities not only with the camels but also with the Anoplotheriae,
Anthracotheriae, and Xiphodontidae. However, they probably are
closest to the Tylopodes.

The color plate on p. 32 gives an idea of how, according to the latest
findings of paleontological research, the single evolutionary branches
of the even-toed ungulates are phylogenetically related.

Dietrich Heinemann
Erich Thenius

4   Swine and Peccaries

Among the present nonruminants, the pig-like mammals (super-family Suoidea) are the oldest and the most primitive even-toed ungulates. Most of the species have teeth characteristic of omnivorous animals. Their menu is extremely varied. They eat the parts of plants under ground as well as grass, leaves, sprouts, fallen fruit, seeds, juicy herbs, and mushrooms. Food below the surface is dug out with the snout. Because of their extraordinary sense of smell, they are able to find food quickly even deep in the ground. Depending on the season, they may also become carnivorous. Then they will eat worms, the larvae of insects, eggs and young birds, snakes, lizards, and small rodents. They may even attack injured and disabled larger mammals and eat carcasses. We distinguish between two families: 1. OLD WORLD PIGS (Suidae); 2. PECCARIES (Tayassuidae).

With their relatively large head, short neck, and mostly short legs, the Old World pigs (Suidae) appear compact. The HRL is from 50 to about 200 cm; the TL is 3.5–50 cm, the BH is 25–110 cm, and the weight exceeds 300 kg. The head is a more or less stretched, oblique cone-shape, terminating in a movable snout with a cartilage disc that is practically hairless and contains many glands, and the nostrils. In many species, strange looking skin structures are found on the facial part of the head: warts, humps, or pads. The small eyes are situated higher or lower. The ears are from small to medium size, usually ending in a point but in some cases they are round. The rump is straight and is higher than it is wide. They are predominantly of one color. In some species, the head, a streak down the back, the tassel, and the legs are of a different color than the rest of the body. The legs are short and graceful. The weight is carried on the third and fourth digits. The second and fifth digits are short, touching the ground only when it is soft. The end of the tail is often flattened, sometimes with a tassel. The thick skin is firm with a few sebaceous and sweat glands. When they are on a rich diet fat is often found under the skin. Most

Old World pig family

Distinguishing characteristics

species are covered by a more or less dense coat of bristles with or without fleece. In most cases, the neck or back mane stands erect when the animal is excited. The skull skeleton is bulky; the body skeleton is rather light and delicate. The scapula is long and rather wide.

Pigs are omnivorous. Originally they had 44 teeth: $\frac{3 \cdot 1 \cdot 4 \cdot 3}{3 \cdot 1 \cdot 4 \cdot 3}$. In some genera, there are fewer teeth. The canines, bent upward and hollow like a tube, are especially well developed in the boar. They grow continuously. The incisors, premolars, first and second molars usually come out later. The gestation period lasts from four to five months; from two to fourteen young are born, most with a brightly striped and dotted coat, depending on the species. There are five genera (bush pig, wild boar, wart hog, giant forest pig, and babirusa) with eight species and seventy-six subspecies.

Pigs have few special adaptations

Most species of pigs are relatively unspecialized. This made it possible for them to spread over an enormous area and to adjust to extremely varied habitats and climatic conditions. They live in large parts of Europe, Asia, and Africa. One species, the wild boar, has been introduced by man to North and South America, Australia, Tasmania, and New Zealand. Their habitats are the plains as well as alpine country. They live in forests, steppes, savannahs, swamps, and even in densely populated human settlements. Their habitat must have water—lakes, puddles, and banks of rivers and ponds are necessary because bathing in the mud is of great importance for their well being. They find shelter in brush, bushes, reed, and high grass, in dens, between the roots of trees, or in caves. They regularly visit places where they rub their bodies, and they also visit salt licks.

Many species use regular paths which connect their resting places, feeding areas, water holes, and wallows. When they find enough food, they remain in that locality. Otherwise, they will roam over great distances in search of food. Their usual form of locomotion is a trot; over short distances, however, they also can do a fast gallop. However, they do not like to jump. Being excellent swimmers, they even cross rivers and channels of several kilometers. This way they were able to populate islands quite distant from the coast. Only a few species are genuinely diurnal animals. Their main activities take place at dawn or during the night, especially in areas where they are pursued by man.

All pigs are gregarious animals, living together in pairs or in groups. The focus of a group is usually a female with her young. Older boars often live alone except during the mating season. At preferred feeding places or on migration, the groups may form herds of considerable sizes, but even then the single groups keep some distance from the others. All members of a group stay closely together. Since pigs vocalize readily, squeaking and grunting sounds play an important part in communication. Furthermore, one can distinguish various calls for maintaining contact, calling young and others, alarm, attack, mating, and distress. They also have many different snorting and sneezing sounds, as well as other sounds they produce with their teeth. The sense of smell is important in recognizing conspecifics. Eyesight plays a lesser role, since most species have poor vision and can see at short distances only.

In minor conflicts, pigs bite the opponent or push him away with their snout. Some species show much competition for food and defend delicacies against members of their group. Serious fights occur mainly between adult boars during the rutting season. The males then often use their sharp canine teeth for tearing and slashing. These canines are also used as defensive weapons against predators or as a lever. Boars attack each other either frontally or from the side.

Probably all pigs like close bodily contact, and, therefore, they snuggle up to conspecifics as closely as possible when resting. Then social grooming may be observed. They run their teeth and lips through their neighbor's mane or massage him with the disc of their snout, all of which the other readily tolerates. He even may demand it in no uncertain terms.

While pig species in moderate climates have a rutting season, tropical pigs may mate all year round. In the tropics reproduction may be determined by the rainy seasons. Prior to copulation a special ritual occurs where the boar causes the female to stand still by making special sounds and by massaging her with his disc-like snout. Usually wild pigs have one litter per year. Under especially suitable foraging conditions they may have two.

Before giving birth, the females usually separate from the group. They seek a sheltered, dry place and dig a cave or a depression which they may line with various plant materials. Sometimes they even build a roof over their "nests." In most cases, young are born in rapid succession; the piglets are small and "rounded" in shape and so pass the birth channel relatively easily. After reaching the ground, they immediately start crawling. At first they are rather clumsy, but after a few hours they are quite sure of themselves. The movements result in the severing of the umbilical cord; it may tear anywhere along its length. Pig mothers belong to the "passive" parental type. As a rule they do not eat the placenta, nor do they clean or lick their young. Immediately after birth, the young pigs begin searching for the nipples of the lying mother, massaging her udder with their snouts. Depending on the species, pigs have between two and twelve or even more nipples. While suckling the little pigs quarrel frequently at first because each tries to obtain the nipple with the most milk for itself. In so doing, they may seriously injure each other with their pointed canine teeth. The more young that are born in a litter, the harder they fight. However, after a while they establish a stable hierarchy in which each animal has his own specific nipple.

Fig. 4-1.
Wild boars are contact animals which like to huddle closely together. For resting or sleeping they prefer to lie side by side with their heads pointing alternatingly in opposite directions.

Young pigs have only a few hairs and for this reason require an extraordinary amount of warmth. During their first days, they do not leave the nest but lie together near the mother, who usually stays with them. Female pigs with young are more alert and more aggressive than many other ungulates. They remain in the immediate vicinity of their young as long as they are still small, calling them and warning them in case of danger. Such protection is necessary because the young could neither defend themselves successfully, nor flee over a long distance. Many predators prefer young pigs as prey. Older pigs, however, are so full of fight that they can defend themselves against superior predators and often cause them serious injuries. In some species, the infant mortality rate is rather high during the first year. Many piglets die from diseases, parasites, or inclement weather. Generally, they are nursed for two to three months, but they soon start to take other food in addition to the mother's milk. The earliest date of puberty in wild pigs is nine months. They may live fifteen to twenty-five years.

The bush pigs

Distinguishing characteristics

The BUSH PIGS (genus *Potamochoerus*) can be recognized by the strongly developed canine teeth (Apophysis) of the boar and the conspicuously formed nasal bones. The HRL is 100–150 cm; the BH is 55–80 cm, and they attain a weight of up to 80 kg. In ♂♂, the nasal bones extend up to one centimeter beyond the ridge of the nose, covering it like a plate. It is also covered with firm cartilage tissue which fills the space between the two protuberances of the nasal bones. Therefore, old animals appear to have two horns that point backwards between the eyes and the nose. The canine teeth are not too long, but they are extremely sharp since the upper ones constantly grind against the lower ones. The rump is flattened at the sides. The coat is without fleece. The body of the young is horizontally striped. The ears are long, ending in a point, and the tail has a tassel.

A species with many subspecies

This pig genus is found all over the African continent south of the Sahara and on the island of Madagascar. Therefore, it is not surprising that it has split up into many subspecies. The Red River hogs of West Africa and West Central Africa look so different from their relatives in South Africa and Madagascar that a further splitting into two species with several subspecies would seem justified were it not for intermediary forms. For this reason all bush pigs are considered to be one species *(Potamochoerus porcus)*.

The most striking subspecies, the CAMEROON BUSH PIG *(Potamochoerus porcus pictus;* Color plate, p. 93), which closely resembles the other bush pigs, is one of the most colorful mammals. The white hairs on the back are so long at the croup that they sometimes look like a brush in profile. The ears have a long brush on their pointed tip which effectively enhances expressive postures. Bush pigs display their striking head mainly when they want to impress a conspecific. Then they arch the back so that the white back mane becomes conspicuous, they turn the head towards the opponent as much as possible, and extend their long brushed ears horizontally. In this position, seen from above, the whole body is bow-shaped, and the hind quarters are also displayed, with the long tail with its tassel at the end clearly visible.

In the Frankfurt Zoo, a boar used to fight playfully with his female. Then the animals would stand facing each other, pressing their foreheads together, crossing their noses and snouts like crossed swords, and pushing each other vehemently to and fro, or boxing each other alternately with their snouts with vigorous butts. Then they would swish their tails like whips, although these are normally only used to ward off insects. Between those two animals, who usually got along well with each other, serious conflicts occurred only when they were fed. While the boar dashed for the food, greedily swallowing as fast as he could, the female did not dare to come close. If the keeper did not interfere, she would have to settle for what he left. She would hastily grasp as much fruit and vegetables as she could carry in her mouth and retreat to the remotest corner of the pen. While she was eating, she kept looking over her shoulder at her voracious mate, squeaking piercingly as soon as he approached.

All bush pigs like to eat animal food. In the zoos they accept ground meat and dead chickens. In the Berlin Zoo, they regularly receive a few salted herrings as a tidbit. Since bush pigs constantly turn up the ground, their pens are transformed into something analogous to the crater landscape on the moon, and each evening those craters have to be leveled. In Africa they are severely persecuted near plantations and other cultivated areas, not because they feed in the fields, but because they plough and burrow in the ground everywhere. As regards their habits in the wild, little is known so far. This is true for other forest dwelling mammals also. They are all very shy and difficult to observe. According to A. Jeannin, the bush pigs in Cameroon come out to search for food only at dusk and during the night, and then they do not roam very far. One usually encounters them in smaller or larger groups.

Strangely enough, the first report of these handsome pigs did not come from their African home, but from Brazil. In 1648, the German naturalist Georg Marcgrav wrote a paper on the natural history of Brazil and included the picture which distinctly was that of a bush pig. It is possible that the Portuguese had transported some bush pigs to Brazil. However, in spite of similar climatic conditions they did not become established there. The first live animals reached Europe relatively late. The London Zoo obtained the first ones in 1852. There they bred six times between 1857 and 1902, and each litter had from one to four young. Lately only the Antwerp Zoo has been fortunate enough to be able to breed bush pigs. There, in 1965, a seven and a half year old female had two young which had distinct horizontal stripes. Due

Observations from the Frankfurt Zoo

Habitats and food

to difficult import regulations, the bush pigs are rarely seen in European zoos. However, they are no longer considered to be "the most delicate charges," as Ludwig Heck once wrote. The bush pig pair who are presently (1967) kept in the Frankfurt Zoo have survived remarkably long. The boar arrived in 1956 from Spanish Guinea and the female in 1959 from Leopoldville (Kinshasa). He has been living in the zoo for eleven and she for eight years. Another male lived there for seven years. In New York, a female bush pig reached the age of twelve years, and another one in London lived for fourteen years and eight months.

Although other species of the bush pig are of the same size, they look bulkier because their long hair does not cling to the body but stands loosely from the rump. The color of the coat may vary greatly. Generally, they are not as colorful as the Cameroon bush pig because the colors in the face are rather blurred. We have little information concerning the habits of these subspecies.

At the present time in South Africa, farmers and hunting authorities are taking a special interest in the CAPE BUSH PIG (*Potamochoerus porcus koiropotamus*; Color plate, p. 93). Since leopards have been severely limited in many places, the pigs have so greatly increased that they now endanger the crops. In a survey of their "crimes" made in order to deal with the problems adequately, the authorities consulted numerous farmers about their experiences. There they discovered that peanuts are apparently the favorite food of the Cape bush pig. Losses of this crop have been up to one hundred per cent. Furthermore, they eat corn, pineapple, grapes, pumpkins, and watermelons. The Cape bush pig is able to smell carrion over a distance of many kilometers. It also attacks small, ailing domestic animals, and newborn sheep and goats, eating parts of them. It also digs its way under "jackal-proof" fences which protect the farms and breaks through barbed wire fencing.

Hunting them with dogs was generally not very successful because the pigs often killed the dogs in self defense. Besides, these rather inexperienced dogs were easily distracted in the forests by the tracks of other animals. Poison also was not effective in most cases because the pigs, ironically, can smell it in time and do not eat the bait. Traps were effective in some cases, but many other animals also fell victim to them. In South Africa the strangest "home remedies" have been recommended to combat the bush pigs. Farmers put out a bowl of native beer, wait until the pigs get drunk, and then kill them with a hammer. Three other farmers recommended climbing a tree with a domestic pig and making it squeak. The pig's cry of distress then attracts the bush pig within range of the farmers' rifles. As strange as these suggestions may appear, they give a clear impression of the problem, of how difficult it is to cope with an animal of the wild whose natural predators have been destroyed by man.

The WILD BOARS (Genus *Sus*) have the widest distribution of all Old World pigs. We distinguish four species: 1. BEARDED PIG (*Sus barbatus;* Color plate, p. 94). The HRL is 100–160 cm; the BH is up to 85 cm, and the weight is 150 kg. They are relatively long-legged with a body flattened at the the sides. The long head has a face surrounded by a light beard on the cheeks which extends from the corner of the mouth almost to the ear. In one subspecies, the CURLY-BEARDED PIG *(Sus barbatus oi)*, a kind of beard also covers the ridge of the nose. One wart each is located between the eyes and the canine teeth as well as under each eye; they are larger in ♂♂ than in ♀♀. There is little hair growth without fleece; the gray or pink-gray colored skin can thus be seen. The tail is either round or compressed at the sides with a large two-rowed tassel. The young are striped in most cases. ♀♀ have five pairs of nipples. There are six subspecies. 2. JAVAN PIG (*Sus verrucosus;* Color plate, p. 94). The HRL is 90–160 cm; the BH is up to 85 cm; and they weigh up to 150 kg. There are three warts at each side of the head: one in front of the eye, the second below the eye, and the third at the angle of the lower jaw. They are of different colors with a more or less dense hair. Eleven subspecies are distinguished. 3. WILD BOAR (*Sus scrofa;* Color plate, p. 94). The HRL is up to 180 cm; the BH is 55–110 cm; and the weight is 50–350 kg. The large head makes up about one third of the HRL; they have a short and bulky neck, and the rump is relatively short, massive, and compact. The skull is massive and cone-shaped. The short upper canines have three edges; they emerge first towards the side and then bend upwards and back with advancing age. The saber-shaped lower canines are much longer, bent backward with sharp·edges ending in a pointed tip. In the ♀♀, the canine teeth are much weaker. They are not used as weapons for slashing and tearing but as levers or cutting tools for digging up food. In the winter they have short dense fleece overlaid with long bristles which are extremely long on the neck, on the withers, and on the back, where they may form a kind of dense crest. The coat of the newborn young is soft with horizontal yellowish-white stripes on the back and sides. There are thirty-two subspecies. 4. PIGMY HOG (◊ *Sus salvanius;* Color plate, p. 94) is the smallest species. The HRL is 50–65 cm; the BH is 25–30 cm. This compact species is slightly higher in the rear with the sides of the body flattened. The snout is pointed. Some have white bands on the cheeks where the hair grows thick into a beard. A crest is formed on the back by the bristles.

The BEARDED PIG *(Sus barbatus;* Color plate, p. 94) is easily recognized by the striking beard on his cheeks. These large pigs of Southeast Asia live in family groups or in large herds. Their habitats are the tropical rain forests, the secondary forests, or mangrove jungles. Even though their food consists predominantly of fruits found in the woods, insect larvae, roots, and the shoots of young Sago palms, once in a while they

The wild boars

The bearded pig

Fig. 4-2.
1. Original distribution of the wild boar *(Sus scrofa);* in addition, the wild boar has been introduced into North, Central, and South America. 2. Bush pig *(Potamochoerus porcus).*

Fig. 4-3.
Javan pig *(Sus verrucosus).*

Fig. 4-4.
1. Pigmy hog *(Sus salvanius).* 2. Bearded pig *(Sus barbatus).*

may intrude on yam and manioc fields, doing some damage to them. Bearded pigs sometimes form symbiotic relationships with other animals. They will permit the crowned wood partridge *(Rollulus roulroul)* to pick worms right before their nose and allow the birds to pick ticks from their skin. As soon as the birds give their alarm call, the pigs will flee also. Often they follow moving groups of gibbons and macaques to collect the fruits which the monkeys drop. According to P. Pfeffer, the conditions are almost ideal for them on the island of Borneo. There are neither leopards nor tigers, and the only larger predator, the clouded leopard, becomes rarer all the time. Since the Muslim population along the coast does not eat pork, the bearded pigs are hunted only by a few tribes in the interior.

Many of the BORNEO-BEARDED PIGS *(Sus barbatus barbatus)* remain throughout the year. The animals on the northeastern part of the island, however, migrate in August/September and in January/February to the south. Thousands of the animals are seized by this migratory drive like lemmings. Old and young, using special paths which often are trodden into the ground so that they resemble a gorge, migrate singly or in herds of twenty to thirty animals. They move on without stopping, eating very little, skillfully overcoming all obstacles in their way. Neither rushing mountain creeks nor wide streams can delay them. Although they are basically rather shy and avoid man, now they almost ignore danger. The Dajaks in the interior of Borneo know the bearded pig's migration routes quite well. As soon as a large herd of them arrive in the middle of a stream, the natives, screaming and shouting, approach them closely in their boats and butcher the pigs with their knives. Animals who try to escape by swimming are killed with spears; those who are swept away by the current or are wounded are collected by hunters and women downstream from the hunting place. The number of bearded pigs who are killed in this manner is unimaginable. In some areas in 1956/1957 there were not enough containers for the lard from the pigs, so even boats were filled with it. In 1954 the migrating herds of bearded pigs were so large that the massacre lasted for weeks. The rivers were blocked by the carcasses and water became unpalatable. This hunt almost caused another bloody catastrophe. The Muslims, who consider pork impure, refused to use the water and declared war on the Dajaks for having caused the mess. Such migrations of bearded pigs are known in other areas of their habitats, such as Sumatra and Malakka.

Bearded pigs apparently have no specific reproductive season. They have young only once a year. After a gestation period of approximately four months, the pregnant female leaves the herd and builds a nest for her litter which greatly differs from the resting places these animals prepare daily. The mother-to-be, who prefers a mound in the jungle, brings ferns, branches, and dry palm leaves together to form a pile about one meter in height and one and a half to two in diameter. There are generally four to eight young born, although on the island of Borneo there are usually only two or three; they stay with their mother for about one year, according to P. Pfeffer. The little pigs' main enemies, in addition to man, are the clouded leopard, the Malayan sunbear, and the python.

The bearded pigs have always been rare in zoological gardens. Presently they are being kept, among others, in the Gelsenkirchen Ruhr Zoo in West Germany. In the zoo for domestic animals of Halle in East Germany, two females of this species were crossbred with a European wild boar. The resulting hybrids were then bred with each other, but their offspring were either born dead or died soon thereafter. The experiment, according to Erna Mohr, "does not suggest a very close relationship between the bearded pig and the European wild boar."

The most striking features of the long-headed JAVAN PIG (*Sus verrucosus*) are the three warts on each side of his head. The wart in the far corner of the lower jaw may grow very large and, as Erna Mohr put it, look like a big, slack blister dangling from the jaw. It is especially well developed in old boars.

The Javan pig

The Javan pigs usually live in family groups or occasionally in small herds. They live in the grasslands of their native island, in valleys, on lower elevations with dense plant cover, and sometimes in swamps. They probably give birth once a year to from four to eight young. So far Javan pigs have rarely been kept in zoological gardens. Little is known about their life in the wild. According to Erna Mohr, their average life span is eight years with a maximum of fourteen years.

No other species of pigs has such an extended distribution as the European-Asiatic WILD BOAR (*Sus scrofa*; Color plate, p. 94). While its relatives are restricted to India, East India, and Indonesia, it is distributed with many subspecies in Europe, North Africa, the temperate tropical zones of Asia, and the Malayan Archipelago. The Southeast Asian pig *Sus scrofa vittatus* was long considered to be a separate species. Today it is classified only as an Eastern subspecies of the wild boar, since all transitory forms between the Western and the Eastern specimens have been determined. The descendents of the African wild boars are not clearly defined as yet. The East Asian island pigs may possibly have come from domestic pigs who returned to a wild state or from crossbreeds between the wild boar and the domestic pig. The wild boar was introduced in North America for hunting. In many places, domestic pigs probably have returned to a wild state. There is hardly another domestic animal more inclined to return to the wild state than this pig. Within a very short span of time, it behaves like an animal of the wild and easily mixes with his free ranging conspecifics.

The wild boar

The food of the wild boar's diet differs not only with the area but also with the seasons. Analysis of their stomach contents in Central Europe has shown that, in addition to considerable quantities of acorns and beechnuts, the wild boar likes to feed on the fern *Pteridium aquilinum*, on rosebay and willow herb *(Epilobium)*, hogweed *(Heracleum sphandilium)*, goutweed *(Aegropodium podagria)*, and plantain *(Plantago)*. Various grasses also form an important part of their diet. Even though their diet is predominantly vegetarian, it is amazing to see the variety of animal food wild boars will eat. They eat carrion as well as wounded rabbits, roe deer or red deer, various rodents, and they devour the eggs and young of ground-nesting birds, and young egrets and cormorants that have fallen out of their nests. They eat lizards, snakes, frogs, and fish, dead or alive. Their diet also includes clams, all kinds of insects and their larvae, and especially grasshoppers in some areas. In addition they eat crabs and worms. H. B. Oloff even watched some wild boar skillfully catch mice.

There is no doubt that the wild boar may cause considerable damage to fields and plantations in densely populated settlements, although the damage is due less to what he actually eats than to his tireless rooting in the ground. In Germany where the wild boar no longer has any natural predators, it became a serious pest in the first years after World War II when hunting still was prohibited. On the other hand, foresters could clearly demonstrate that the wild boar may be extremely useful in the forest. It does eat considerable quantities of tree parasites, mainly the larvae of cockchafers and sawflies, the caterpillers of the night moth and pine-beauty. The wild boar's digging and rooting is useful in another way in the forest; it loosens the soil in large areas and actually buries seeds at times, thus contributing to the regeneration cycle. This may even cause a change-over of certain species of trees. Observations showed that where wild boars occur, oak forests turn into birch woods or fir trees replace oak.

Like most of his relatives, the wild boar has poor eyesight. Hunters, knowing that it can barely recognize a quietly sitting person, allow it to approach well within range. In contrast, their sense of smell is all the better. This is not surprising since the wild boar must be able to find a large part of its food below the surface of the ground. The domesticated pig has retained this ability. Formerly, pigs in Southern France were trained to search for truffles in the manner of tracking dogs. Since vocal contact within a group of wild boar is important, their sense of hearing is also well developed. The faintest sound from a breaking branch may cause them to flee before a hunter. In areas where they are limited they become so shy that they remain concealed and one can hardly ever see them. All who know them testify to their considerable intelligence; for example, they are well able to distinguish where and from whom danger threatens.

Wild boars are anything but defenseless. Their sharp-edged,

pointed canine teeth are excellent weapons for attack as well as for defense. This becomes obvious when two boars fight. "When boars of equal strength meet during the rut," according to K. Snethlage, "bitter fights will result. They leave tracks of blood in the snow from wounds which the furious rivals have inflicted on each other. They strike each other with their heads and try to ward off the blows with their shoulders. They may also lunge towards each other, trying to run over the opponent." If their shoulder parts were not especially well protected, such violent fights would often cause the death of one of the animals. The protection is due to a "shield" that develops on the boar before the rutting season. On both sides of the chest, growths of connective tissue begin to grow and from a rugged plate several centimeters thick which reaches from the shoulder to the last ribs. It is not elastic like the normal skin and it slides forward and backward, clearly visible on the moving animal. This plate may even become reinforced by a layer of resin, caused by the boar's rubbing his shoulders against trees.

The habitat of the wild boar is unusually varied. When they find shelter and food, they live as readily in the plains as in the mountains at elevations of up to 4000 meters, in swamps as well as in relatively dry steppes, and in remote areas as well as in densely populated areas where they actually thrive close to civilization. Their weight and size apparently increase from West to East. The animals in wet areas are usually heavier than those in dry regions. Continental forms grow larger than those living on the islands. According to Theodor Haltenorth, the wild boars on the island of Sardinia have a body height of approximately 55 cm and weigh up to 50 kg, while animals from the Carpathian Mountains, the Caucasus, and the Ussuri area grow up to 110 cm high and reach a weight of 350 kg.

As long as the wild boars are not forced by unfavorable conditions to migrate or to roam about the animals remain in a specific home range whose size will vary. According to the studies of Heptner, Nasimovic, and Bannikow, the territory is largest for solitary individuals, and smallest for females with young. The size of the home range also changes with the seasons. It is especially large in the fall when the pigs have to walk for several kilometers in order to reach the fields with crops. But in the winter, when high snows keep them from moving about, they stay in an area of from one half to two and a half square kilometers. When snow is more than 40 cm deep, they use deep-cut paths and walk in single file. During the cold season, they rest in shallow beds which are padded with plant material. The beds of the females and the young animals are usually round in shape. In these up to fifteen pigs may sleep together. The boars generally stay by themselves. Their beds are long and narrow and have a thicker padding. During the summer months, the beds are less elaborate than in

Fig. 4-5.
In serious or playful encounters wild boars often fight with the sides of their heads together or the side of the head against the other's body (lateral fight).

Fig. 4-6.
They try to slash their opponent from below with their heads which are armed with tusks.

Fig. 4-7.
The sides of the frontal part of the rump in the male wild boar are protected against injuries from lateral fights by a "shield" of especially tough, densely coated skin.

winter. The animals usually leave their den at sundown to start searching for food. Then they always walk against the wind or across it. In the same manner, they return to the den. There are one or several wallows in each home range which are used throughout the year. They are used more at high temperatures, during shedding, and in the rutting season. If the animals do not go to special waterholes, they may also drink at these wallows. In the vicinity of such a wallow, there are "marking trees" where the animals rub their skin and frequently leave distinct cut marks from their canine teeth.

Except for the adult boars, wild boars live gregariously during the whole year. A group may consist of one or several females, their young, and last year's offspring of more than one year of age. Usually there are about six to ten animals in a group, but there may be more. As soon as the male yearlings become pubescent in the fall, they leave the group and become solitary. However, adult boars join the estrous females during the mating season from November until January. In order to reach the estrous females they may cover long distances, rarely resting, and eating very little. Having arrived at their destination, they first chase away the yearlings, if they have not already left on their own, and start violent fights with their rivals. One boar usually manages to acquire three females, although sometimes he may win up to eight.

The reproduction in the wild boar

A distinct courtship ceremony precedes actual copulation. The boar usually drives the female in circles, boxing and massaging her roughly with his snout. He urinates frequently and he may utter strange rhythmic sounds. Scientists J. F. Signoret, F. Du Mesnil Du Buisson, and R. G. Busnel found that in this manner the domestic boar causes the female to stand still. When the French zoologists played a tape of the boar's mating sounds to young virgin females, seventy-one percent of them would stand rigid and even allow people to sit on their backs. The mating may last for quite some time and is repeated several times. Towards the end of the rutting season, the initially well-fed boars are covered with serious wounds and are emaciated because they have lost up to twenty percent of their original weight. Again they leave the groups and go back to their original solitary way of life.

Male and female wild boars become pubescent at the earliest at the age of from eight to ten months. But whether they will then reproduce depends mainly on the availability of food. H. B. Oloff found that in very favorable years in Europe about half of the females of less than one year reproduce; in average years, only about ten percent do so; and when food is scarce none breed that year. In years when food is plentiful, wild boars may have two farrows which results in the herd having two groups of young. Before farrowing, which in Central Europe usually occurs at the end of March or the beginning of April, the pregnant female separates herself from the group and finds a quiet place with dense plant cover where she has good cover. Shortly before giving birth, she digs out a spacious flat depression and pads it with all kinds of plants which she bites off or gathers in the vicinity. Those plants from nearby she scratches into the depression walking backwards; those from farther away she carries in her mouth. In this manner she builds a thick, relatively soft bed. Sometimes the females use branches to put up a roof-like structure. The largest nests are built by the East Asiatic wild boars. This maternal behavior can be observed in our domestic pigs. It guarantees the offspring's protection against water, wind, and cold temperatures. This is essential for the young pigs because they are born with little hair and, compared to other even-toed ungulates, they are very small. They would quickly cool off without the appropriate protection.

According to A. Bannikow, the gestation period in old females is 133–140 days, and in primipaous females, 114–130 days. In Central Europe, the average farrow is five to six young. The younger the mothers are, the fewer young they have. During the first days after birth, the young will not leave the den. Sometimes they stay in it for a week. Lying side by side or piled up, the siblings keep each other warm and sleep over long periods in the first days. The mother lives with them, but when she leaves for a short while she often covers the young with nesting material. Soon the young accompany her on her trips. As long as they are small, they like to return to the bed to rest; eventually it disintegrates. The bond between the mother and her young is very close. Even though the young may engage in wild fighting and chasing, they never go far from their mother. She is constantly on the alert, ready to defend them. As soon as she gives her alarm call, the young remain motionless.

Birth and raising of young

When nursing the mother lies down on her side and calls the young with soft grunting sounds. The young gather in a row along the mother's abdomen and violently massage the udders. They begin to suckle hastily as soon as the milk begins to flow. When the milk ceases to flow, the mother grunts again and gets up. As they grow older, the young literally coerce their mother into nursing. Squeaking loudly, they follow her and butt her abdomen so strongly that she finally lays down in response to this stimulus. One can make pigs (and also tapirs) lie down just by massaging their bellies. The unused nipples soon become small again, and of the others, some have much more milk than others. While initially searching for the fullest nipple, the young frequently fight until they have established a "suckling order" (see also p. 78). Since they already have their upper and lower canines and third incisors, the newborn often injure each other on the sides of the heads.

As soon as the young wild boars leave the nest for the first time, they start rooting in the ground. At the age of two weeks they eat, in addition to soil, small amounts of solid food, such as small parts of plants and soft animal food. However, until the age of two and a half to three months, milk is their main food. At this time the characteristic striped coat of the newborn begins to fade. Finally in the fall the juvenile coat is replaced by the bristled coat of the adult. The young pigs shed their coat for the first time about one year after they are born.

Defending the young

The younger the piglets are the more aggressive is their mother. Many people have come into dangerous situations because they unsuspectingly came too close to a wild boar's den with young. K. Snethlage reports of a young shepherd who had climbed a tree to escape from an attacking female wild boar. There he was besieged for quite some time by the excited animal. Also, females with young will not tolerate conspecifics in their vicinity and they will chase them away unless they are other females with young. With these they may often form large herds that live together with little conflict. After several weeks when the young become more independent, the yearlings may join the herd and they are tolerated until they leave on their own. In some cases, however, the yearlings already have formed groups of their own after their mother has left them to give birth to a new litter.

Infant mortality and longevity

Infant mortality in the wild boar is extremely high. Many of the young die before they reach the age of one year. S. I. Ognew has observed in a conservation area in the Caucasus that twenty percent of the young died during the first three months of their life, due to cold weather, severe rain storms, hail, predators, or parasites. Ognew writes further: "During their first seven months, 55 percent of all young die. Thus an average of only two young for each female with young survive. Only a very small percentage of the litter ever reaches puberty." The average life span of the wild boar is fifteen years.

Natural enemies

In the Soviet Union, the most important natural predators of the wild boars are the wolves. In the Caucasus Preserve, Russian scientists S. S. Donaurow and W. P. Teplow found that wolves may occasionally destroy entire groups of wild boar with young and yearlings. Females are less frequently attacked by the wolves, who try to avoid the well-armed adult boars. Lynx also take young wild boars, and so do swamp lynx *(Chaus)*, clouded leopards, and leopards. In contrast, bears rarely attack live wild boars. In many places of Southeast Asia these pigs are the tigers' main food. Since all these natural predators are absent in Central Europe, man must assume the role of the predator. The birth rate of the wild boar is comparatively higher in temperate climatic zones than in colder countries. Without controlled limiting, the wild boar would become too numerous for Central Europe and do serious damage to the farmland. In times when hunters are prevented from killing them, the number of wild boars increases rapidly. The pressure of overpopulation causes the yearlings to emigrate from their place of birth to other, more suitable habitats.

Because of their gregariousness and their need for close body contacts, all species of pigs would be suitable for domestication. However, to our knowledge, the DOMESTIC PIGS (*Sus scrofa domesticus;* Color plates p. 91 and p. 94) all stem from the Eurasian wild boar. Pigs did become domesticated in different areas of Europe and Asia. According to Pira, they came to Sweden only in 3000 B.C. The Austrian zoologist Otto Antonius emphasizes that typical nomadic people do not have pigs: "Pigs cannot be driven over long distances like other domestic mammals; therefore, their presence is a clear sign that these people led a rather sedentary life and practiced intensive or extensive animal husbandry. Genuine pastoral tribes never raised pigs."

In those areas where people do not selectively breed pigs for a specific purpose, they often live in a half-tame state, especially in East Asia. They spend the day in the woods searching for food and in the evening return voluntarily to the settlements. Grzimek reports from New Guinea that women nurse orphan pigs at their breast and raise them. Therefore, it is not surprising that the pigs consider the settlements their home and even remain there when they would have the chance to return to the wild. Since in these areas domestic pigs frequently crossbreed with wild boars, it sometimes is difficult to tell the difference between the two. In South Asia, those crossbreeds that had become wild were occasionally described as a new species of wild boar. In Northern Australia, near Port Darwin, Grzimek found many pigs in the wild who were derived from domestic pigs and were homogeneously black.

A breed of domestic pig that was bred primarily in Northern China is the MASKED PIG. It has large, drooping ears, has black or black-slate color, and has thick skin folds in its face. In the northern parts of its habitat, it has a coat with long, black bristles. The races in the southern countries of East Asia have predominantly erect ears. This group includes the Vietnamese pig, which is known for its sagging abdomen, an early maturing and fertile breed which, soon after its introduction to Europe, became a favorite zoo and laboratory animal. In Vietnamese villages, these animals are often kept in any huts made of bamboo and palm leaves. Hans-Georg Petzold of the Animal Park Berlin-Friedrichsfelde has seen these funny looking small pigs in their native country. "They are usually black all over and spotted ones are rare. Adult animals do not weigh more than forty kilograms. They are characterized by erect ears; their body looks 'pushed together' from front and rear because the backbone is bent like a saddle. Therefore the abdomen almost drags on the ground, even in males, while in pregnant females it actually does scrape the ground. The skin has many folds and hair growth is scarce. The head looks like a pug's and is greatly exaggerated, which is a characteristic of many domesticated animals. The extremely short nose and the deep wrinkles on the

The descent of the domestic pig

▷

The domestic pig (*Sus scrofa domestica,* top) originates from the European wild boar (*Sus scrofa,* bottom).

▷▷
Entelodontidae

1. *Daeodon*
Pigs:
2. Warthog (*Phacochoerus aethiopicus*); subspecies:
(a) Cape warthog (*Phacochoerus aethiopicus aethiopicus*)
(b) West Sudanese warthog (*Phacochoerus aethiopicus africanus*)
3. Babirusa (*Babyrousa babyrussa*)
Peccaries:
4. Collared peccary (*Tayassu tajacu*)
5. White-lipped peccary (*Tayassu albirostris*)

Asiatic breeds of the domestic pig

▷▷▷
Pigs

1. Bush pig (*Potamochoerus porcus*); subspecies:
(a) Cameroon bush pig (*Potamochoerus porcus pictus*)
(b) Cape bush pig (*Potamochoerus porcus koiropotamus*)
2. Giant forest pig (*Hylochoerus meinertzhageni*)

1a ♂ ♀

1b

2

European breeds of
the domestic pig

Pigs

1. Wild boar *(Sus scrofa)*;
subspecies:
(a) Central European wild
boar *(Sus scrofa scrofa)*
(b) Japanese wild boar
(Sus scrofa leucomystax)
(c) Southeast Asian pig
(Sus scrofa vittatus)
2. Domestic pig *(Sus scrofa
domestica)*; breeds:
(a) Bavarian country hog
(b) German pasture hog
(c) German improved
country hog
(d) White German im-
proved hog
(e) Mangaliza hog
(f) Masked pig
(g) Vietnamese pig
(h) Papuan pig
3. North Celebes Javan
pig *(Sus verrucosus cele-
bensis)*
4. Borneo bearded pig *(Sus
barbatus barbatus)*
5. Pigmy hog *(Sus sal-
vanius)*

Domestic American
breeds, by Leslie
Laidlaw

animals' faces gives them a roughly original appearance, especially in the young."

In Europe, too, breeds of pigs were developed, but few of practical value in our day. Until the eighteenth century, pigs were kept mainly in the pastures, in fields, and in forests of oak, beech, and chestnut trees. They were tended by swine herdsmen and they searched for their own food. When this food became scarce, because of the increasing settlements and cultivation of the land, people began to keep them in sties. But man's needs for their lard and especially for their meat increased continuously. So it became necessary in the middle of the last century to selectively breed from the original country hogs which grew slowly and were tall, new improved types that would reproduce faster and produce quantities of muscle and fat. This first took place in England where breeders crossed small, early-maturing Indian-Chinese and Roman pigs. From the crosses between the WHITE ENGLISH YORKSHIRE HOG and the unselected GERMAN COUNTRY HOG came the WHITE GERMAN IMPROVED HOG. Presently, this breed is still widely distributed in Germany. It has large, erect ears and a dense, smooth white coat which clings to the body. It matures early, is lively, and grows fast. Another important breed is the similarly built GERMAN IMPROVED COUNTRY HOG with drooping ears that has come from crosses between the white German improved hog and the long-eared country hog. The GERMAN PASTURE HOG is a late-maturing, hardy animal which does well even on sparse foods. It comes from the area of Hannover and Braunschweig in Germany; it strongly resembles the wild boar with its long, pointed head, its long-legged body, and erect ears. The SCHWÄBISCH-HALLISCHE HOG, which resembled the improved country hog, is characterized by great fertility, a short fattening period, and good health. It has drooping ears and is black, except for a white band of varying width. The ANGLER SADDLE HOG is also black except for a white "saddle" on the shoulders. This droopy-eared, improved country hog is well suited to the pasture. It matures quickly, is fertile, and is appreciated for its meat, especially in Schleswig-Holstein from whence it originates. Finally the original English Cornwall and Berkshire hogs are of certain significance for the German hog breeders.

Many of the current European hog breeds are certain to become extinct. Just to mention one more breed from this large number is the curly-haired MANGALIZA HOG from Hungary, whose young are horizontally striped like the young of the wild boar during the first weeks. However, this unusual pattern soon fades in domestic pigs. Also remarkable are the so-called SINGLE-HOOFED HOGS of several European countries in which two center toes are enclosed at the tip by one large hoof.

The BERKSHIRE was imported from England to America in 1823. The Berkshire is black with white feet, a white strip on the face, and a white tail tip. The face is dished and broad, with a short snout. The ears are of medium size and are carried erect. The jowl is prominent, but is not flabby. The Berkshire produces a high quality carcass with little excess fat. The breed is not exceptionally prolific, averaging nine pigs per litter. Berkshires have excellent dispositions. They are better grazers than most of the breeds developed in the United States. The breed is not as large as some, with boars in show ring condition weighing around 400 kg, and sows weighing about 360 kg. The feet and legs of the Berkshire could be more desirable. Often both front and rear pasterns are long and sloping, and they have a tendency to spread at the toe.

The AMERICAN LANDRACE was developed from Danish Landrace which were imported in 1934. The American Landrace also carried from one-sixteenth to one-sixty-fourth Poland China breeding. The Landrace is a white, long-bodied hog lacking the arch of back desired by many modern breeders. The head is long and narrow, and the ears are carried close to the face. The sows are prolific, but reach top milk production later than many other breeds. The Landrace has a plump but trim ham. The long, weak pasterns often seen in the Danish Landrace have been eliminated to enable to American Landrace to do well under pasture conditions as well as in confinement. The Landrace hangs a high quality carcass. It is used extensively in the United States in crossbreeding programs.

The HAMPSHIRE breed is uncertain, but it was probably developed in England. The breed acquired its name in 1904, but was known as early as 1895. The Hampshire is black with a white belt encircling the body. The white includes both front legs and feet. There may also be some white on the hind legs. Hampshires produce a high quality carcass. They are smooth-sided and lack excess fat on the ham. The face is long and straight, and the ears are carried erect. Hampshires are active, and are good grazers. They have a somewhat nervous disposition and require careful handling. The breed is medium-sized with boars conditioned for show weighing about 400 kg and sows weighing about 320 kg. The Hampshires are fairly prolific, averaging eleven pigs per litter, and are among the best mothers. The breed does well on pasture or on grain feeds, showing efficient gains. The Hampshire is the leading breed in registration numbers, and is frequently used in crossbreeding programs.

The POLAND CHINA was developed in southwestern Ohio in the early 1800's, and achieved breed association status in 1860. The Poland China is ebony black with white on four feet, the tip of the nose, and the tip of the tail. The Poland is a good-sized breed with boars fitted for the show ring weighing about 445 kg, and sows weighing about 385 kg. The Poland China breed is noted for maximum weight at any age. They are good breeders and show rapid gains. Polands have rugged constitutions with substantial bone to provide sound feet and

The Berkshire

The American Landrace

The Hampshire

The Poland China

legs. They have very quiet dispositions. The sows are quite prolific and are very good mothers. Polands are well-muscled and therefore produce a good carcass. These hogs are often sluggish and lack the grazing ability of many breeds. Their hair coat lacks the quality found in some of the breeds, and they may be faulted for uneven arch of back.

The Spotted Breed

The SPOTTED breed of swine was developed in central Indiana from hogs of Poland China origin, registered Poland Chinas, and "Gloucester Old Spots." The breed was officially recognized in 1912. Spotted swine are ideally fifty percent white and fifty percent black, with well-defined black spots. Hogs with at least twenty percent of either color and not more than eighty percent of the other are eligible for registry. The type is similar to that of the Poland China. The head is of medium length with a slightly dished face. The ears are medium-sized and drooping. The spotted, like the Poland, is a large breed with boars weighing about 430 kg or more, and sows weighing about 340 kg. They also show the constitution and sound feet and legs of the Poland. With the exception of color, there is very little difference in the breeds. The spotted is popular in the Corn Belt, and has shown little increase or decrease in popularity.

The Duroc-Jersey

The DUROC-JERSEY breed originated in the Eastern United States and in the Corn Belt. It was originally developed from two strains, the Jersey Red of New Jersey and the Duroc of New York. Durocs vary in color from a light golden red to a dark mahogany red. The breed has a medium-length face with a slight dish. The ears are drooping. Mature boars fitted for the show ring usually weigh around 430 kg and fitted sows usually weigh around 340 kg. The breed has good length of leg and depth of body, but is often lacking in muscling in the lower part of the ham. The jowl is clean and smooth. Durocs have a mild disposition and are fair grazers. They are an early maturing breed. The sows are excellent mothers and lay on fewer young pigs than equal-sized sows of other breeds. Durocs occasionally have too much fat for the amount of lean meat present. This is being improved by present day breeders. The grain of the meat is somewhat coarse on occasion. Durocs are sometimes faulted for having front legs too close at the knee, and hind legs with too much angle at the hock.

The Chester White

The CHESTER WHITE was developed in Chester County, Pennsylvania. The breed was developed from hogs of English origin and hogs from New York. The Chester White is solid white and may sunburn if not kept in confinement. The face is of medium-length with an occasional slight dish. Chesters have clean jowls and medium to large drooping ears. They are smooth-sided and have a strong topline. The ham is deep and full. The hair is fine textured, so that the carcass shows up well when hung up. The Chester White hangs a high-quality meaty carcass. Boars in breeding condition weigh around 420 kg, and sows weigh around 330 kg. The sows are very prolific, and are excellent mothers. The breed has a very good disposition, and is fair at grazing. Some Chester Whites, especially those of the Ohio Improved Chester strain, are too short and carry too much fat cover. The opposite problem of length and height without muscling is also seen.

The YORKSHIRE originated in England. Of the three types found there, only the large white type has been popular in the United States, where it was introduced in 1893. The Yorkshire has a medium-length face with a broad nostril and a medium dish. The ear is medium-sized and is inclined forward. The jowl is very clean. The neck is smooth and of medium length. The hog is white in color, and thus may sunburn if pastured. Boars in show ring condition weigh around 320 kg and sows weigh about 270 kg, but individuals have weighed well over 450 kg. Yorkshires are active at foraging. They are of a somewhat nervous disposition. Yorkshires show good rates of gain and efficient feed conversion. The sows are quite prolific and are excellent mothers that probably suckle their litters better than any other breed. The Yorkshire produces a high quality carcass, although they do not reach as great a market weight as do some of the other breeds. The hogs are used extensively in crossbreeding programs.

TAMWORTH hogs were imported from Britain in 1882. They vary in color from a light golden red to a dark red. The hair coat is straight. The head is long with a moderately long, straight snout. The ear is medium-sized and is held erect. The Tamworth is a long-bodied, deep-sized breed, but lacks the width over the top desired by many breeders. The neck is long and clean, and the jowl is trim. The ham is muscular and firm, but lacks the size seen in many of the other breeds. The sows are excellent mothers, though not exceptionally prolific. It is a rugged breed and very active when foraging. Tamworths show good substance of bone and stand correctly on their legs.

The HEREFORD breed of hogs was developed in Iowa and Nebraska, and achieved recognition in 1934. Durocs and Poland China hogs were used to a great extent, but other breeds were also included. The Hereford has a white face and at least two white feet. It must be at least two-thirds red to be eligible for registry. The Hereford has a medium-length face with a slight dish. The ear is of medium-size and is drooping. It lacks trimness in the jowl area. Boars fitted for show generally weigh around 365 kg, and fitted sows weigh about 270 kg. It is an early maturing breed. Hereford sows are prolific and are good mothers. The hogs of this breed are fair grazers and good feeders. This breed lacks the uniformity of the longer established breeds. Weak toplines and weak pasterns are prevalent, but its primary fault is lack of muscling. Breeders have not kept up with packers demands, so the breed does not show up well in performance tests or in carcass competition.

Domestic pigs are raised only for man's use. They are not "pets" in the sense of dogs or cats at which one enjoys looking. Therefore,

The Yorkshire

The Tamworth

The Hereford

the number of carefully selected breeds is decreasing steadily in favor of a "standard hog," which meets the requirements of the farmers. Nevertheless, several scientists are currently working on the development of another new breed, a dwarf domestic pig which would be used mainly as an important laboratory animal in medical and pharmaceutical research. Being omnivorous, this animal is closer to man with respect to its metabolism than most of the other laboratory animals.

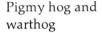

Pigmy hog and warthog

The small PIGMY HOG (*Sus salvanius*; Color plate, p. 94) from the Southern Himalayan countries looks like a toy pig. Unfortunately, nothing is known about the habits of this apparently very rare species. For quite some time now, no one has seen pigmy hogs in the wild. In spite of all fears, however, they do not seem to be extinct in Assam. The animals are said to leave their hiding places only at night. "While the old males of the Southeast Asian pig *(Sus scrofa vittatus)* always stay by themselves, joining the females only during the rutting season," Erna Mohr reports, "the pigmy hog males stay permanently with the herd to protect and to defend it. Even when a male pigmy hog is encountered during the annual brush burning in the forest, he energetically manages, even when surrounded by the flames, to escape from the pursuing men across uneven and still covered ground. When attacking, he snaps with unbelievable speed and agility, being capable of causing serious injuries with his sharp canine teeth. The herds of the pigmy hog are not large. Usually there are only five to six, although in rarer cases fifteen to twenty animals may be seen together in a herd. It may be that the animals live in pairs for a few days during the rutting season. The rut is said to occur once a year and the litter consists of three to four young." The young of the pigmy hog are, according to Sanderson, "tiny little fellows that look like toys." They have distinct yellow-brown stripes. The life span of this smallest species of wild boar is presumably ten to twelve years.

The most unusual pigs are the WARTHOGS (genus *Phacochoerus*). They have a barrel-shaped body, a big, wide head ornamented with odd skin warts, and huge canine teeth. The L is 145–190 cm; the BH is 65–85 cm, the weight is 50–150 kg. There are skin warts behind and below the eye, between the corner of the mouth and the eye, and at the side of the lower jaw. These may grow in ♂♂ into cone-shaped protuberances, while they are much smaller, and sometimes hardly visible, in ♀♀. There is a distinct mane on the neck and back, and often a white beard on the cheeks. The tail has a tassel; otherwise, they are mostly hairless. Old pigs have only sixteen teeth because some of the incisors, the premolars, and the anterior molars have fallen out. All the chewing is finally done with the last molar in each jaw which grows so large that it almost covers the entire jaw bone. There is only one species (*Phacochoerus aethiopicus*; Color plate, p. 92) with seven subspecies.

Fig. 4-8.
Warthog *(Phacochoerus aethiopicus).*

The warthog is the only species of pig who moves on the wrists while searching for food; wide calluses have developed on their wrists as an adaptation to this unusual method of locomotion. Although the warthogs must lower their heads while grazing they are able to see over a much larger area since their eyes are higher and further back in the head than in any other species of pig. This enables them to look over the grass. The food of the warthog consists mainly of different kinds of grass and its teeth are used primarily for grinding. Therefore, the special formation of the teeth, as described in the former paragraph, is another characteristic of the warthog.

Fig. 4-9.
Warthogs prefer to fight by pushing each other with their foreheads (frontal fight; see preceding page for description of fighting behavior).

The upper canine teeth may grow especially long. They start out growing to the side, then they arch upwards, and finally they bend up and inward. The straight and much shorter lower canines grind against the lower edge of the upper canines to form sharp edged, pointed stilettos. With these weapons warthogs can seriously injure not only lions, leopards, and wild dogs, but also people. Many predators are afraid to attack adult warthogs, and in zoological gardens, keepers sometimes have been injured; for example, in the Frankfurt Zoo these animals had been completely tame with their keeper until one suddenly attacked. In the Duisburg Zoo a very experienced keeper was even killed by a warthog.

However, when warthogs fight with each other they generally do not actually harm each other with these sharp tools which are used for slashing and tearing. These encounters usually are much more harmless. The attacker runs towards the opponent with his mane erect. The opponent lowers his head, catching the blow with his broad forehead. Now both animals press the ridges of their noses and foreheads together and, while trying to push one another back or to upset the other's balance, butt each other with upward thrusts of the snout and hit sideways. The opponents always fight in such a way that neither of the animals is seriously injured in the unprotected sides of the body. Constantly growling, the animals push each other backward and forward. They may "kneel" down on their wrists and flick their tails in excitement. This continues for a while; then the animals separate by slowly backing up. If an inferior animal is cornered, however, it may attempt to escape by whirling around. In such a situation it may happen that the opponent may hit its flank with one of the lower canines when his mouth is half opened. But only very few warthogs bear such scars in the wild.

Warthogs prefer an open habitat like savannahs, light bush and grass steppes. Where the dense forest begins, their habitat ends. Except for the rutting season, they live in smaller groups with usually one or two females and their young. During the day adult boars may join these groups, but generally they stay by themselves. In contrast to most of their relatives, warthogs are active only during the day. As soon as the sun sets, they retreat into their hiding places and do not leave them again until dawn. Even though they are well able to dig and to move large amounts of earth in a short time with their wide noses, they prefer in the wild to move into the abandoned dens of the aardvarks (Oryc-teropus afer). The aardvarks, who belong to the Tubulidentata (see Vol. XII), are shaped so similarly to the warthogs that the pigs hardly ever have to enlarge the tunnels of their builders. Adult warthogs enter the den sliding in backwards. This is of advantage because they can keep an eye on a predator as long as possible. Young animals, however, enter the den head first. The last animal to enter is always an adult and its large head almost completely plugs up the entrance.

The den is not only a hiding and resting place for the group, but it also serves as a nursery. During the rutting season, the boar persistently follows the female in heat, often in circles, making sounds which are reminiscent of a clattering motor. When he smells at a place where the female has urinated, he will step over it and mark it with squirts of his own urine. When the female becomes pregnant, she leaves her young from the previous year and goes into an area where no other warthogs live. In a den in the ground, she gives birth to an average of three young. Since the female has only four nipples, she would not be able to nurse more than four young. If more young are born (reports are of up to seven), they usually die.

The young are a grayish-pink color without stripes. They are sensitive to cold and so do not leave the den during the first days after birth. R. Geigy found temperatures in the den of up to thirty degrees centigrade and a humidity of approximately 90 percent. In such an "incubator," the young warthogs have much better protection against bad weather than their relatives in nests of often careless construction. Therefore, warthogs do not need to carry nesting materials into the aardvarks' dens. In contrast to other mother pigs, the warthog females leave their young alone during the day for an amazingly long period of time. They leave in the morning, remain outside the den until noon, returning briefly for nursing before leaving again until evening. At the end of the first week, the young accompany their mother on short excursions, but they regularly return to the den. With increasing age, the nursery is used only at night. In the wild the young stay with their mother for about one year before leaving her. In the Frankfurt Zoo the second generation of warthogs is now living (1967). The first female came to the zoo in 1954 as a subadult and remained there until 1963. One of her female offspring died upon giving birth for the first time. Altogether a total of twenty-nine young were born in this group, fifteen of which were raised. The number of young per litter averaged between two and seven. The females were able to reproduce by the age of one year. The average life span of the warthog is ten to twelve years. A subadult warthog male, who came to the Berlin Zoo in 1952,

Fig. 4-10.
Professor Geigy found an abandoned aardvark den in a termite nest which had been occupied by warthogs and used as a nursery. Some bats also lived in this cave.

died there in 1966. A female of the same age arrived there in 1952 and she is still (1967) alive and well. Her age seems to be well above average.

One of the few African game animals to be discovered in the beginning of this century is the GIANT FOREST PIG (*Hylochoerus meinertzhageni;* Color plate, p. 93). The HRL is 155–180 cm; the TL is 30 cm; the BH is up to 110 cm; and the weight is up to 250 kg. The ♂♂ are taller than ♀♀. The East African form is larger than the West African. The head is large and wide, and the nose disk of the trunk reaches 13 cm in diameter. The facial wart below the eye is very large in ♂♂. The rump is round, and the slate colored skin is more or less densely covered with long black hair. Many animals have a well developed mane on the forehead and neck. The tasseled tail is carried straight up when the animal runs fast, just as the warthogs do. Three subspecies are distributed from the African Tropical Forest Belt in Liberia to East Africa.

In spite of the considerable size and its conspicuous appearance, it was not until 1904 that the giant forest pig was discovered by Meinertzhagen, a British officer, in Kenya. Two years later, skulls of the same species were found in the Congo area. Its distribution is not completely known to this day. This may be because the animal lives in very inaccessible habitats. According to Duke Dönhoff, "The animals show a preference for areas with alternating dense and impenetrable jungle and more open forest. The jungle is their fortress, full of tunnel-paths and enlarged dens made by generations of forest pigs. They are actually very stationary and usually remain within their home range. But since the range of each family is very large, the families move to another den every day." However, Bernhard Grzimek saw giant forest pigs several times in the Albert National Park in the Congo on open grass plains which had scattered bushes. With some luck they may also be seen near the famous Treetops Hotel in Kenya.

The animals rest and sleep in thickets. Early in the morning and in the afternoon, they leave their hiding places to search for food in the forests or in clearings. Their main diet is grass, leaves, fresh sprouts, and fruit. They hardly ever dig and, therefore, they do not eat roots or bulbs. Usually they live in family groups consisting of a male, a female, and two to four young. Apparently, the young may be born at any time of the year. The parents are so vigilant and defend them so determinedly that very few of their young fall prey to predators.

The huge facial wart especially of the strong males gives them a rather strange appearance. It reaches from the ridge of the nose to the base of the ear and looks, according to Erna Mohr, like a sponge growing on a tree trunk. Serious fights in the wild rarely occur, according to Erna Mohr, because each family occupies its own specific area and hence rarely comes into contact with other conspecifics. Like the

The giant forest pig

Fig. 4-11.
Giant forest pig (*Hylochoerus meinertzhageni*).

warthog, the giant forest pigs fight head-to-head. A report by the Uganda Game Department, mentioned by Bernhard Grzimek, shows how strong the bond is in a group of giant forest pigs. Two game wardens in Uganda had found a giant forest pig which apparently had been wounded by a spear some days prior. "He could not walk well because half of his hind quarters were paralyzed. Four of his group stayed close to him, making a 'terrible' noise and trying to help him out of a mud hole." Similar cases of coming to each other's aid are also known from other species of pigs.

Giant forest pigs rarely come into zoos. They turned out to be rather difficult to keep. Since in many cases they were fed ordinary hog food instead of fresh greens, they did not last long. So far the Frankfurt Zoo has had the most success in keeping them. There a male lived for almost three years and a female for more than two.

The babirusa

An especially unusual representative of the Southeast Asian pigs is the BABIRUSA (*Babyrousa babyrussa;* Color plate, p. 92). The HRL is 90–110 cm; the TL is 20–30 cm; the height is 65–80 cm; and the weight is up to 100 kg. The head is relatively small, ending in a point. The strongly arched back is joined to slightly higher buttocks. The long legs are rather graceful. There is no tassel. The skin is either smooth or wrinkled, while the coat differs in density depending on the sub-species. Sometimes from a distance, the gray or gray-brown skin seems to be completely naked. The upper canine teeth in the ♂♂ are very long (up to 31 cm); in the ♀♀ they are much shorter. The canines may break through the upper cheeks and grow backward in an arch. There are four subspecies, which differ mainly in size, type of coat, and teeth formation. The babirusa is not closely related to any of the other presently living pigs; it is considered a rather primitive form.

The large, bizarre canines have given this animal its German name, "hirscheber," which means "deer boar." In old males these canines may arch back so far that they grow back again into the upper jaw close to where they originally emerged. In exceptional cases they grow straight forward. We do not know why these animals have such strange "weapons" or whether they in fact are "weapons." In contrast to the lower canine teeth, they are hardly suited for damaging the opponent. Rather, they are more useful in warding off an opponent's attack. According to Erna Mohr, they are "rather brittle and, therefore, museum collections contain more skulls with broken and damaged upper canines than intact ones." Naturally, these structures have inspired many tales and legends among the natives. According to the Malayans, the babirusa can hang from branches with his teeth. The babirusa can easily crack hard nuts with his ridged teeth. In the wild they feed mainly on leaves, shoots, and different fruits. While searching for food, they do not root in the ground like other species of pigs do, but on occasion they dig insect larvae out of rotting trees. Their vocalizations consist of grunting and squeaking sounds. When excited, they chatter with their teeth like peccaries.

Babirusas live singly or in small family groups in swamp forests and reed thickets. They like to wallow and to swim. While searching for food, they not only cross rivers, but also wide inlets into the sea. Because of their unusual structure villagers used to keep them as pets. Such handraised animals are said to be quite tame. On the other hand, they were hunted for their meat in many areas throughout their range, and so today the species is probably endangered. The clearing of the forests undoubtedly contributes to their decreasing population.

Fig. 4-12.
Babirusa (*Babyrousa babyrussa*).

Babirusa have reproduced several times in the Berlin Zoo. There it was found that the gestation period is approximately five months. The female has only two nipples and so she can nurse no more than two young. At birth the young are from fifteen to twenty centimeters long, and, in contrast to most other wild pigs, they do not have a striped coat. The mother defends them vigorously. According to the few available records, their twins are always of the same sex. They are rare in zoological gardens, although babirusas do rather well on ordinary hog feed and have reproduced several times in addition to the Berlin Zoo, also in Chicago, San Diego, and London. Their average life span is about ten years. The oldest known animal was a female who died at the age of twenty-four.

Seen from a distance, the PECCARIES (Family Tayassuidae) resemble the Old World pigs. But, as Ludwig Heck once remarked, "They deserve the name 'pig' only, so to speak, because of their appearance, i.e., their bristled coat and their trunk. In all the other characteristics that are important for classification such as the teeth, stomach, and feet, they are clearly closer to the majority of the present-day even-toed ungulates, the ruminants." Located in the center of the croup is a submerged pouch which contains the opening of a gland lying in the connective tissue, which already develops in the embryo. This gland secretes an oily substance with a musky odor. Usually it remains hidden under the long bristles of the back in front of the tail. An inexperienced veterinarian at the Frankfurt Zoo even diagnosed the gland as an abscess which he intended to cut out. When the animal flares the bristles apart, the bald area with the gland's opening can be clearly recognized.

Family: peccaries

The PECCARIES (*Tayassu*) are one genus. The HRL is 75–100 cm; the TL is 1.5–5 cm; the BH is up to 55 cm; the weight is from 18 to 30 kg. The short head on the short neck is pointed. The body is flat at the sides and rises slightly higher toward the rear. There is no bone in the snout. The eyes are small and the rather round ears are only lightly covered with hair. The dense coat of bristles is especially long on the crown of the head, the neck, and the back. The tail is rudimentary and not visible externally. The short and graceful legs have the second and

Distinguishing characteristics

the fourth metacarpal and second metatarsal more developed than in the Old World pigs. The fifth metatarsal is reduced to a small rudiment. In the COLLARED PECCARY, the third and fourth metatarsals are fused into one bone in the upper part, thus forming a type of bone which closely resembles the hollow bones of the ruminants. This is also true of the third and fourth metacarpals of the white-lipped peccary. There are thirty-eight teeth: $\frac{2 \cdot 1 \cdot 3 \cdot 3}{3 \cdot 1 \cdot 3 \cdot 3}$. The upper canines are not directed upward, as in Old World pigs, but point downward as in predators. These teeth form strong, three-edged daggers, whose front edge is continually being sharpened by the lower canine teeth which grow upward. Externally the stomach appears sectioned; internally there are three clearly distinguishable gland regions. The frontal section has two sausage-like cul de sacs. There is no gall bladder.

There are two species: 1. COLLARED PECCARY (*Tayassu tajacu*; Color plate, p. 92). This is a smaller species, almost 100 cm long. The disk of the snout is flesh colored and only 3-5 cm in diameter. The coat is brown-black with a black stripe on the back which is clearly visible in the sub-adult's coat from the nose to the end of the back. The young have a similar coat pattern, but it is usually lighter. The bristles have yellow-black rings and a cross section is flat or oval. There is no fleece. A band which is of a dirty yellow-white color runs from the withers to the throat on both sides, making the relatively small head appear larger. There are fourteen subspecies. 2. WHITE-LIPPED PECCARY (*Tayassu albirostris*; Color plate, p. 92), is a larger subspecies which is up to 113 cm in length. The gray-black or brown-black coloring is augmented by a striking white spot on the lower part of the face. The young are a reddish yellow with a dark brown-black stripe on the back from the ears to the hindquarters. There are five subspecies.

The gland on the peccaries' back seems to play an important role in their lives. Erna Mohr once observed a male collared peccary walking around quietly and then standing still. He spread the bristles into all directions, thus exposing the gland. "With a slight but sudden movement, he lowered his back a little, and a two to three centimeter long white plug came shooting out of the gland at an angle of about sixty degrees, accompanied by the drizzle of little milky-white drops. The females did not pay any attention to it, either before or after." Ludwig Heck assumed that the gland was of importance for their sexual life. According to Hediger, it is used for marking their territory, among other things. He describes how the collared peccaries in the Zurich Zoo rub the milky secretion on grass and the roots of trees with up-and-down circling movements. The odor and sometimes the brown marking spots remain visible for a long time. In Hans Krieg's opinion, the glands are used "to facilitate the meeting of the sexes by leaving a scent-track by applying the secretion to plants or the entrance of their dens." However, he sees another function for the gland. "It was strik-

Fig. 4-13.
Collared peccary (*Tayassu tajacu*).

Fig. 4-14.
White-lipped peccary (*Tayassu albirostris*).

ing to me, as well as to other observers, that both wild and tame animals, as well as the subadults, often rubbed the sides of their heads on each other's glands. They did this with such persistence and seeming devotion that one is inclined to believe that it was a pleasurable sensation for them." Hans Krieg assumes that this behavior is both a stimulus to increase the secretion and a means of impregnating each other with personal odors that facilitate the group's or the family's cohesion.

The structure of their upper canine teeth suggests that the peccaries do not defend themselves by blows with the head in a sideward-upward direction, but rather that they actually bite the attacker. Characteristic for both species is a conspicuous threatening gesture where the teeth are displayed accompanied by an equally threatening clapping and rattling of their teeth which can be heard over a long distance. In their habits, the peccaries closely resemble their Old World relatives. Like them, the peccaries prefer areas with plant cover, they like to wallow, and they are good swimmers. Their food consists of roots, herbs, grass, and a variety of fruits. But they also like to eat the larvae of insects, worms, and small vertebrates. Usually they live in small groups, but they may also form large herds. As residents of warmer regions, they reproduce throughout the year. Once a year the female gives birth to two unstriped young after a gestation period of 140 to 150 days. The young soon follow their mother around; they probably do not go through a "nestling period" like the young of the Old World pigs.

Both species can be crossed with one another. So far there are no reports of crosses between peccaries and Old World pigs. The New World pigs are not as important to man as are their Old World relatives. However, they are hunted because people like to eat their meat and their skin can be made into straps and bags. Therefore, they have already become rare in many areas. Bernhard Grzimek reports that the town of Iquitos in Peru exported the skins of 129,000 collared and 30,000 white-lipped peccaries in 1965 alone. Even though the peccaries' aggressiveness has often been exaggerated, hunting them can sometimes be dangerous. They are said to surround mounted hunters and to jump up at their horses.

The COLLARED PECCARY (*Tayassu tajacu*) lives in varying habitats and is widely distributed between the Southwestern United States and Central Argentina. Therefore, his diet differs depending on the location. F. Kühlhorn found grass, leaves, and the remains of roots in the stomachs of peccaries in the Mato Grosso area, although they also eat small animals and fruits. By checking the stomach contents and excrements and by observation, T. A. Eddy found out that the diet of peccaries in Southern Arizona consists mainly of the sprouts and the fruit of the opuntias. In areas where succulent plants are available through-

The collared peccary

Habits and food

out the year, peccaries do not need water as long as the plants do not desiccate in the long periods of drought. Captured animals, who were fed exclusively on prickly pears for two weeks and were not given any water, did not show any signs of ill health.

During the hot months of summer, the collared peccaries feed only early in the morning and in the evening; during winter, however, they eat all day long. They dig up mushrooms, roots, and bulbs. Occasionally, they eat grasshoppers, beetles, insect larvae, dead rodents, birds, and reptiles. Some farmers claim that collared peccaries were serious competitors for the cattle. According to Eddy's investigations, this is not true. Collared peccaries are gregarious; one rarely finds a single individual. In Arizona, they live, according to B. J. Neal, in smaller groups which consist of from six to twenty-nine animals of different ages and sexes. The group stays closely together and serious fights over food or for other reasons rarely occur. They use regular paths in their home range which is only a few square kilometers in size. F. Kühlhorn saw these paths in the wet savannah leading from one prime forest region to another. These paths may be tunnels covered by dead plant materials and fresh grasses which may be several meters high: "Often only the peculiar, gnashing noise that is produced by their teeth when their jaws clap together makes one aware of their presence."

Usually peccaries avoid people. Approaching enemies are detected by their keen sense of hearing. Their senses of vision and olfaction are less well developed. Adult animals become primarily the prey of jaguars and mountain lions; the young are killed by mountain lions, lynx, other cats, and coyotes, in spite of the mother's vigorous defense. While tame collared peccaries may become very attached to people, they have a distinct antipathy to dogs. H. Sick describes how a tame young male would always fight with dogs and how the animal reacted to an aggressive opponent: "Chico stood still and threatened with a quick sequence of opening and closing of the jaws, which finally turned into a kind of yawning. Then he lifted his head slightly, turning it aside. The eyes were closed to narrow slits. He did not utter a sound. With increasing excitement, the mere opening and closing of the mouth changed into a loud clacking noise. After threatening in this manner for quite some time, Chico walked off very slowly, backing off to one side."

Reproduction, raising of the young, and longevity

According to L. K. Sowls, the females become sexually mature at 33–34 weeks and the males at 46–47 weeks. There is no specific rutting season, although most births take place during the months of high precipitation. Within one herd females in heat may copulate with several males, yet fights between the males are rare on these occasions. After a gestation period of 142–149 days, the female leaves the herd for a short while to give birth to usually two, but occasionally three or even four young in a sheltered place, often in a cave. Twins seem to be most frequent; according to Sowls, they comprise seventy-nine percent of all births.

A few hours after birth, the young are already up and around and are able to follow their mother. The next day the mother rejoins the herd with her young, although she stays somewhat apart from the others in spite of their usually friendly acceptance of the young. Collared peccaries have four pairs of nipples, but only the two in the rear produce milk. The mother nurses her young while standing. Because the milk supply at any one time is small, the young drink frequently. Hediger observed in the Zurich Zoo the strange case of two young who not only nursed at their own mother but also on another female who produced milk without having given birth. The life span of collared peccaries in the wild is not known. A female in the New York Zoo reached the remarkable age of twenty-five years.

The WHITE-LIPPED PECCARY (*Tayassu albirostris*) has a life style very similar to that of the collared peccary, although, as a rule, it seems to live in large herds of up to one hundred or even more animals. In some areas, collared and white-lipped peccaries share the same habitat. The white-lipped peccaries are much more aggressive than their smaller cousins, and their food consists of a larger proportion of meat. The young of this species are much more conspicuous than those of the collared peccary. Hans Krieg writes on this subject: "During their first year, these pigs are of a more or less rusty red, which caused bad observers to take them for a different species." The coat of the young changes into the adult's during the second year. In contrast to the collared peccaries, white-lipped peccaries are seldom on display in zoological gardens. In the Berlin Zoo, they reproduced regularly between 1928 and 1937. Among the ten births recorded during this period, nine were twins. According to C. G. Roots, the gestation period of the white-lipped peccaries is 158 days. The longest life span of a zoo animal was, according to L. S. Crandall, more than thirteen years.

The white-lipped peccary

Hans Frädrich

5   Hippopotamuses

Distinguishing
characteristics

The HIPPOPOTAMUSES (family Hippopotamidae) have a plump and bulky build. The ♂♂ are usually larger than the ♀♀. The head is thick and heavy. The mouth is wide and can be opened wide. The small ears are movable. The neck is short and the body is barrel-shaped. The short legs, each with four only slightly spreadable toes, are webbed. The pseudo-claws are well developed. The short, round tail is partly flat near the tip. The thick skin has mucous glands and practically no hair. The bristles are evident only around the mouth and near the tip of the tail. The newborn young are grayish-pink in color; the adults are brownish. The underside is usually lighter than the back. There are seven neck, fifteen chest, four lumbar, six croup, and twelve to thirteen tail vertebrae. There are fifteen or sixteen pairs of ribs. In spite of their heavy weight, they are good climbers. Since the nostrils can be closed, they are relatively good divers.

They are herbivorous, eat grass, aquatic and reed plants, leaves and fruit. They have 38–40 teeth: $\frac{2 \cdot 1 \cdot 4 \cdot 3}{1\text{-}2 \cdot 1 \cdot 4 \cdot 3}$. The canines are very long, especially those of the lower jaw. The canines and incisors, which have open roots, grow continuously. The large stomach has three sections; there is neither appendix nor gall bladder. The liver is simply constructed, the kidneys are lobulate, and the feces and urine are also used for marking. Copulation occurs predominantly in the water. Usually one young is born. The ♀♀ do not eat the placenta; they have two nipples. The young are usually nursed while the mother lies stretched out on her side, as pigs do. The nursing period lasts about four to eight months. The life span is thirty-five to fifty years.

The two species of
hippopotamuses

The family has only two species: the HIPPOPOTAMUS (*Hippopotamus amphibius*) and the PIGMY HIPPOPOTAMUS (*Choeropsis liberiensis*). People are quite familiar with the bulky shape of the hippopotamuses and they are not confused with any other group of animals. They are different from all the other even-toed ungulates because they are not true terrestrial animals, but spend a great deal of their life in the water. Besides these characteristics, there are few similarities in the habits of the two species. The one species, the large hippopotamus, is a gregarious animal who lives in the open and is well adapted to life in the water. People have taken an interest in it since ancient times. In contrast, the pigmy hippopotamus is a solitarily living, shy forest dweller of limited distribution, which has been discovered only recently and about whose life in the wild little is known.

The first scientific description of the pigmy hippopotamus appeared in 1841. Twelve years later, the American zoologist Joseph Leidy classified it as a special genus because of the difference in teeth characteristics and size. Decades later other important scientists believed the pigmy hippopotamus to be just an extremely small species of hippopotamus. As late as 1879 and 1886, the Swiss zoologist Johann Büttikofer made observations on living pigmy hippopotamuses in Liberia. After a few unsuccessful attempts to capture this almost unknown animal for European zoos, the German explorer Hans Schomburgk, in 1912 was able to bring back alive five pigmy hippopotamuses for the animal trader Carl Hagenbeck in Stellingen. Since then we know a little more about this "legendary creature" of yesterday, usually from observations in zoos. Now they are very rare, even in their home range in West Africa, but they are still hunted by the natives for their meat.

PIGMY HIPPOPOTAMUS (*Choeropsis liberiensis;* Color plate, p. 111): The L is approximately 150 cm; the BH is 77–83 cm; and the weight is 180–260 kg. The smooth black-brown skin is kept moist by a crystal clear mucus secretion. The body is completely naked, with hair only at the ears, on the upper lips, and on the tassel, which in the ♂♂ is bushy. The body is almost torpedo-shaped; the legs are short and sturdy. The head is comparatively smaller and rounder than in the large hippopotamus. In the mouth, which can be opened widely, there are only two incisors in the lower jaw. They can feed under water as well as on the ground, and their diet consists of aquatic plants, leaves of bushes, algae, and short grass. The gestation period is 190–210 days. The life span is approximately thirty-five years.

Like other animals of the prime forest, the pigmy hippopotamus is difficult to observe in the wild. Some researchers have stated that the animal, when in danger, would flee from the water into the dense bush. More recent information, such as the reports of two collectors who, in the beginning of the sixties, managed to capture several pigmy hippopotamuses alive in Liberia, indicates the contrary. The pigmy hippopotamus uses deep, tunnel-like paths to reach the water, and silently dives into it. When encountering people, it flees immediately into the nearest river or swamp. This also has been observed in zoos. Since it has not yet been possible to get close to this shy animal in the wild, we have to be satisfied with zoo observations on its behavior.

Since they were first imported in 1912 by Schomburgk, pigmy

The pigmy hippopotamus by E. M. Lang

Fig. 5-1.
Pigmy hippopotamus (*Choeropsis liberiensis*).

▷
Hippopotamuses

1. Pigmy hippopotamus (*Choeropsis liberiensis*) (a) young
2. Hippopotamus (*Hippopotamus amphibius*) (a) young

▷▷
Egyptian geese (*Alopochen aegyptiaca*) on the backs of hippopotamuses (*Hippopotamus amphibius*).

hippopotamuses have done very well, reproducing in several zoos in Europe and North America. In the Washington Zoo from 1931 to 1956, sixteen female and seven male young have been born. They have also reproduced in the zoos of Berlin, New York, and Tokyo. The breeding of the Basel Zoo has become world famous. It began with a pair that arrived in 1928 and 1931. By 1967, a total of thirty-eight young had been born, ten of which were males and twenty-eight, females. Pigmy hippopotamuses today would be even rarer than the Okapi in zoos if it had not been for the Basel Zoo, which sold and traded pairs and single animals to other zoos. The old male Sämi died there in 1961 after thirty-three years. Five months before his death one young was born of which he was the sire. From these breedings all zoo requests could be filled so that there is no need for further imports from the home country of this rare species.

In a zoo the keeper usually becomes aware when a female is in heat from the male's behavior. The male remains in the female's stall and smacks his lips. When the animals are allowed to come together in the common outdoor pen, the male approaches the female from behind and places his head on her back. The animals spread the mucous from their skin glands and they often look as if they were covered with a white foam. At first the female runs away a few times; then she stands still, finally allowing the male to push her down on her stomach. In this position, copulation takes place, often lasting up to twenty minutes.

In twenty-three cases the gestation period averaged 199 days. The newborn weighs between 4.5 and 6.2 kg. When giving birth, the female usually stands on her legs, although she may also lie down with either the front or the rear part of the body. It is a rule in zoos that the birth in pigmy hippopotamuses must not take place in the water, in contrast to the large hippopotamus, because often the young that were born in water have drowned. The young have to learn how to swim and to dive before they are allowed to go into the deep water. Therefore, in Basel any female advanced in pregnancy is carefully watched when she is taking a bath. She spends most of her time in a stall to which she has become accustomed some weeks prior. An unusual observation has been made in the Basel group: the first young of each of the three breeding females was born dead or died shortly thereafter.

Birth is easy in pigmy hippopotamuses. The young has a cylindrically-shaped body and is pushed out usually in one or two minutes. The umbilical cord breaks immediately. The young soon begins moving and tries to get onto its legs. About half an hour later, it starts searching for the mother's udder at her head, the sides of her body, and the joints of fore and hind legs. The nipples are short and blunt and sometimes the young has difficulty finding them. It drinks only twice or three times a day, but then it takes a rather large quantity of

The male hippo opens his mouth widely in an impressive threatening gesture. The teeth are presented to the opponent to intimidate him. Nevertheless, serious fights are not unusual among male hippos.

milk. It grows rather fast; the daily increase in weight is from three hundred to six hundred grams. Within two or three weeks, the young has doubled its weight at birth; after five months, it weighs ten times as much as at birth. Although first it is nursed on the ground, later it nurses under water. It becomes sexually mature at four or five years.

The pigmy hippopotamuses usually defecate against a wall or a fence. The rapidly moving tail spreads the excrements. The animals probably mark their territory in this manner, as the large hippos do, but there are no records from the wild which would confirm this observation. In any case, pigmy hippopotamuses in the wild have never been seen in large herds or groups like their large cousins, but only in small families. In a zoo, at most one pair or a mother with her young can be kept together. Even related animals will fight each other as soon as they are pubescent and, naturally, the males do too. The animals can injure each other seriously with their sharp lower canine teeth. The teeth, which do not grow out of the mouth as they do in large hippos, reach into pouches in the upper jaw.

The pigmy hippopotamuses become rather tame in captivity, even when they have been captured in the wild; they tolerate the keeper in their pen, and allow him to touch them. The care and breeding of pigmy hippopotamuses is another proof of the importance of zoos in the study of the biology and behavior of wild animals. We hope that these experiences may be useful for the protection and conservation of this species, which is threatened by extinction in his natural habitat.

The HIPPOPOTAMUS (*Hippopotamus amphibius*; Color plate, p. 111, p. 112-113, and p. 114) is much larger: the HRL is 400–450 cm; the BH is up to 165 cm; and they weigh up to 3200 kg. They are copper brown with a darker back, and an underside of a light purple. The skin has glands which secrete a red liquid. The gestation period lasts from 227–240 days.

The hippopotamus by H. Frädrich

The first part of the name hippopotamus indicates a relationship to the horse (Greek: *hippos*), which is misleading. As even-toed ungulates they are not closely related to horses. About four thousand years ago, the hippos were so numerous in Egypt that they caused serious damage to the crops and consequently were vigorously hunted. The last hippos in that region were destroyed in the beginning of the last century.

Hippopotamuses from the Nile were seen regularly in Rome since 58 B.C., where they died in fights in the arenas together with other wild African animals. After this period, it took a long time until hippopotamuses again came to Europe. In 1850 one animal came into the London Zoo; in 1853 another one arrived at the Paris Menagerie. Both of these animals were presents from the Viceroy of Egypt. After the initial problems of keeping them had been overcome, the hippos soon became part of the permanent inventory of each zoological garden, where they reproduced well. Many remarkable details of their

Fig. 5-2.
Hippopotamus
(*Hippopotamus amphibius*).

Fig. 5-3.
The hippo's tracks show that the animal steps evenly on all four toes and that there is a common cutaneous pad behind the toes.

habits were observed for the first time in zoos. Thorough field research began only about two decades ago. Many wrong ideas from reports of travelling adventurers could now be corrected.

In historic times, the hippopotamus had been widely distributed over large parts of Africa and Palestine. Now this species has disappeared in many areas due to unrestrained hunting and the cultivation of their habitat. Hippos live as well in the lowlands as at elevations of up to two thousand meters. The low night temperatures in these altitudes seem to bother the animals only a little. Therefore, in zoos they may be kept in outdoor pools during the day, even in fall, provided that the animals are used to it. Hippos are sedentary animals. They stay in areas where they have ponds or slowly-moving streams with foliage-covered banks. They do not migrate even in time of temporary food shortage. Since they are mammals with an amphibian way of life, the water habitat is as important to them as the terrestrial part. In areas where they are hunted they leave their water refuge only after dark. However, their activity periods are very regular even in natural parks and wildlife preserves. During the day, as long as it is not too hot, they sun themselves on the sandbanks or spits of land; otherwise, they rest in the water. Most of the night is spent grazing. R. Verheyen observed in the Albert Park in the Congo that they emerge from the water after sundown, and earlier on evenings with cloudy skies than on clear nights. The younger males are the first to emerge while the old males will not come on land until 8 or 9 P.M. After extensive foraging, most are back in the water by four o'clock in the morning although some laggards may return as late as five o'clock.

Adaptation to life in the water

One would expect that an animal who spends more than half of his life in the water would be well adapted to this habitat, but in comparison with the streamlined seals and whales, the hippo does not do very well. However, the hippo is not a high performance swimmer but rather, as H. Hediger put it, a "fresh water buoy." It prefers depths of only about one and a half meters and places where there is either no current or only very slowly moving water. In such places it does not even have to swim; it may walk floatingly over the bottom. Unlike other better adapted water animals, the hippos could easily drift off because of their size and round shape. Upstream from Murchison Falls in Uganda, they predominantly stay near the banks of the Nile or in lagoons, avoiding the center of the stream. But occasionally an animal is caught in the rapid current and hurled fifty meters down. Even though most of the hippos live in the interior, they do not avoid brackish or sea water. In several cases, they have reached the islands of Zanzibar and Mafia off the East African coast, which means that they had to swim a distance of about thirty kilometers.

The sensory organs

The sensory organs are very appropriately placed. The nostrils, eyes, and ears are on one level on the head. Often they are all that shows above the water of such a colossus. H. Hediger writes on this subject: "After rising to the surface, both ears are whirled around independently of one another, doubtlessly to clear them of water so as to enable the animal to detect possible enemies." Although precise tests have not been made, field observations indicate that the hippo's senses of hearing and sight are well developed. Its voice is quite impressive. They do not often make sounds, but their calls belong to the most striking among the large African animals. In addition to the short grunting and sniffing sounds, they roar, especially early in the morning and in the evening. This roar can be heard over long distances; the meaning of it is not yet understood. Their sense of smell does not seem to be lacking either. When they invade plantations during the night they do not eat all the fruits they could, but select among them carefully.

When they dive, the slit-shaped nostrils are closed. There was no agreement among investigators for a long time about the duration of the dive. The present opinion is that infants remain under water for about twenty seconds, while adults remain two to four minutes, with a maximum of about six. When hippos feel threatened or bothered, they prefer to surface under floating acquatic plants or low branches hanging near the edge of the river. Then they take a quick breath and disappear again, barely moving the water surface.

How well their skin is adapted for the water can be observed when the animals are kept from going into their pools for a longer period. The glands under the surface of skin then produce a thin mucus of a red-brown color that has a high salt content. Such a "perspiring animal" then looks almost as if it were bleeding all over its body. Only when they bathe regularly do hippos have the oily shine which is the unmistakable sign of their well-being. The toughness of the hippopotamuses' skin is also famous. The skin is thickest in places where it is most exposed to the attacks of conspecifics. In order to make the dreaded "Kiboko" whips, the skin was cut into four to five hundred thin strips which were soaked in oil and then twisted together and dried. Bernhard Grzimek reports that tanning of a hippo skin takes at least six years. After that, it is as hard as rock and about four and a half centimeters thick. Therefore, it is not surprising that it even may be used for cutting diamonds.

The skin of the hippopotamus

The open mouth of a hippo is especially impressive. If a bull wants to intimidate a rival, he opens his mouth and remains in this position for a moment. Then he raises his head and opens his powerful jaws as far as possible. This exposes his huge lower canine teeth, his weapons for attack and defense. They are used for biting and tearing in fights. The largest of such canine teeth ever found is in the collection of the Duke of Orléans—and measures 64.5 cm in length.

Threatening and fighting

"The large canine teeth of hippos are covered with a yellow dental enamel that is hard as glass," writes Bernhard Grzimek. "This has to be dissolved in acid in order to obtain the actual ivory. The tooth loses about one third of its weight in this procedure. But the large hippo tusks are more expensive and more in demand than elephant tusks because they do not turn yellow. Therefore, for many years dentures were made out of hippo teeth." Quite frequently abnormally-grown hippo teeth are found. Some grow out of the mouth to the sides and others bend in such a way that they are not ground down by the opposing tooth, causing injury to the upper jaw. Because of their life-long growth, they may cause great pain to the animal if it is not fortunate enough that they break off in time. When a hippo opens its mouth, it is not always yawning. On the contrary, it may be a sign of the highest degree of aggressiveness. In actual yawning, the mouth is not opened quite as wide. In the zoo this mouth opening may become a gesture for begging, although here, too, the angle of the opening is smaller.

Besides the threatening yawning, the hippo has other means of intimidating his opponent. According to Hediger, "The lead bull rises well above the surface of the water in a so-called counter-dive and then disappears into the water with great speed in the direction of the opponent who may be standing at the bank. Because of the opaqueness of the water, it is difficult to tell whether or not the hippo will surface right in front of the observer. In most cases, the animals swim off under water in another direction. In other variations of this behavior the bull will suddenly rise high out of the shallow water, and look at the enemy with his goggle eyes. Another animal may splash unexpectedly so that the water bubbles up or he exhales under water sending foaming fountains up into the air."

A fight between two equally matched opponents may last up to two hours. This is quite a frightening spectacle. The combatants dash towards each other with splashing bow waves, trying to damage one another with their mouths wide open. They usually stand parallel to each other facing in opposite directions. With vigorous sideways blows from their heads, they drive their canine teeth into the other's body. In spite of the thickness of their skin, deep wounds, holes, and lacerations often occur which may cause a considerable loss of blood. Between "rounds," the opponents retreat by stepping backwards, but after a short pause they again roar and rush each other. Not rarely does one animal kill the other in such fights. On examination of dead hippos, it was found that an opponent's canine tooth had pierced the other's heart. Old "warriors" or inferior animals are often covered with deep scars all over the body. This shows that adult hippo males do not fight together in a ritualized manner, as do many of the other armed ungulates. The smaller the area in which they must live, the more often serious fights occur. Therefore, in the densely populated

Fig. 5-4.
Fighting hippopotamus bulls often injure one another seriously with their tusks. Their fights are not at all according to "fair rules," in contrast to most of the other mammals that have weapons with which they could injure each other.

preserves of Uganda, more injured animals are found than in other places. One also frequently finds pieces of their teeth which have been broken off in fights.

Even though hippos in many areas eat aquatic plants and reeds, the main part of their diet in Uganda consists exclusively of grass, which they bite off close to the ground with their slightly horny lips. The stomach contents of 122 animals shot in Uganda contained twenty-seven identifiable species of grass. No traces of aquatic plants were found. In the Queen Elizabeth National Park, R. M. Laws and Ch. R. Field found that adult female hippos daily eat about 1.3% of the weight of their bodies, while adult males eat only 1.1%. A subadult animal, who was kept in a small pen with plenty of food, ate about 1/100 of its body weight per day. Females eat 9.4% more than males, while nursing females consume 16.8% more than pregnant females.

The long and voluminous stomach consists of three sections: the anterior part which is lined with smooth furrows and two secondary pouches, a second chamber, and the short, thin third stomach. The hippos as genuine herbivores have comparatively long intestines. While the elephant's intestinal tract attains a length of only about thirty meters, the hippo's may well extend to fifty or sixty meters. Those who have seen the hippo's sparse feeding grounds in East Africa, where often only dry, short and tough grass grows, are amazed at how well-nourished and how round the animals appear. They must utilize their food very well. There are large quantities of droppings because their diet is high in cellulose. The excrements too play an important part in the balance of nature. These animals continually add organic materials to the water which are necessary for developing the microscopic plants on which the fish feed, especially the delicious *Tilapia*. The fishermen of Lake George in Uganda annually catch an average of three million fish, each of which weighs about one kilogram. If it were not for the numerous hippos, their catch certainly would be much less.

The excrements also are of importance in other aspects of the hippos' life. They use them to mark the terrestrial part of their territories. They emerge at distinct places along the banks and where their paths begin, leading up the steep slopes which are up to fifteen meters high. "They are actual sunken roads, trodden by generations of hippos over the centuries," Bernhard Grzimek says of such a path. "It is hard to believe how steep such a road is, with thick roots across it about half a meter above the ground, and with high steps stamped into the soil and rock. The ground has been pounded by the wide soles of the hippos' feet, which have to be placed close together on this narrow path, thus forming two rows with a center line between them. It is hard to imagine how these colossi weighing forty or sixty hundred weight climb up these steep slopes." Because of these observations, narrow,

Food, digestion, and marking

steep accesses were built into the Frankfurt hippo pool which allowed two animals to be kept where previously there had been space for only one because of the space-consuming flat steps. This is proof of how well, in some cases, observations from the wild may improve the methods of keeping zoo animals. If one has climbed such a sunken road with its smooth walls that are polished by the animals' bodies, it is easier to follow it further, sometimes over several kilometers.

Frequently along the sides of the path, there are strange, cone-shaped or prismoidal manure piles of one to two meters in width and one meter high. Usually they adjoin small bushes, piles of rock or soil, or termite hills. There the adult males distribute their feces and urine, using their short, flat tail as a catapult. The tail movements are so vigorous that at times the excrements are splattered over a distance of several meters. The younger and more subordinate male hippos, however, drop their feces in the manner of females without distributing them. The significance of these markings is not yet clear. In any case, these manure piles with their scent, which are renewed every time such a bull passes by, are "calling cards" which make known the presence of one or more bulls. Whether or not such landmarks actually keep other bulls from entering the "occupied" territories, as has been suspected, is not quite clear. This could be determined only by individually marking all animals in a specific area. However, this is not easily done with hippos.

As with so many other unusual habits of animals the marking behavior of the male hippopotamus has attracted the interest of the Africans and led to a rather peculiar interpretation. G. Schaller relates the following interesting legend: "God created the hippopotamus and ordered him to cut grass for the other animals. But when the hippo arrived in Africa and found out how warm the climate was there, he asked God's permission to stay in the water during the day and to cut grass only at night. God was reluctant to grant his permission because the hippo might eat fish instead of cutting the grass. Consequently, the hippo spreads out his excrements with his tail to show God that there are no fish bones in them." When a male hippo wants to intimidate an opponent or to impress a female, he sprays his feces intentionally. The one who is able to produce the largest quantity in short bursts is at once superior to a less "productive" conspecific. This innate distributing of excrements may be easily seen in zoos, much to the unhappiness of the keepers. This explains why the water in their pools looks muddy and brown most of the time despite the keeper's best efforts to keep them clean by regularly changing the water.

In contrast to their smaller cousins, hippos are gregarious. In Uganda, R. M. Laws and G. Clough found groups of ten animals as an average. Adult females, young subadults of both sexes, and a few adult males live together. In Albert Park, the Congo, R. Verheyen

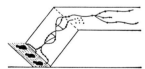

Fig. 5-5.
Every strong male hippo has his own, often rather small, area in the water which he defends against rivals. Each has his own exit on the shores which leads to paths leading to their grazing grounds. The black dots along the paths are marking places where the males distribute their feces.

found large concentrations of hippos consisting of females and numerous young. He calls these mother and child communities "nurseries," because "aunts" tend the children when the mother leaves for a short while. Not only are the young protected in such "nurseries," they also find peers with which to play. In the vicinity of the nurseries, adult males live together with younger males, old single males who bitterly defend their specific bathtub-shaped territory, and, finally, young single adult males. Since the males try to establish their territories as close as possible to the females', there are always fights about such choice spots. Mothers with infants live apart from the large group. When the young are seven weeks old, the females return to the large groups which they had left prior to giving birth. According to Hediger, a large bull seems to hold a leading position in herds of twenty to thirty animals. The denser the population in an area, the harder it seems to be for obviously weaker males to find appropriate living conditions. Instead of living near the females, they have to settle for flat puddles or pools further inland. During the dry season, the water level of such depressions is often very low, hardly enough to cover the animals' backs.

The best way to determine the age of pubescence in hippos is from zoo reports. "Knautschke," a male who was born in the Berlin Zoo in 1943, sired for the first time in 1949. The female "Jette" at Berlin had her first young at the age of almost five years; a female in the New York Zoo was five years and seven months old when she had her first young. In the wild, females seem to become pubescent later. According to R. M. Laws and G. Clough, they become sexually mature at an average age for females of nine years and for the males, seven and a half years. Apparently both sexes are reproductive until old age. Examination of their sexual organs showed that the males can sire young throughout the year. Females in the wild, however, have an estrous cycle. Most of the copulations in the Elizabeth Park take place in February and August, always at the end of the dry season. A reproductive cycle that is determined by the climate has its advantages; the young are born after a gestation period of approximately 240 days in the months of October and April which have the largest amounts of precipitation. This guarantees that the females will find an abundance of high protein grass food which helps to produce the milk they need for nursing. Copulation takes place in shallow water, probably because of the more favorable weight conditions. It lasts a long time, and the female is almost completely under water. She must repeatedly raise her head to breathe. The sex ratio of unborn young of all stages of development in Uganda was rather balanced; 48.95% were males and the rest were females. A similar sex ratio was also found in the adult population.

Female mammals always give birth in places where they feel abso-

Social life

Reproduction and development

lutely safe. Therefore, it is not surprising that the female hippo goes into the water when her time comes because that is her usual refuge and resting place. When labor starts, the animal lies down on her side. In most of the cases, the young emerges almost explosively, head first, followed by a jet of blood. The umbilical cord may break anywhere. The first task of the young is to quickly struggle to the surface in order to fill its lungs for the first time. Nursing also takes place in the water. After an initial search on all parts of the mother's body, the young soon finds the two cone-shaped breasts filled with milk. It is remarkable that the milk is not squirted into the young's mouth; instead they must actually suckle under water. According to Hediger, for a long time they were not thought to possess this ability. "It was obvious that the little hippo, in spite of all theories to the contrary, did suckle, changing from one nipple to the other." Unlike whales, hippopotamuses do not have the circular muscles which make possible the injection of milk under pressure. In a bottle-raised hippo, L. Dittrich could clearly see that it was able to perform suckling movements. Although fed in the pen during the first days, this animal showed an interesting behavior: "With the nostrils closed and the ears clinging to the head, the young drank a few times and then stopped to breathe deeply several times. It raised its ears and blew as if it were in the water." After a few days this behavior ceased; the young had adjusted to being fed out of the water, and the innate drinking pattern thus had adapted to the changed conditions.

Even though little hippos, which are precocial animals, are able to move about right from the first day, they do not have enough strength to follow their mother over long distances on the ground. The difference in size between the adult and the newborn animal is striking: two of three very young animals, which had been weighed in Uganda immediately after birth, weighed 45.5 kg, while one weighed only 25.4 kg, a mere fraction of the adult's weight. The average weight at birth is about 50 kg; the length at birth is 127 centimeters. At first, the mother does not move far from the edge of the water. She avoids all conspecifics and is very aggressive when another hippo comes too close to her young. Only in very rare cases have hippo mothers been seen with two young; usually in those cases it was a newborn and the previous offspring which was not yet independent. In Uganda only two females out of 276 were observed with what possibly were identical twins. In 1963 in the St. Louis Zoo, hippopotamus twins were born and one of them had to be raised with the bottle.

In the wild it is striking to observe how closely the young walk to their mother; they almost appear to be attached to her neck or shoulder. R. Verheyen speaks of an actual training of the young: If they move too far away from the mother, she will punish them by butting them with her head. The mother carefully sees to it that the young is

Fig. 5-6.
Each hippopotamus mother makes certain that her young stays close to her. In this way she can best protect it.

always on the side of her which is away from danger. She acts as a buffer between her young and a possible attacker. According to Verheyen, the old, irritable, and unpredictable bulls are usually the ones who attack and crush the young. Zoo keepers usually allow males back with the females only after the young have grown and are more skilled in their locomotion. After four to six months, the male usually behaves peacefully again toward the mother and young. Often he even seems to appreciate the agile young as a playmate. Wild, playful fights and extended water battles between father and young may ensue. One is amazed to see how carefully the giant bulls treat the young.

When hippos in a zoo do not have conspecific playmates, unusual animal friendships may develop. H. Petzsch reports about an adult female hippo in the Halle Zoo who made friends with a female boxer. The hippo seemed to enjoy it when the dog crawled halfway into her mouth and licked the inside. "Both animals seemed to enjoy this unusual kind of fun which was repeated several times on several days. Neither of the two ever showed any sign of bad intentions." C. Van Doorn saw a similar episode in the Rotterdam Zoo. There a six year old male lived with a female more than thirty years old who apparently did not care to play anymore. Therefore, the male tried to make friends with the Indian and African elephants, from whom he was separated by a trench. He planted his front legs on the rim of the trench, opened his mouth, and shook his huge head invitingly. The elephants seemed to understand this gesture, came close, and leaned far over the rim of the trench. The game which followed looked rather peculiar. The elephants put their trunks into the hippo's open mouth, or they rubbed their tusks on the hippo's canines, or they allowed him to gently nibble at their foreheads and the bases of their trunks. Some of the animals offered a hind leg, an ear, or even the tail. The hippo grasped these parts very carefully with his jaws without ever actually biting.

During the first weeks and months of life, the young may sometimes become the prey of predators. According to R. Hoier, young hippos under one year of age in Albert Park were killed mainly by lions and leopards. These casualties occur most frequently when the young become more independent of their mother and start roaming with peers. Hyenas and wild dogs, too, are said to follow the tracks of young hippos and to attack them on their way to the feeding grounds. It is rather doubtful whether crocodiles actually capture young hippos as frequently as older reports say, although it probably depends on local conditions. Adult hippos are comparatively safe from predators, but they too may occasionally be attacked. C. A. W. Guggisberg reports of how several lions attacked two adult hippos, throwing them on their backs and killing them with bites in the chest and the throat. "But those incidents usually are accompanied by vicious fights where occasionally a hippo not only saves his own life, but also seriously injures one or more of his attackers."

With other large herbivorous mammals, for example buffalos or elephants, the hippos usually live in peace. But accidents may happen. In the Tsavo National Park Guggisberg found the dead body of an almost full-grown rhinoceros which had been grabbed, pulled into the water, and then drowned by a hippo.

Some waterfowl seem to take advantage of the hippos. They like to sit on their backs, catch water insects and small fish, or just rest. As they do with many other African game animals, the ox peckers perform a kind of grooming on the hippos by picking bothersome insects and their larvae from their skin. They also appear when hippos have been wounded in a fight, except instead of picking insects, they peck small pieces of flesh out of the wounds. Even under water, hippos seem to attract other animals. H. Hediger reports on the carp-like fish *Labeo velifer,* which swims about hippos and picks something from their skin: "We were not able to find out actually what it was. It could be the red-colored drops of perspiration or another secretion, but I rather think that the algae or other matter adhering to the skin are involved." The Africans appropriately call this fish the "cattle egret of the water."

Famous zoo hippopotamuses

Many hippopotamuses living in a zoo have become famous. In the animal collection of the Emperor Augustus in Rome a hippo allegedly lived from 29 B.C. until A.D. 14. Furthermore, the male hippo "Knautschke," who was born during a bombing raid, became very famous. He was one of the ninety-one animals of the Berlin Zoo who survived World War II and then he became the ancestor of all the Berlin and Leipzig Zoo hippos. Another is the male "Toni" at Frankfurt who dived under water when his house burned down above him and remained so tame, even when an adult, that his keeper could ride on his back. Hippos in captivity often owe their popularity to their longevity and their considerable number of offspring. In 1914 a one year old male hippo came to the Memphis Zoo; he was still living there in 1963. The female who had come with him died in 1955. In the New York Zoo, one male hippo reached the age of forty-nine years, six months, and nineteen days. In his old age he was seriously handicapped by arthritis, his molars were worn down, and his jaw muscles were so slack that he lost most of the food out of his mouth. When he finally could no longer move, he was painlessly put to sleep.

Judged by the amount of wear of these teeth, the average life span of hippos was found to be approximately forty-one years in Uganda. During the course of such a long life, a considerable number of offspring may be produced if conditions are favorable, even when only one young is born at a time. In the Frankfurt Zoo, one pair had twelve young between 1952 and 1967, all of whom were successfully raised. In many zoos hippos reproduce so regularly that sometimes it is difficult to find a place for the offspring. From the Frankfurt Zoo, two young hippos went to African zoos and others to Japan and North and South America.

English scientists in Uganda found that the mortality rate among young hippos during the first year of their lives is around twenty percent. Once beyond this critical period, their chances of survival are considerably higher. Between the end of the first year and the thirty-third year the mortality rate declines to six percent. When hippos reach the age of forty years, the rate increases to forty percent. It is not known whether the high rate during the first year is due to accidents, predators, or diseases. Compared to other animals, hippos are not very susceptible to diseases. In the zoo cases of sickness among hippos are extremely rare, and they probably are as hardy in the wild: for example, they are able to develop immunity against the dangerous cattle plague.

The infant mortality rate

Since hippos usually reach a rather advanced age, give birth to many offspring, have few predators, and are rather hardy, one may ask what prevents an overpopulation in a certain area? One of the reasons may be that they can exist only in places with enough water. But especially the shallow ponds, as well as the small lakes or rivers in Africa, are strongly dependent on climatic factors. Often the lack of one rainy season may dry them out. Other animals that are less specialized may migrate to find other places with water. But hippos find this hard to do. E. Huxley reported in the 1930s in North Kenya that elephants and hippos died by the thousands because they became stuck in the mud and were too exhausted to get themselves out. The second reason may be that the hippos have been hunted by man since time immemorial. The meat was just as desirable as the skin and the ivory from the tusks. Tribes who depended on agriculture probably tried their best to keep the huge animals from their fields. Because of their size and fighting prowess, it is much more difficult to hunt hippos with primitive weapons than any of the other large African animals which damage the crops, with the exception of the elephant. Consequently, the number of hippos killed by Africans was probably not very high. Most of the hunting methods were, and still are, very cruel. Thus, pits lined with pointed spikes are dug, and harpoon-like tools and complicated traps suspended from trees above the hippo paths have been invented.

How can an overpopulation of hippopotamuses be prevented?

With the appearance of the Europeans, however, the hippopotamus population decreased considerably in many parts of Africa. The first discoverers, soldiers, and explorers felt challenged by the mere sight of these giants to try their modern firearms on them. The number of animals shot just for the fun of it during the middle and the end of the last century must have been unbelievably high. Especially in areas where white men settled there was no further room for hippos. It is characteristic that hippos were destroyed in all those areas in which intensive, modern agriculture was practiced. Along the lower Nile, they disappeared by 1815; they are gone forever from the Cape Province in South Transvaal, and Natal, with the exception of Zululand.

Decrease due to persecution from man

Although large numbers of hippopotamuses still live in many African wildlife preserves, they are seriously endangered in all the other areas of their habitat. This seems to be due to not only the damage they do to crops, but also to their fighting powers and unpredictability.

How dangerous are hippopotamuses?

Of course in many cases when they attack people it is their own fault. Animals with spear or arrow tips in their bodies are usually irritable and aggressive. However, they may also attack without provocation, so that coexistence between them and man can be dangerous. During the last few years, it happened frequently in Uganda that cyclists or pedestrians on their way home to their village in the evening inadvertently crossed hippos' paths. When the animals, which have the tendency to flee towards water, found their escape route to safety cut off, they felt threatened and attacked. It is amazing what speed these clumsy fellows may attain while trotting. Consequently many people found themselves face to face with a hippo as soon as they first became aware of its presence. It is known that old males defend their territory in the water against any intruder. It is probably they who attack, turn over, and destroy the light boats of the fishermen with a vigorous movement of the head.

Hippos in protected areas

Presently the largest populations of hippos live in the national parks of western Uganda and the eastern Congo. In this relatively small area live the majority of all wild hippos. Tourists are often thrilled to see the large concentrations of these huge beasts everywhere. On the other hand, this population density presents serious problems to zoologists and ecologists. In the Victoria-Nile of Murchison Park, 6077 animals were counted by plane. This means there are ten animals per square kilometer. On the shores of Lake George, Lake Edward, and the Kazinga Canal, the count was 26,000. In Western Uganda and the Congo, R. Sachs reports on the effects: "The bare shores of Lake Edward were caused by the hippo overpopulation, according to statements from the authorities. They estimated that about two animals lived on each hectare of pasture. However, the conditions of the plant life led biologists to the conclusion that even ten hectares would probably not be enough for two animals in this type of soil around the lake. The hippos were also blamed for the migration of other game animals."

After a long discussion of whether or not it was justified to destroy animals in a preserve, the authorities finally decided on an experimental cropping program in the Elizabeth Park in 1958. According to R. Bere, some overgrazing cannot be helped when a large population of hippos inhabit the shores. The situation becomes dangerous only if this damage spreads further inland. However, it was especially complicated in Elizabeth Park because many of the hippos had settled in ponds and puddles farther inland. At first, they attempted to make the hippos dislike the pools. One wallow was drained, and in another which contained about one hundred hippos, the population was scared off with firecrackers and rocks, and then the area was fenced off. But it turned out that these places had not only been the hippos' sleeping places, but were also the favorite water holes for other game animals. A controlled shooting, especially at the visitor's quarters in the vicinity of the Mweya Peninsula, seemed inevitable. The pastures were totally barren, offered no food for the hippos or any other animals, and were subject to erosion. The few remaining plants could not protect the soil from weather influences. Therefore, all hippos were killed off in limited areas and others were prevented from coming into those areas which were especially endangered.

The results of these measures were surprising, especially to those who had been against this human interference. The plant life recovered remarkably well. Five surveys taken in 1957 on the peninsula found that in addition to hippos, only 89 other species of larger mammals, from the elephant to the duiker, were living there. A few years later, the grassy regions again attracted more herbivores; 208 species were counted. In spite of this success, there is no proof that cropping by shooting will have such favorable results in every case. The actual benefit of this experiment was not so much in the improvement of the living conditions in a limited area, but in the opportunity it afforded to scientists who, for the first time, were able to thoroughly examine a significantly large number of hippos. Only when extensive knowledge on age, reproductive rate, diseases, and mortality is available is it possible to determine the specific quota of animals which need to be cropped.

The hippos were shot both during the night and the day. To kill them in the water was somewhat complicated. Since only the eyes, ears, and nostrils are above the surface, it is hard to aim correctly. According to C. Willock, the bullet has to enter between the eyes in order to hit the brain. When a hippo is killed in the water it sinks down immediately. After one to three hours, when the grass in the stomach and the intestines has produced enough gas, which in turn depends on the quantity of food the animal has eaten and the temperature of the water, the carcass returns to the surface. The carcasses were pulled ashore with a winch and dissected according to a precise plan. First, the hippo is measured, his skin parasites are determined, the scars from fights are recorded, and blood samples are taken for tests. The remnants of food from between the teeth are collected and the lenses are cut out from the eyes. Then the legs are cut off with a panga (a kind of machete) and the chest, stomach, head, and back are opened. The thyroid gland, testicles, mammary glands, adrenal glands, ovaries, and sperm are collected and separately conserved. The contents of stomach and intestines are weighed and analyzed. Not only does this give excellent material to the zoologists, but it is of interest also for nutritionists and veterinarians.

Research on killed hippopotamuses

On this occasion Dr. I. Mann from the Veterinary Institute of Kabete studied the possible utilization and profitableness of the hippopotamus as a food source. The results were surprising, as R. Sachs reports: "The weight of the dressed and skinned carcass is sixty-eight percent of the live weight, a much higher percentage than in other wild animals. Also surprising is the fact that the short, thick muscles are reminiscent not at all of pork, but rather of lean beef. The expected fatty layer was missing; this is quite in contrast to assertions in old Africa books that big game hunters killed the hippopotamus mainly for lard. From one specimen weighing 1450 kilograms, about 520 kilograms of meat were obtained. The exceptionally high protein content makes hippopotamus meat an extremely valuable food item."

Whether or not the Uganda hippos are so lean due to overpopulation on the bare pasture is not known as yet. In other parts of Africa, animals were found with a fatty layer eight to sixteen centimeters thick which yielded about ninety kilograms of lard.

Hippopotamuses for human nutrition

It is not easy to preserve the meat properly in Africa. R. Sachs, one of B. Grzimek's assistants, reports on this subject: "Methods for the preservation of meat are still primitive, but nevertheless effective. Large chunks of hippo meat, which perishes easily in the African climate, are placed on a wire grill over a simple fire and covered. The heat and the smoke preserve the meat and at the same time cover it with a crust that keeps flies from the meat. Meat which has been treated in this manner keeps for several days and is sold on the market as fresh meat." The carcasses from the controlled cropping operation were not only studied scientifically, but the meat was also sold on the market in the nearby villages. Since in many areas of Africa the people lack high protein food, the "cropping" of overpopulated areas offers a good opportunity to meet the needs of these people. However, this contains the danger that national parks will be considered as a kind of farm from which a regular meat supply would be expected. Therefore, it would be better to establish hippo farms out of the national parks where they could be raised like pigs. They could be bred, raised, and butchered on demand. The hippo is well suited for this purpose. It requires no special food, needs little space, and can be kept together with many others in a relatively small enclosure.

Hans Frädrich
Ernst M. Lang

6   Camels and Llamas

▷

For thousands of years in the deserts of the Near East, dromedaries were vital to the people. Settlement of these arid regions would not have been possible without them. Only lately have automobiles begun to assume the task of the "ship of the desert."

▷▷
Large camels

1. Dromedary *(Camelus dromedarius)*
(a) Mehari (riding camel)
(b) Pack-dromedary
2. Two-humped camel *(Camelus ferus)*
(a) Wild camel *(Camelus ferus ferus)*
(b) and (c) are different color phases of the domestic camel *(Camelus ferus bactrianus)*

While most of the other contemporary even-toed ungulates touch the ground only with the hoofed tips of the last (third) toe of the front and hind legs, the Tylopodes (suborder Tylopoda) walk on the plantars of their center digits. The small nails on the hooves protect only the tip of the end limbs, but the plantar part of the digits is padded with thick, elastic cutaneous pads. In the present Tylopodes, all of which belong to the camel family (Camelidae), only the two center bones of the front and hind legs remain, while the lateral bones have completely disappeared.

Tylopodes are ruminants and, accordingly, have a four sectioned stomach, but, however, an only slightly developed esophagus (see Ch. 7). However, certain characteristics in the structure of the camel's stomach indicate that they developed into ruminants independently from the ruminants of the suborder Ruminantia.

Like the horses described in Volume XII, the Tylopodes are extinct in their original habitat. The four surviving species are restricted to Asia, North Africa, and South America, unless they were re-introduced by man into other continents. In the prehistoric past camels lived mainly on the North American continent, which was the original habitat of the whole suborder.

Phylogeny
by Erich Thenius

The geologically oldest remnants of camel-like animals were found in the Early Eocene strata of North America. Their age is estimated at approximately forty to fifty million years. They were small even-toed ungulates with a complete set of teeth and "seledont" molars, that is, they had the same crescent-shaped surface pattern as the contemporary camels and the ruminants of the suborder Ruminantia (see Ch. 7). *Protylopus,* an ancestor of the camels who was only the size of a hare (Color plate, p. 32), still had four digits on the front foot, two on the hind foot, and his metacarpals and metatarsals had not become fused. The fusion of the forearm bones, which is significant in camels, does not begin until a later age. The roe-sized *Poëbrotherium* from

1

2

3 ♀   3a

4

Contemporary
camels
by Bernhard
Grzimek

Distinguishing
characteristics

the North American Middle Oligocene (approximately thirty-five million years ago) was much more camel-like. Its longer skull still had a complete set of teeth, yet these had already somewhat higher crowns (high crowned molars are typical of camels). Its forearm bones were fused and the lateral digits had disappeared almost completely. Still in the Oligocene, a branch of slender, light-footed GAZELLE-CAMELS (Stenomylus) and another kind of long-legged, long-necked GIRAFFE-CAMELS (Alticamelus) had branched off from the main phylum of the camels. This branch reached its prime during the Miocene period (approximately ten to twenty-five million years ago), was extinct by the Pliocene (about two to ten million years ago).

The main stem of the camels evolved from the PRE-CAMEL (Procamelus) into its present forms. They only recently came into their present area of distribution; the actual camels migrated during the end of the Tertiary (about two million years ago) to Eurasia, while as late as the Glacial Period (Pleistocene), llamas migrated over the Central American land bridge, which was then in the process of forming, into South America. By the end of the Glacial Period the camels, with the genus Camelops, became extinct in North America, their original habitat.

The present-day camels (Family Camelidae) are either medium-sized or very large even-toed ungulates. The HRL is 125–345 cm, and the BH is 70–230 cm. The slender head has a cone-shaped snout with nostrils that can be closed. The eyes are medium-large, the ears are small to medium long, and the hairy upper lip is cleft. The rim around the nostrils is naked and the neck is long. Camels have an ambling gait, since they lack connective skin between their underside and thigh. The legs are long and slender, and the forearm bones are fused, not only at the base but, in some cases, also at the upper end. The rudiments of the fibula exist only in the upper and lower ends of the splint bones. The first, second, and fifth parts of the limbs have completely disappeared, while the third and fourth metacarpals, respectively, and the metatarsal bones have fused except for the lower ends which are spread apart. For the structure of the toes, see p. 72. The dense fleecy coat has few bristles and often there are manes on several parts of the body. There are 30–34 (up to 38) teeth (for tooth formula see genera); while the milk teeth still contain three upper incisors, only the third is present in the permanent set. The canines and first premolars are hook shaped. The long lower incisors are inclined forward, while the molars have high crowns (hypsodont) with crescent shaped surface humps (selenodont). For a description of the stomach, see Ch. 7. There is a short, primitive appendix and no gall bladder. The red blood corpuscles are oval-shaped and occur in no other mammal; the red blood cell count is very high. The sheath of the penis is directed to the rear, so the ♂♂ urinate in that direction. Copulation takes place while the animals are lying down. There are two genera: the CAMELS *(Camelus)* and LLAMAS *(Lama),* each made up of two species.

The CAMELS (genus *Camelus)* are among the largest of all even-toed ungulates. The HRL is 225–345 cm, the BH with the hump is 190–230 cm, and the weight ranges from 450–650 kg. The long skull has an arched crown. The lips are pronounced and hang down to some extent; the small ears are egg-shaped. There is one pair of skin glands on the back of the head in both sexes (sexual glands) which secrete a black, creamy substance. The long neck is arched; the rump has one or two humps. The legs are especialy long, and the cutaneous pads which form the soles are slightly clefted only on the tip and are very broad. There are callouses on the chest and the joints of elbows, hands, knees, and heels which are already present in the embryo. The soft fleecy coat is rather short, but is longer on the crown, neck, throat, elbows, hind legs, humps, and tail. There are 32–38 teeth (usually 34): $\frac{1 \cdot 1 \cdot 3 \cdot (2\text{-}4) \cdot 3}{3 \cdot 1 \cdot 2 \cdot (3) \cdot 3}$. The females have one pair of nipples. The two species are: 1. The TWO-HUMPED or BACTRIAN CAMEL *(Camelus ferus;* Color plate, p. 132) has two humps on its back which are rather small and upright in the WILD CAMEL *(Camelus ferus ferus),* and larger and heavier, often bending to one side in the Bacterian camel *(Camelus ferus bactrianus).* During the summer the coat is short and dense with thin manes at the chin, shoulders, hind legs, and on the humps, while in the winter the coat is longer and denser, with longer manes, including those on the elbow. The coat color in the wild camel is uniform and darker in the winter than in summer. The color of the Bactrian camel varies. There are no callouses on the wrist in the wild camel. 2. The DROMEDARY, or ONE-HUMPED CAMEL *(Camelus dromedarius;* Color plate, p. 131 and p. 132) is known only as a domestic animal. There is only one hump on the back, and the coat is rather short, and it is longer only on the crown, neck, throat, hump, and the tip of the tail. The ♂, during the rutting season, is capable of inflating the soft palate of his mouth to form the so-called "goulla" out of the side of his mouth. It can reach the size of a human head, and a gargling roar is made in the process.

It is strange how differently various persons who knew camels have, in the past, felt about and interpreted their nature. Alfred Brehm, the famous "Father of Beasts," thought they were dull, stubborn, stupid, apathetic, and cowardly animals. He was annoyed by their odor, the earsplitting roaring of the dromedary, and even the mere sight "of its unbelievably stupid-looking head." However, Sven Hedin, the great explorer of Asia who travelled on camel's back across the arid deserts of Central Asia, praised the majestic deportment of a particular male camel: "He carried his head with solemn gravity and his quiet eye searched the horizon with an expression intimating that he felt he was the unlimited and sovereign master of all the deserts of Asia." Perhaps the contradiction between Brehm and Hedin may be explained in part

The camels

Fig. 6-1.
1. Domestic camel
(Camelus ferus bactrianus).
2. Dromedary *(Camelus dromedarius).* In overlapping ranges of the two species of camels (1 + 2), hybrids occur.

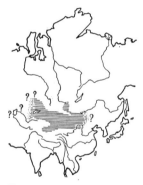

Fig. 6-2.
Original distribution of the wild camel *(Camelus ferus ferus).*

Is the camel ornery?

Fig. 6-3.
The facial expression of the dromedary seems to be "arrogant" and "'stupid." Our innate understanding, which is adapted to the expressive movements of our fellow men, in this case misinterprets the appearance of the animal's face. Indeed, its appearance has nothing to do with arrogance.

The two-humped camel

by the fact that Brehm predominantly speaks about the dromedary, the one-humped camel of Africa and Southwest Asia, while Hedin's experience was with the two-humped camel of Central Asia. However, Brehm also has a low opinion of this camel. He calls it "a camel in the truest sense," whose mental talents "are on as low a level as the dromedary's." (In German a dull-witted person is called a "Kamel," perhaps reflecting Brehm's prejudice.) In contrast, Sven Hedin tells how, during a later expedition where he purchased by chance the same male camel that had once accompanied him on his way through Tibet, it was unable to stand the hardships of the trip and died in the snow-covered mountains. Sven Hedin ends by writing: "Thus, during many years of travelling in Asia, my life has been closely connected with camels and, even though I now live far away from the Gobi Desert, I imagine hearing the waning echo of caravan bells."

To some people certain features of an animal may appear "arrogant" or at least unpleasant owing to a possible "Innate Releasing Mechanism" (IRM) which in man would be considered an expression of disdain or rejection. In the dromedary and the llama, for example, the corners of the mouth are usually pulled downward a little bit, the head is carried slightly raised so that the nostrils are on the same level or even higher than the eye, and the dromedary's nostrils are narrow slits which appear pressed together as a protection against dust. However, all this tells us nothing about the animal's feelings. If one wants to know whether a llama is about to spit or to be friendly, one should watch the ears, which are the indicators of his mood. Nevertheless, these animals have facial expressions which may seem to us to be arrogant and rejecting, or at least unpleasant.

The two-humped camel first became known only in his domestic form, which Linne described in 1758 under the name of *Camelus bactrianus,* because he thought that this form originated in Baktria, the name of an area near the present Afghan-Soviet border. However, in this area not the two-humped camel but the dromedary is kept. One hundred and twenty years later the famous Russian Asian traveller, Przewalski, found the wild camel in the deserts at Lob-nor, naming it *Camelus ferus* in 1878. The wild camel once was distributed widely over the arid areas of Central Asia. "Today this habitat has diminished greatly," reports the Moscow zoologist Heptner, "but because of the scarce information, it is difficult to outline its actual habitat. Besides, it may change rapidly. In Mongolia, the wild camel occurs only in the Transaltai Gobi Desert, where it is protected by law, and in nearby areas of China. It is possible that there are more wild camels at Lob-nor and to the north, east, and south in the desert of Takla-Makan (Kaschgaria)." Unfortunately, even in these last refuges and in spite of legal protection, the wild camels become rarer all the time, and their total extinction may be just a matter of time.

The last specimens of the wild camel live in deserts, semi-deserts, and grass-steppes in the plains, and in the mountainous regions, where they are found at elevations between 1500 and 2000 meters. During the summer, they stay in the vast dry valleys and on the hills where desert grass, herbs, and scanty shrubs grow. In the winter, they prefer the dried-up creeks in the vicinity of oases, where they live in small groups of about six to a few more than twenty animals. These groups consist of females and young led by a male. The surplus males usually live singly. The wild camels, who search for food in the morning and evening, eat grass, herbs, and thin branches from bushes and trees. In the fall, they especially like the fallen poplar leaves.

During the mating season in February, the male chases off the young animals. The gestation period is thirteen months, and thus the young are born in March.

Since tame camels are used for riding, the terms used for horses, with respect to sex and age, like stallion, mare, and foal, are also used for camels.

For a long time the wild camels were thought to be the descendents of domestic camels which had returned to a wild state. Now we know that they are the true wild form. This does not exclude the possibility, however, that occasional runaway domestic camels may have joined groups of wild camels, so that characteristics of the domestic camel could have passed into the wild populations.

The camel became domesticated in Central Asia as early as the fourth and third century B.C. From here man distributed the domestic form of the two-humped camel to extend over wide areas of Asia, to the east as far as North China, to the west as far as Asia Minor, and to Southern Russia. In contrast to many other domestic animals, the domestic camel has changed little from its wild ancestors. It is somewhat plumper and heavier with larger humps which may tip over to one side. Its foot pads are wider and its hair grows longer, especially in the manes and "cuffs." The most noticeable difference is that the domestic camels are not as uniform in color as the wild form, although no very distinct breeds have been developed.

The temperature in the range of the two-humped camel may reach fifty degrees centigrade during the summer, while in winter the temperature often sinks to twenty-seven degrees below freezing. However, this does not seem to affect the camel, although it does not tolerate a wet climate very well.

Camels are mainly used for carrying loads although they are occasionally used for riding and for pulling wagons. In addition, the wool, milk, and meat are utilized, and dried camel's manure is an indispensable fuel in steppe and desert areas which are poor in timber.

In contrast to the cold-adapted two-humped camel, the dromedary lives in a hot, dry climate. In the past the wild dromedaries presumably

The dromedary

lived in North Africa and Arabia, but presently only the domesticated forms exist there. The one-humped camel probably was domesticated first in Central or South Arabia, perhaps as early as 4000 B.C. The oldest written report on the domestic dromedary is in the Bible. It says that Abraham sent his servant with ten camels from Palestine to Mesopotamia to find a bride for his son Isaac. To our present knowledge, this occurred in or around 1800 B.C. Accordingly, there must have been domestic camels in Palestine at that time. Later dromedaries were introduced to India and North Africa for riding and carrying loads. The conquering Muslim Arabs were especially responsible for the camel's wide distribution. They brought them to Spain and later to East Africa. Ferdinand di Medici brought 1622 dromedaries to Italy and planned to use them near Pisa. This herd was preserved until recent times, although by the nineteenth century it consisted of only about 200 animals. However, these Italian dromedaries did not survive World War II, when they were allegedly butchered by soldiers. Eighty dromedaries were introduced into Spain in 1829, where they were released in the Coto Donana in the Guadalquivir Delta. There their descendants were living—strangely enough, in a swamp—until 1950, when the last five of these Spanish dromedaries were stolen. In colonial and pioneer times, dromedaries were also brought to Australia, Southwest Africa, and the United States-Mexican border regions, where their descendents that had reverted to a wild state are still living. The exploration of Central Australia was made possible only by the use of the dromedaries. During the campaign of the British against the Italians in Ethiopia, the 20,000 dromedaries, which the British army brought from the Sudan, died there in the cold alpine air.

Camels for riding and as beasts of burden

Two principal types of dromedaries are always found among the many breeds: the plumper, heavier, and slower animal used for carrying loads, and the light, long-legged and fast running dromedary. An especially "noble" breed of the riding dromedary is the North African MEHARI (Color plate, p. 132).

Dromedaries with unusually shaped humps often bring fantastic prices. For example, in the Arabian town of Hodeida, a four-humped camel brought approximately ten thousand dollars.

Like the two-humped camel in the arid areas of Central Asia, the dromedary is an indispensible aid for man in the hot desert and dry steppe countries of Africa and southwest Asia. In adventure stories, there are often reports of travelers about to die of thirst in the desert who slaughter one of their camels and drink the water supply which the animal has in a special section of his stomach. This "emergency store" allegedly is the secret of how the animal manages to cross the desert for days and weeks without ever drinking, while men and horses could not have survived.

In such books, also, miracles are told about how fast and far a camel can travel. A trained riding camel with a rider on his back once walked a record eighty kilometers a day, covering four hundred kilometers in five days. But this took place in the "winter," when it does not get too warm even in the North African desert and when the plants the animal feeds on are somewhat green and contain water.

Caravans travel comfortably at a speed of four kilometers per hour. Because the animals need resting periods, the daily performance is about twenty kilometers per day. If horses and dromedaries (in Africa they are always one-humped dromedaries) are pitted against each other in a race, the horse will usually win over short distances, while the dromedary will excel on trips of several days. In any case, in making bets on the outcome, much depends on the individual horse or camel in question.

In modern biology books, the story about the "emergency store" of water in the camel's stomach is no longer found. Whoever observes the slaughtering and cutting up of a camel in North Africa may well find the mushy contents in the stomach, but this contains less liquid than a cow's or other ruminant's stomach. Of course, the liquid may drip through a cloth. Its salt content is like that of blood. Therefore, one would really have to be close to death to drink this rotten-tasting green soup. Raswan, a German who has lived for a long time with the Rualas in central Arabia, once was in this situation. During a campaign, the valuable horses were close to death. "In order to get something for our mares to drink, Raschejd killed fourteen of our reserve camels. Rumen and intestines of these camels yielded enough liquid to fill eleven containers. When it was filtered through the shepherd's coats and mixed with ten liters of milk from the female camels, this strange brew became drinkable for our mares. With blood in their beards and tangled hair, the butchers bent over the carcasses of the killed camels, put their bare arms feverishly into the intestines, tore out the rumen, and poured the sour tasting liquid into the containers."

No matter whether a man or a horse can really drink the contents of a camel's stomach, the camel itself can only use the water it has actually drunk, no more and no less. People can also get along without drinking in cool weather if they have to, provided that they eat fresh vegetables and juicy fruits, so it is not too surprising that a camel in the North African "winter" does not need to drink for months. However, scientists are curious to know how the camel manages to last ten times as long as man and four times as long as a donkey in the summer blaze of the desert.

For the most part, all terrestrial animals consist of water, and they all utilize the water in their body in the same manner. The water passes through the kidneys where the urea and salts diluted in the water are eliminated, through the lungs during breathing, and finally water evaporates continuously on the skin and the mucous tissue of the

How far and how fast can dromedaries travel?

Can one drink their stomach juices?

mouth in order to cool off and to maintain constant body temperature. Some desert rodents are so economical with the water in their urine that it becomes solid immediately after leaving the body. They spend the hot time of the day in their moist caves deep underground.

The camel, however, is not able to go underground. So how does it manage to stay alive? In the latest publications on the subject, the secret seems to have been convincingly explained. The camel makes use of the fat in the hump on its back. All the components of our body, such as protein, starch, and fat, contain hydrogen. When they metabolize, the hydrogen combines with the oxygen of the atmosphere resulting in water. Thus, computations show that 100 grams of protein are metabolized to 41 grams of water and 100 grams of fat will yield 107 grams of water. A 40 kilogram hump on a dromedary thus yields more than 40 liters of water. The explanation seems convincing and, at the same time, answers the question of why camels have humps on their backs.

But this beautiful theoretical solution contains a flaw. The animal must inhale the oxygen through his lungs in order to "burn" up the fat. However, in exhaling the body loses more moisture than it gains when metabolizing the fat. Two American scientists, Dr. Schmidt-Nielsen and T. R. Haupt, in collaboration with Dr. Jarnum of the University of Copenhagen, have studied this fascinating problem in the oasis of Beni Abbas in the Sahara Desert, south of the Atlas Mountains.

The first difficulty was rather unexpected: It was not possible for them to get camels. No one was willing to sell them any. In the summer the climate in the oasis Beni Abbas becomes unbearable. At this time no Europeans are there. The air temperature reaches 50°C, and where the sun blazes on the rocks, it even reaches 70°C. Humans lose 1.14 liters of perspiration during one hour and become very thirsty. After having evaporated more than 4.5 liters of perspiration which is 5% of the body weight, a person is unable to see correctly or to judge his environment. At 10% loss of weight, he is unable to hear, is in terrible pain, and becomes insane. When it is cool, a person may go without drinking for quite a long time and he dies only after his body weight has decreased by more than 20%. But in the heat of the desert, we die of a heatstroke by the time we lose 12% of our weight due to a lack of water.

A camel is able to endure more. When the scientists kept one of the animals during the desert summer for eight days without allowing it to drink, it lost one hundred kilograms of its weight, which was 22% of its normal body weight. It looked terribly emaciated; the stomach was drawn in, the muscles had shrunk, and the legs appeared longer than they actually were. Naturally, it would not have been able to work or to walk very far, but it did not look seriously ill. What I liked about the three scientists was that they did not try to see how much weight a camel must lose until it dies of thirst. After it had become so thin and light, the three camel researchers gave it something to drink. It drank one bucket after another and became increasingly round and normal again. A camel may lose a fourth of its body weight in the blazing sun without being in danger of death from thirst.

One explanation for why camels can endure what humans cannot depends on the difference of the water content of the blood. Actually there is the same quantity of water in the camel's as in the human's blood: approximately one twelfth of all the water in the body. But when the camel has lost one fourth of its body weight due to evaporation, only one tenth of the water content of the blood has been lost; hence, the blood's viscosity is almost as it was before. Our blood, on the other hand, would have lost one third of its water in the same time; it becomes very thick, flows slowly, does not circulate well through the capillaries, and moves only sluggishly through the body. Thus, it cannot bring the increasing heat from the inside of the body to the skin where it is normally released. The temperature inside the body rises rapidly and we die from heatstroke.

A camel, on the other hand, can tolerate having very little water in its body. But this is not all of it: A camel also has the ability to evaporate much less water than humans do. When temperatures rise beyond that of our own body (36.5°C) we begin to perspire. Only through evaporation of water from within our bodies can we prevent the inside from becoming increasingly warmer. But perspiration requires water and lots of it.

This is different from the dromedary. During the day its body temperature increases up to 40°C and only then does the camel begin to perspire. Of course, this saves a lot of water. Furthermore, the camel's body temperature drops down to 34°C during the very cold nights in the desert. Due to the daily variation of body temperature of about six degrees, it takes much longer during the morning and noon hours before the bulky body of the camel warms up again to the point where perspiration begins. This wide variation in body temperature of camels occurs only during the summer, while during winter and in the Mediterranean area it varies much less.

Donkeys too are animals of the desert, and in contrast to humans, they too can lose up to a fourth of their body weight through lack of water. However, they lose the water in their bodies three times as fast as the camel. While a dromedary, even in the blaze of the desert, could endure to be without water for seventeen days, the donkeys had to be watered every fourth day. Their body temperature may also vary much more than that of humans, but not as much as the camel's. Thus, donkeys begin to perspire sooner. One of the reasons is their thin coat.

Camels lose most of their hair in the summer, but they keep a fleece on their back which may be as thick as ten centimeters, which of course is an excellent protection against the sun's rays. Using the same principle, the Bedouins of the desert wear woollen burnooses on their heads, often one over another. Therefore, it is of special advantage that camels and dromedaries store their fat on their backs. It is a bad heat conductor; thus, it serves as an additional protection against the blazing sun. If the fat were distributed evenly throughout the body between the intestines and muscles, it would prevent the heat flow to the surface.

However, the donkey is superior to the camel in one regard. A dried-out dromedary will drink 135 liters of water in ten minutes, replacing the body weight it has lost. It is almost frightening to see an animal empty ten buckets of water in this short span of time. But a donkey is able to replace one fourth of its body weight in only two minutes. Humans, after a day in the blazing desert, can replace their lost weight only after several hours of drinking, and they often have to eat in between. The wild animals' hasty method of drinking has many advantages. Water is rare in the summer, and there are generally only a few water holes, at which predators lie in wait. If the drinking takes only two minutes, the time of danger is correspondingly decreased.

These mysterious abilities of the camels, about which we knew little until recently, have for centuries made trade possible through North Africa and large parts of Asia. Because of them, kingdoms could thrive and dreary wastelands come under man's control. The importance of the dromedary has diminished in recent years, since it has become possible to cross the deserts faster and more safely by car or plane. The camel is being replaced by the car not only in its native Arabia and North Africa, but also overseas, where Europeans had introduced the camel as a means of transportation. The man-made machine is even more economical in regard to water needs than these living animals which came into being in their own right rather than because of man.

H. Gauthier-Pilters has closely studied the threatening and fighting behavior of dromedaries. At a distance of one hundred meters, male camels will try to intimidate one another with display behavior, raising of the head, swishing of the tail, and presenting the bushy beard at the throat. When they come closer, the animals gobble, sometimes protruding the "goulla" out of their mouths. With each "blo-blo-blo," the male lowers his neck and head, but then he throws them back, salivating heavily. He spreads his hind legs apart and slowly beats his tail down on his penis and up on his back while urinating and defecating. The attacker first snaps at the head, throat, neck, and legs of the opponent. He keeps trying to press the opponent's neck down with his own, attempting to knock him over and hold him down with

Fig. 6-4.
Fighting dromedary males snap at the opponent's legs. If one of them falls he often brings the other down with him. It sometimes happens that the opponents choke each other to death. Such casualties are usually rare among conspecifics.

the weight of his body. Since 1967, these cruel fights between male camels are prohibited by law in Turkey where they were a popular entertainment.

The domestic forms of camels and dromedaries have remarkably limited mating seasons in wide areas of their distribution. The domesticated camels mate mainly in February and March and the dromedary, at least in the northern parts of its distribution, mates from January until March. Then the males are very aggressive and may seriously injure or even kill people with their powerful teeth. The gestation period lasts between 365 and 440 days, and the female gives birth while standing to only one young. After two to three hours the young begins to walk. At first, the young moves in a combination of a trot and an amble, but by the second day it has the sure-footed amble of the parent. The camel female nurses her young for more than one year. When milked, a female may yield four and three quarters liters daily; one female dromedary even gave eight to ten liters. According to Grzimek, camel's milk consists of 6.4% fat, 4.5% lactose, 6.3% nitrogen substances, and 0.9% ash.

In those countries where the two species occur, they are regularly crossbred. These hybrids are often larger and stronger than their parents, yet they are either sterile when bred to each other or else produce only weak offspring. Therefore, the original species are always crossbred, or the hybrids are bred to purebred animals.

The LLAMA or CAMELIDS (genus *Lama*) are medium-sized. The HRL is 125–225 cm, the TL is 17–25 cm, the BH is 70–130 cm, and the weight can reach 75 kg. The ♂♂ are taller than the ♀♀. The profile of the head is straight. The large eyes have long lashes on the upper lid, the ears are long and pointed, and the lips are not too large. The long, thin neck has a slightly arched base, and is usually erect. The body has no humps, and the back is level. The round tail is rather thick, with an almost naked underside, and it is usually carried bent down and away from the body. The dense, woolly, and smooth coat has a few thin bristles which do not protect against the rain. The cutaneous foot pads are smaller than in camels, and there is a deeper cleft between the toes. The outer and inner sides of the metatarsus have a vertically stretched, callous, and hairless area containing glands ("chestnuts"). There are thirty teeth: $\frac{1 \cdot 1 \cdot 2 \cdot 3}{3 \cdot 1 \cdot 1 \cdot 3}$. The shape of the teeth is like the camels', but in the vicuna the lower incisors have smaller crowns, with open roots and continuous growth. The females have two pairs of nipples. There are two species:

1. GUANACO (*Lama guanicoë*; Color plate, p. 133). The HRL is 180–225 cm, the TL is 15–25 cm, the BH is 90–130 cm, and the weight is 60–75 kg. There are dark and light parts of the coat which contrast strikingly. The "chestnuts," in most cases, are clearly visible. The mating season

The llamas
by D. Heinemann

Fig. 6-5.
Guanaco (*Lama guanicoë*). The crosses mark areas where the guanaco has been destroyed.

Fig. 6-6.
Vicuna *(Lama vicugna)*.

Fig. 6-7.
While fighting guanaco males try to bite into each other's front legs, their necks are often crossed. The "neck fight" which is also found in many other even-toed ungulates probably evolved from this behavior.

lasts from November until February, and the gestation period is eleven months. This is the original form of the DOMESTIC LLAMA (*Lama guanicoë glama;* Color plate, p. 133) and the ALPACA (*Lama guanicoë pacos;* Color plate, p. 133).

2. VICUNA (*Lama vicugna;* Color plate, p. 133). The HRL is 125–190 cm, the TL is 15–25 cm, the BH is 70–110 cm, and the weight is approximately 50 kg. These animals are smaller and more graceful than the guanaco. The head is shorter and the ears are longer. The dark and light colors of the coat are not distinctly separated; at the base of the neck and the front part of the chest they have a 20–35 cm long white mane. The "chestnuts" are often hidden under the hair. The mating season is from April until June, and the gestation period is ten months.

While the Old World camels are animals of the plains, adapted to the endless desert and steppe belts of North Africa and Southwest Asia and the plateau of Central Asia, the llamas live partly on level terrain but mainly in the mountains. Therefore, the cutaneous foot pads of their toes, which are not as wide as the camels', are small and movable, helping the animals to move securely over narrow rocky trails and gravel slopes.

The guanaco, the larger of the two wild species, originally had a wide distribution. Although the general opinion is that the guanaco and the vicuna are genuine alpine animals, Hans Krieg discovered that they prefer dry areas, regardless of whether they are on the plains, at the coast, or in the mountains. As long as it is dry, this large, wild llama can tolerate the heat as well as the cold. In the areas near the equator, therefore, it lives only in the dry highlands; it does not go into the hot and humid plains. At the beginning of this century, the guanacos lived only a little to the south in the hot and dry lowland plains of the Gran Chaco, and in southern Patagonia it lived in the savannahs and deserts of the plains and coastal areas. It even reached some of the off-shore islands and Tierra del Fuego.

In large areas of its original distribution, the guanaco became either extinct or very rare. They do best in the impassable highlands of the Andes where they live at elevations of up to more than four thousand meters. The animals live in small groups of about twenty individuals which are kept together and defended against rivals by a male. The fighting method of male guanacos corresponds to the original fighting behavior of the even-toed ungulates from whence, according to Walther, the different "ritualized" fighting and courtship gestures have evolved, especially in the horned ungulates (see Ch. 11): "Fighting llamas (and camels too) frequently try to snap at the opponent's front legs in order to knock him down and to force him 'on his knees.' If the two opponents do this at the same time, they cross their necks. Then, consequently, one tries to press the other down with his neck."

In addition to the biting and the neck-pushing, there are several other fighting methods in the guanaco and its relatives, such as beating with the front leg, jostling the opponent from the side, and spitting.

The mating season is during the late summer and early fall, and the males are then especially aggressive. Often an excited male guanaco spits saliva and gastric juice into the face of his reluctant female. Copulation, as in all camels, takes place in a prone position and may last for several quarters of an hour. After approximately eleven months, the female gives birth to only one young; twins occur rarely. The young are nursed for about four months. After birth the guanaco female does not lick her young like most other ungulate mothers. Young guanacos are delightful and graceful. Ingo Krumbiegel once had to lock up a young guanaco that was born in a zoo pen because it was romping about so high-spiritedly at the age of two hours that it was in danger of slipping on the wet ground and getting harmed.

Fig. 6-8.
A guanaco female kicks her front leg at a sub-adult young. F. Walther thinks that the ritualized "leg-kick," which occurs in many horned ungulates (see Ch. 11), is derived from this original fighting behavior.

For a while people thought that the llama was derived from the guanaco, while the alpaca came from the vicuna. A zoologist and specialist on domestic animals, Wolfgang Herre, of Kiel, found proof that the derivation of the alpaca from the vicuna is rather doubtful and that this second domesticated form comes also from the guanaco. Supporting this opinion are certain characteristics of the skull, and the size of the alpaca's brain. For quite some time it has been known that all domestic animals have a smaller and lighter brain than their wild ancestors. Thus, even the largest and heaviest breeds of dogs have a smaller brain than does their ancestor the wolf. This is easy to understand, because man freed his domestic animals from much of the necessity to struggle for life so that the domestic animal may well function with a smaller brain than the wild animal which lives in constant danger. The brain of the alpaca is smaller than that of the guanaco, but larger and heavier than that of the vicuna. This finding could not be explained if one were to continue holding the view that the alpaca came from the vicuna.

The descent of the llama and the alpaca

In the South American Andes, the llama is kept for carrying loads and for its meat, while the alpaca is valued for its wool. Thus, the differences between the two may well be explained as a consequence of breeding them for these different purposes. Although the llama is tall and strong, the adult males cannot carry more than fifty kilograms per load for a distance of no more than twenty-five kilometers per day, which they carry safely and steadily over even the most difficult gravel slopes and mule tracks. Ingo Krumbiegel writes on this subject: "The pack-llama males are not sheared because the thick fleece on their back serves as a natural blanket for loads that they have to carry. The pace of such a caravan is not usually fast but it is steady, and the llamas hardly show any signs of exertion. They walk with a strikingly light and graceful gait, constantly watching on all sides, and they have a

The utilization of the domestic forms

Fig. 6-9.
Young guanacos (and other species of llama) stretch both legs forward while they are reclining, bending them under their body as do most other ungulates.

good relationship with the animal-loving Indians. A friendly attitude towards these animals is extremely important because they would immediately become stubborn and difficult to handle if treated unkindly or brutally." The female llamas are not used for carrying and therefore they are sheared regularly. But, in contrast to the extremely fine wool of the alpacas, their wool is of no great value.

Alpacas have much more a mind of their own than do the llamas; therefore, it is rather difficult to shear them. This domestic form is smaller than the llama and has a shorter head. Alpacas are frequently plain black or brown-black, while the llamas are of all different shades, even piebald, except plain black. Behavior and reproduction in the alpacas and llamas are hardly different from the guanaco's, and often the domestic animals graze together with their wild relatives on the slopes. However, the vicunas also like to join herds of tame llamas and alpacas; hence, no conclusions can be drawn from this behavior about the relationships among the forms. Since all four forms of the genus crossbreed in captivity and have fertile offspring, this does not aid in determining their origin.

Presumably, it was not the Incas who were the first to keep and to breed the two types of llamas as domestic animals, but already their predecessors in the highlands of Peru. According to Frederick E. Zeuner, bones of llamas have been found in a stratum dating back to 2550 to 1250 B.C. in the Virzu Valley of Peru. In recent decades, the llamas and the alpacas have been of decreasing importance to man. Today, even in the highlands of the Andes, the transportation of goods is made easier by cars, airplanes, and trains, and the wool production is more economical with sheep.

The vicuna

The vicuna is the second, smaller wild form of the genus *Lama.* It has a smaller distribution than the guanaco and is restricted to the highlands of the Andes. Like its larger relative, it prefers the dry climate. However, its more specific environmental needs apparently not only prevented its wider distribution but also made it less suitable for domestication. Vicuna groups consist of a male and five to fifteen females. The young excess males gather in groups of their own consisting of twenty to thirty animals which often do not get along with each other.

The Indians use the wool of the wild-living vicuna. They drive the animals along stone walls into pens, where they shear them and then release them again.

Whether this species has several subspecies is just as doubtful as in the guanaco. The GREATER VICUNA *(lama vicugna elfridae),* described by Ingo Krumbiegel as a subspecies and later as a species of its own, is so far known only from a few specimens in zoos. It probably is only a cross between the vicuna and guanaco or one of their domesticated forms.

While all types of llamas are easy to keep and breed in zoological gardens, the alpacas and vicunas are considered somewhat more particular than the llamas and the guanacos. It is very important not to feed these animals too richly. With good care they may become as old as twenty-four years.

Bernhard Grzimek
Dietrich Heinemann
Erich Thenius

7   The Cud Chewers

During their phylogeny herbivorous mammals have developed various methods to escape from their enemies. Many rodents became small with the ability to disappear quickly into hiding places, literally able to "crawl into a mouse hole;" the elephants developed into almost unassailable giants; the rhinoceros grew into valiant, armed knights; and the horses became fleet runners.

Among the RUMINANTS (suborder Ruminantia) we also find small brush dwellers who lead a concealed life, like the mouse deer (Tragulidae), the muntjacs (Muntjacinae), the pudus (Pudu), and the duikers (*Cephalophus* and *Sylvicapra*); among cud chewers we also find large, well-armed creatures like the wild cattle and the musk ox. There are also a large variety of light-footed fleeing animals like gazelles, antelopes, and deer.

All these animals which differ so much in so many ways belong to the third suborder of the even-toed ungulates; however, they have one characteristic in common which gave the name for the entire group: It is the special structure of the stomach and esophagus which gave them the ability to ruminate. A plant diet is hard to digest because vertebrates do not have the ferments and enzymes necessary to dissolve the cellulose from the walls of plant cells. However, certain bacteria, which live in a symbiosis of mutual advantage with the ruminants, have this ability. They live in large numbers in the stomach of the ruminant and feed on the cellulose of the plant cells, thus making accessible to their "hosts" the nutritious contents of the plant cells, most of which have remained intact after chewing. "Only in connection with this arrangement," writes zoologist Karl von Frisch, "does the strange construction of a ruminant's stomach become comprehensible (see fig. 7-1). The huge rumen, which first receives the food, represents the living quarters and the workshop of the bacteria. The reticulum, which is a protrusion of the rumen, is the ladle of this kitchen in which the bacteria pulp is mixed with the food by repeated

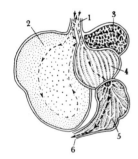

Fig. 7-1.
Cross Section of the Ruminant's Stomach.

The coarsely chewed food reaches the (2) rumen through the esophagus (1). There it is partly digested by bacteria and transported in small portions into the reticulum (3) where it is predigested and regurgitated back into the mouth. In ruminating, the food is chewed thoroughly, mixed with saliva, and then swallowed through the esophagus into the omasum (5), where digestion continues. Finally the food passes into the abomasum (6).

contracting and slackening." After the bacteria have sufficiently broken down the food mash, it is regurgitated and once again chewed thoroughly. The second swallowing of the food does not return to the rumen but now passes through a part of the esophagus formed by two folds, into the third section of the ruminant's stomach, the omasum, from where it finally passes into the abomasum. The rumen, reticulum, and omasum are protrusions of the esophagus; they are without glands and are partly lined with a horny tissue layer. The digestive glands are only found in the abomasum, and this is the ruminant's actual stomach, where the actual digestion begins.

The ability to ruminate not only provides for a better utilization of the food but also carries with it another important advantage. It reduces to a minimum the time the animals have to spend grazing in a dangerous environment with little cover. The thorough chewing can be done later in a safe hiding place. Therefore, it is not surprising that such a practical "invention" within the class of mammals has been made several times. Ruminating has been discussed in Volume X with the kangaroos, with the camels (see Ch. 6), and in some early development stages of the structure of the peccaries' stomach (see Ch. 4). But the greatest efficiency is reached in the actual ruminants of the suborder *Ruminantia.*

The present-day members of this suborder are classified into two infraorders, the chevrotains (Tragulina) with only one family, the mouse deer (Tragulidae), and the infraorder Pecora with four families whose members are characterized by horns and antlers.

The MOUSE DEER (Family *Tragulidae*) are only of rabbit to hare size. The L is 45–100 cm, the BH is about 20–40 cm. The head is small, and the narrow snout has a hairless nose region and nostrils which are long slits. The large eyes have no pre-orbital glands, and the ears are small and pointed or medium sized and round. They have no horns on the forehead. The neck is short. The arched back ends in a compact rump which is humped towards the rear. There are four toes, and the pseudo-claws are distinctly weaker than the center toes. The fine dense coat clings to the body. The hair on the back is longer and in some parts is bushy. There are thirty-four teeth: $\frac{0 \cdot 1 \cdot 3 \cdot 3}{3 \cdot 1 \cdot 3 \cdot 3}$. The upper canines in the ♂♂ are long and saber-shaped, while in the ♀♀ they are only very small studs. The lower canines are like the incisors. The stomach has four sections, but the omasum is reduced and small. There is a gall bladder, and a simple appendix. The females have four nipples. In males, the scrotum is not sharply defined and the tip of the penis is spiral-shaped with lobes on the sides. There are two genera with four species and fifty-seven subspecies.

The AFRICAN WATER CHEVROTAIN (*Hyemoschus aquaticus;* Color plate, p. 134) is an excellent swimmer and diver. The HRL is 75–85 cm, the TL is 10–15 cm, the BH is 35–40 cm, and the weight is 10–15 kg. The sturdy

Family: mouse deer (chevrotains) by L. Heck

Fig. 7-2.
African chevrotain *(Hyemoschus aquaticus).*

Fig. 7-3.
1. Spotted mouse deer
(Tragulus meminna).
2. Large Malay mouse
deer *(Tragulus napu).*

Fig. 7-4.
Lesser Malay mouse deer
(Tragulus javanicus).

legs have strong pseudo-claws. There is no chin gland. The intermaxillary does not reach the nasal bone. Metacarpals 3 and 4 are not fused. The gestation period, which is not known, is estimated at four to six months; one young is born.

In contrast to this species, the ASIATIC MOUSE DEER (genus *Tragulus*) live in dry or rocky habitats, in prime forests, and in mangrove jungles. They have thin legs with small pseudo-claws. The chin gland may be present or not. The intermaxillary bone reaches the nasal bone. Metacarpals 3 and 4 are fused to form the cannon bone. The gestation period, which lasts about six months, usually produces two young. There are three species: 1. LARGER MALAY MOUSE DEER *(Tragulus napu).* The HRL is 70-75 cm, the TL is 8-10 cm, the BH is 30-35 cm, and the weight is 5-8 kg. There are twenty-eight subspecies. 2. LESSER MALAY MOUSE DEER *(Tragulus javanicus;* Color plate, p. 134). The HRL is 40-47 cm, the TL is 5-8 cm, the BH is 20-25 cm, and the weight is 2-2.5 kg. There are twenty-eight subspecies. 3. SPOTTED MOUSE DEER *(Tragulus meminna;* Color plate, p. 134). The HRL is 45-55 cm, the TL is 2.5-5 cm, the BH is 25-30 cm, and the weight is 2.25-2.70 kg. In some characteristics, there is more similarity to the African chevrotain than to the other mouse deer (no chin glands, etc.) and, therefore, it has been placed into the separate subgenus *Moschiola.*

The mouse deer have certain physical characteristics in common with the ancestors of the pigs and the ruminants. Theodor Haltenorth aptly calls them the tip-toeing creatures of the dense brush and edge of the prime forests. These rabbit-sized ungulates are active mainly at dawn. They live singly and form pairs only during the mating season. In addition to all kinds of plants, the chevrotains also eat insects, crayfish, fish, small mammals, and carcasses.

Some fossil chevrotains from the Tertiary period, which were found in Eppelsheim near Worms, Germany, and in the Indian Siwalik Mountains, resembled the contemporary African chevrotain to such an extent that the extinct and the contemporary species were classified for a time into one genus. Erich Thenius is right when he calls the African chevrotain "a genuine living fossil which probably would be indistinguishable from the *Dorcatherium* of the Tertiary. Like them, it is primarily a forest dweller. The chevrotains lead such a secretive life that they are rarely seen in their habitat."

Therefore, it is not surprising that even during my long stay in the Cameroon prime forest I never saw a chevrotain in the wild. Only once did I see one and that was an animal killed by a leopard who, when fleeing from me, dropped it from a tree. During the day, chevrotains like to hide in caves underground. They skillfully disappear into the water and disappear by diving when they are pursued. Very little is known about the biology and the behavior of these small primitive ungulates. In the resting position, they place both front legs under their body and sit on them. It is said that they occasionally climb hanging tree trunks or trees covered with lianas in order to sun themselves or to escape from predators. Even though the natives in many areas pursue them with dogs, sling traps, and nets, they do not seem to be threatened by extinction. They are much harder to keep in captivity than their Asiatic cousins. Some chevrotains, which were captured in traps by the natives and then given to me, have, however, been kept for years in the Berlin Zoo.

The ASIATIC MOUSE DEER, whose Malayan name is Kantschil, are much better known. The natives of Southeast Asia consider them especially intelligent; they have a saying, "cunning like a kantschil." When pursued, the mouse deer hides in the bush and remains motionless and plays dead. When a person comes near to pick up his prey, the animal makes one or two jumps and dashes off. In spite of the fact that there are many mouse deer in South and Southeast Asia, they have rarely been seen since they are extremely shy, coming out of cover only at night. Because of their small size, these animals are called "mouse deer" in English.

After World War II, the Frankfurt Zoo kept two spotted mouse deer for some time. Rosl Kirchshofer reports on them: "Our pair retained their natural activity cycle in captivity. During the day, the animals slept, dozed, and ruminated while hidden in their little house. But at night and in the early morning, they became quite lively. They are agile and quick, always on the alert, and shy even with their keepers. They are cautious about everything strange and thoroughly investigate, with great patience and often with one front leg raised and their head and neck stretched forward. However, they usually flee anyway and take off in quick, long, arching jumps that are reminiscent of a fleeing rabbit."

Rosl Kirchshofer also described their "death-feigning" by an example from one of the Frankfurt animals: "It slowly walks in front of the pursuer, then at last stands still, ducks down a little, and lets him approach. As soon as the person stretches out his hand, the animal darts off. If it should be caught, it bites hard and tries to escape in this manner."

The chevrotains are a geologically very old group of animals. They were already living during the latter part of the Eocene, about fifty million years ago, long before the first deer or horned ungulates appeared. We may well imagine that the ancestors of all ruminants were similar to the chevrotains. In addition to many primitive characteristics, we also find some that are more highly evolved than in the other ruminants, especially in the structure of the metacarpal and the digestive organs.

The closest relatives of the chevrotains, and perhaps the direct ancestors of the horned ungulates, were the Gelocidae (Family Gelocidae)

Phylogeny
by E. Thenius

from the earlier Middle Tertiary (Oligocene). The two other families that played an important role during the Tertiary period, especially in North America, were the Hypertragulidae and the Protoceratidae. An especially primitive ruminant was the *Archaeomeryx optatus* (Color plate, p. 134), that is considered by some specialists to be a Hypertragulida, while others consider it the ancestral chevrotain. Perhaps this species should be considered the immediate ancestor of the horned ungulates, but certainly it was only slightly different from them.

The Protoceratidae, whose horns were covered by velvet, must have been fantastic looking creations. Pictures of *Protoceras* and *Syndyoceras* are on p. 134. Their long skull has three pairs of protuberances, one on the upper jaws, another on the forehead, and one on the crown. Presumably, these were bony extensions, covered with fur and similar to the giraffe's. Therefore, for some time, those animals were thought to be relatives of the giraffe. However, they seem to be related to the chevrotains. Some of these species lived in North America until the end of the Pliocene, which is the latest Tertiary.

The classification of the ruminants

The last infraorder of the ruminants includes four families, 69 genera, 141 species, and 682 subspecies; this includes the majority of the even-toed ungulates and the horned ungulates altogether. All members of this group, with few exceptions, have, at least in the males, bony weapons on the forehead; they form the infraorder *Pecora*, which is characterized by the horns on their foreheads. They range from the size of a fox to that of a giraffe. All the upper incisors are lacking, while the lower canines are incisor-shaped and join the incisors. The soles lack cutaneous pads and have hooves instead. The omasum is fully developed. They are classified into four families, according to the type of horns on the forehead:

1. DEER (Cervidae). Some have no horns on their foreheads; in others, there are antlers which consist of bare bone that is dropped every year and then grown again. The velvet covering is polished off.

2. GIRAFFES (Giraffidae). The weapons on the forehead consist of bony knobs that are covered with fur.

3. PRONGHORNS (Antilocapridae). The horns on the forehead consist of bony knobs that are covered by forked horny sheaths that are dropped and regrown annually.

4. HORNED UNGULATES (Bovidae). The horns consist of bony cores covered with hollow, unbranched horn sheaths that are permanent or are changed only once in the subadult animal.

Lutz Heck
Dietrich Heinemann
Erich Thenius

8 Deer

The DEER (Cervidae) are the first of the families with weapons on their foreheads. They range in size from the rabbit to the horse. The BH is 30-150 cm and the L is 85-300 cm. The head size is from short to long. The nose region varies from hairless to being almost totally covered by it. The eyes are rather large and the small to large ears are more or less pointed. There are no antlers in some species; otherwise, they are present only in the ♂♂ (with the exception of reindeer). The neck varies from medium length to compact. The shape of the body is small and graceful, medium sized and slender, or bulky to large and slender or heavy. The rump may be elevated in the rear or it may be straight; it may be slender or massive. The tail is from rudimentary to long, from plain to bushy, or with a tassel. The legs range from graceful to compact, and from rather short to long. The coat, which is from short-smooth to dense-long, often changes with the seasons. Sometimes there is a tuft on the forehead, a mane, or a beard under the throat. The coat of the young is usually spotted, and sometimes this also occurs in adults. There are numerous skin glands. The females have four nipples. Bile is not collected in the gall bladder but empties directly into the duodenum (an exception is the musk deer). There are 32-34 teeth: $\frac{0 \cdot 0\text{-}1 \cdot 3 \cdot 3}{3 \cdot 1 \cdot 3 \cdot 3}$. The gestation period lasts from five to eight months.

The distinguishing characteristic of the deer is their antlers. Except for two primitive species (musk deer and Chinese water deer), all males have these weapons on their foreheads. Another exception to this rule is the reindeer where the females have antlers too. The antlers are more or less branched rods which grow on the bony structure of the frontal bone and are, in most cases, dropped annually. The new antlers then grow within a few months; they are covered with a thin skin—the so-called velvet—which is well supplied with blood vessels. When growth is complete, the velvet is shed by rubbing and the animal again possesses his head ornaments. The antlers have made deer the

The development of the antlers, by Th. Haltenorth

▷
Fig. 8-1.
The relationship of hormones and growth of antlers in male roe deer. 1-12. During the months of January to December. The arrows indicate when the hormones are effective. 1-3. Stimulated by the anterior lobe of the pituitary gland (FLP), the testes secrete a hormone to develop the antlers (DH). The antlers grow. 2-3. The stimulating hormone from the Leydig's cells (ICSH) stimulates the testes to produce increasing quantities of the male sex hormone testosterone (T). 4-8. Increased production of T inhibits FLP and hence DH. Therefore, the antlers cease growing, calcify, and are polished (5). The testes produce sperm. 9-12. T and sperm production decreases and lower DH production causes the antlers to be dropped and the wound to be covered with new tissue.

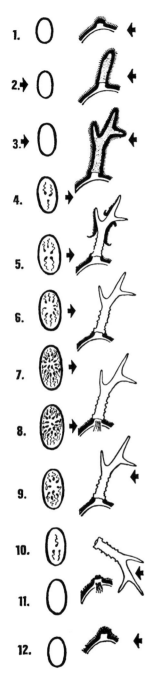

ICSH Testes T Antlers DH

1.
2.
3.
4.
5.
6.
7.
8.
9.
10.
11.
12.

Deer antlers as
aphrodisiacs, by
B. Grzimek

most desired game animals by hunters in many countries. Because of extensive hunting, some species of deer have been threatened by extinction or were actually exterminated. However, others owe their continued existence to the desire of hunters to ensure a supply of trophys, and therefore, hunting of the protected game is well controlled.

In deer, the basic structure of the antlers cannot be changed during the new growth. Again the reindeer is an exception; it is the only species of deer which is able, in both sexes, to add new points during the growth, thus changing the basic structure from base to tip. To what extent the antlers of other deer are sex-related is clearly shown when a deer is deficient in testosterone, the male sex hormone, due to loss or destruction of both testicles. This hormone is produced by Leydig's cells which are in the interstitial tissue of the testicles. In all species of deer, where the development of antlers is sex-related, it is possible to prevent any growth of antlers if the animal is castrated soon after birth. Castration after pubescence and during the growth of the antlers leads to a continuous growth which covers the head like a wig. However, when the testes are removed after the velvet-shedding phase, the deer will drop his antlers immediately and then grow small, permanent antlers covered with velvet. Only in the reindeer is castration without much effect. Small antlers may also grow in females of other species as a result of hormonal disorders.

Besides the testes, the thyroid gland and the anterior lobe of the pituitary gland also have an effect on antler formation. The pituitary gland is responsible, so to speak, for the growth of the antlers. Initially the pituitary provides merely the stimulus to the testicular hormone testosterone that the pubescent animal should grow the secondary sex characteristics of his species. This stimulus by the pituitary can be suppressed with an injection of the female sex hormone, which causes already present cartilage tissue to calcify in the budding antlers, thus stopping any further growth. In the normal process, the testosterone from the testes effects a more gradual calcification of the cartilage, then the drying up of the velvet, and finally the dropping of the antlers. In this process, the thyroid gland directs the rate of growth of the antlers. There is then a very complicated relationship between the growth of the antlers, the activity of the hormone glands, and the annual rutting season of the deer.

For centuries, the people of Siberia, Kazakhstan, and the Altai Mountains sold antlers in the velvet stage of the spotted East Asiatic sika or Dybowski's deer to the Chinese, neatly bundled and packed in boxes. This was one of the reasons why the deer disappeared in large parts of Russia. The peasants had a second source of income by keeping the deer in pens and cutting off the plethorial antlers each year. To the East Asiatic peoples, the antlers in the velvet phase, especially those of the sika and red deer, are considered a stimulating and restorative remedy. The hormones and their components present in the velvet antlers can be extracted by a process of careful drying, grinding, or bleaching. The extract has been named "Pantokrin" after the Russian word "panty." The blood from the cutting also is collected, dried, and used as a remedy. Even shed antlers are ground into powder in East Asia and used for abcesses and rashes.

In their pharmacies, the Chinese used to sell many ineffective remedies, for example, rhinoceros horns, teeth of extinct animals, and preserved snakes. Due to this, they are mainly responsible for the extermination of the Asiatic species of rhinoceros. Some species of Asiatic deer also have been nearly or perhaps completely destroyed in the wild for the same reason. Fortunately for the deer, Pantrokrin presently is being produced in state farms which have been established in the Soviet Union. After cutting, the antlers are dipped into almost boiling water repeatedly at regular intervals. In the drying process they lose 40 to 60 percent of their weight. The Academy of Sciences in Moscow has done clinical tests with such antlers. They found that, in contrast to the medically completely ineffective rhinoceros horns, the deer antlers indeed contained many sex hormones and basic substances which have beneficial effects on old age symptoms and the healing of wounds. Therefore, many subspecies of the sika deer and of the Asiatic red deer are still seen on such farms, whereas they are nearly extinct in the wild.

The earliest ancestors of the deer probably had no antlers, like the present musk deer and Chinese water deer. Beginning at the end of the Middle Tertiary, gradually and over millions of years, the different shapes of antlers have evolved. The oldest fossils suggest that the evolution of the deer began in Asia. The earliest forms, such as the *Eumeryx* from the Oligocene of Asia (approximately forty million years ago) were similar in body structure to the mouse deer. A number of primitive deer can be derived from them, including the *Amphitragulus* from the European and the *Blastomeryx* from the North American Miocene (approximately ten to twenty-five million years ago.) They are already true deer, but they still had a low skull without antlers, a complete set of molars, and low crowned teeth. The canines in the upper jaw were elongated. From such forms all of the other deer seem to be derived.

The TRUE DEER (Cervinae) are derived from ancestors who were similar to the present muntjacs (subfamily Muntiacinae). In the late Pliocene they went beyond the earlier six point stage of the antlers. The common ancestor of all true deer was closely related to the *Cervavitus tarakliensis* of the late Pliocene.

Among the best known deer of the Glacial Period were the GIANT DEER (genus *Megaloceros*). They were not related to the presently living

Phylogeny, by E. Thenius

Fig. 8-2.
The Giant Deer (*megaloceros giganteus*) lived in the glacial steppes of Europe.

fallow deer or even the moose, as was formerly assumed, but were derived from large-bodied deer with ordinary antlers. They finally developed huge antlers with large flattened surfaces, which surpass in size any known deer antlers. For example, the EUROPEAN GIANT DEER (*Megaloceros giganteus*) of the early Glacial Period had antlers with a span of more than three and half meters! It lived in open country, and became extinct when its habitat reverted to forest at the end of the Glacial Period. These extremely large antlers have often been blamed for the extinction of the giant deer, i.e., they were said to be too heavy, and the annual growth would have been too much of a burden on its metabolism. But these interpretations are almost certainly incorrect. The sizable antlers corresponded to the animal's size, and so they could be expected under the natural growth conditions.

Suddenly, during the Glacial Period (Pleistocene) in the Old World, the moose, roe, and reindeer appeared, and so far none of their ancestors are known from the Tertiary.

Subfamily: musk deer, by L. Heck

The MUSK DEER (subfamily Moschinae) have a special place among the deer. Some of their characteristics are reminiscent of the chevrotains. Like them, they are without antlers and have a gall bladder. The males have tusk-like upper canines and a thread-like extension at the penis; they also lack glands on their face and feet. However, the well-developed four-sectioned ruminant's stomach of the musk deer indicates a closer relationship to the deer than some zoologists had presumed. Its foot structure, which is like the deer's, is a special adaptation to alpine life. The main digits are spread far apart, and the pseudo-claws touch the ground in walking. This gives the musk deer a larger surface to stand on and represents, at the same time, a protection against slipping. A special characteristic of the males is a musk pouch on their abdomen which is situated in front of the sex organs. The glands in this pouch secrete a strongly scented paste called musk during the rutting season.

There is only one genus and species: MUSK DEER (*Moschus moschiferus*; Color plate, p. 209). The HRL is 80–100 cm, the TL is 4–6 cm, the BH is 50–70 cm, and the weight 7–17 kg. The head is small, and the front part of the body is more slender, weaker, and lower than the rear. The tail is very short and almost triangular. The front legs are shorter than the hind legs. The pseudo-claws of the front legs are telemetacarpal. The gestation period lasts five to six months; one or two young are born. There are seven subspecies, four of which are endangered.

In ancient India, the musk deer was named "muschkas" because of his sexual glands; the word means "testicles." This secretion has been used for thousands of years in Chinese perfume production. Before the invention of artificial aromatic substances, musk had also been used in Europe as the most expensive ingredient of fine perfumes.

The shape of a musk deer reminds one of a large hare because it

Fig. 8–3.
Musk Deer (*Moshcus moschiferus*).

is greatly enlarged in the rear and has large, round ears. The head is especially striking; it is very small in relation to the rest of the body. In their Central and East Asiatic native habitat, musk deer prefer dense forests and the large rhododendron bushes on the mountain slopes. Because of their foot structure, they are able to climb steep slopes; sometimes they will even climb inclined trees and feed in their tops. Their main diet is young grass, moss and lichen, tender shoots, the leaves of the dwarf oak, and occasionally cedar nuts. They spend the day time in a depression scraped in the ground in the manner of rabbits. They use fixed paths and are sedentary. During the mating season in late fall, the males have intense fights. Each tries to wind its neck around the other's and to use its long tusks which may be up to seven centimeters long and are capable of tearing deep wounds in the hide and muscles. All adult males bear scars from such fights. During this time, they carry the penetrating musk odor. To mark their territories, they rub a secretion from their tail glands onto small branches and rocks.

Little is known about keeping musk deer under human care because these animals rarely reached European or North American zoos. The first musk deer in Germany was kept in the Berlin Zoo in 1889.

Lately musk deer have been kept and studied in the Frankfurt, the Kronberg, and other zoos. Rosl Kirchshofer reports on the Frankfurt pairs which "were extremely shy at first, but now even strangers may enter their pen without causing the animals to climb the walls. It is interesting that on such occasions the male appears to be more lively, 'fearless,' and hence more curious. He continues to come out of his house, peers at the visitor, and utters an alarm call periodically which also seems to be meant as a threat. The call resembles a strong human sneeze and is difficult to describe. The animal raises its head up and back and then thrusts it forward while it is vocalizing. I never heard this sound from the female, but, in the same situation, she urinated long and repeatedly in several places as a fear reaction. It is striking that after even a slight stress both animals breathe heavily with their tongues hanging from their mouths, but this is not surprising considering that they are alpine animals which now live in the climate of a Frankfurt summer." The Frankfurt female at first did not show her advanced state of pregnancy. She had a female fawn in June, 1962, which she raised successfully. The only recorded case of previous reproduction had been the musk deer of the Duke of Bedford in Woburn Abbey (England). In 1965, two more female young were born in Frankfurt but, unfortunately, they were so seriously injured by magpies immediately after birth that they died.

Hans Frädrich has thoroughly studied the behavior of the Frankfurt musk deer. Inspired by reports from the wild that musk deer like to climb inclined or fallen trees and to feed on the lichens in their tops,

Musk deer in the zoo

a platform was put up 1.8 meters high in the Frankfurt antelopes' pen, enabling the animals to reach it by climbing an inclined tree trunk, and soon this became their regular resting place. A short while later the musk deer also occupied a second platform on a somewhat lower level. Although they did not seem to mind snow and severe frost, wet and cold weather as well as the heat and humidity of the summer did not bother them. Obviously in captivity these animals definitely need a shady place and sufficient shelter against rain and wind. Hans Frädrich reports: "The strikingly high rump of the musk deer's body determines most of its movements. Its movement is like a deliberate glide with diagonally opposed feet on the ground at a time and with the neck stretched forward horizontally. When they move faster, they carry it slightly higher and, in a trot or gallop, the neck is raised up even higher. When in flight the animals usually gallop and the hind legs are placed in front of the forelegs. They rarely leap for a longer distance like a gazelle, with all four feet raised off the ground at the same time. The animals tire quickly, at least in our latitudes, after a short trot or gallop, and their tongues hang out of their mouths as they breathe heavily. They are able to jump up over the tree trunks to their platforms with two or three leaps without difficulty, and they often jump straight down to the ground from the platform. They can easily stand upright on their hind legs for about half a minute. Sometimes they support themselves by touching vertical surfaces with their front legs pointing upwards, in the manner of gerenuks."

The long neck with its extremely elastic vertebrate column enables the musk deer to do extensive grooming. Frädrich found that they can reach a large part of their own body either while standing up or lying down, with nibbling or licking movements of their lips. They scratch the regions of the upper neck and their ears with their hind legs. After prolonged rest, they shake their body and often many hairs fall out of the long, brittle coat. An animal when caught showed how easily the hair comes off. Where the bent arm of the keeper touched the animal, all the hair broke off and spots of considerable size of naked, dark skin became visible.

According to reports from the wild, musk deer are said to be unsocial, living singly or in small groups consisting of females and their young. The Frankfurt animals also lived side by side for a long time without paying much attention to each other. Only after they had used the same platform did they occasionally sniff or lick each other. On cold evenings in winter, they would chase each other in the freshly fallen snow. In the young born in 1962, Frädrich observed two different methods of suckling. When the mother lay down, the young drank with its head lowered, the front legs bent back, and the hind legs spread apart. Sometimes it would then lie on its belly. But when the mother was standing, the young pushed searchingly with its head and neck erect in the direction of the udder, frequently kicking up a stretched-out front leg with each push of the head. Such blows, which usually miss, are reminiscent of the treading movements of some predators and also of different species of deer and antelope.

Depending on the development of the metacarpals of the front legs, the other deer are divided into two large groups. In one group only the lower parts of the second and fifth digit metacarpals are present (Telemetacarpalia) while in the others, only the upper parts (Plesiometacarpalia) remain. To the plesiometacarpal deer belong the subfamilies of the muntjacs (Muntiacinae) and the true deer (Cervinae); to the telemetacarpal deer belong the water deer (Hydropotinae), the American and roe deer (Odocoileinae), the moose deer (Alcinae), and the reindeer (Rangiferinae). We shall begin the description of the plesiometecarpal deer with the primitive muntjacs (Muntiacinae) which have antlers and extended tusk-like upper canines. They are the size of a duiker. There are no gall bladder and no thread-like attachments on the tip of the penis. There are two tear duct openings. There are two genera (muntjacs and tufted deer), with two species and twenty-three subspecies.

Subfamily: muntjacs, by L. Heck

1. MUNTJAC *(Muntiacus muntjak)*. The HRL is 89–135 cm, the TL is 13–23 cm, the BH is 40–65 cm, and the weight is 15–35 kg. The base of the antlers is long. The short antlers occasionally have additional branches and one to three tines. The short and shiny coat is thicker and denser in subspecies from moderate climates. The gestation period is about six months and only one, rarely two, young are born. There are twenty subspecies, among them the NORTH INDIAN MUNTJAC *(Muntiacus muntjak vaginalis)*, the CHINESE MUNTJAC *(Muntiacus muntjak reevesi;* Color plate, p. 209), and the JAVAN MUNTJAC *(Muntiacus muntjak muntjak:* Color plate, p. 209). Some of the endangered subspecies are the TENASSERIM MUNTJAC *(◊Muntiacus muntjak feae)* and the BLACK MUNTJAC *(◊Muntiacus muntjak crinifrons)*.

2. TUFTED DEER *(Elaphodus cephalophus;* Color plate, p. 209) is larger. The HRL is 110–160 cm, the TL is 7–15 cm, the BH is 52–72 cm, and the weight is 40–50 kg. The tail is somewhat shorter. The insignificant antlers are unbranched and almost invisible in the forehead tuft. The base of the antlers is short. The lachrymal fossa of the preorbital gland is very large. The ♂♂ have a dark tuft up to seventeen centimeters long on the forehead and around the base of the antlers. The gestation period is about six months, and only one young is born. There are three subspecies.

The muntjacs are deer with a slightly larger rump. They live in the bushes and dense plant cover of their East and Southeast Asiatic habitat. Fossils of deer similar to the muntjacs from the Tertiary Period have been found in different places in Europe. The bones and skull fragments hardly differ from those of presently living forms. Millions of years ago these small deer also lived in our latitudes. Out of these

Fig. 8-4.
Muntjac (*Muntiacus muntjak*).

Fig. 8-5.
Tufted Deer (*Elaphodus cephalophus*).

Subfamily: true deer
by L. Heck

primitive species of deer with tiny antlers came the bigger animals with the large antlers. In the true muntjacs, the small antlers sit on a high, hair-covered bone structure on the forehead which is as long as the antlers. In the tufted deer, they are short and dagger-like. While these weak structures may be used in the rut for fighting, the long upper canines play a more important role. I once saw a muntjac, who was about to be caught in a small pen, tear a gaping wound with these hook-like teeth in the keeper's upper arm.

The muntjacs are also called "barking deer." When they are excited, they utter a series of short, hard sounds, similar to the barking of a dog, of a loudness which one would not expect from an animal of this small size. The famous tiger hunter Jim Corbett reported that the small deer had often announced the approach of a tiger by their alarm calls. In the tropical subspecies, the rut is not restricted to one season; however, in the northern species it begins in January and February. The rivals fight vigorously, using their canines rather than their small antlers. About six months after copulation the female gives birth to one or two spotted young. With her long protruding tongue, the female licks her young clean carefully. The tongue is used, as in giraffes, to strip off twigs and leaves from bushes. It is so long that when grooming it can reach over the face up to the eyes.

The CHINESE MUNTJAC (*muntiacus muntjak reevesi*) in its habitat lives in climatic conditions similar to those of Europe. Therefore, these winter-adapted muntjacs were introduced several times into France and England. I often saw them flee from some bush with quick jumps at Woburn Abbey on the Duke of Bedford's estate, and in Clere (Normandy) in the park of Jean Delacour. They crossed the clearings with large leaps and disappeared into the closest cover. It is interesting that these primitive deer do not form groups as the more highly evolved species do. They live singly or in pairs in their territories, which they hardly ever leave, preferring certain places and paths. When pursued, they do not flee very far, but rather hide and wait for the pursuers to leave.

Muntjac deer have successfully been kept for years in the Berlin Zoo; in 1961 a young was born there. Today the Javan and the Chinese muntjacs are kept and bred in many zoos, where some have reached an age of more than sixteen years. The much less known tufted deer of the Central Asiatic mountains, however, has seldom reached a zoo. One had lived in the London Zoo for almost seven years from 1932 until 1939.

Like in the muntjacs, in the TRUE DEER (subfamily Cervinae) the front legs are "plesiometacarpal," which means that only the upper ends from the reduced metacarpals of the side digits remain. The base of the antlers is rather short and the antlers are long. There are two lachrymal duct openings. The head varies from short to long, with a

Forest in Central Europe 1000 years ago

Artiodactyls: 1. Aurochs *(Bos primigenius*, see Ch. 13). 2. Roe deer *(Capreolus capreolus)*. 3. Wild boar *(Sus scrofa*, see Ch. 4) 4. European bison *(Bison bonasus*, see Ch. 13). 5. Red deer *(Cervus elaphus)*. ☐ Predators (see Vol. XII): 6. Pine marten *(Martes martes)*. 7. Brown bear *(Ursus arctos)*. 8. Fox *(Vulpes vulpes)*. 9. Lynx *(Felis lynx)*. 10. Badger *(Meles meles)*. 11. Wolf *(Canis lupus)*. 12. European wild cat *(Felis silvestris)*. ☐ Rodents (see Vol. XI): 13. Dormouse *(Glis glis)*. 14. Squirrel *(Sciurus vulgaris)*. ☐ Insectivores (see Vol. X): 15. Common shrew *(Sorex araneus)*. ☐ Birds of prey (see Vol. VII): 16. Hawk *(Accipiter gentilis)*. ☐ Owls (see Vol. VIII): 17. Eagle owl *(Bubo bubo)*. ☐ Cuckoos (see Vol. VIII): 18. Cuckoo *(Cuculus canorus)*. ☐ Woodpeckers (see Vol. IX): 19. Black woodpecker *(Dryocopus martius)*. 20. Green woodpecker *(Picus viridis)* catching ants. ☐ Sparrows (see Vol. IX): 21. Chaffinch *(Fringilla coelebs)*. 22. Crested tit *(Parus cristatus)*. 23. Coat tit *(Parus ater)*. 24. Golden oriole *(Oriolus oriolus)*. 25. Song thrush *(Turdus philomelos)*. 26. Jay *(Garrulus glandarius)*. 27. Nuthatch *(Sitta europaea)*. ☐ Reptiles (see Vol. VI): 28. Viper *(Vepera berus)*. ☐ Insects (see Vol. II): 29. Garden tiger beetle *(Arctiea caja)*. 30. Ant lion (larva of *Myrmeleon formicarius*) in its crater catching ants. 31. Red forest ant *(Formica rufa)* with nest mound and "ant road".

straight or humped profile. The tip of the nose region is hairless and delicately patterned. The eyes are of medium size. The neck ranges from short to medium long. The rump is from compact to long with legs ranging from rather short and sturdy to long and slender. The tail varies from short and smooth haired to medium-sized and bushy (in Père David's deer, long with tassel). There are 32–34 teeth:$\frac{0 \cdot 0\text{-}1 \cdot 3 \cdot 3}{3 \cdot 1 \cdot 3 \cdot 3}$ The four genera (fallow deer, axis deer, red deer, Père David's deer) comprise thirteen species and sixty-two subspecies.

The FALLOW DEER (genus *Dama*) differ from the other true deer in their flattened palmate antlers with many points. The HRL is 130–235 cm, the TL is 15–20 cm, the BH is 80–105 cm, and the weight is from 35–200 kg. The legs are medium-long to long and the body is compact and usually spotted. The rather short head has well-developed preorbital glands. The neck is compact and the larynx protrudes. The summer coat is smooth, thin, and clinging, with little fleece, while the winter coat is rougher and thicker, with dense fleece. The upper canines are lacking. The gestation period lasts seven to eight months; usually one young is born, rarely two or three. They live from twenty to thirty years.

The subspecies are: 1. EUROPEAN FALLOW DEER (*Dama dama dama*; Color plate, p. 210). The antlers have strong anteocular and brow branches, and at the upper end of the antler is a flat palinate expansion. The tail is medium long. 2. PERSIAN FALLOW DEER (⊖*Dama dama mesopotamica*; Color plate, p. 210). The antlers have a very short anteocular branch, a very strong middle brow branch, frequently an additional point, and a flat palinate expansion of the antler. The tail is short. The coat is especially woolly and dense on the neck and withers.

Before the last Glacial Period, the fallow deer were part of the fauna of Central Europe. After the ice had receded the European fallow deer was found only in Asia Minor. By ancient times it had been introduced throughout the Mediterranean countries by the Phoenicians and, later, the Romans. The Romans also brought it to Germany and into the northern part of Western Europe. Today the fallow deer is the most popular and widely distributed animal in the European parks. Since this species of deer is extremely adaptable, it has been successfully introduced to other continents.

Because the fallow deer has been kept and bred by man for centuries, it occurs in many different colors. Besides the natural color, there are animals that are white, black, reddish, and the so-called "porcelain-colored" fallow deer. In earlier centuries, such mutations were preferred by the breeders of the princely game parks, according to the prevailing fashion. The white animals are yellow at birth.

In the European forests, the fallow deer survived mainly because of their senses. Their sense of smell and hearing are as good as the red deer's, but their sense of sight is much better. Most species of deer are not able to see a motionless person, but the fallow deer can. Their eyes are constructed so that they can be focused on certain objects. Man and the related primates are also able to do this. While the fallow deer in the game parks becomes very tame and is often seen during the day in clearings, it is extremely shy in the wild. Hunting the fallow deer requires the same skill and endurance as for the red deer.

A population of fallow deer has been living for a long time in the Berlin Grunewald Park. On a previous visit to Berlin, a friend of mine was quite surprised to see a fleeing fallow deer when he stepped off the bus at the city limits. With its peculiar jumping manner and long tail, the fallow deer is a little like an antelope. In the zoo, I have often received phone calls from people who told me that some foreign antelopes had escaped and were running around in the Grunewald.

Usually fallow deer live in large herds, which consist of females and their offspring. The adult males live in groups of their own. The rutting season begins during the second half of October, when the males join the females and begin their fights. They utter a rough, burping sound which sounds like coughing from a distance. When the antlers clash together, it sounds quite serious, but the antlers are shaped in such a way so as to prevent serious injuries. The females give birth to their beautifully spotted young from the end of May to the beginning of June. During the first two to three weeks, the mothers leave them hidden in the dense brush; later they follow the adults.

The PERSIAN FALLOW DEER still had a wide distribution at the end of the Glacial Period. It became extinct in North Africa and wide parts of its Asia Minor habitat when the land became increasingly arid, as well as through heavy persecution. It is remarkable that the ancient Egyptians, Sumerians, Assyrians, and other civilized nations of pre-Hellenic times only knew the Persian fallow deer, using it as the most important sacrificial animal in their moon and hunting cults. Later, during the era of the Phoenicians, Grecians, and Romans, the Mesopotamian species disappeared from these cultures, and the European fallow deer, which then lived in Asia Minor, took its place.

Today the Persian fallow deer is one of the rarest and least known mammals. In 1875 this forgotten animal was rediscovered in Western Persia by the English Vice Consul Robertson and Sir Victor Brooke, who described it as a new species. During the following years, Robertson sent some living animals to the London Zoo, and they reproduced there. Some of them came to the Duke of Bedford's estate in Woburn Abbey, and were still there at the beginning of the twentieth century. In 1951, the Persian fallow deer was considered extinct since no one had seen or heard of this animal for decades.

The American scientist Lee Merriam Talbot travelled to Asia Minor in 1955 on behalf of the International Union for Conservation of Nature and National Resources. There he learned about deer in the

Fig. 8-6.
Fallow deer *(Dama dama)*. Black: Original distribution. Hatched: Present distribution of European fallow deer. Arrow: Present distribution of Persian fallow deer.

Persian fallow deer

area of the Persian-Iraqui border. After his return home, he mentioned this to two German zoologists, Theodor Haltenorth and Werner Trense, and the two suspected that it might be the Persian fallow deer. Soon after, Trense had received information from North Persia on a population of deer in the area of Chusistan. The two zoologists contacted Georg von Opel, founder of the Opel Zoo in Kronberg (Taunus), in order to obtain the necessary funds for a trip in search of the Persian fallow deer. On the first scouting trip in 1957, Werner Trense found about two dozen Persian fallow deer in a small wooded area along the banks of the Karcheh River. This probable last remnant of the species was obviously in extreme danger from environmental conditions as well as animal and human predators. Trense, who brought a young pair of these animals to Kronberg, tried to establish conditions for the preservation and reproduction of this species.

According to Theodor Haltenorth's reports, the Persian fallow deer has retreated into the dense jungle of brush where it has developed the art of escaping persecution by moving around silently: "The Arabs in this area, whose water buffalo graze in the steppe next to the forests and on clearings along the river, hardly disturb the fallow deer, but they endanger the species because they take the young which they find in the jungle and eat them." According to the information from these Arabs, the rutting season is in August and lasts for about one to one and a half weeks. The fawns are born from mid-March to mid-May in the dense jungle.

There are only estimates about how many Persian fallow deer now exist in the small remaining range in Southern Persia. Presumably, after the area is cultivated, there will only be a few dozen left. Haltenorth estimated that in 1960 there were still two to four hundred animals. Therefore, Georg von Opel's efforts in his zoo in Kronberg and those of an Iranian landowner to preserve this rare species of deer are all the more praise worthy. Because of Haltenorth's research and his report to the Iranian government, the Persian fallow deer now is fully protected.

Axis deer and hog deer

The SOUTH ASIATIC INDIAN SPOTTED or AXIS DEER (genus *Axis*) have a heavily spotted, and thus presumably phylogentically old, coat. The head is short, the body is compact, and the tail varies from long to medium long. The antlers are lyre-shaped and widely separated at the ends, with usually six and rarely seven or eight points. The long anteocular branch begins immediately above the base of the antler and the second branch is shorter than the anteocular branch. The upper canines are usually lacking and the lower canines are similar in shape to the incisors. The subgenera (*Hyelaphus* and *Axis*) each contain one species.

Distinguishing characteristics

1. HOG DEER (*Axis porcinus*; Color plate, p. 211). The HRL is 105–115 cm, the TL is approximately 20 cm, the BH is 60–75 cm, and the weight

is about 50 kg. The animal is low-built and is reminiscent of a hog; its tail is bushy. The gestation period is 220-235 days; one young, but rarely two, are born. They live twelve to fifteen years, but may on occasion reach twenty. There are four subspecies, some of which are unspotted. The BAWEAN-HOG DEER (*Axis porcinus kuhli*) was almost extinct, but has recovered nicely, and the CALAMIAN-HOG DEER (⬦ *Axis porcinus calamianensis*) which remains endangered.

2. AXIS DEER (*Axis axis;* Color plate, p. 211). The HRL is 110-140 cm, the TL is 20-30 cm, the BH is 75-97 cm, and the weight is 75-100 kg. These animals have longer legs and a longer body, and they have a rather long tail. The eyes are large and the upper lids have long lashes. The ears are shorter than the hog deer's. The gestation period is from seven to seven and a half months; one to three young are born (♀♀ sometimes give birth twice a year). Their age is the same as in the hog deer. The two subspecies, which are always spotted, include the INDIAN AXIS DEER (*Axis axis axis*) which has been introduced into New Zealand, Australia, Hawaii, Brazil, and Argentina.

While the hog deer usually live singly, some individuals may form a small group. Their preferred habitat is grassland with bushes near the edge of the forest. Although the hog deer does not grow to the withers' height of the European roe deer, it is much heavier. These animals rest during the day in the jungle and feed only at night. When scared, they flee with a peculiar gait which some observers have compared to a pig's.

While hog deer have been successfully introduced to European game parks several times, keeping them presented some problems because individual deer may become rather pugnacious. During the rutting season, the male attacks and fights with all kinds of objects. He dashes against trees and fences, ploughs the ground with his short antlers, and threatens every one in his vicinity by tilting his head to the side and approaching in a diagonal direction, often inflicting damage with his antlers. He even attacks deer species very much larger than his own.

The closest relative to the hog deer, the true axis deer, probably has the most vividly spotted coat of all deer, and it retains this pattern during its entire life. This species prefers to remain close to rivers, taking to water to escape predators. Axis deer live more frequently in larger herds than hog deer.

As in all tropical deer, the antlers are not dropped in any specific season. The ruts occur irregularly, occurring throughout the year with single individuals. The females usually have their young during the cooler time of the year. Axis deer are very fertile and, after an interim of six months, the female may have another young.

These beautiful, colored deer already attracted the attention of the Romans. Since ancient times they were brought to Europe from time

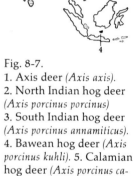

Fig. 8-7.
1. Axis deer (*Axis axis*).
2. North Indian hog deer (*Axis porcinus porcinus*)
3. South Indian hog deer (*Axis porcinus annamiticus*).
4. Bawean hog deer (*Axis porcinus kuhli*). 5. Calamian hog deer (*Axis porcinus calamianensis*)

to time. In recent times they come especially to English game parks where they do well and reproduce readily. They have become established in areas where the climate is not too cold, but they are very sensitive to snow and extended periods of subfreezing temperatures. In Germany, a herd has lived successfully for 150 years in the Royal Wuerttemberg Game Park Solitude near Ludwigsburg.

In some cases, the axis deer may reproduce too much in their new habitats. They were introduced to New Zealand because there were no endemic mammals and hence no four-legged game animals. Soon they became so numerous that they represented a serious danger to crops and forests. They also threatened the birds of New Zealand, which were already endangered. Therefore, the authorities had to hire professional hunters as "deer exterminators." They have orders to kill as many deer as possible. Without this action, New Zealand's national parks would have been damaged and their plant life, which is well worth preserving, would have been lost. This is another clear example of the fact that the introduction of new species to places where the fauna and flora are not adjusted to such aliens may cause more damage than good.

Most of the true deer belong to the RED DEER (genus *Cervus*). They are of a medium to large size and have more or less heavy antlers. Frequently, there is a well developed mane around the neck. The coat varies according to the season. Upper canines are present, but are reduced in both sexes of the sika deer. There are five subgenera: SAMBARS, BARASINGHAS, SIKA DEER, RED DEER, and THOROLD'S DEER.

Sambars

A large number of South Asiatic continental and island deer, which grow antlers with only six tines, are included in the subgenus SAMBAR *(Rusa)*. They range in size from that of a roe to that of a horse. The hair is coarse and the neck often has a little or a strong mane. The tail has either medium long hair or it is bushy with coarse, woolly hair. The antlers are strong thick bars which are pearled up to the crown-fork. The frontal bone knobs are short to medium long and the strong anteocular spiller rises from the bar at an acute angle. The gestation period of 249-284 days yields one young, seldom two. Longevity is twelve to twenty years (in captivity sometimes more). There are three species with eighteen subspecies.

1. INDIAN SAMBAR *(Cervus unicolor)*. The HRL is 170-270 cm, the TL is 22-35 cm, the BH is 120-150 cm, and the weight is 150-315 kg. These large animals are long-legged. The tail is wide and bushy, and the coat has long, firm, woolly hair. There are six subspecies, which include the CEYLON SAMBAR *(Cervus unicolor unicolor;* Color plate, p. 212) and the SUMATRAN SAMBAR *(Cervus unicolor equinus;* Color plate, p. 212). Two of the subspecies are endangered.

2. SUNDA SAMBAR *(Cervus timorensis;* Color plate, p. 212). The HRL is 130-215 cm, the BH is 80-110 cm, the TL is 10-30 cm, and the weight

Fig. 8-8.
1. Indian sambar deer *(Cervus unicolor)*. 2. Sunda sambar deer *(Cervus timorensis)* 3. Philippine sambar *(Cervus mariannus)*.

is 80–125 kg. They are smaller, more graceful, and longer necked than the Indian sambar. The tail is slender and not bushy. The dense coat is shaggy and rather long. The ♂♂ often have a mane around the neck. There are eight subspecies; three of them are endangered.

3. PHILIPPINE SAMBAR (Cervus mariannus). The HRL is 100–115 cm, the TL is 8–12 cm, the BH is 55–70 cm, and the weight is 40–60 kg. This is the smallest species of the subgenus. It is distinguished by a higher croup and short ears and head. The ♂ is more compact and bulkier than the graceful ♀. The tail is not bushy. There are four subspecies, among them the LUZON SAMBAR (Cervus mariannus mariannus; Color plate, p. 213) and the ALFRED'S SAMBAR (Cervus mariannus alfredi; Color plate, p. 213).

Already Alexander the Great encountered the SOUTH INDIAN SAMBAR (Cervus unicolor niger) on his Indian campaign. Thus Aristotle, the famous Greek philosopher and zoologist, learned about this animal. He mentioned in his *History of Nature* that this deer's antlers regularly had only six endings. Because of this early and accurate observation, this subspecies is also named Aristotle's deer. The sambar seemed so large to British hunters in India that they gave it the confusing name "Elk," in the same way that the North Americans named the wapiti. In size and weight the Indian sambars cannot compete with the moose, but they can with the European red deer. Some individuals reach a weight of 250 kilograms.

South Indian sambar

Depending on their habitat, the sambars vary extremely in color and size. The subspecies of East India, Sumatra, and Borneo (Cervus unicolor equinus, a.o.) are almost plain black with very short antlers and bulky bars. The anteocular spiller joins the bar of the wide frontal bone knobs. These knobs, therefore, are oval and often quite large. It is so large that *objets d'art* and even whole hunting scenes may be carved on it and the basal part of the bar. Because of their especially long, thick, and coarse-haired tail, these Southeast Asiatic forms in England and North America are called "horse-tailed sambars."

For their habitat, they prefer dense forests and bamboo jungles. When alarmed, they utter a loud, whistling call and raise their tail, which reveals the white underside.

Even though the sambars are fully protected in several Indian preserves, the many cattle which graze along the boundaries of these areas may endanger them. In the Kaziranga Conservation Area, in 1949, forty great Indian rhinoceros, many wild buffalo, and many deer died of anthrax, a contagious disease which doubtlessly had been brought by the cattle to the game animals. Therefore, during the last years, efforts have been made to decrease the number of cattle near the conservation areas and to vaccinate the rest against contagious diseases. But such procedures are resisted by many Indians to whom the cow is sacred.

In the wooded parts of India, the sambars occur rather frequently, but they are rarely seen. It is amazing how soundlessly these large and heavy animals can move in the jungle. They feed mainly on grass, leaves, and wild fruit. They like water; when they swim, only their heads and antlers are visible above the surface. During the rutting season, each male fights for his territory and keeps those females which enter his area. After the mating season, he leaves the females and stays by himself until the next rut begins. In Central and Southern India, the young are born mainly at the beginning of the rainy season, by the end of May or the beginning of June. However, the mating and reproduction seasons vary according to the habitat. Sambars do well in zoological gardens, and have been introduced to Australia, New Zealand, and Florida.

Sunda sambar

On the islands of Java, Celebes, and the Lesser Sundas, there are somewhat smaller relatives of the Indian sambar which were formerly considered to be a species of their own. Presently, all these forms are included in one species, the SUNDA SAMBAR (Cervus timorensis). These island deer have a longer, denser, and shaggier coat than the continental sambars. They are said to be much more gregarious than their larger relatives and they often live in herds with many animals. Their preferred habitats are grass covered plains with trees and bushes. They go to the forest only for resting, and they are not as dependent on water as is the Indian sambar.

Previously the sunda sambars have been severely hunted. One method was to encircle them with tame buffalos, while groups of riders using swords and spears killed hundreds of them. The German naturalist Justus Karl Hasskarl described such a mass hunt which occurred in Java in 1855: "It is a touching sight when a female and her young are pursued. She tries over and over to cover it with her body and to protect it, and, consequently, makes the oddest jumps in all directions, until finally a rider gets between the two. Then it tries to flee, but often it is too late." No wonder that the sunda sambar has become rare on the rather small and densely populated islands.

Sunda sambars from the islands of Java, Timor, and the Moluccas have been introduced by man into many other places in the world, such as South Australia, West New Guinea, Madagascar, Mauritius, and the Comoro Islands. They are seldom found in zoos. One female placed in the New York Bronx Zoo after World War II reached an age of more than twenty-six years. This is the all time record for any Indian sambar in captivity.

Philippine sambar

The small Philippine sambars seem to be the most primitive form of the subgenus. One of the subspecies, the ALFRED'S SAMBAR (Cervus mariannus alfredi), is the only sambar to wear a spotted coat as an adult animal, which is generally a significant sign of a lesser evolved deer. Strangely enough, the Philippine sambars had first been described scientifically on the South Sea Island of Guam in 1822 by the French zoologist Desmarest. The Spaniards had introduced them to Guam long ago. Most of the Philippine sambars have short, six-tined antlers with a flat-surfaced end-tine.

These beautiful and graceful deer are rare inhabitants in our zoos. Little is known about their behavior in the wild. They live in prime forests and swamp jungle, climb mountains of altitudes up to 2500 meters, and have lately adjusted to cultivated land with trees. Alfred's deer have reproduced several times in the London and Berlin Zoos.

Closely related to the sambars are the BARASINGHAS (subgenus *Rucervus*). The animals have a long body and long legs. The head is medium long and sturdy. They live in the water and the swamp. Their coat has hardly any fleece and appears to be somewhat shaggy. The ♂♂ in rut have a more or less distinct throat mane. The reproduction season depends on the climate in their habitat. After a gestation period of eight months, one young is born, seldom two.

Various barasingha forms may have antlers of varied shape. According to the latest studies, there are so many similarities in the structure of their bodies, coats color, and habits that we now distinguish only two species:

1. BARASINGHA (♂ *Cervus duvauceli;* Color plate, p. 213). The HRL is about 180 cm, the TL is 12–20 cm, the BH is about 115 cm, and the weight is 230–283 kg. There are three subspecies, one of which is the now-extinct SCHOMBURGK DEER (*Cervus duvauceli schomburgki;* Color plate, p. 213).

2. ELD'S DEER (♂ *Cervus eldi;* Color plate, p. 213). The HRL is about 180 cm, the TL is about 20 cm, the BH is 107–115 cm, and the weight varies from 80 to 150 kg. Of the three subspecies, one is extinct and the other two, including the THAMIN (♂ *Cervus eldi thamin),* are endangered.

Barasingha

Trophy hunting has turned out to be disastrous, especially for these South Asiatic species of deer. While the white man in his native habitat used to take good care of the deer by increasing their numbers and "improving" the breed with "fresh blood," just for trophies, he exhausted the deer population by uncontrolled killing in his former colonies. He ruthlessly destroyed those species whose antlers were considered especially "beautiful," and one of these is the barasingha. Of the six forms, only the North and the Central Indian barasinghas are still safe. Since barasinghas avoid the forest and prefer pastures with trees as their habitat, they easily fall victim to man in comparison to sambars, which are able to live concealed in the jungle. Schomburgk's deer, which differs from the barasingha by its chocolate-brown coat and its greatly branched, sometimes basket-shaped antlers, was frequently found, until the last century, on the open, swampy grass plains of Thailand. Due to cultivation and hunting, it was forced from

Fig. 8-9.
1. North Indian bara-singha *(Cervus duvauceli duvauceli)*. 2. Central Indian barasingha *(Cervus duvauceli branderi)*. 3. Schomburgk's deer *(Cervus duvauceli schomburgki)*. 4. Manipur Eld's deer *(Cervus eldi eldi)*. 5. Thamin *(Cervus eldi thamin)*. 6. Siamese Eld's deer *(Cervus eldi siamensis)*. Schomburgk's deer is extinct and the other Indian forms are almost destroyed.

its habitat into the jungles of West Siam, an unsuitable area where the subspecies finally became extinct in the nineteen thirties. There are hardly any specimens in museums and only a few photographs exist of this animal.

With its summer coat of gorgeous amber to golden brown with pale spots, the Indian barasingha is, to our eyes, among the most beautiful deer. Its antlers, too, are of a strange beauty; reaching a length of more than one meter, they look better proportioned and more "elegant" than the "basket" shaped antlers of Schomburgk's deer. Barasinghas prefer living in the swamps and often spend time standing in the water. Therefore, the English call them "swamp deer," which may cause one to confuse them with the South American swamp deer. The firm, water repellent hair and the widely spreadable toes enable the barasingha to live in the swamp. While they prefer reeds and other plants of the swamp, they also eat grass, herbs, leaves, buds, and, in cultivated areas, rice plants. In the spring and summer, the males live singly, while the females form smaller groups of two to three animals. In winter all gather into larger herds.

In Assam and North India, the males are said to be in velvet during the rut and for this reason, do not fight. Although the mating season may occur at different times of the year, which is common in all tropical deer, the young are regularly born during the monsoon or shortly after when enough food and shelter are available. The mating call is very melodious; it is a long sound with alternating high and low notes. According to S. H. Prater, barasinghas are much more diurnal than sambars. When alarmed, all members of the group utter a piercing bark which they continue during their flight.

Man has also introduced the barasingha to Australia. They do well in zoos and reproduce there. In a zoo, the gestation period was found to be about 250 days. Several of the animals in the London Zoo reached an age of more than twenty-one years; one female even attained twenty-three years.

The situation of the East Indian eld's deer is much worse, and the subspecies in Assam may already be extinct. If the laws for their protection are not strictly enforced, they may soon share the fate of Schomburgk's deer. The habits of these very shy deer are similar to the barasingha's. Avoiding hill country and dense forests, they like flooded plains and spend the hot time of the day hidden in the brush. Because much of their habitat has been changed into fields and plantations, they often cannot find sufficient natural food and, consequently, they "harvest" the farmers' crops. This has led to such ruthless hunting that extinction of all subspecies is anticipated. Only Burma has strict laws for the protection of the local subspecies, the thamin, so that their population which numbered only a few hundred animals some decades ago has now increased to 3500 head.

In zoological gardens, eld's deer are rare. Presently (1967) this species may be seen in Europe in only three zoos. A handsome breeding group of the Siamese eld's deer *(Cervus eldi siamensis)*, whose first progenitors came from what was formerly called Indochina, lives in the zoo of Paris-Vincennes. In the Berlin Zoo, Burma thamins are presently kept and bred. Heinz-Georg Klös reports: "Even though the animals have access to a heated room during the cold season, they spend even cool fall nights in the open. Since the handsome male is absolutely peaceful towards his two females even in the mating season, they have never needed the so-called 'wedding fence' (a separation across the pen with small openings to enable the females to escape from the obtrusive males), which is so important for many of the other deer."

The SIKA DEER have become a popular game park animal in many places. They are of medium size, compact, small-headed, and mostly spotted. The tail is medium long, and the simple antlers have eight to ten endings. There is only one species, the SIKA DEER *(Cervus nippon)*. The HRL is 105–155 cm, the TL is 10–20 cm, the BH is 75–110 cm, and the weight is 25–110 kg. Of the nine subspecies, some are endangered in the wild, but preserved on game farms; among them are the JAPANESE SIKA DEER *(Cervus nippon nippon;* Color plate, p. 214) and the DYBOWSKI'S DEER *(Cervus nippon dybowskii;* Color plate, p. 214). Seven subspecies are endangered or extinct, including the FORMOSA SIKA DEER (⊹ *Cervus nippon taiouanus;* Color plate, p. 214).

Sika deer

Tame Japanese sika deer are kept in several temple regions of Japan, and families come to visit these park deer and offer them rice cookies for food. Since many sika deer live in a semi-tame state in parks and on farms, they are easily introduced into other countries, especially since the northern subspecies does well in the climate of Europe. In addition to many European forests and parks, the Japanese sika deer now also inhabit the island of Madagascar. Dybowski's deer had been brought to New Zealand by 1875, and they were later introduced into Australia. In 1950, the Russians introduced this subspecies to Aserbeidjan. Thanks to their minimal requirements and endurance, all sikas have successfully survived such transfers. Presently in Germany, there are several populations living in the wild where the red deer cannot be tolerated. Included in the German game laws, they have the same closed seasons as do the fallow deer.

Sika deer are amazingly hardy. They thrive even in winter on scarce food and without additional feeding. Actually, they are not really such strangers in Europe as may be assumed. They occurred there in older geological times as well. They probably had been forced out gradually by the stronger red deer.

Originally the preferred habitat of sika deer was large forests with clearings, but they have adjusted well to the cultivated, man-made

Fig. 8-10.
The greatly reduced Sika deer *(Cervus nippon)* have been introduced into some areas of Europe and other countries.

Fig. 8-11.
Red deer *(Cervus elaphus)*
in Europe and Asia.

The red deer, by L.
Heck

Fig. 8-12.
Red deer *(Cervus elaphus)*
in America *(Wapiti)*. The
former distribution is
shaded and the present
distribution is black.

landscape. The rut of the northern subspecies is in November and December. The males then cry with high notes which often change into piercing sounds. The females' gestation period is seven and a half months, and the young are born by the end of June or the beginning of August. Usually only one young is born, although twins occur on occasion. All sikas do extremely well in zoos. Even though some subspecies have become rather rare in the wild, sikas, which are considered by some zoo people to be "too common," have had to make room for other species of deer. Therefore, it is very commendable that some zoos and game parks lavish much love and care upon the breeding of the endangered sika deer subspecies. For many years, large herds of FORMOSA SIKAS *(Cervus nippon taiouanus)*, which in their native habitat are close to extinction, were kept in the English zoo of Paignton, in the Duke of Bedford's Park, in the Whipsnade Zoo, and on the Catskill Game Farm. Some sikas in captivity reach a very old age. In Washington, one animal lived for more than twenty-five years.

The Frankfurt Zoo, which has had a breed of Dybowski's deer for decades, has kept records on them since 1946. They had 159 young up to 1967. 11.5% of them were born in May, 57.5% in June, 21.5% in July, and 7% in August. Only 1.4% were born in September and 0.7% in October, among them the only recorded female twins.

For Europeans, the prototype of the deer family is the RED DEER (subgenus *Cervus*). The HRL is 165-265 cm, the TL is 10-22 cm, the BH is 75-150 cm, and the weight is 75-340 kg. The head is medium long and the antlers have many tines, with a crown or fork on the tips. The long, square rump is higher in the rear of the Szechuan deer. The coat is firm and somewhat shaggy. The ♂♂ have a well developed throat mane, with the exception of the Szechuan deer. The gestation period is 225-262 days (average 235); one young, rarely two, are born. The nursing period is nine months to one year. The life span is 17-20 years, rarely longer.

The European red deer, the marals and other Asiatic forms, as well as the Northeast Asian-North American elk, have been grouped in one species, the red deer *(Cervus elaphus)* because geographically they occupy the same niche and are linked by many transitionary forms throughout the range. There are twenty-three subspecies, six of which are threatened by extinction: the Tyrrhenian red deer, the Barbary stag, the hangul, the yarkand deer, the Szechuan deer, and the wapiti.

The basic shape of red deer antlers, as in all species of the genus *Cervus*, is formed by antlers with subsequent branches which all lie in one plane. In the larger subspecies, especially in the elk, very large antlers with long endings are formed. As a rule the European red deer, and especially their western forms, develop a "crown." The fourth or fifth spiller is not on the same level with the rest of the branches, but protrudes to the side. At this point or near it more tines are added with increasing age of the deer. In this manner, huge crowns with many tines may develop.

The NORTH AMERICAN ELK or WAPITI consists of six subspecies and among them are smaller forms, like a dwarf of the California Sacramento and San Joaquin Valleys, and also real giants, the largest of all red deer. Originally, elk were not native to North America. Like many other animals of Europe and Asia, they migrated over the formerly existing land bridge between East Asia and North America. Bison, brown bear, wolf, and beaver belong to these immigrants. Because of its size, the English settlers in North America called the animal "elk," which is, of course, zoologically incorrect. Therefore, American zoologists and zoo people have begun to use the original Indian name, wapiti.

The wapitis prefer semi-open woodlands. In the summer, they migrate up into the mountains and in the winter they return to the valleys. Their very strong symmetrical antlers usually have twelve powerful tines, rarely more, which only form crowns in exceptional cases. They may weigh from twenty to more than twenty-five kilograms. Strong males may weigh up to 450 kilograms. During the rut, these deer utter a long, pleasant, yodeling sound which is very different from the roaring of the European red deer. It is quite an experience to hear these elk calls echoing in the valleys of the Rocky Mountains.

Originally, elk were more widely distributed than all other North American ungulates. Their habitat ranged from Northern Canada to Southern Mexico and from the Atlantic coast to California. They lived on the sea level plains as well as in the high mountains up to the timber line. In this wide habitat they were more and more reduced by increasing cultivation. The advancing settlers, who killed them for their meat and skins, caused the gradual disappearance of the Eastern subspecies *Cervus elaphus canadensis;* (Color plate, p. 217). Probably the last animal was killed in 1967 in Pennsylvania. But since these animals had been introduced to New Zealand in 1870, they live on in the red deer population of that country.

Presently, the most numerous among the surviving subspecies is the ROCKY MOUNTAIN ELK (*Cervus elaphus nelsoni;* Color plates, p. 188/189). Smaller groups of these alpine deer have been brought to some of the eastern woodlands in order to replace the extinct wapiti. There are other subspecies in Canada and California which vary in size, coat, and antler shape. Thanks to the large conservation areas established in time in North America, they have been saved. If it were not for the U.S. National Parks, there probably would be no elk left.

The decrease in the elk population was caused not only by hunting, but also by farming. In winter the elk live in the valleys and on the hills near the high mountains, which have so little snow that they can easily get to the grass. But lately these areas have been claimed by man.

American elk or wapiti

Fig. 8-13. Wapiti, maral, and Central European red deer. The red deer (*Cervus elaphus*) steadily decrease in size from North America over east, central and west Asia to Europe.

They have been fenced, cattle are kept on the land, or the grass is cut for hay. Thus, the elk are forced more and more into the narrow space between the high, snow-covered peaks and the cultivated valleys. Many of the deer died of starvation during the recent severe winters. Presently, the herds can be kept alive only with additional feeding in the above-mentioned preserves.

In Jackson Hole, Wyoming, the famous preserve for elk, 20,000 elk are fed during the winter on former farm land every year. Strict hunting regulations prevent both a decrease and an uncontrolled increase of the deer population. If there were too many of them, there would not be sufficient food. When spring comes, the herds begin their migration through the valleys up to their summer quarters in the mountains.

Among the Central and East Asiatic subspecies the ALTAI MARAL (*Cervus elaphus sibiricus;* Color plate, p. 216) most closely resembles the American wapitis. Therefore, when the red deer and the wapiti formerly were considered to be different species, it was called the Altai wapiti. "Maral" is the Siberian name for deer. The Altai maral is a huge animal with antlers similar to the wapiti's. Because of these heavy antlers with many tines, it has been crossbred with the European red deer. These offspring had antlers weighing up to fifteen kilograms, and often with the shape of the European red deer's.

Asiatic red deer

Other Asiatic subspecies are the ISUBRA MARAL (*Cervus elaphus xanthopygus;* Color plate, p. 217) from Manchuria and the Amur region, the BACTRIAN DEER (♦ *Cervus elaphus bactrianus;* Color plate, p. 216), the HANGUL (*Cervus elaphus affinis;* Color plate, p. 216), the YARKAND DEER ♦ *Cervus elaphus yarkandensis)* and the SZECHUAN DEER (♦ *Cervus elaphus macneilli).* Most of these forms are endangered in the wild, but they are kept on farms for panteen production. Sometimes they are also bred in zoos and game parks.

The CAUCASIAN MARAL (*Cervus elaphus maral;* Color plate, p. 216) from the Caucasus, Asia Minor, and North Persia is smaller than the Central and East Asiatic Deer. Its long antlers often have only a few tines. Very large marals are able to develop a crown.

In the Balkans, the Carpathian Mountains, and Hungary are deer which might be described as a transition form, or a cross-breed, between the maral and the CENTRAL EUROPEAN RED DEER (*Cervus elaphus hippelaphus;* Color plates pp. 215, 185, and 186/187). However, this Central European subspecies is not pure in many places because several different subspecies have been crossed with them to "improve" and to "touch up" the blood line. Besides this Central European form, there are other subspecies, like the SCANDINAVIAN RED DEER *(Cervus elaphus elaphus)* and the nearly extinct MEDITERRANEAN RED DEER or DWARF RED DEER (♦ *Cervus elaphus corsicanus;* Color plate, p. 215), whose habitats were the islands of Sardinia and Corsica.

"There is no doubt that the most striking, and at the same time the most impressive, part of this deer are its antlers," writes the director of the Zurich Zoo, Heini Hediger, about the European red deer. "The better a person gets to know this strange structure in all its phases, the more wonderful it appears to be. The antlers, which the hunter sees on the killed animal, are dead bones. This in itself is peculiar. What other animal would carry around dead bones? Usually a dead bone would be eliminated from the body, but this is not so in deer; here, the antlers are functional and serve their function only after they have died. As long as they are "alive," which means in the process of growing, the deer could not possibly use them because they are soft and very delicate. But one day, after the dead antlers have served their purpose as a weapon and a meaningful symbol, they are finally dropped, only to be replaced by others which possibly may grow even larger."

It takes about 100 days for the new antlers to reach their final shape and size. In this surprisingly short period, the deer's body produces several kilograms of bone material and forms a beautiful, ramified structure with it. In this short growth period, the deer needs more "building material" than he could possibly take from his food. Therefore, he has to use the calcium reserves stored in his body. Growing bones contain radioactive material to a much greater extent than other bone material. Therefore, skeletons of human infants have more of the hazardous radioactive strontium in them than do adults' skeletons. The antlers of a Scottish deer, analyzed in 1957, contained ten times as much radioactive strontium than others examined in 1952. Thus, the antlers distinctly reflected the increasing pollution of the atmosphere from nuclear bomb tests.

But man has not only affected the growth of the deer's antlers since the invention of the atomic bomb. By changing the original habitat of the red deer through cultivation, he has varied the nutritional base which is also important for the development of the antlers. In our forests, which are now mainly used for the production of timber, we have to offer additional food, since otherwise the red deer could not survive the severe winters.

Depending on the season, the male red deer has a completely different behavior. During the summer, he leads a concealed life in the forest. But in the fall rutting season, he stays with the herd and announces his presence by roaring, which can be heard over long distances. After dropping his head ornaments, his behavior again changes completely. Before shedding, he would intimidate any lesser opponent, but now he becomes so cautious that he even avoids younger conspecifics. Except for the rut, the male deer will form groups with males of all ages and a female will hardly be tolerated. These "men's clubs" begin to break up in mid-July after the new antlers have grown.

European red deer, by B. Grzimek

First antlers (one and a half years).

Second antlers (two and a half years).

Fourth antlers (four and a half years).

At first the skirmishes are rather harmless, but gradually the fights increase in intensity until the rut when each rival is considered a "deadly enemy." Then the deer may cover long distances in order to reach the traditional rutting areas. Each year one particular deer was known to move the forty kilometers from his summer range to the rutting area in one night.

During the growth of the antlers, the deer are sexually uninterested. They stay away from the females and do not seek their company. But this behavior changes completely and rapidly in the rut. The male seeks out the females, who in turn move to an area which a strong male deer has occupied. Such a "territorial deer," who is now anything but uninterested in the females, announces his presence by his roaring. This roaring is a loud, growling sound which indicates that the animal is ready for mating. In this phase, the deer is aggressive, jealous, and ready to defend the females and his territory against any rival. The usually cautious deer turn into wild daredevils; they almost "forget" about their environment and danger. When the rut is over in the first half of October, the deer no longer pay attention to the females. In remote places, exhausted single deer may be seen who are hardly interested in what goes on around them. They recover gradually and, in trying to regain their lost strength with good food, they go to feeding stations. There they join mixed herds or groups of males for the winter.

Hediger reports on the new growth of the antlers: "Let us take a look at these bloody, cup-shaped dropping scars. By the next day, they are covered with a light scar tissue, and the rims are somewhat enlarged. During the following days, these rims grow toward the center, and the originally recessed scar bulges distinctly. Within one week, a bulbous structure may have developed which almost pushes out of the basic bone structure and which becomes covered with a velvety material. In this phase the deer is known as a 'knobber.' On the velvet covered mound, the initial dropping scar may be visible as a small crust, and large blood vessels may shine through the velvet.

"The blood flows to the tips from a circumventricular artery and each pulsation transports building material to the incredibly fast growing ends. Soon the knags begin to show protrusions to the sides: the anteocular branch and the bay branch, the two lowest tines of the future antlers, are about to be formed. The third knob contains the center bar on the summit where the crown will later develop its fan-shaped propping up of four or more tines. Gradually, the final shape and number of ends become clearly visible. But still the antlers are very delicate and may not be used as weapons. For his defense now, the deer must use a different method, which is the same as that of the female. With the muscular front legs he may hit violently when he rises on his hind legs.

"Finally, in July the antlers are ready in form and size, but they are

Twelfth to fourteenth antlers.

"Regressed" antlers in old deer.

Fig. 8-14.
In the first years the Red deer's antlers become gradually stronger and grow increasingly more points. The full-grown deer develops antlers with strong crowns every year. At an older age, the antlers become weaker, and the deer "regresses".

still covered with the velvet (the living tissue). This phase lasts for several weeks until this huge, barely completed structure begins to die. The large blood vessels and the nerves dry up, and the velvety skin comes off in bloody pieces from the dying bone which appears pale underneath. This process must cause a vehement itching. The deer has an urgent need to remove these shreds as quickly as possible, especially since they attract masses of flies and hang down over his eyes. The tormented animal then looks for brush and thin trees and rubs the velvet off of his antlers. Often he not only rubs off this material, but also greedily eats it. Strangely enough, this completely herbivorous animal becomes a meat eater during this time, eating dead parts of his own body. Apparently nothing of the precious material must go to waste but rather must be stored, for seven months later the deer's system again has to undergo the strain of building up new antlers."

Of course, this applies to all species of deer with antlers. But it is especially impressive in the red and other large deer with their huge large head weapons. Such red deer antlers may weigh 14-16 kilograms. It must not be too convenient to carry this head structure around during a great part of the year until it finally is dropped. In red deer this occurs mostly in February. Right above the frontal bone, under the basal bone structure, special cells begin to dissolve the bone material. One day the two antlers fall off at almost the same time. The loss of these weapons must come as quite a shock to the owner. He probably suddenly feels defenseless or inferior. It occasionally happens in a zoo that a dominant deer then allows himself to be beaten up by some female who strikes out with her front legs and drives him away, even though he could fight back in the same manner. Both temporarily "bald" males and those in the "velvet phase" use this method in quarrels with each other; they stand erect on their hind legs. The dropped antlers are soon eaten by mice, squirrels, and other rodents; the Scottish deer are said to do it themselves.

In some countries, the ruling houses and the aristocracy saw to it that a number of deer remained for the "noble" pleasure of hunting. They alone reserved the privilege to hunt the game in their forests. Perhaps this was a substitute for war but, in any case, it was a change from the boring life of the palace. Formerly, deer hunting was anything but a fair sport. Chasing deer on horseback and with dogs was the usual procedure. Often deer were driven into large pens where they could not escape; there they were butchered with spears, crossbows, and muskets. The indentured peasants erected walls many kilometers long so that at any time the master would have a sufficient number of deer available in his game park for shooting by himself and his guests. The deer were also driven along hedges which had openings where nooses and nets were concealed. In order to keep this equipment in good shape and to preserve it from rotting, they built special houses

Hunting as a privilege of aristocracy

for storage. The Zeughaus in Kranichstein near Darmstadt, Germany, which is 150 meters long, is as long as the net which was hung in there.

The time of rut in the deer

Every fall in Central Europe the roaring of the red deer resounds in the forests, because this is the time when the male deer are interested in their females. Their mating calls are part of the romantic atmosphere of the European woods. It is so attractive to tourists that some innkeepers pay lumbermen to imitate this call on watering cans in the dense forest. The guests are most impressed; the game wardens are too, because it keeps the curious crowds away from the actual mating grounds of the deer. Such a male deer may gather fifty females of all age groups around him. But this is no "harem," since the male is not their leader, nor does he watch or defend them. He will quickly run away when hunters or dogs come to the area. His main concerns are to keep other males away and to see that none of his females leave him.

Fights in the rut

When two male deer fight with each other, their antlers are clearly more of a "tournament" weapon than a tool for killing. The multitude of tines prevents the antler from injuring his opponent's body because they usually get caught in his antlers. Once they stand head to head, they fight; this consists mainly of pushing and pressing, and it is a contest of strength, like a tug-of-war or Indian wrestling is for people. It may happen that the antlers become so locked that the two opponents cannot separate any more and so they die miserably of starvation. It is bad when a deer has two long, plain, pointed bars instead of the branched antlers; he easily passes through the opponent's antlers and stabs him to death. This is not dangerous in young deer whose antlers, although straight, are just beginning to grow since they do not dare to participate in these fights. However, older deer with such antlers are called "murderers." Deer antlers can also be used as tools. It was observed how a deer in captivity, unable to reach ripe apples in a tree, stood on his hind legs, beat the branches with his antlers, and so knocked down the apples.

Birth

Eight and one half months after copulation, some time in May or June, the female deer will separate herself from the herd, which consists of six to twelve animals, and prepare to give birth. Before doing so, she will chase away her young from the previous year. The actual birth hardly lasts longer than ten minutes, and the mother moans frequently during the process. For the first few days, the young remains hidden, and the mother comes to it only for nursing.

Red deer and man

In some areas where the deer are not being hunted, they will become so tame, especially in winter at the feeding stations, that people may come close to them. One hunter had put up a feeding station twenty meters from his house. The deer came the whole year round at night and did not mind a light bulb shining four meters above their heads in a tree. He even had a device connected to a small lamp on his desk in the house which lit up when the deer were there and they could be watched then. Around seven P.M. on Easter Sunday in 1943, in the Saxon town of Zahna, a herd of about twenty deer paid a visit which turned out to be very exciting. They jumped over the Zahna Creek, swam through the Pine Pond, and went into the city. When they saw all the people walking in the streets, they became terrified; some of them ran to the railroad station, others jumped fences, and one destroyed a store window and suffered serious injuries. Some even wound up in hallways of houses. The total damage was more than $500.

Two hundred, five hundred, or even a thousand years ago the deer had much larger antlers than they have now. One of the reasons may be that the deer have been forced into the forest even though their natural habitats are open areas with scattered trees.

Ostentatious princes continually tried to hitch deer before coaches. The first to do this was the Roman Emperor Aurelian, who, after his victory over the Empress Palmyra, used a team of four tame deer in his triumphal procession. They were said to have belonged to a Gothic king at one time. Afterwards, these animals were sacrificed to the god Jupiter in the capital. The Duke of Pückler was pulled by deer over the famous Berlin boulevard "Unter den Linden." These deer had been castrated and were easier to handle. The French "King Lustik," who reigned in Kassel during the Napoleonic era, enjoyed exaggerated and expensive festivities. On August 15, the birthday of his brother Napoleon, he displayed a team of four deer. The animals bolted, dashing along the avenue; the terrified passengers jumped off. When they returned on foot to the stables, the deer, who had only the torn harnesses and the tongue of the cart with them, were already there. Ludwig VIII of Hesse-Darmstadt also had a team of six strong deer which he used to travel from his hunting lodge, Kranichstein, to Darmstadt.

Deer as draft animals

Since deer hunting had long been considered the sport of sovereigns, the aristocracy used to exchange deer as presents. As early as 1661 and 1662, 159 Neumark red deer were brought by boat to London. Many estate owners abroad, especially Europeans, introduced red deer into their forests. There are large populations of European red deer now in Argentina, Australia, and especially in New Zealand. Unfortunately, the imported red deer in Argentina gradually displaced the endemic species of deer. They took over their habitat, ate their food, and drove them out. But the hunters of Argentina prefer the trophies of the red deer over the lesser antlers of the deer of the Andes and the pampas. Therefore, in the beautiful Nahuel-Huapi National Park of the Argentine Andes are large herds of the European red deer, which grow even better antlers there than they do in Central Europe. The original, endemic species of deer, the guemals and the pudus, have totally disappeared from this area.

There were endemic red deer in North Africa as well. Today the BARBARY DEER (♀ *Cervus elaphus barbarus*) are restricted to a small area north of the Sahara Desert, at the border between Algeria and Tunisia. Their antlers rarely grow more than ten tines. The Barbary stag has a relatively low box shape, and the fawn has striking dots along the spine. Not long ago this subspecies was about to become extinct; only a few Barbary deer live free in North Africa. Then about one dozen deer were moved, at great cost and difficulty, to a well controlled preserve in Tunisia.

Thorold's deer and Père David's deer, by L. Heck and H. Wendt

Among the rarest and least known of the Central Asiatic deer are THOROLD'S DEER (subgenus *Przewalskium*), which has only one species, (♀ *Cervus albirostris*; Color plate, p. 218). The HRL is 190-200 cm, the TL is 10-12 cm, the BH is 120-130 cm, and the weight is 130-140 kg. The high legged animal is compact and the croup is slightly higher than the withers. The short head has a short, wide muzzle. The antlers, which have ten to twelve ends and no crown, are white or light brown. The wide hoofs are like those of the oxen, and the pseudo-claws are especially well developed and long. The dense coat, which is rough and without a fleece, is twice as long in winter as in summer. The ♀♀ have a short tuft between the ears. The gestation period is nine to ten months.

The Tibetian name for Thorold's deer is "Shou." This deer lives on the alpine meadows above the timber line in the Asiatic Mountains. There they may climb up to elevations between 3500 and 5000 meters. They feed mainly on grass and herbs. Thorold's deer are gregarious, usually living in groups of five to twenty animals. Herds consisting of as many as forty Thorold's deer have been observed. The old males are loners, and the young ones join in small groups. The females tolerate their offspring even if they are about to have another young.

Fig. 8–15. Thorold's Deer (*Cervus albirostris*).

Today Thorold's deer are rare and extremely endangered. They probably never have been in a European or a North American zoo. The Peking Zoo, with its efforts to preserve and to breed many rare species of East Asiatic animals, is the only zoo in the world to keep a herd of these large alpine deer. They have already reproduced.

The last genus of true deer, PÈRE DAVID'S DEER (*Elaphurus*) are, in many ways, the most extraordinary of all. There is only one species, Père David's deer or MILU (♀ *Elaphurus davidianus*; Color plate, p. 218). The HRL is about 150 cm, the TL is about 50 cm, the BH is about 115 cm, and the weight is 150-200 kg. They are high legged and long headed with short pointed ears. The large eyes have very large preorbital glands. The almost hairless nose region is delicately sectioned. The neck is rather short, and in ♂♂ appears to be compact. The long, square rump has a tail which is unusually long for a species of deer, with the tassel reaching to the heels. The hooves are large, the pseudo-claws are long, and the toes, like the reindeer's, spread far apart and make a cracking noise in walking. The soft coat, which has a thick fleece in winter, is longer on throat, chest, and withers in ♂♂. The antlers have a unique shape among deer. The main bar has tines which are directed backward, and decrease in length from the base to the top. The lowest and longest of the spillers begins only a few centimeters above the frontal bone structure and may itself have up to a half a dozen tines, all pointing backward on the rear edge. At first sight, the antlers appear to be "reversed" on the head, with the short, strong main bar pointing forward, the tined ends and the long, arched anteocular branch point backwards. The rut is in June and July. The gestation period is about 250–270 days, and one or two fawns are born in April and May. Sexual maturity is usually attained at two and a quarter years, rarely at one and a quarter years.

Père David's deer has a special place among all actual species of deer. Its odd antlers with their ends pointing backwards, the long skull and unusually long donkey-like tail, and hooves which make that cracking sound when walking, are all characteristics which essentially distinguish it from its relatives. However, Père David's deer is not that different from red deer. In the Berlin Zoo, even crosses of Père David's deer and red deer were fertile.

The original habitat of this strange swamp deer is still debated. It probably lived in the marshes of North and Central China. Fossils of Père David's deer have been found in Manchuria and South Japan. It has been extinct in the wild since times unknown. Large herds were kept by the Chinese Emperor as game animals in his park in Nan Hai-tsu, south of Peking. A high wall, seventy-two kilometers long, surrounded this game park. Guarded by Tartars, it was strictly forbidden for anyone to enter it or even to glimpse over the wall. In 1865, the French Jesuit priest Armand David, who was an excellent zoologist, managed to bribe one of the guards to allow him to climb the wall. At this moment, a herd of about 120 deer passed by.

At first, Père David took them for an unknown species of reindeer. He was quite surprised that there had been no scientific report on these tame creatures which were grazing in the Emperor's park. For several months, he collected information on the Imperial deer, which, of course, had to be fragmentary since hardly anybody in Peking knew about these animals. It seemed that they had been living for ages in the park of Nan Hai-tsu and they did not occur any place else. With great difficulties Père David managed to get two complete skins of this new deer species. Based on these skins, the French zoologist Milne-Edwards described Père David's deer in 1866.

Soon after that, by royal order, the Chinese minister Hen-Tchi presented three living Père David's deer to the French Ambassador in Peking. Naturally, they did not survive the long voyage to France, but now Milne-Edwards was able to study these dead specimens of Père

▷
Roaring Red deer (Cervus elaphus hippelaphus) in Central Europe.

▷▷
The sonorous roaring is heard only in the European subspecies.

▷▷▷
Rocky Mountain wapitis (Cervus elaphus nelsoni) at the Norris Geyser Basin in the Yellowstone National Park.

▷▷▷▷
Moose (Alces alces) like to graze on aquatic plants while they are wading or diving.

▷▷▷▷▷
Like in most other species of animals, white forms occur occasionally in roe deer (Capreolus capreolus)

David's deer. He immediately saw that they were not reindeer and he classified the deer as belonging somewhere among the relatives of red deer and wapiti. Unfortunately, his "Notes on *Elaphurus davidianus*, a new species in the family of deer" contained a verbal error. Père David had found out that the Chinese name for this animal was *Sse-pu-hsiang*, which means "four-in-one" because the Chinese felt that this deer had the characteristics of the deer, the reindeer, the ox, and the donkey. Milne-Edwards, however, named the species "milu." Actually this name refers to sika deer; still the name was accepted in the zoological literature.

After the publication of Milne-Edward's paper, demands arose from European diplomats directed to the minister Hen-Tchi. Since the Emperor had presented the French with the deer, he could hardly refuse such a gift to the English and the Germans. In 1869, the English diplomat Sir Rutherford Alcock brought two live Père David's Deer to the London Zoological Society. Some years later the German Ambassador von Brandt and Consul von Möllendorf received several of the animals for the Berlin Zoo. Until 1890, more Père David's Deer went to England and France, reproduced, and soon made up a sizable population. Strangely enough, the offspring of the Berlin specimens, which had been crossed with red deer, resembled more closely the odd Chinese rather than the familiar European form, although they had only one fourth of the Père David's deer's blood. Altogether the large herd in the Imperial Gardens of Nan Hai-tsu and the approximately two dozen of the animals in European zoos seemed to guarantee that the species would not die out.

But soon Père David's deer became extinct in China. One of the many flood catastrophes which regularly occurred, in 1895 drowned not only the inhabitants south of Peking but also the game animals of Nan Hai-tsu. The Hun-Ho River flooded villages and fields, which caused starvation and misery among the population, and it also destroyed a part of the old, rotten wall around the game park. Thousands of deer, gazelle, and other park animals were carried away by the current. Most of Père David's deer escaped through the openings in the wall and fled to the country where they were killed and eaten by the starving Chinese. After this catastrophe, only about twenty or thirty of Père David's deer remained alive in Nan Hai-tsu. They lasted for only five more years. During the Boxer Rebellion, they were totally destroyed. One single female survived in the Peking Zoo and died of old age in 1920.

As soon as the news of this destruction reached Europe, some far-sighted directors of several zoos decided to give all their Père David's deer as breeding stock to one of the most famous animal breeders of his time, the Duke of Bedford. Thus, a breeding population could be established in Woburn Abbey. For many of these zoo people, it meant

Strong male reindeer
(*Rangifer tarandus*).

quite a sacrifice to give up this rare animal from their collection. But the preservation and the survival of this strange Chinese deer depended on it.

In the large game park of Woburn Abbey, the Duke of Bedford supervised their breeding program. It was a great success. In 1932, there were 182 animals. Saving Père David's deer established a classic example of prudent conservation and well organized breeding in human care.

Only twenty years later, the first animals from the steadily increasing population in Woburn Abbey could be given to zoos in Europe and abroad. The Duke of Bedford gave some of his Père David's deer to the Whipsnade Zoo near London. Here, too, they reproduced well. Some of Père David's deer were even returned to their original habitat, the Peking Zoo in China.

The population of this deer, which only a few decades ago was nearly extinct, has increased in several zoos to such an extent that some of the animals were released into the wild as an experiment. Thus, several of these deer came into the game park of Dr. Heinrich Prince Reuss III of Styria/Austria. Others were brought to South America to the estates of Consul Vogel from Munich. The preservation of Père David's deer is another proof that species of animals which are threatened by extinction may well be saved in zoological gardens and eventually be reintroduced into the wild.

Like the musk deer, which is also the most primitive contemporary deer of the Telemetacarpalia, the WATER DEER (subfamily *Hydropotinae*) have no antlers. Instead, their males have very long, tusk-like upper canines. There is only one species, the CHINESE WATER DEER (*Hydropotes inermis;* Color plate, p. 219). The HRL is 75–97 cm, the TL is 5–8 cm, the BH is 45–55 cm, and the weight is 12–15 kg. It has a small, compact body with an arched back and long legs. The short head has a slightly bulging forehead. The preorbital glands are small. The neck is short and the rear is higher. The front legs are shorter than the hind legs. In summer, the coat is rather thin and clinging, while in winter it is thick, brittle, and curly. The fleece is very fine and sparse. The upper canines in the ♂♂ are very long, protruding up to six centimeters from the jaw, sword-like, flat and slightly bent, with a sharp rear edge which is used as a weapon. The gestation period is approximately six months. They live ten to twelve years. There are two subspecies, one of which is the rather unknown KOREAN WATER DEER (*Hydropotes inermis argyropus*).

Subfamily: water deer, by L. Heck

With its red-brown summer coat and its darker winter color, the Chinese water deer resembles the European roe deer. However, the habits of this animal are more similar to those of the chevrotains and musk deer than to the American or roe deer. This nocturnal animal lives singly or in pairs in the jungle and the high plains grass along rivers. In the mating season, serious fights between the males precede

Fig. 8-16.
Water deer *(Hydropotes inermis).*

copulation in the same manner described previously for the musk deer. "Thus our two males continuously chased each other," reports Rösl Kirchshofer from the Frankfurt Zoo, "uttering strange trilling and drumming sounds. At the beginning of the fight, they stand opposite each other, head to head. Each animal tries to put his head over the other's neck. Then they use their teeth. This neck fighting ends by tearing out pieces of the coat and skin. Our males used to show white, hairless scars for a long time after these encounters. Finally, the father proved to be superior over the son. Consequently the son, badly damaged in the fight, repeatedly jumped the fence which was one meter high." This confirms observations from the wild that only one pair live in a specific area.

During the last decades, several Chinese water deer have been brought to Europe. Since the animal adjusts well to the European climate, it has been released in several large parks. Now there are good populations at the Duke of Bedford's Park in Woburn Abbey, in the zoos of Frankfurt and Whipsnade, and in the private game park of the French naturalist Jean Delacour in Cleres, Normandy. I observed feral Chinese water deer in Whipsnade, Woburn Abbey, and Clere. When in danger, they, like rabbits, remain quiet and motionless in a cover, and take off only at the last moment. The Chinese water deer that were released have reproduced especially well in England. Near Whipsnade there are now several hundred of them. They have spread all over the adjoining areas and have actually become naturalized.

For a long time, it was believed that the female regularly had four to seven young. At the Frankfurt and the Berlin Zoos, each female had only one or two young per birth; only one in Frankfurt had three. In the Frankfurt female the number of young born each time decreased as she became older. However, this may well be due to conditions in captivity. The fawns, which are brightly spotted, stay put in the dense brush during the first weeks, as do roe deer young and other small deer.

Subfamily: American and roe deer, by L. Heck

In formerly Spanish regions of America, the WHITE-TAILED DEER, the best known of the North American deer, is called simply "deer," which in Europe refers to the roe deer. In Walt Disney's movie "Bambi" is depicted as a roe deer. While European hunters and naturalists have often been upset by this "mistake," they are wrong from the zoological point of view, as the European-Asiatic roe deer is more closely related to the New World deer. Hence, modern classification places them into the subfamily of the American and roe deer.

The AMERICAN and ROE DEER (Odocoileinae) range in size from that of rabbit to red deer. The ♂♂ have antlers which are medium-sized to long. Rudiments of the metacarpal pseudo-claws remain only at the lower end (telemetacarpal). The upper canines are reduced or missing. The ♀ gives birth to one to four young. There are two tribes: ROE DEER

(Capreolini) and AMERICAN DEER (Odocoileini) which have five genera, thirteen species, and eighty-seven subspecies.

Presently all roe deer (tribe Capreolini) are classified in only one species, the roe deer *(Capreolus capreolus)*. The HRL is 100–140 cm, the TL is 1–2 cm, the BH is 60–90 cm, and the weight is 15–50 kg. The animal has a graceful head, long legs, and a tail which is hardly visible. The forehead and nose form a straight line. The large eyes have upper lids with long lashes. The rump is higher in the rear, and the slender legs have pointed hooves and well developed pseudo-claws which do not touch the ground in a standing animal. In contrast to the American deer, there are skin glands at the eyes, the forehead, and the anus. The metatarsal glands, which are high on the hind legs, are marked by especially bristly hair. The surface of antlers has ridges and small, "knotty" protuberances ("pearls"), and the base is enlarged. There are three points, although in rare cases there may be four or even more. The summer coat is thin and smooth, while the winter coat is thick and shaggy. The upper canines are lacking or reduced in size, and the lower canines are therefore like the incisors. The gestation period is prolonged by an extended embryonal phase, which, depending on the subspecies, may last as long as nine months. Occasionally, there is a delayed rut with five months' gestation. There are usually two young. The nursing period lasts two to three months. They live approximately fifteen years. There are three subspecies: 1. EUROPEAN ROE DEER *(Capreolus capreolus capreolus;* Color plate, p. 219); 2. SIBERIAN ROE DEER *(Capreolus capreolus pygargus;* Color plate, p. 219), which is larger with stronger antlers; 3. CHINESE ROE DEER *(Capreolus capreolus bedfordi),* which has been destroyed by man in some areas.

Like the red deer, the roe deer has become a "follower of civilization." Instead of its original diet of foliage and buds, it now prefers the juicy, nutritious plants raised by farmers. The European roe deer presently is the most frequent species of deer. The annual take of roe deer by hunters, which is strictly controlled by the game authorities, has reached since 1960 in West Germany about 500,000 animals, and the population itself remains constant at about that level. Roe deer are abundant everywhere in the plains, the mountains, the forests and brush, and in fields and meadows. It is one of the animals which enlivens the central European countryside. Generally, the summer coat is a blazing red and the winter coat is a gray-brown. In some regions there occur a striking number of darker, almost black, roe deer. These black or white mutations (Color plate, p. 191) differ only in color from the normal color roe deer; their shape, weight, antlers, and social life are the same. Even piebald animals are not too rare.

Hunters incorrectly call the small antlers of the roe deer, which usually have six points, "horns." While this may lead to confusion with the horns of the horned ungulates (see Ch. 11), but it is hard to

The roe deer

Distinguishing characteristics

Fig. 8-17.
Roe deer, *(Capreolus capreolus).*

Fig. 8-18.
The Siberian roe deer (below) is not only larger than the European roe deer but it also has much stronger antlers which may have more than the six prongs that are characteristic of the roe deer.

Fig. 8-19.
In cases of hormonal disorders a "wig" will grow.

Habits and food

change such traditional usage.

Basically, the antler development in the roe deer is the same as in all other species of deer. "Naturally, the roe deer's antlers are much smaller than those of the red deer," reports the director of the Zurich Zoo, Hediger. "Usually it has no more than six ends; males with eight ends are very rare. The roe deer male sheds his antlers in a different season than the red deer. He does not drop them in March, but in November or December, and he sheds off the velvet in spring. The peak of the rut is at the end of July and the beginning of August, and this important weapon has to be ready for use then. Some of the male roe deer fawns grow their first pair of antlers, which vary in size, during their first year. It is rather a mere indication of antlers, sometimes only the size of a pea." By the age of three to four years, the male has developed his best antlers. The basic shape of the antlers may be affected by climatic influences or injuries. In cases of hormonal disorders, the antlers' bone material may form a growth called a "wig" by hunters.

The daily activity rhythm of the roe deer in the densely populated countries of Europe has become adjusted to that of the people's. The roe deer search for grass and herbs mainly early in the morning. Then they insert a short, hasty feeding period at noon when people are at lunch. Ph. Schmidt has observed that this feeding period depends on whether the farmer's noon breaks are at eleven or twelve A.M. When daylight is longer in the spring and summer, the animals appear correspondingly later in the meadows; they wait until the farm hands have left for the day. On fields where the farmers work late with lamps on their tractors, the roe deer do not leave the cover of the forest.

The Swiss theologian and conservationist Ph. Schmidt has observed that the roe deer often graze until after dark. Then they lie down in the field or meadow to sleep, rarely getting up during the night. They begin grazing again with the first daylight and finish eating before the grass and herbs become too wet from the dew. They do not like the wet grass, which is not good for them. In early spring when nights in the fields are too cold for them, they rest overnight in the brush or the forest. In summer the grain fields are the safest since people hardly ever penetrate them. When a farmer mowing his fields finds their beds he is not at all happy about it.

When grazing at noon in the summer, roe deer are plagued by insects. They often have to flee to their hiding places after a few mouthfuls; soon after, they return to the meadow to graze for a few more moments. When blood thirsty insects and disturbing people are absent, roe deer seem to "enjoy" grazing when they do not have to return immediately into their hiding places.

Grooming does not play as important a role for roe deer as it does for many other horned ungulates. Usually they only remove ticks and other parasites by licking and biting their coats; the males scratch their backs and the inner sides of their thighs with their antlers. Only during spring shedding when the winter coat comes off in bunches do they like to rub against trees in order to remove it. In this period, they also nibble each other until some areas of the skin are almost bare. The coat then looks mangy, when the animals in fact do not have this disease.

When at ease, roe deer do not make sounds. When disturbed or scared, they make soft, monosyllabic "teeping" sounds. The fawns "teep" more piercingly than adults. Their alarm call is a rough, short "Bo" or a longer "Ba" which usually is repeated several times; it almost sounds like the barking of a dog. Somewhat softer is the call with which a female marks her resting place or her grazing area. In extreme danger roe deer make loud, piercing cries.

Vocalization of the roe deer

In 1966, Fred Kurt, from the Zoological Museum at the University of Zurich, found that female roe deer, when they have their young and during the rut, have territories which do not overlap at all or only at the boundaries. When two females with young meet, they may attack each other. Presumably, the females mark their territories by urinating repeatedly on the same spots. Usually, roe deer live singly or in small groups of two to ten animals; in the winter, however, they may form larger herds. In the spring, these herds split up, and so do the family groups. The young males finally separate from their mother, and siblings and from this time either remain alone or live in loose contact with peers or older males. In early summer, the males separate and avoid each other, trying to attract the females without young.

Territorial behavior

In late spring and early summer, the home range of the adult males become territories. Each territory has only a narrow boundary line in common with its neighbor, and it is marked with odors. Along boundaries, neighbors have frequent quarrels. Only the one year old males have no territories as yet. Their home ranges overlap and are in part situated within territories of adult males.

In the rutting season from mid-July to mid-August, the males abandon their territories. They sniff the grass and the forest ground for a substance from the females' interdigital glands which they are then secreting. "Suddenly, the female flees," Ph. Schmidt says of the courtship. "He follows. This hectic chase may go across cornfields, uphill and downhill, until the female tires of it. It ends with the female suddenly standing still, and the male also standing still at a distance of about fifteen to twenty meters from her. Both animals stand motionless. Once in a while, the male will utter soft blowing or snorting sounds, which is sometimes reciprocated by the female. She turns her head towards the male. And now one can see that a roe deer would never attack a conspecific from the rear. 'Driving' the female is not initiated by the male, although it may appear that he was attempting

The rut in summer

to coerce her. Driving just is part of the courtship, and that is all there is to it. Now the female starts to walk or trot slowly in order to give the following male the chance to catch up with her. The closer he comes, the slower the female moves, until she finally walks at a quiet pace. Now the male touches her conspicuous white tail-patch with his nose and the female does a quick turn and then she stands still. The corners of her mouth are pulled backward in a strange manner; the mouth is slightly opened and her expression is typical of females that are ready to copulate. Mounting quickly he copulates, the corners of his mouth are also pulled backward. Then the female turns around, ready to walk off, but the male licks her and again she stands still. Now a second copulation follows. Finally, the male glides down from the female, and both animals lie down in the grass for a short rest."

Since the female flees from the male in a narrow circle, the pair make the so-called "witches circles" in meadows and grain fields; they may be round as well as serpentine. The male stays close to the female during the three to five days when she is in heat; he will not go farther than five to six meters from her. When the browsing female moves on, he jumps to his feet and follows her. He smells at all places where she has urinated and shows "flehmen" (lip curl) afterwards (compare with Ch. 11). Roe deer females are still with their fawns even during the rutting season. On the days of copulation the young stay by themselves. When they come close to their mother, the male, threatening with his antlers, will chase them away. Schmidt saw two such youngsters repeatedly enter the mating area and they were apparently trying to snatch a drink of milk in between. But each time the male chased them off in a fairly rough manner. Occasionally a female may return to her offspring between copulations and nurse them while the male is watching.

Fertilization and development of the embryo

For a long time people disagreed about the question of when fertilization actually takes place in roe deer—whether it is during the summer rut or during the so-called "pseudo-rut" in the late fall. In the bodies of females killed in October and November an embryo was never found. In addition, it seemed improbable that the small roe deer should have a gestation period of nine and a half months, while that of the much larger red deer is only eight months. It turned out that two different types of gestation are possible in roe deer. Those females who have been fertilized in the summer have a so-called "pre-gestation" period from August to mid-December. In this period, the impregnated egg rests after the first cell divisions and then continues to develop during the winter. Usually, the pseudo-rut has nothing to do with reproduction, because the female's ovaries have eggs ready to be fertilized only in the summer. But it may happen that for some reason some of the females did not come into heat in the summer. These animals are then in heat in November and December, when they may then become fertilized. They give birth after five and a half months, without the "pregestation" period.

Annually in May and June, the female usually has two fawns, although she may sometimes have one, three, or in very rare cases, four. When the female is about to give birth, she retreats to forest meadows or high grass where she can find quiet, cover, sun, and sufficient food for the young. Roe deer fawns are the "stay put" type (compare Ch. 11). In the first days, the mother leaves them alone and comes back only for nursing. Many people then think that the mother has deserted her fawn and take it home for bottle raising. But such artificial ways of upbringing seldom work. Such fawns who stay motionlessly pressed against the ground are without any odor. It is said that touching such a fawn could mean its death because the mother is frightened by human scent and abandons it. Therefore, it is better not to touch it.

The "stay put" period lasts for about three to five days, after which the fawn starts to follow its mother around. By the age of ten days, it begins to browse, to play, and to practice "fleeing." The flight maneuver begins with a sudden standing still posture. Schmidt reports that "the fawn pretends that some flower or grass was the enemy. Then the fawn quickly turns and dashes off in a wild chase, but then comes strolling back to the mother to graze for a few minutes and then to begin another 'flight exercise'." When the fawns utter a soft, high-pitched 'teeping' sound, the mother comes immediately. Does with fawns may be easily attracted by imitating this sound.

A remarkable behavior of roe deer mothers and many other ruminant females is "comfort nursing." Schmidt describes it in the following way: "When a fawn has a frightening or scary experience, it will be nursed 'for comfort'. One day a friend of mine was mowing a pasture when he found a fawn which he carried aside. When he picked it up with a bunch of hay so that it would not become permeated by his scent, it was striking to see how the little animal was paralyzed from terror. Its legs were dangling so weakly that when he put it down they were not able to support the little body and it collapsed. At this moment, I knew that it must be this same reaction that causes fawns to be injured by mowers; since they are paralyzed by terror, they are not able to get up and flee. When my friend held the fawn a little too long, it uttered a distress call. The next moment, the mother appeared at the edge of the forest, her eyes wide open with terror. Then she approached to within fifteen meters and stood motionless. Now the fawn, which had overcome its paralysis, struggled violently to hurry to its mother. All of a sudden it was able to walk beautifully. She led it for only thirty or forty meters to the edge of the forest and nursed it right there." It is known from human mothers, too, that they nurse their crying babies, or, as a substitute, give them a pacifier.

The birth of the fawns

"Staying put" and "comfort nursing"

Roe deer in west and east

Roe deer increase in body and antler size from west to east in Europe. The Asiatic subspecies are even larger, especially the animals from the Altai and Tian-Schan Mountains. Some of the Central Asiatic and Siberian roe deer may reach twice the weight and antler size of the Central European roe deer. Male Siberian roes may reach a weight of fifty kilograms. The largest antlers from a Siberian buck bred in the Munich Zoo were forty-two centimeters and weighed 1100 grams. Such a strong antler can develop in the wild only in an environment where the original food is available at all seasons and is not spoiled by farming, cattle grazing, forestry, or overpopulation of roe deer.

In the past, European cultivated countries probably had not as many roe deer as there are now in well-tended areas. Roe deer, which avoid red deer and wild boar, are controlled in numbers by strong predators. Years ago, when the Central European forests were inhabited by many red deer and wild boar, as well as by bear, lynx, and wolves, there were fewer roe deer, but they were younger and stronger. Since the decreasing density of the woods due to farming and forestry, which drove off the larger animals, the roe deer have become the favorite game for hunting. In Germany and other European countries without predators, the roe deer population has to be controlled through planned cropping by hunters. The hunting authorities determine how many males and how many females may be killed. This kind of management is designed to weed out weak males with poor hereditary characteristics and to let "good" bucks become as old as possible so that they may reproduce over many years. In all countries with game laws, there is either only a short shooting season or only the killing of specific animals is permitted.

In contrast to the red and fallow deer, the roe deer in captivity are most difficult and hard to keep. In the wild, they feed mainly on herbs, tips of branches, foliage, blackberries, raspberries, hawthorne, and multiflora rose. This diet is difficult to replace in captivity. In additon, roe deer are very susceptible to a variety of parasites, for example a kind of fly, and lung and intestinal worms, which also are difficult to combat in the wild. Nevertheless, roe deer are being kept and bred successfully in many zoos. Even the difficult and troublesome bottle raising has been done often by zoo people, as well as by expert and dedicated amateurs. K. Borg found that roe deer's milk contains 6.7% fat, 8.8% protein, and 3.9% lactose (sugar), which is about twice as much fat and protein as in cow's milk. If cow's milk is diluted as is usual in artificial diets, a roe deer fawn being bottle raised hardly obtains the appropriate nutrition. In most cases, infections of the intestines and diarrhea follow which often lead to death. Goat and sheep milk is more similar to roe deer milk than is cow's milk. According to Grzimek's findings, sheep's milk contains 5.3% fat, 6.3% protein, and 4.6% lactose (sugar) and, therefore, is best suited for raising roe

Fig. 8-20.
Attacking male roe deer.

deer fawns.

All bottle-raised male roe deer usually become very aggressive when older. Since they have no fear of man and may even consider him a rival, they often injure their foster parents badly. Such dangerous tame animals cannot be turned loose in the wild successfully, either. Therefore: stay away from "deserted" fawns which you may "find" on a walk.

The AMERICAN DEER (tribe Odocoileini) belongs to the main genus *Odocoileus*. These animals are medium to large in size, slender and high-legged. They have large ears and a tail of medium length. The antlers have many tines and curve forward. The preorbital gland is large. Usually there are no upper canines. There are two subgenera (*Odocoileus* and *Blastoceros*) with four species and fifty-three subspecies.

Through documentary films and zoos, the WHITE-TAILED or VIRGINIAN DEER *(Odocoileus virginianus)* has also become well known in Europe. The HRL is 85–205 cm, the TL is 10–35 cm, the BH is 55–110 cm, and the weight is 22–205 kg. The antlers have a weak lower spiller on the inside and many tines on the upper part in an equally spaced sequence (record of ends, 28; record of width, 79 cm). The gestation period is 196–210 days, one to four young are born. The nursing period is only one to one and three quarters months. They live fifteen to twenty years. There are thirty-nine subspecies, among them the VIRGINIAN WHITE-TAILED DEER (*Odocoileus virginianus virginianus*; Color plate, p. 220) and KEY'S WHITE-TAILED DEER (⟡ *Odocoileus virginianus clavium*; Color plate, p. 220).

The most striking characteristic of the white-tailed deer is its large tail which is best seen during flight and then looks like a waving white flag (Color plate, p. 220). The observer usually does not get to see anything else but this white flight signal. It aids in keeping group members together in covered areas. In the wild the white-tailed deer lives a rather concealed existence. Where it is not pursued by man, it prefers the daylight hours before dawn and night. They live singly or in small family groups and may form larger herds in winter. Its paths are not used as regularly as the roe deer's. Similar to the roe deer in Europe, the white-tailed deer in North America has become a follower of civilization. They represent the major game for hunters. Wildlife biologists estimated in 1967 that there were more than 5,000,000 deer in America and they believe that today there are more white-tailed deer in America than at the time since discovery of the continent by the white man. For example, in the state of New York alone, the annual take by hunters of white-tailed deer is 70,000 and in Pennsylvania, 165,000. But it is also a fact that several subspecies in the Rocky Mountains, in Mexico, and on some of the islands in the Caribbean are endangered and probably threatened with extinction.

The heaviest specimens of the Virginia deer live in the north; towards the south, the animals' weight decreases. White-tailed deer

White-tailed or Virginian deer

Distinguishing characteristics

Fig. 8-21.
White-tailed deer (*Odo-coileus virginianus*).

Fig. 8-22.
A white-tailed deer mother licks her young clean. Thus, it remains almost odorless, hence predators can discover it in its hiding place only by chance.

Mule deer

in the state of New York weigh up to 200 kilograms, while the small subspecies on the islands around Florida weigh only twenty-five to forty kilograms. The population of this pigmy form diminished in 1947 to about forty animals. Then a part of their habitat on the island of Big Pine Key was declared a protected area in 1954. Twelve years later, there were 300 animals and thus the danger of extinction has been decreased.

The females of the northern white-tailed deer have their young regularly in spring. They often are twins; sometimes there are even more than two fawns. The tropical and subtropical forms may have offspring at any time of the year. In moderate climates, the rut begins in November or even sooner. If at this time the male meets a rival, fights follow. Then it may happen that two deer of equal dominance are caught by each other's antlers so that they cannot pull apart and thus die. Several times hunters and naturalists have attempted to separate such locked antlers but always in vain.

Like the roe deer, the white-tailed deer rarely make sounds. The young call their mother with a soft bleating; the mother calls her fawn with a murmuring sound. Disturbed deer snort briefly and may utter a shrill, whistling sound at night. Only badly injured animals scream loudly. Their vision is rather poor, but they hear well and have a keen sense of smell. Audubon reports: "One morning in fall we saw a female in heat passing at a close range. Ten minutes later a male followed her with his nose on the ground, precisely on her tracks. After another half hour, a second deer came along, and some time later a third one came, all of them following the same track."

In order to prevent a decrease in the population of white-tailed deer, the hunting season in the United States has been restricted to a few days. Each of the hunters in this period may kill only one deer. A group of hunters together may kill only one female. Under the legal restrictions, only the member of this group who wears an official band around his arm is entitled to the shot. Due to these conservation methods, the white-tailed deer presently is the most abundant ungulate in the wilds of North America.

White-tailed deer have been successfully introduced to New Zealand and Europe (Finland) as useful game. They have adjusted very well to the new habitats. In Czechoslovakia around Dobris and Brdy, there are approximately 200 white-tailed deer. They are descended from two males and four females imported from the United States in 1852.

While the white-tailed deer has a very wide distribution in North, Central, and South America, the MULE DEER (*Odocoileus hemionus*) occurs only in the mountains and the deserts of the North American west. The HRL is 100-190 cm, the TL is 10-25 cm, the BH is 90-105 cm, and the weight is 50-215 kg. The tail is not bushy but has a modest hair growth. The ears are unusually large. The antlers are up to 78 cm long, with many points. The gestation period of 182–210 days produces two young. There are eleven subspecies, among them the ROCKY MOUNTAIN MULE DEER (*Odocoileus hemionus hemionus;* Color plate, p. 221) and the BLACK-TAILED DEER (*Odocoileus hemionus columbianus;* Color plate, p. 221). Six of the subspecies are endangered.

At the beginning of their development, the mule deer's antlers resemble those of the roe deer. The younger males have either unbranched prongs or forks or small antlers with six points. With the continuing division of the ends into two spillers, after a while in mature males a tremendous multitude of tines may develop. While the largest antlers I ever saw had forty-two points, I have heard about antlers with fifty or more points.

Fig. 8-23.
Mule deer (*Odocoileus hemionus*).

In some regions, for example, Manitoba, this animal is called the "jumping deer" because of his habit of jumping high in flight and then landing on the ground with all four legs at the same time. The tail is, in most cases, white with a dark tip. However, in the black-tailed deer, black covers the entire tail. From this comes the name "black-tailed deer." In the northern regions of its habitat, the mule deer lives among cedars and firs, while in the south, it lives among mammoth trees. They also exist in dry, arid semideserts, where they feed on succulents which store enough water so that they do not need to drink for several days.

In contrast to the white-tailed deer, the mule deer do not follow civilization. Avoiding large forests, they prefer open clearings which have sufficient cover. In the winter, they migrate from the mountains to the lower elevations where they may form large herds which include the males. On their way back to the mountains in late spring or early summer, the females give birth to the young. As in other species of deer, the mothers leave their young alone during the first days and come back to them only for nursing. When a predator approaches, she paws the ground with her hooves and tries to lure it away.

In their home ranges the older deer are able to conceal themselves rather well. When we had our first camp in the Canadian wooded mountains, I tried in vain for ten days to see the large mule deer whose tracks I had spotted several times in the soft ground. Only one time did I catch a glimpse of the white tail patch of one of these animals from a distance between the brush on a slope opposite my place. But before I could make sure whether it was a male or a female, it had disappeared between the dense cover of young fir. This amazing shyness may have been caused by the many wolves in this area which had preyed heavily upon the deer population the winter before.

In the Jasper National Park of the Canadian Rockies, I was finally lucky enough to observe mule deer from a very close distance. There was scattered brush and hill country so that I could easily approach. The calves, which have a spotted coat like the roe deer fawns at birth, were changing. The adults also had their light gray winter coat. There the graceful animals were not disturbed, and it was a pleasure to watch them for a long time roam all over the place.

Marsh deer

The SOUTH AMERICAN MARSH DEER (*Odocoileus dichotomus*; Color plate, p. 222) is of a very different color. The HRL is 180–195 cm, the TL is 10–15 cm, the BH is 110–120 cm, and the weight is 100–150 kg. The tail has long, bushy hair. The rather long coat has tangled hair. The shape is somewhat arched, since the animal lives in bush and reed jungle. The hooves are very long, about 7–8 cm, can be spread wide and have long pseudo-claws. Their tails are kept erect during flight like the white-tailed deer's. They have a long gestation period, allegedly about twelve months, with one young born.

This large and conspicuous red-brown deer lives in the swamp forests, jungles along rivers, creeks, and other waters. Feeding mainly on swamp and aquatic plants, it is active predominately at dawn and during the night. The marsh deer live singly or in small groups of up to six animals, which include usually one older male and two females with their young. Occasionally larger herds may be found. Females advanced in pregnancy and mothers with young stay away from the herd. In marsh deer, antler-shedding and the rut are not dependent on the season but may occur the whole year round. Consequently, rival fights among the males hardly ever take place.

In spite of its size, marsh deer belong to the "bush crawlers," those species which skillfully move in the high grass and jungle. Marsh deer hide rather than flee. The extremely spreadable inner edges of the hooves are connected by strong membranes. This enables them to live in the swamp where other ungulates with a comparable weight would sink in too deeply.

On his travels throughout South America, Hans Krieg has often observed the marsh deer. He says: "In the flood region of the Paraguay River, I rode one day for about an hour across a swamp which reached up to my horse's hocks. Suddenly, when I turned around a bush, two marsh deer stood in the water a few meters from me. In spite of the noise of the approaching horse, they had waited in order to find out what this turmoil was about; they then took off with steep jumps."

In mule deer and marsh deer, the antlers develop so many points that they might be called clusters of tines. This type of antler is quite irregular. They never show the harmony and balanced beauty which is to be found even in the uneven antlers of red deer. Many of their antlers are conspicuous because of their bright color. Obviously, there is a lack of tannin in the trees and bushes while they are polishing their antlers. In the collection of my brother, Heinz Heck of the Munich Zoo, there is an especially beautiful marsh deer with antlers that have the color of amber.

My father, Ludwig Heck, liked to compare the marsh deer to the Sitatunga antelope who also live in the swamp: "Everyone who sees

Fig. 8-24.
Marsh deer (*Odocoileus dichotomus*).

both animals every day like I do cannot help but be puzzled by the similarity of shape: the same thin, extremely high legs which carry the body in a strange, staggering manner, the same pointed, long hooves which spread apart while the animal is walking, and the same rugged, somewhat tangled coat. All these characteristics distinguish the deer and antelopes from their respective genera and make these animals similar to each other, not because of a close phylogenetic relationship, but because they share the same habitat and have accordingly developed the same appearance and habits. Even more strange and inexplicable is the similarity in coat color with the peculiar predator of this region, the high-legged, large, red-maned wolf. There is the same fox-colored coat and the same black pattern on the legs, which confuses even experienced hunters who are not able to say for sure which one of the two animals they have seen at a glance."

Presently the marsh deer is among the endangered species which are urgently in need of protection. It is already extinct in Uruguay, the densely populated areas of Brazil, and probably also in Paraguay and northern Argentina. Since its habitat is restricted to moist, swampy plains, it actually lives in only a few areas of its wide distribution. Often the swamps are separated by wide areas which are inhabitable for the deer. The increasing human settlements in these areas cause further obstacles for deer migration between those swamps. Therefore, marsh deer have rarely been displayed in zoological gardens because they do not last long in captivity.

Even more critical is the situation of the second large South American deer. This species, the PAMPAS DEER (⚲*Odocoileus bezoarticus;* Color plate, p. 223), has been classified by zoologists into a subgenus of its own *(Blastoceros).* The HRL is 110–130 cm, the TL is 10–15 cm, the BH is 70–75 cm, and the weight is 30–40 kg. The slender animals are long-legged. The ears are medium-sized. The weak antlers have few points. The ♂♂ and ♀♀ either do or do not have upper canines. The fine, clinging coat has undulated bristles with a dense fleece. A large whirl of hair is on the center of the back. The gestation period is unknown, but one young is born. There are three subspecies, one of which, the PATAGONIAN KAMP DEER *(Odocoileus bezoarticus color),* has been destroyed by the farmers except for a few herds on private estates. The other two subspecies are also endangered.

The pampas deer, as the name indicated, lives on the dry, open steppes, which presently are being steadily turned into grain fields. Therefore, it has retreated into areas which are less populated. It is distinguished from its relatives, aside from several physical characteristics, chiefly by a strong garlic scent which comes from a secretion of the interdigital glands of the adult male's feet. Such a deer can be smelled over a distance of more than one kilometer. The scent is especially strong when an animal is disturbed in his bed or during the rut.

Pampas deer

Fig. 8-25.
1. Pampas Deer (Odocoileus bezoarticus).

The gland secretions may aid in keeping group members together. Pampas deer live in pairs or small family groups. During the rut, the male usually remains with only one female. The family stays together after the young has been born, and both the parents watch it. This makes the pampas deer one of the few deer which have a kind of "family life."

Like marsh deer, pampas deer also drop their antlers during the rut and have their young throughout the year. The young are chocolate-brown with four horizontal rows of white dots. When in danger, the mother tries to distract the intruder by fleeing in the opposite direction from the young. Some cases have been reported where she feigned injury or limping, thus attracting the pursuer's attention to herself.

Thirty years ago, Hans Krieg observed pampas deer in Central Brazil. He reports on this: "It happened several times that when I suddenly came around a bush, a deer would stand motionless a few meters from me, seeing and hearing with his ears widely spread. Then in sudden terror, it would jump off with wide, flat leaps. After 100 or 200 meters, the deer would stand still again. But then, naturally, it is hardly possible to get closer to this animal. A resting kamp deer may, like the European roe deer and the South American brocket, jump a few meters away from the approaching hunter. Several times I also saw single animals trying to sneak away inconspicuously. Many different species of game animals employ this method of sneaking away. In this case, it is necessary that the animal become aware of the disturbance early and does not feel immediately threatened. A kamp deer in flight is a rather strange sight to the European hunter who is not used to seeing deer with long tails. In flight, the kamp deer raise their tails so steeply that the white under side is directed backward. In open country, these white signals may be visible after the deer itself can no longer be recognized."

Hans Krieg learned then that the number of kamp deer has been extraordinarily reduced during the past years. This population decrease cannot be blamed on extensive hunting; the South American cattlemen despise the odor of the deer and say that the meat reeks of it. In Krieg's opinion, the deer suffered severely from hoof and mouth disease, which is almost completely ignored by the large cattle ranches of Central America instead of being treated. Krieg once talked with an old Brazilian who was very sorry about the disappearance of the deer and for a rather strange reason. He claimed that these animals were very useful because they killed snakes. This opinion, which is widespread in the South American pampas, was considered at first to be only a hunter's story. Now, Hediger states that deer in North America kill venomous snakes by kicking them with their hooves, and that according to ancient sources deer were introduced into the Mediterranean island of Rhodes to control the snakes.

Explanation of the following illustrations. The deer family *(Cervidae)* includes a total of thirty-one species, twenty-eight of which are illustrated in color on the following pages. There are forty-seven subspecies.

Plate I
Musk deer (subfamily Moschinae)

1. Musk deer *(Moschus moschiferus)*, a) skull of a male with the long upper canines.

Muntjacs (subfamily Muntiacinae)

2. Javan muntjac *(Muntiacus muntjak muntjak)*, a) skull of a male with the long upper canines and the small antlers on the high forehead structure.
3. Chinese muntjac *(Muntiacus muntjak reeves)* in velvet.
4. Tufted deer *(Elaphodus cephalophus)*.

Plate II
True deer (subfamily Cervinae)

Genus fallow deer *(Dama)*:
5. European fallow deer *(Dama dama dama)*. The variety of colors are due, in part, to selective breeding by man. a) Wild color, b) Blacks, c) Whites, d) "porcelain"-color.
6. Persian fallow deer *(Dama dama mesopotamica)*.

Plate III
True deer (subfamily Cervinae)

Genus axis deer *(Axis)*:
7. Axis Deer *(Axis, axis)*.
8. Hog deer *(Axis porcinus)*.

Plate IV
True deer (subfamily Cervinae)

Genus red deer *(Cervus)*:
Sambar (subgenus Rusa):
9. Ceylon sambar *(Cervus unicolor unicolor)*.
10. Sumatran sambar *(Cervus unicolor equinus)*.
11. Sunda sambar *(Cervus timorensis)*.

Plate V
True deer (subfamily Cervinae)

Genus red deer *(Cervus)*,
Sambar *(subgenus Rusa)*:
12. Luzon Sambar *(Cervus mariannus mariannus)*.
13. Alfred's sambar *(Cervus mariannus alfredi)*. Barasinghas (subgenus *Rucervus)*:
14. North Indian barasingha *(Cervus duvauceli duvauceli)*.
15. Schomburgk's deer *(Cervus duvauceli schomburgki)* now extinct.
16. Eld's Deer *(Cervus eldi)*, greatly endangered.

Plate VI
True deer (subfamily Cervinae)

Genus red deer *(Cervus)*:
Sika deer (subgenus *Sika)*:
17. Formosa sika *(Cervus nippon taiouanus)*, almost extinct.
18. Japanese sika *(Cervus nippon nippon)*.
19. Dybowski's deer *(Cervus nippon dybowskii)*, endangered in the wild, a) male and female in their winter coats b) summer coated female with young c).

Plate VII
True deer (subfamily Cervinae)

Genus red deer *(Cervus)*:
Red deer (subgenus Cervus in the strict sense), subspecies of red deer:
20. Central European red deer *(Cervus elaphus hippelaphus)*, male deer in winter coat, female in summer coat with young a).
21. Mediterranean red deer *(Cervus elaphus corsicanus)*, almost extinct.
22. Barbary stag *(Cervus elaphus barbarus)*, greatly endangered.

Explanations to the plates VIII-XVI are on page 225.

I

1

1a

2a

♂

♀

♂   ♀

2

♀

3

♂

4♂

♀

5 a

♂

5b♀

5c♀

5d♂

6

♂

♀

7 ♂ ♀ a

♀

a ♀ 8 ♂ ♀

V

14 ♂

16 ♂

15 ♂

12 ♂

13 ♂

17 ♂

♀

18 ♂

19 a ♂

19 a ♀

19 b

19 c

20 ♂

20 ♀

20 a

21 ♂

22 ♂

VIII

23

24

25

26

♀

27

♂

♂

28

a

♀

X

♂

30

♀

30 a

29 ♂

31♂ 31♀ XI 31b 31a ♀ 32 ♂ 33♂ 33b♀ ♀ 33 a

b♂

35 ♂

34

♀

♂

a

36 ♀

♂

a

♀

37 ♂

XIV

♀

38

♀

♀

♂

40♂

41♂

43 ♂

♂ 42

♀

♀

♂ 39

♀

♀ 44 ♂

XVI

45 ♂

♀

a

46 b

46 a

47

Explanations to the preceding illustrations.

Plate VIII
True deer (subfamily Cervinae)
Genus red deer *(Cervus):*
Red deer (subgenus Cervus in the strict sense), sub-species of red deer:
23. Maral *(Cervus elaphus maral).*
24. Bactrian deer *(Cervus elaphus bactrianus).*
25. Hangual *(Cervus elaphus affinis).*
26. Altai maral *(Cervus elaphus sibiricus).*

Plate IX
True deer (subfamily Cervinae)
Genus red deer *(Cervus):*
Red deer (subgenus Cervus in the strict sense), subspe-cies of red deer:
27. Isubra Maral *(Cervus elaphus xanthophgus).*
28. East wapiti *(Cervus elaphus canadensis),* a) young (The Rocky Mountain elk *Cervus elaphus nelsoni).*

Plate X
True deer (subfamily Cervinae)
Genus red deer *(Cervus):*
Thorold's deer (subgenus *Przewalskium):*
29. Thorold's Deer *(Cervis albirostris)* genus Père David's deer *(Elaphurus):*
30. Père David's deer *(Elaphurus davidianus)* exists only in zoos and game parks.

Plate XI
Chinese water deer (subfamily Hydropotinae):
Chinese water deer (subfamily Hydropotinea):
31. Chinese water deer *(Hydropotes inermis)* male and female with six young, a) skull of a male with the long upper canines, b) American and roe deer (subfamily Odocoileinae), tribe roe deer (Capreolini).
32. Siberian roe deer *(Capreolus capreolus pygargus).*
33. European roe deer *(Capreolus capreolus capreolus),* a) young, b) winter coat.

Plate XII
American and roe deer (subfamily Odocoileinae)
34. Virginia white-tailed deer *(Ococoileus virginianus vir-ginianus),* a) young, b) in flight with the tail erect as a signal to follow.
35. Key's white-tailed deer *(Odocoileus virginianus calvium).*

Plate XIII
American or roe deer (subfamily Odocoileinae), tribe American deer (Odocoileini):
36. Rocky Mountain black-tailed deer *(Odocoileus hermianus hemionus),* a) young.
37. Black-tailed deer *(Odocoileus hemionus columbianus).*

Plate XIV
American or roe deer (subfamily Odocoileinae), tribe American deer (Odocoileini):
38. Marsh deer *(Odocoileus dichotomus).*
39. See plate XV.
40. Peruvian Guemal *(Hippocamelus antisiensis).*
41. Chilean guemal *(Hippocamelus bisulcus).*

Plate XV
American or roe deer (subfamily Odocoileinae), tribe American deer (Odocoileini):
39. Pampas deer *(Odocoileus bezoarticus),* more closely re-lated to marsh deer (plate XIV).
42. Red brocket *(Mazama americana).*
43. Grey brocket *(Mazama gouazoubira).*
44. Southern pudu *(Pudu, pudu).*

Plate XVI
Elk deer (English), moose deer (American)
(subfamily Alcinae):
45. Elk resp. moose *(Alces alces).*

Reindeer (subfamily Rangiferiane):
46. North European reindeer *(Rangifer tarandus tarandus),* a) wild form, b) pie-bald domestic form.
47. Barren ground caribou *(Rangifer tarandus arcticus).*

At the turn of the century pampas deer were occasionally displayed in zoos. The last specimen to come to Europe lived in the forties at the Duke of Bedford's Woburn Abbey.

South American deer or guemals

The South American deer or GUEMALS have been classified by some zoologists in the genus *Odocoileus.* But they have so many different characteristics from the species described so far in this chapter that we consider them a separate genus *(Hippocamelus).* The HRL is 140–160 cm, the TL is 10–15 cm, the BH is 75–85 cm. The rather short-legged animals have large ears. The short, triangular tail ends in a point. The dense coat is long and thick with brittle bristles; the fleece is short and sparse. The antlers usually have a simple, compact fork; sometimes the rear bar has several spillers on the front and other formations may also occur. The ♂♂ and ♀♀ have upper incisors, but they are very small. These are alpine animals which live in altitudes of 3000–4000 meters. There are two species: 1. PERUVIAN GUEMAL *(Hippocamelus antisiensis;* Color plate, p. 222), where antlers are forked close to forehead. 2. CHILENIAN GUEMAL *(Hippocamelus bisulcus;* Color plate, p. 222), whose antlers are slightly higher with several spillers.

The Peruvian guemal lives in small herds above the timber line and migrates in winter to lower altitudes. He is considered to be very shy. An older male usually joins two or three females with their young. These deer may be found grazing at any time, night or day.

In 1899, a guemal came to the Berlin Zoo. My father, Ludwig Heck, writes about it: "This animal was of special interest to me because of its strange, but fascinating, combination of deer and ibex characteristics. When the keeper came into his pen, this deer would playfully attack him, raise on its hind legs and shake his head pugnaciously. I have never seen this behavior in any other deer. Standing still, he would have looked like an ibex if it had not been for the typical deer's head." A couple of Peruvian guemals, purchased for the Berlin Zoo in 1931, reproduced there regularly until 1940. This species also has been kept by American zoos.

A rather deformed picture of the guemal is found on the Chilian coat of arms, along with the condor. Strangely enough, this animal's scientific name has been changed no less than twenty-one times between 1782 and 1902. At first it was described as a horse, then a camel, then a "horse-camel," then a llama—probably all by people who knew about it only from hearsay. In its native countries, it is now extinct except in a few tiny, remote habitats.

Hans Krieg saw this deer in the area where the Argentine Nehuel-Huapi National Park is now located. He said: "In summer the guemal prefers to stay near the timber line, although it may also cross it and enter the rocky and gravel slopes. The creeping *Nothofagus* trees of this region supply sufficient cover for it. As soon as the heavy snow begins to fall, the mountain guemals start to migrate slowly into the forest

Fig. 8-26.
1. Peruvian guemal *(Hippocamelus antiensis).*
2. Chilean guemal *(Hippocamelus bisulcus).*

regions. In undisturbed areas, some animals may remain permanently in the lower forest regions. The guemal seems to like variety in its food, like many ruminants whose main diet are leaves. At times it will prefer the leaves of the Maytén tree and then it will change to the juicy shrubs of the yellow-flowered *Alstroemeria.*"

Krieg also found that the cattle rancher's dogs chased some of the guemals to death. He reports: "These deer are not enduring runners. The guemal who is being chased will go downhill in utmost distress to the nearest lake. From this behavior, the natives have developed the rather distasteful method of killing the deer in the water." Some cattle diseases are also contagious for the guemals and reduce their population. But the most disastrous consequences came from the European red deer, the fallow deer, and the Indian axis deer which were introduced as game animals into the Southern Andes of Argentia and Chile. These imported newcomers reproduced to such an extent that they forced the guemal from its habitat so that it disappeared from the Nahuel-Huapi region. Several attempts to reintroduce it to this area have been in vain. Presently, only a few small groups of guemals remain near Los Alerces and Los Glaciares in Southern Argentina and in the eastern section of the Chilian province of Aisen.

Brockets

In the BROCKETS (genus *Mazama*), the antlers are unbranched, short prongs. These animals range in size from a roe deer to a rabbit. The head, neck and tail are short, and the ears are wide. The lumbar region is higher with an arched back. The hairless nose region has small wrinkles. The preorbital glands are usually large. The upper canines are either reduced to small, pointed hooks or are missing. There are four species with twenty-nine subspecies:

Distinguishing characteristics

1. RED BROCKET (*Mazama americana;* Color plate, p. 223). The HRL is 95–135 cm, the TL is 10–15 cm, the BH is 65–75 cm, and the weight is 20–25 kg. The shining coat has a yellow-red to fox-red color on the back. The head is carried rather low. The preorbital glands are small or missing. This animal is a forest dweller.

2. GREY BROCKET (*Mazama gouazoubira;* Color plate, p. 223). The HRL is 90–125 cm, the TL is 10–15 cm, the BH is 60–70 cm, and the weight is 17–23 kg. The coat is dull, the back is gray-brown or other colors. The head is carried high, and the eyes are more in the side of the head. The preorbital glands are small or missing. Their preferred habitat is the bush and the savannah. This animal follows civilization.

3. LESSER BROCKET or BORORO (*Mazama nana*). The HRL is 90–100 cm, the TL is 10–12 cm, the BH is 45–50 cm, and the weight is 10–20 kg. The shining coat has a dark red-brown back and a yellow-red underside. The head is carried rather low. The tips of the antlers are short and compact. The preorbital glands are large. This animal lives in forest and hill country and likes to bathe and to swim.

4. GREY DWARF BROCKET (*Mazama bricenii*). The HRL is 72–92 cm, the

TL is 8–10 cm, the BH is 35–40 cm, and the weight is 8–12 kg. The back is red-brown and the underside is white. They have medium-sized preorbital glands. They live in forests and brush in mountains up to 3000 meters. First described in 1908, the habits of this species are virtually unknown.

These small deer take the place in the South American fauna which the duikers occupy in the African fauna. They are typical "brush creepers" for which the thickest plant cover does not present an obstacle. According to Sanderson, they like to rest during the day in the dens between the huge roots of prime forest trees or under fallen tree trunks. At night they come to the edge of the forest and graze. The gray brocket, which also visits the plantations, has adjusted well to human settlements.

Fig. 8-27.
1. Red brocket (*Mazama americana*). 2. Grey dwarf brocket (*Mazama briceni*).

Hans Krieg found tracks of the gray brockets repeatedly in the settlements near the Alto Parana. There the animals remained in the same areas. At night they went into the yerba and orange plantations or into the fields of corn, manioc, and sweet potatoes. The red brocket leaves its hideout at sundown and begins to graze on the plantation's mellons, green corn, and other produce. Therefore, landowners chase it with dogs. The lesser brocket, which is the size of a half year old roe deer fawn, is pursued in the same manner.

Krieg found that the marking of the red brocket males is similar to that of the roe deer. Territorial marking consists of a seemingly senseless polishing of the already velvet-free lower parts of the antlers against a small trunk or shrub. According to S. V. Schumacher, the deer applies the secretion of the scent and sebacious glands in the skin between the bone structure on the forehead with this antler polishing behavior, thus the animal indicates that this territory is occupied. The deer paws the ground vehemently with his front legs while he is marking. The question of whether brockets are original forms or whether they have acquired their seemingly primitive characteristics in adaptation to life in the brush has occupied zoologists for a long time. The small size of these deer is not important in this regard, for there are other small deer which are otherwise highly evolved. One primitive feature is the head shape, especially that of the "full profile." Another feature is the short spikes which branch only in rare cases. Krieg comments on this subject as follows: "It is justifiable to believe the brockets have the essential characteristics of the original telementacarpal deer. However, this does not answer the question of whether a part of their initially primitive-looking features, especially the poor development of their antlers, are not the result of a reduction which could have taken place in a relatively short time. I presume that their ancestors were medium-sized deer from the savannah with prongs or antlers with four, six, or at the most, eight points."

During the last years, Helmuth O. Wagner, formerly with the

Overseas Museum in Bremen, has thoroughly studied the biology and the behavior of the Mexican red brocket *(Mazama americana temama)*. Wagner assumes that these deer either never or only irregularly change their antlers.

Being sedentary has caused the brockets' adjustment to cultivated areas in many areas. They live concealed in plantations and in the brushwood between them. Therefore their number may easily be underestimated. The two sexes are together only briefly during the mating season. Never was a female found with more than one young. Depending on regional conditions, the brocket's females may have young at all seasons. The young, which "stay put" are found occasionally on the coffee plantations when the plant growth is removed from the clearings of former prime forests.

In their native habitat, brockets are caught at a young age and then kept tame on farms and in gardens. These pleasant animals often come to zoological gardens through the natives. Before World War II, the Berlin Zoo recorded several births of the red brocket, and Philadelphia later repeated this success. A gestation period of seven and a half months was observed there. Kenneth found in 1953 that the gestation of the lesser brocket was 217 days. On the island of Barro Colorado in the Panama Canal, a captured male liked to eat fruit and banana peels. In Berlin a red brocket lived for seven years, and one in Philadelphia lived for more than nine.

The Berlin Zoo has had gray brockets since 1962. The director, Heinz-Georg Klös reports: "This species, too, reproduces well in our zoo, which recorded the probable first birth in a zoological garden. By 1967, three young had been born. Coming from the tropics, the brockets need more warmth than the pudus. Low temperatures are apparently unpleasant for them. In their food preferences, they are choosy, resembling the pudus and the roe deer."

The pudus

The smallest of all American and contemporary deer are the PUDUS (genus *Pudu*). They have delicate legs, a short neck, an arched back, a compact rump, and a short head with a narrow facial structure. The short, oval ears have a thick hair covering. The coat is thick and dense. The brocket-like antlers are small, short, and almost invisible in the forehead tuft. There are no upper canines. There are two species: 1. NORTHERN PUDU *(Pudu mephistopheles)*. The BH is approximately 35 cm; no other measurements are known. It has a colorful coat, no tail, small preorbital glands, and a flat skull indentation. The hooves are narrow and pointed. Two almost unknown subspecies exist in the northern Andes. 2. SOUTHERN PUDU (⚥*Pudu pudu*; Color plate, p. 223) is smaller. The HRL is 80-93 cm, the BH is 30-35 cm, and the weight is 7-10 kg. It has simple coloring, a small tail, large preorbital glands, and a deep skull indentation. The hooves are wider than in the northern species. A seven month gestation period precedes the birth of only one young.

Formerly the southern pudu was rather widely distributed in the Andes of Chile and West Argentina and on the islands off the Chilian coast. This smallest form of the deer is an animal of the unpenetrable jungles in these southern rainy regions. "Few people know it in the Chilean Andes," reports Hans Krieg. "Its unbelievably small tracks are found only in the remotest corners of the valleys and on the shores of hidden bays where its favorite food, the fuchsia, grows in abundance and where the Colihue-bamboo offers sufficient cover. Even though it is extremely shy in summer, it seems to go close to settlements after heavy snow falls in winter." Remorseless persecution has led to the destruction of this tiny deer in many areas of its native habitat. Some of them remained on the island of Chiloe and in forest and lake regions of Southern Chile, but their number is decreasing steadily. Unfortunately, they suffer from so many parasites, mainly tapeworms, that it is difficult to keep them in captivity.

Fig. 8-28.
1. Southern pudu (*Pudu pudu*). 2. Northern pudu (*Pudu mephistopheles*).

The Berlin Zoo has a long tradition in keeping these tiny deer. The first pudus were displayed there as early as 1896. During the first four decades of this century, they reproduced several times in Berlin. While a pudu young at birth is only fifteen centimeters high, after three months it has reached adult size. There are two rows of light spots, the usual "camouflage" of the young deer, on its back. The little male, who starts to grow the first spikes after three months, rubs off the velvet tissue only six to eight months later, and then sheds them after twelve months. Then he begins to grow the first "real" antlers. A pudu becomes sexually mature at about one year.

Pudus were displayed in the Berlin Zoo at intervals until 1940. They are rather hard to keep. "Pudus are very choosy in their food preferences, as are roe deer," reports Heinz-Georg Klös. "They nibble a leaf here, a berry there, and maybe a juicy shoot, and it takes a lot of experience to make them adjust. In 1966, another pair was imported from Chile and added to the world famous Berlin Zoo deer collection which now has 22 species of deer. Shortly after they arrived, the little fellows became so tame that they fed from a person's hand and could be touched. The zoo of Cologne also has had pudus since 1961."

The largest and probably the most conspicuous of contemporary deer are the MOOSE DEER (subfamily *Alcinae*). There is only one species, the MOOSE (*Alces alces*; Color plate, p. 224). This animal is the size of a horse, with long legs, a long head, and a short neck and tail. The HRL is 240–310 cm, the TL is 5–10 cm, the BH is 180–235 cm, and the weight is 300–800 kg. The short antlers form great flattened surfaces armed with numerous tines or snags, often more than forty. The total weight of the antlers is more than twenty kilograms. Antlers with several forked bars are rather common. The eyes and preorbital glands are small. The nose region is extended, and the animal has a very wide

Subfamily: moose deer, by B. Grzimek and L. Heck

muzzle with a highly movable upper lip. The neck, which is compact, has in ♂♂ a sausage-like skin pouch which grows into a dewlap with beard as the males age. The rump is lower than the humped withers. The large hooves are long and narrow; the pseudo-claws are large and wide. The toes may be spread far apart, thus making it possible to walk in the swamp and marsh. It is a good swimmer. The rather long-haired coat clings to the body in the summer, and in the winter the animal is covered densely with brittle, undulating bristles and little fleece. There are 32 teeth: $\frac{0 \cdot 0 \cdot 3 \cdot 3}{3 \cdot 1 \cdot 3 \cdot 3}$. They eat soft wood, branches and twigs with juicy bark, swamp and aquatic vegetation, grass, herbs, heather, etc. The 35–38 week gestation period ends with the birth of one or two, but rarely three young. The young nurse for nine to twelve months. They live twenty to twenty-five years. There are seven subspecies, among them the NORTH and EAST EUROPEAN MOOSE *(Alces alces alces)*, whose distribution reaches as far as East Prussia, the COMMON or AMERICAN MOOSE *(Alces alces americana)*, and the especially large ALASKAN MOOSE *(Alces alces gigas)*. The CAUCASIAN MOOSE *(Alces alces caucasicus)* has been destroyed and the YELLOWSTONE MOOSE *(Alces alces shirasi)* is endangered outside the national parks.

I, Lutz Heck, saw the first moose on the Courland seashore. Some moose lived there in willow brush and forest. Each time it was an exciting experience to see these giants appear suddenly near the sea, in the surf, on high, bare sand hills, or in the forest. Since then I have seen hundreds of moose in the wilds of Germany, Finland, and Canada. Several of them I could keep for years. When they reproduced, I turned their offspring loose in the wild. From their habits which I studied, it became quite clear that the moose was an animal originally from the forest who has excellently adjusted to special environmental conditions. Formerly, many of his characteristics were misinterpreted and resulted in the strangest reports, even up to the present.

Although moose were found in the cave drawings of the Early Stone Age, their first description appears in Caesar's "Gallic War." However, it sounds so strange that it might be considered the world's first hunting tale. Caesar reports that the moose were hornless and that their legs had neither ankles nor joints. Therefore, they could not lay down, but once they fell, they could never get up again either. To sleep they would just lean against trees. The Teutonic hunters would chip notches into these trees so that when the tired moose tried to rest, both it and the tree would fall down together, thus becoming an easy prey for the smart hunters. The only essence of truth which exists in this tale is the observation that moose straddle rather strong trunks in order to break them to obtain twigs, buds, and leaves from the soft parts of the tree.

Later, the famous Roman naturalist Plinius stated that the moose

Fig. 8-29.
Moose *(Alces alces)*. In the dark-shaded areas of Europe, the moose has been exterminated. In the dotted tundra regions, they pass through only on migrations.

"had to graze backward because of his enlarged upper lip." This too, of course, is wrong. The moose's upper lip is indeed huge and extremely well suited for breaking off branches or for stripping the leaves from them. The striking short neck of the moose is another adaptation to his environment, which makes it impossible for the animal to feed readily from the ground. For this purpose, the moose must spread its front legs widely apart or "kneel" on his metacarpals. Therefore, it cannot graze like other deer. The natural method of feeding is to stretch its neck and head upwards to obtain his food from bushes and trees or to feed on aquatic plants (Color plate, p. 190), in the swamp, where it even may totally submerge.

Of course, a large deer with such gigantic antlers has always been a temptation for hunters. However, moose hunting in the prime forest habitat is quite difficult due to the amazingly appropriate behavior of the animals. In battues, the strongest males may sneak through the rows of the beaters without being sighted by the hunter. Thus, the moose's intelligence and abilities, which are absolutely appropriate to their habitat, appear to be peculiar only in a strange environment.

Even though the moose may come to cultivated areas and feed on oats or beets, its actual habitat is the forest and its preferred diet is branches and leaves. Therefore, moose used to live in all those regions of the Northern Hemisphere with deciduous forest. Where the deciduous trees and the swamps disappeared, the moose population subsequently decreased. In all countries with appropriate game laws, they have recovered beautifully. This trend has been furthered by the frequent twins born to the females.

The moose roam irregularly in their large territories. They migrate often, sometimes over hundreds of kilometers. They may cross channels of the sea; moose have been seen many kilometers off shore between Sweden and the Aland Islands in the Baltic Sea. The drive to migrate seizes them predominantly in the rutting season which occurs in early September. Then the moose's call, interrupted by the clacking blows of the antlers against trees and branches, sounds with a muffled roaring through the swamp forest. Searching moose roam about and often have vehement fights. Occasionally fatal accidents may occur, as in red deer. Dead moose with their antlers locked together have been found, as well as ones that were pierced by antlers. Generally the moose rut is much more quiet (as in roe deer) than that of most other deer. Moose in rut do not try to gather dozens of females or to defend them against rivals. A male moose joins a female temporarily. If there are two or three females with him, it will usually be an adult with her young of the last two years. Finally the pair will separate, and the male seeks out another mate.

The uniformly red-brown young which may be born anytime from April to the beginning of June, are usually born in mid-May. They

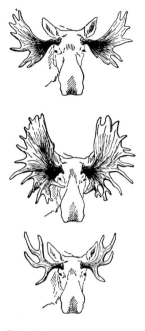

Fig. 8-30.
Moose have usually
stronger or weaker
palmate antlers, although
lesser branched bar-
shaped antlers also occur.

Preservation of
moose deer in East
Prussia

grow up quickly. At birth their height is about seventy to eighty centimeters, but by New Year's, it is one and a half meters. They are pubescent by one and a quarter years of age.

The moose's antlers are the strongest in all presently-living species
of deer. In the fall, protuberances of only two centimeters develop on
the young's forehead. In the following spring, these grow into antlers
without having been shed. At first the moose has only six spikes; occasionally it may have forks with six, ten, or, in exceptional cases, even
more ends. All these have a distinct flattening in the center, indicating
the formation to follow. The male moose is fully grown by the age of
five years. At eight years he has his best antlers. They drop every year
no sooner than November, while those of the younger animals drop
one to two months later. The time varies, according to the age of the
animal, like in the red deer, where the older males drop them earlier
and the younger ones later. The phase during which the velvet is shed
from the new-grown antlers lasts from the end of June until August.
Thus, the moose has his perfect antlers only for three or four months.

In all moose habitats, there are animals with bar shaped antlers
which never develop the typical flat surface in the center. The antlers
have long, round ends and are often strong and bulky, but they lack
the typical palm. For some time, it was believed that the moose was
in the process of evolving and the bar-shaped antlers represented the
"modern" type. Meanwhile, many fossils have been found which
prove that such formations have existed at all times.

The East Prussian hunters did not like such bar-shaped antlers. So
for decades they killed all moose which had them in order to insure
that the offspring came only from those males which had the significant antlers with the flat surfaced center. Under the influence of
Manfred von Kobylinski-Korbsdorf, the moose became the first game
animal in Europe to be hunted under restrictions established by the
authorities concerning the size and age of the animals to be killed. Due
to this selection, the antlers of the East Prussian moose were better
developed in comparison with those of the Scandinavian and east
European moose.

The East Prussian moose was preserved by methodical care over the
decades. Many specialists, from the foresters of moose habitats to
well-known scientists, worked together on the project and used all
their knowledge in studying this giant deer. In 1849, after the short
period of unlicensed shooting which followed the East Prussian Revolution of 1848, only eleven moose were left. While this population
seem destined to become extinct, careful protection led to a steady and
remarkable increase in their numbers. This long period of indulgence
made the East Prussian moose completely tame. It was possible to walk
within twenty meters of a resting moose; a horse carriage could approach even closer. This lack of shyness was thought by many ob-

servers of this time, among them Brehm, to be stupidity. But this really has no correlation whatsoever with intelligence. An inhabitant of the dense willow and alder forests, the moose feels secure in its cover. Therefore, he remains motionless and flees only at the last moment. He does so in an unbelievably secretive and soundless manner. In Sweden and Canada, huge moose unheard and unseen disappeared from my sight as if swallowed up in the ground.

The East Prussian moose often lost their young from flooding which caused the death of many. In the early thirties more than eighty young moose drowned. Therefore, the dams were improved and "moose hills" were established as refuges. The moose kept increasing in number. Therefore, shortly before World War II moose were to be resettled elsewhere, as in the Schorfheide, a wildlife refuge near Berlin. It failed after initial successes, because the area was too small and already overstocked with red and roe deer. However, due to this failure much was learned about the life of moose, which should prove valuable in future attempts at resettlement. Probably no moose survived World War II in East Prussia. Meanwhile the Soviet administration has introduced new moose into their former habitat. This means a new start for a moose population in Central Europe and gives hope for their preservation.

Bottle-raised male roe and red deer regularly turn vicious and dangerous when they grow up and come into rut. This has rarely been experienced with the largest and strongest of deer, the moose. The reason may be that male moose in rut are much more quiet and gentle. (see p. 232). They are especially easy to handle when they swim in the water. Conservationist D. W. Simkin drove them into the lakes in Ontario, Canada, with his helicopter and then he applied ear-marks with special pliers to the animals. With these marks they can be recognized later, and they will give information about the age and migration of the animals.

Taming of moose deer

Since the moose become tame so easily, their young have often been bottle raised in the Baltics, Sweden, and Russia. When they are familiar with the house, they walk about in the rooms and even climb stairs. In Livonia a forester by the name of Harry Walter had a male moose, Tschuk, who skillfully jumped through an open window into and out of his room. During meals it would lie in the living room, and it followed him in the forest like a dog. While it could be ridden, it would not tolerate a saddle. This tame animal had only one bad habit: it liked to eat mushrooms. Therefore, it would surprise mushroom-hunting women and empty their baskets. The question, then, is why the gentle, tame, and strong moose has never been domesticated like the reindeer. Indeed, such attempts have been made.

About ten years ago, a biologist, Dr. Peter Krott who then resided in Finland, tried to ride two bottle-raised young moose and drive them

Moose deer for riding and as draft animals

before a wagon. When they were almost full-grown, they were strapped into a special harness, bridles, and halters without any resistance on the first attempt. This could not be done with any young horse, no matter how tame, since it is more ticklish. Because a metal bit would have become too cold in winter, they had one of rubber which they seemed to enjoy chewing. They did not pay any attention to the empty sleigh behind them. Pussi and Magnus, the two moose, walked around with the harness and sleigh just as if they were on their way to the pasture. There was no way to direct them by pulling the reins or shouting; their hard mouths did not respond to the bit, But, without any problems, they would follow when their master went ahead of them. After twenty minutes, they would browse in the bushes, and after another half hour, they would lay down on the road and remain there for three quarters of an hour to recover from the strain. With the attached sleigh, they could jump over the meter wide creeks or one meter high fences.

But as soon as Mr. Krott put a load on the sleigh, even if it were only branches and greens, the harness rubbed the skin off their bodies, no matter how well or on how many different places it was padded. Later Peter Krott travelled with Pussi and Magnus to several towns in Finland to show off his team. The two large animals would follow their master, who lured them up the ramp onto the truck with a cabbage. There their halters were tied to the front wall and, while the truck was moving, they would lie down quietly. When they turned their heads, their dangerous antlers were only millimeters from Mr. Krott's eyes—proof of how precisely they could control their movements.

When they stayed in their truck overnight in a town and heard their approaching master's familiar voice in the morning, they jumped through the canvas two and a half meters down onto the road. The crowds of people fled in all directions. When Krott called them, they immediately put their muzzles under his arm and followed like dogs. They were not bothered in their shows by the crowds, the many cyclists, cars, and buses. Only at the station were they terrified when a train dashed past them, although after the train was gone, they calmed down and returned without any difficulties to the exhibition grounds. The children were delighted when, amidst the crowd, the animals lay down in the road to rest and ruminate. Even a private plane flown right above their heads did not bother them. Since all attempts to make a useful working team of them were unsuccessful, Pussi and Magnus were given to zoological gardens in Hamburg and Denmark.

In the Soviet Union, they were more successful with training moose for riding and pulling carts. Many prehistoric drawings on rocks show moose tended by humans or hitched to a sleigh, suggesting that the inhabitants of Siberia during the Upper Stone Age kept moose which were later replaced by horses and reindeer.

As early as 1938 in the experimental institute of Serpuchowsk near Moscow, thirteen tame moose were kept who were hitched daily to a sleigh. They would pull one to one and a half cubic meters of logs for fifteen kilometers per hour, once for a record distance of eighty kilometers. Unfortunately, this experimental station was ruined during the German occupation. It was later rebuilt in Siberia in the state nature preserve of Petschora-Iljitschewsky where the moose's main diet grows plentifully. There the moose feed mainly on leaves and branches, and only in the summer will they also eat grass and mushrooms. In some mountain forests in winter, they eat mainly fir needles. The moose easily adjusts to totally different environments. In the summer, they migrate to the coast to escape from the horseflies; they can also be found on the steppe of Kazakhstan in sunflower fields. The moose in the Siberian breeding farm were not afraid of predators and dogs. A large male moose, without even using his antlers, smashed a bear's skull with blows from his front legs. Apparently the antlers are weapons mainly for intraspecific tournaments in the rutting season. Young moose may be killed by bears. In 1962, two bears invaded the moose farm in broad daylight and killed two young moose in sight of the keepers.

From 1946 until 1948, moose were first captured when one to three days old for the moose farm. They immediately drank cow's milk from the bottle. Such animals became extremely tame and attached to humans. The stomach of a young moose can take one and a half liters of milk, and during one day it drinks up to two liters. A female moose in the wild gives an average of 150 liters of milk per year. Due to regular milking, the lactation period may be extended from four to six months. After six years, Maika, the best female, gave almost six liters daily, 402 liters per year.

Already at an age of two to three months, the animals are put into a halter and attached to a post with a rope. Within a few weeks, they learn how to walk on a leash. It is important that they not only obey the familiar person's voice, but also that they respond to the sound of a horn. Otherwise, they would follow only the people they know; with the horn anyone can make them follow. With the horn signal, the herd follows the herdsman to the farm where each of the animals receives several kilograms of potatoes or beets as a reward. The young moose when used to the leash will follow its keeper around as it would its mother in the wild. It is more difficult with pulling, when the animal is expected to walk in front of the person. The easiest method to start this training is during the trip from the pasture to the farm. There the animals walk the familiar road without needing the keeper in front of them.

As soon as the moose have reached a weight of 130 to 150 kilograms, which may be by the age of six months, they are trained to carry loads

Experimental stations in the Soviet Union

and to pull loaded carts. The most difficult moment, as it is for horses, comes when the animals are supposed to leave their herd. In the second winter, the sleighs will be loaded with up to 150 kilograms, and the animals walk over eight to ten kilometers at a speed of eight kilometers per hour. At three years, their training is completed and they are ready for work. They may be ridden and pull weights of several hundred pounds. Full-grown moose can carry up to 250 pounds, which is one third of their live weight, on their backs. The moose Akwi could pull a sleigh with a load of 1860 kilograms. In contrast to farm horses, the moose are not scared of airplanes; they will bring the sleigh right up to the plane. Neither do they shy from cars and trucks with running engines.

Thus the moose may turn out to be of use for labor in the northern Taiga. It needs no special food supply, as horses, or gasoline as cars. Deep snow, swamp, or fallen trees are no obstacles for it. Even in the Siberian winter it does not need a stable but can lie down on the snow. After a heavy snowfall, only its head will stick out of the snow the next morning. Perhaps, at least in Siberia, people will ride increasingly on mooseback and in sleighs pulled by moose.

Moose deer and wolf

As for the role that the wolf plays in the moose's habitat, the American biologists D. L. Allan and E. L. Mech have done extensive research on Isle Royal. This island, located in the north of Lake Superior, belongs to the state of Michigan. It is about twenty kilometers long and two to six kilometers wide. Formerly it was populated with reindeer which had come from Canada. When the reindeer population disappeared from southern Canada, the migration to the island ceased. Then a small group of moose, who either swam the twenty-five kilometers from the mainland or had migrated over the ice, came to the island. Since these animals were totally protected, their reproduction increased the population so that soon 1000 to 3000 moose were on the island. They gradually destroyed the forest, and since there was not sufficient food, parasites and disease overtook the moose. The game biologists and conservation authorities were deeply concerned about the situation. Then wolves moved in over the ice between Ontario and Isle Royale, where they found bountiful food in the moose. Since the island is a national park, they were not pursued. The present sixteen or eighteen wolves reduced the population to a rate of zero growth. In 1958, a ten year research program on the relations between moose and wolves was begun. It showed that wolves established a biological balance. The records showed that the remaining moose population of 600 head had 250 young per year, the same as the number of moose killed by the wolves. Mostly they were either the weak, the young, or the older animals over five or six years of age. The strongest moose in their prime could escape the wolves by retreating into the water where the predators were so occupied by swimming that they were not a threat any more. Many of them did not flee at all but defended themselves against their pursuers. In all cases, the wolves stopped their pursuit as soon as they met a healthy, strong moose. Only weak animals fell victim to them. Thus, the wolves destroyed the sick moose of the population; the consequence was that the island now supports about 600 moose, who are in good health and have sufficient food.

The REINDEER (subfamily Rangiferinae) differ from all the other deer because their females also have antlers. There is only one species, the reindeer *(Rangifer tarandus)*. The HRL is 130-220 cm, the TL is 7-20 cm, the BH is 80-150 cm, and the weight is 60-315 kg. The ♂♂ are considerably larger and heavier than the ♀♀. The animal has a long head and rump, rather long legs (shorter only in Spitsbergen reindeer), and a short tail. The short bone structure on the forehead supports a bar of long antlers with many ends. The hooves are very broad, round at the sides, capable of being spread far apart, and connected with a membrane. The long, strong pseudo-claws are located to the sides and touch the ground when the animal walks. The walking feet make clattering sounds. In the ♂♂, an inflatable fist-sized skin pouch on the throat serves as an amplifier. The very dense coat is of a different color in the winter than in the summer. There is a throat mane. The females have four nipples; sometimes there are two more which are then reduced. The gestation period is between 192 and 246 days; usually one young is born, although two, or in rare cases, three or four, may occur. The nursing period lasts five to six months. They live twelve to fifteen years, although there are rare cases of reindeer living twenty and even more years. There are twenty subspecies, among them the NORTH EUROPEAN REINDEER *(Rangifer tarandus tarandus;* Color plate, p. 224), the only deer which has actually become domesticated; the SPITSBERGEN REINDEER *(Rangifer tarandus platyrhynchus);* the EURASIAN TUNDRA REINDEER *(Rangifer tarandus sibiricus);* the WEST CANADIAN WOODLAND REINDEER or CARIBOU *(Rangifer tarandus caribou),* whose ♂♂ sometimes have a pouch-like fur at the throat; and the BARREN GROUND CARIBOU *(Rangifer tarandus arcticus;* Color plate, p. 224), the northernmost American form. Some of the woodland reindeer grow into giant forms, and some tundra subspecies are dwarf forms. Three subspecies have been destroyed and six more are seriously endangered.

The reindeer inhabits the far north of the earth in a belt around the pole. Its distribution ranges over the tundras and the northern woodlands of Europe to Asia and America. In this region, the topsoil thaws for only a few months and the water does not drain away but creates swamps. The reindeer's feet are well adapted to these environmental conditions. The toes may be spread far apart, the hooves are broad, and the pseudo-claws touch the ground and help carry the animal's weight. This prevents it from sinking and facilitates locomotion in the snow and the swamp. When the animal walks, a strange clatter is

Subfamily: reindeer, by L. Heck

Fig. 8-31. Reindeer *(Rangifer tarandus).* In the dark-shaded regions, wild reindeers are extinct.

audible from the tarsal joints. An earlier explanation was that the hooves clicked together, but Erna Mohr discovered that this sound is caused by a movement of the ligaments. This clatter is also found in Père David's deer.

Naturally, with such a wide distribution various forms of reindeer have evolved. For a long time zoologists classified two species, the woodland reindeer and the barren ground caribou. The woodland forms, which are larger than the barren ground caribou, often have a darker coat. They also differ from the steppe types in their behavior. But presently all woodland, barren ground, and island forms are included in one species.

The reindeer's antlers are especially conspicuous. They differ from all other deer antlers by their weak base, the flat bars, the light color, and the asymmetrical shape. The tines are along the arched bars and the first and second spiller grow closely together.

The anteocular branch, which begins near the base only at one bar, grows into a flat surface in the adult deer. It reaches down to the nose and forms a kind of protective shield over the skull. Formerly this formation had been called a "snow shovel" because people thought that the animal would remove the snow with it to reach their main diet, the reindeer lichen. But many observers have reported that the reindeer scratch the snow from their feeding grounds with their front legs. The protective shield may be handy in the male reindeer fights in the wild, which are often hard and vehement. So far there is no explanation as to why female reindeer have antlers.

Their barren habitat forces the reindeer into seasonal or even year-round migrations over very long distances. The barren ground caribous may cover distances of up to 1000 kilometers. There are only a few stationary forms, some in Newfoundland. During the summer at noon, the reindeer would roam high in the mountains where they were not bothered by the millions of mosquitoes which inhabit the plains of the tundra. In the fall they migrate to lower elevations. Depending on the habitat, the rutting season is from August to early November. After lively fights, the males gather groups of ten or twelve females. In the winter the herds move to the valleys where the snow is not frozen. On their migration, these excellent swimmers not only cross large streams but also wide sea channels. Since they are adapted to life in the coldest regions where other ruminants could not find food, their number has greatly increased in some places. At least this was true until man with his modern firearms came along and destroyed them in many areas of their original habitat.

Wolf Herre reports on the reindeer's habitat in the far north: "It is dark here for many months in the tundra, and there is snow, ice, and cold weather. Then comes a short period of light with a twenty-four hour day and pleasant temperatures. An abundance of plant food is

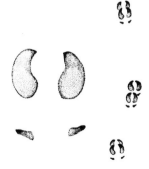

Fig. 8-32.
On the right are the tracks, and on the left is one single footprint of reindeer. Behind the imprints of the hoofs, the psuedo-claws are clearly visible. They prevent the animal from sinking into the snow or swamp.

provided by the 760 species of flowers, the 330 species of moss, and the 250 species of lichen. Suddenly freezing temperatures are back again, followed by snow, and the landscape is thickly covered. Now only those animals survive in the tundra which can tolerate darkness and cold, are able to find food and can live on a frozen diet. This is the reindeer's secret: they can do all these things. With their front legs, they paw narrow paths in the snow which may often reach up to their bellies. Under the snow, they find their lichen, many other kinds of brush, and grass. The reindeer eat a variety of food in the winter; they pass up only moss. Long migrations are necessary in order to find enough of the appropriate food."

During the Pleistocene of the Glacial Periods, the reindeer, which were widely distributed in Europe, were the most important game animal to the people of that era. Not only did they supply meat and skins, they also furnished antlers for the production of various tools and art objects. From this time, which prehistorians call the "reindeer era," countless pictures of reindeer are preserved. On cave walls in Dordogne (France), 121 engraved pictures of reindeer were counted. In all these artistically decorated cave dwellings made by the people from the late Glacial Period, drawings or carved figurines of these northern deer have been found. For the hunters of the Lower Stone Age, who were following the receding ice from the then barren Central Europe, the reindeer were an important game animal. Now the large herds of wild reindeer are of equal importance to the peoples and tribes in North Asia and northernmost North America. Since Indians and Eskimos have modern firearms, their hunting has greatly reduced the population of the North American caribou.

The reindeer in the Glacial Period

"But part of the success in hunting depends on chance," writes Wolf Herre. "People learned how to improve their severe living conditions by domesticating the reindeer. Presently, large herds of reindeer are still the essential basis of existence of people in the far Northeast of Eurasia." Often reindeer are the only wealth of the northern nomads. They wear reindeer coats with the hair outside, which gives them almost complete protection against the cold. The female reindeer's milk is processed into cheese and butter. From the reindeer's antlers and bones, the inhabitants of the polar regions manufacture knives, scrapers, and other household utensils.

Reindeer as domestic animals

During their year round migration, the reindeer nomads live in pointed tents of reindeer skin or in small, quickly established huts, the roofs of which are made of soil. They must continuously follow the herds and cannot settle in one place. The migratory drive, which is an inherent part of reindeer behavior since their early existence, could not be changed by domestication. Therefore, the breeders have adjusted to the animal's habits since it was the only possible solution. Other species of deer are generally not well suited for domestication, except for the moose (see p. 234). The male deer may become danger-ous, especially in the rut. Therefore, according to Herre, all male rein-deer are castrated right after pubescence. "Only the young males, which are gentler and have smaller antlers, are used in reproduction. This apparently insignificant procedure had distinct effects on the reindeer's social structure. Instead of small groups, large herds could live together and all the known variations from the "classically" do-mesticated animals were found in the "naturally" living reindeer. There are large numbers of black and white, brown and gray, piebalds and spotted domestic reindeer in addition to those of natural color. There are short legged individuals, which have high meat yields and bulky antlers, and there are long legged, slimmer animals with long, high reaching antlers. Even white reindeer with blue eyes may be found."

In order to study the habits and the keeping of reindeer, I stayed with them in Central Sweden in mid-winter. I walked along the Fjäll for hours before I saw the first herd. A Laplander on skis appeared and, a few meters behind him, was a reindeer. This was the leader, and then, following like sheep, came a huge herd, head to head. Their antlers looked like a forest. Thousands of animals, a huge gray crowd on the white snow, came in a fast trot. They avoided us by making a semicircle around us and they had passed within minutes. Bringing up the rear were two Laplanders with two small, black dogs who, responding to a signal, drove the herd with their barks. I followed behind the herd on my skis and admired the graceful movements as the animals easily trotted through the snow. Some groups tried to stand still and to feed on the dwarf birch, but the dogs quickly made them move on. All of a sudden five animals took off to the side and disap-peared between the birches. A Laplander with a dog dashed after them, but returned after ten minutes. He had not been able to catch up with the animals, which had become too shy. He expected they would be back the same or the next day, although this is not always the case. As a consequence feral herds of reindeer form and behave like genuine wild reindeer.

In the rutting season in September, the Laplanders drive their herds together in one place. Afterwards, the owners pick out their animals, which are marked with different ear notches, from the strange herds. This is a difficult job which often takes several days. I saw a Laplander, after a long search, catch several marked reindeer with a rope which he threw over their antlers from a distance of ten meters. The same procedure was used for slaughtering. The captured animal was drawn near and killed with a quick stab in the neck. Other members of the groups stood a few meters off and watched the slaughtering without shying. The dead reindeer were cut up immediately. The Laplanders hung the meat in the birch trees and it was soon frozen solid.

It is very exciting to see the hitching of reindeer and their habituation to the sleigh. In the evening, my Lapland host caught a strong deer with huge antlers with his rope. He attached a boat-shaped sleigh with only one rope to the reindeer's shoulder strap and two Laplanders held the animal for hitching. Releasing the animal as soon as we sat down in the sleigh, off we went with the deer dashing off. We raced downhill along the narrow path between the birch trees so snow splashed into my watering eyes. At one place we approached a narrow bridge over a creek in a gorge. Until the last moment I was not sure whether the reindeer would actually cross over the bridge or fall into the gorge. The Laplander pulled several times hard on the rope in order to keep the animal in the middle of the road, but it seemed to me to be mere chance that the reindeer finally went over the narrow bridge. Suddenly, right behind the bridge, the animal jumped aside and our sleigh turned over in the snow. The Laplander quickly held onto the animal, we jumped back into the sleigh, and we went on, all within seconds.

The reindeer as a draft animal

The Canadian wild reindeer, the caribous, which become rarer all the time, have been famous for their migrations. In 1947 Farley Mowat, a Canadian, witnessed such a giant march: "At first I saw only a moving line. It seemed as if the slopes were cautiously approaching the bay, as if countless boulders and rocks were growing out of the hills and were suddenly gliding downhill over the icy surface in slow motion. Then the immensely slow avalanche reached the far edge of the bay and began to spread over the lake. In undulating or ragged lines, in bunches, singly or in broad formation, the caribous poured onto the ice of the lake finally moving northward on a mile-wide front. The long lines changed into endless columns of single reindeer who stayed one behind the other, as precisely as pearls on a string. Now and then a young was walking beside its mother, who was carrying the heavy burden of a new life within her again. There were no males among all these pregnant animals who were driven by an irresistable instinct north to the wide Barren Grounds, which are the reindeer's calving grounds.

Migrations of the caribou

While the leaders of the herd now reached our edge of the lake and began to ascend the slopes, the avalanche of bodies moving across the bay was still increasing. For ten kilometers to the east and the west, the surface of the bay had turned into an undulating sea of animals without any end in sight. The animals climbed without haste and without rest, without conscious will except for their instinct, down to the ice of the bay, one following in the other's tracks.

Now the herds began to crowd around our observation station, ten steps away from us, now only five—and then we had to get up and wave our arms so as not to be trampled to death under their hooves. The animals, who looked at us calmly and without curiosity, moved one meter to the side and continued north, straight north, without hesitating for a moment. Hours passed like minutes while crowds of animals moved past us without interruption and in undiminished numbers until finally the sun sunk deep beyond the far horizon."

After several days, the females were followed by the migrating males. Some laggards could be found on the route even weeks later. Canadian and North American scientists found that the caribous did not always use the same routes like the huge herds of bison once did. All they know is that in certain seasons and certain areas huge numbers suddenly appeared and disappeared some months later in the same manner. Gradually the many contradictory reports merged in an approximately correct picture. Most of the large herds migrate in the summer to the barren grounds of the far north and they return in the fall to spend the winters in the cover of the immense Arctic woodlands of the south.

Decrease of the caribou

But the era of the large herds, which had inspired with awe the first French explorers of Canada, are now over. The population of the barren ground caribou *(Rangifer tarandus arcticus)*, which was distributed from the west of Hudson Bay to Alaska, is being reduced at a frightening speed. In the winter of 1947/1948, 600,000 barren ground caribou were still said to be alive; in the winter of 1954/1955, there were about 300,000, and presently there may be only some few ten thousands of them left. For the caribou-Eskimos and the Indians whose life and existence depended on this species of deer, a time of hardship and great need began. Until then, the caribous on annual migrations had been an easy prey for these tribes. The herds could be expected at specific and suitable places. But when these migrations ceased, whole tribes died of starvation. In some areas where twenty to thirty homes had existed, only one remained. The exorbitant destruction of the caribou had caused a catastrophe.

Experiments to introduce reindeer

In order to replace the caribous, tame reindeer from Norway were imported and brought in a large trek to northern North America. The Eskimos in the interior had to learn how to handle tame reindeer. In his book *The Trek of the Reindeer*, Canadian Allen Roy Evans describes the difficulties of driving 3000 tame reindeer from the Alaskan coast to the estuary of the Mackenzie River. Together with a few men and women, a Laplander set out around Christmas from the Buckland Bay in Alaska to drive the animals over 3000 kilometers of Arctic land into this area of permanent famine east of the Mackenzie Territory. The trip lasted five years. Some of the drivers died; even murders occurred. Along the path were the graves of men and the carcasses of reindeer. A number of the animals finally reached their destination. But the project to free the natives from starvation was only partly successful.

In 1963, Bernhard Grzimek visited these reindeer herds in the area of the Mackenzie estuary near the town if Inuvik. They were still in a state farm under the supervision of a European. So far, it had not been possible to train the Eskimos and the Indians to take care of the reindeer as a domestic animal and to tame them like the Laplanders do in Europe. The wild barren ground caribous, which have supported these tribes keep decreasing in number. In spite of all government efforts, it was not possible to replace them with domestic animals.

Even though the reindeer, especially the domestic forms, are displayed in many zoos, it is not that easy to keep them in good health and to breed them over a longer period. In zoos they rarely become more than ten years old. According to the records of Jarvis and Morris in the London Zoo, 33% of twenty-seven reindeer lived up to twelve months, and the others to an average of fifty-seven months. The oldest reached an age of nine and a half years. In Nürnberg and Berlin, reindeer have become twelve years old; one animal in Basel reached almost fourteen years.

Alfred Seitz reports: "Animals imported from their original habitat frequently are infested with all kinds of parasites of the stomach, lungs, liver, and with the larvae of different types of horseflies. Therefore, these animals are in a debilitated condition and cannot be saved. Even reindeer without parasites often lack vitamin A. They do not get their natural food in the zoos. Reindeer in their habitat eat quantities of green plants, certain grasses and herbs, and willow and birch leaves. Even in winter they may eat the green plants which are preserved by snow." They eat lichen, which, according to W. Herre, comprises only about fifty per cent of the total diet. Lichen may be easily replaced by good hay from mountainous areas. Lucerne, grass, bran, pressed cereals, crushed oats, crackers, and large quantities of leaves from willow, birch, and oak are added to the hay.

For eleven years all the reindeer in the Nürnberg Zoo have been injected with vitamins A, D_3, and E twice a year. Perhaps such a supplement may increase the number of long-living, healthy animals in the zoos. The vitamins proved to be especially valuable to debilitated and infested reindeer. Furthermore, they improved breeding results. It was very rare that a female reindeer in a zoo had more than five young during her life. All larger numbers are exceptions. In Copenhagen, a female reindeer had a total of eight young while another one in Nürnberg had seven.

Even more difficult is breeding the caribou from the far north. In 1961, in the Assiniboine Park Zoo of Winnipeg, Canada, a four week old East Canadian woodland reindeer *(Rangifer tarandus sylvestris)* was bottle raised. The zoo's director, Dr. Günter Voss, reports on the matter: "Irk, as we named the little animal, received at first a compound of a powder with the essential parts of dog's milk, enriched with vitamins and minerals. In the beginning, the animal drank six times per day; this decreased later on. Right from the beginning, Irk nibbled at solid food which consisted of lucerne, cattle feed, oats, chopped

Keeping reindeer in the zoo

apples and carrots, and branches with willow, birch, poplar, hazel and, in fall, dry leaves. Irk developed beautifully without any lichen. Occasionally I gave him some dog biscuits which he liked to eat. Irk never had an enclosed shed. Sometimes we turned him loose in the children's zoo. Then he would immediately leap about in a most unpredictable manner. He could play beautifully all by himself as soon as he had enough space for this kind of game. When we later put him with some young moose and a young elk, he jumped around with them so wildly that he even lost some weight."

Even though the reindeer's antlers start to grow after the first year, by the age of four and a half months Irk's antlers were thirteen to fourteen centimeters long. His coat was, according to Voss, "of a fabulous softness and density, and shining like silk, simply beautiful." Caribous are very seldom seen in zoos. In 1967, only the St. Louis Zoo, the Catskill Game Farm near New York, and the Albert Game Farm in Canada, where Irk now is living, had some.

Bernhard Grzimek
Lutz Heck
and contributors

9 Giraffes

Presently, the giraffe family includes only two genera with one species each, and their distribution is restricted to Africa south of the Sahara Desert. Long ago the giraffes were a widely distributed family of even-toed ungulates with a multitude of forms. They evolved relatively late, probably about twenty-five million years ago in the Lower Miocene, from a group of ungulates which included the European genera of *Lagomeryx, Procervulus,* and *Climacoceras,* which had a set of teeth like the deer's. These Lagomerycidae (family Lagomerycidae) had forked, branched, or flat tined, bar-shaped forehead bone structures which are like a deer's antlers, but which presumably were permanently covered with skin and never changed. This would suggest a European-Asiatic origin for the giraffe if it were not for recently found fossils from North Africa *(Prolibytherium)* which indicate an African origin.

The geologically oldest and most primitive forms, which were distinctly classified as giraffes, were the PRIMITIVE GIRAFFES (Paleotraginae) of the Middle and Upper Miocene of Eurasia. They were giraffes with a short neck, about the size of red deer, with paired forehead structures covered with skin. They still had many deer characteristics. It seems to be doubtful whether the contemporary okapi *(Okapia johnstoni)* is only a slightly different, direct descendent of these Paleotraginae, as is the opinion of some scientists. Therefore, it seems to be more correct to classify them as a separate subfamily, the Okapiinae.

Besides the genera *Palaotragus, Giraffokeryx Orasius,* and *Samotragus,* which belong to the original giraffes of the Lower Pliocene (approximately ten million years ago), genuine steppe-giraffes of the subfamily Giraffinae appear which were doubtlessly derived from *Paleotragus* or closely related forms and which had a distribution from West Europe to East Asia. In addition to the extinct genera of *Honanotherium* and *Decennatherium,* the genuine long-necked giraffe of the contemporary genus *Giraffa* was also included.

Phylogeny, by
E. Thenius

Fig. 9-1.
The Sivatherium, a giant short-necked giraffe with a cattle-like physique and fur-covered palmate antlers, lived during the glacial period in south Asia.

Fig. 9-2.
Paleotragus, an okapi-like
short-necked giraffe
which existed more than
ten million years ago, was
close to the origin of all
giraffes.

Subfamily: forest
giraffes

The okapi, by
G. Grzimek

A third evolutionary line of giraffes became extinct in the Pleis-tocene (Glacial Period). The Sivatheriinae were giant, short-necked giraffes with an oxen-like rump and normally developed, sturdy legs, like the *Helladotherium* from the Lower Pliocene of Greece. Some of these giant animals had oddly-shaped, huge forehead protrusions. They were conically-blunt in *Brahmatherium*, but had flat, enlarged surfaces in *Sivatherium*.

The present-day GIRAFFES (family Giraffidae) are large to very large ruminants. The L is about 2.5–5 m, with a height at the crown of 1.8–5.8 m. Head is small, long, with a narrow snout. The very mobile upper lip overlaps the hairy lower lip. There may or may not be a small, naked muzzle. The long flat nostrils can be closed. The eyes are large or medium-sized with long lashes. On the forehead and the crown there are two to five bony knobs covered with skin. The long, sleek, and very mobile tongue is adapted to grasp leaves (its main diet). The neck is moderately long (okapi) or very long (giraffe). The legs are long or very long with the front legs longer than the hind legs and a declining back line. The medium-long tail has a tufted tassel. The psuedo-claws are missing except for tiny rudimentary bones at the end of metacarpal and metatarsal. The hooves are wide and low. The thick skin has an even hair growth. In the giraffe, the flexure joints have bare callouses. There is a bristled short mane in the giraffes and in young okapis. As an ambler, there is no extensive tissue between the rump and the thighs. The animal has an ambling walk. There are interdigital glands only in the okapi, there are no other skin glands. The females have two to four nipples. There are 32 teeth: $\frac{0 \cdot 0 \cdot 3 \cdot 3}{3 \cdot 1 \cdot 3 \cdot 3}$, the premolars and molars have low crowns. There is no gall bladder in the okapi; it never occurs in the giraffes, but all have an appendix. There are two subfamilies with one species each.

The FOREST GIRAFFE (subfamily Okapiinae) has only the species OKAPI (*Okapia johnstoni,* Color plate, p. 262). The HRL is approximately 2.10 m, the TL is approximately 30–40 cm, the BH is approximately 1.50–1.70 m, and the weight is about 250 kg. The neck is only moderately long, and the back line declines only slightly. The ♂♂ have two horns, and the ♀♀ are hornless. The points of the horns are without skin or with a small horn sheath which is changed. The ears are very large and the eyes are medium-sized. The gestation period is fourteen to fifteen months (426–457 days) and the nursing period is six months. There are no subspecies.

Around 1890 Henry Stanley penetrated the unexplored prime forests of the Congo and met pigmies for the first time. In his famous report, he mentions that these forest dwellers were not surprised at his horses, they claimed that they could catch similar animals in covered pits. A few years later, Johnston, the governor of Uganda, checked on this strange statement. After all, horses are animals of the steppe. Should there really be such a thing as a forest horse? In his Congo expedition of 1899, he asked the pygmies about the fabulous large animal. The little people, as well as Lloyd, the Archdeacon of a missionary station, who was the first white man to see the animal, confirmed Stanley's report. They told him that they called it "okhapi," that it was a pale brown or dark gray color on the upper part of the body, with stripes on the belly and the legs. The Belgians in Fort Beni told him that they had a skin somewhere which he might have. Unfortunately, the black soldiers had already cut it into pieces for belts and cartridge pouches. Only two pieces could be found and Johnston took these with him. He then sent them to the Royal Zoological Society in London. It was evident that they could not belong to any of the familiar zebra species. Therefore, in December of 1900 Sclater gave notice of a new species of animal which was given the scientific name "Equus (?) johnstoni," which literally means "Johnston's horse." The question of whether it was actually a horse-like animal was answered a half year later. In June, 1901, a complete skin and two skulls of this animal arrived in London. The friendly Belgians of Beni had kept their promise to send these trophies to Sir Johnston as soon as possible. The skull bones gave evidence that the animal certainly was not related to a horse or a donkey, nor to antelopes, oxen, or any other of the contemporary animals. However, there was a startling resemblance to the extinct, short-necked giraffe *Helladotherium* which had lived more than ten million years ago in Europe and Asia. The zoologists had to classify the okapi into a genus of its own; "Johnston's horse" was named "Okapia johnstoni."

Fig. 9-3.
Okapi *(Okapia johnstoni).*

It was one of the first great sensations of the new century for the daily newspapers all over the world. Such a large mammal with a conspicuous pattern and shape had been unknown for so long to the zoologists. The big hunters competed with each other to be the first white man to kill this shy game, and the leading museums searched for skeletons and skins. Duke Adolf Friedrich zu Mecklenburg, then a famous African traveller and hunter, hoped in 1907/1908 to shoot one of these animals; however, he did not succeed. He only brought home five skins and a skeleton which he had received in trading with the pygmies. Similar things happened to many others who tried.

Fig. 9-4.
A young male okapi uses his nose to push his father's elbow, thus inviting him to a fighting game.

For many years an expedition under the supervision of Herbert Lang struggled to bring live Okapis to the New York Bronx Zoo. But the okapi still remained concealed from the white man. Lang and his crew did not reach their goal. The first okapi was captured near the Ituri River; it was probably a very young animal and lived for only a short while with a Belgian officer. When Lang in 1909 was lucky enough to capture another young animal, it died because he only had enough condensed milk for four days with him. He could not reach the next village in time with his little okapi.

Fig. 9-5.
He goes into a slightly exaggerated display posture.

The first zoo
experiences

Fig. 9-6.
In a playful neck fight, the younger one "defeats" the much stronger opponent.

Fig. 9-7.
Finally, the two lie down and the young continues to press the elder's neck down.

Fig. 9-8.
The young arises in the display posture, while the old one remains in the submissive posture.

It was in December, 1918, that an okapi was captured which arrived alive at a zoo. Natives seized it when it was only a few days old, and the wife of the district commander of lower Uele raised it with condensed and cow's milk. On August 19, 1919, a government official delivered it to the zoo of Antwerp. Unfortunately, it lived there for only fifty days. The next okapi to reach Europe alive came nine years later. Its destination was London, but it died on a stopover of the transport in Antwerp.

The few others who came to Europe in the following years also died soon. The female "Tele," who came to Antwerp in 1928, was the first one to adjust to the zoo. She lived there for more than fifteen years and died of a lack of food during the war in 1943. The male "Congo" lived in New York only a few days less, 15 years and 33 days, until he had to be killed because of irreparable joint damage. He was the first specimen of his species in America. The scientific assistant of the Antwerp Zoo, Dr. Agatha Gijzen, reports that during the first forty years of zoo experience with okapis, thirty males and eighteen females left Africa. Nineteen of them did not survive either the transport or the first year in a zoo. In the beginning of 1966, there were twenty-nine males and twenty-four females in eighteen different zoos. Thirteen of these males and twelve of the females had been born in a zoo.

Efforts to bring okapis alive out of Africa were initially discouraging. The animals were captured more carefully and cautiously than any other game animal ever was. In spite of this, they died from hardships en route in the first decades of our century. From the place of the capture and the acclimatization station, they had to travel for hundreds of kilometers to the Congo River in boxes, and then they travelled for about three weeks on river boats to the capital in Leopoldville, now Kinshasa. They were transferred to trains, and finally they had to wait in the port of Matadi for many days until a boat arrived to take them on the long voyage to Europe or America. Of the ten okapis which were transported in 1949 from the area where they had been captured, only five arrived alive in Leopoldville. Of the four remaining animals who boarded the boat, two died there and another one died in Basel two weeks after its arrival. As soon as okapis could be shipped by plane, the transportation losses decreased. The first okapi to travel by plane came to Copenhagen in 1948. The first okapi to come to a German zoo arrived in 1954 in Frankfurt and was transported in a large passenger plane which had been chartered specifically for this purpose. Since considerable difficulties were encountered on this first air trip, I constructed a transport bow whose weight could be adjusted so that the box did not need to be tipped over when passing through the door of the plane. This method of air transport has been used for okapis ever since.

Since 1933 okapis have been fully protected by law in the Congo;

the former Belgian Colonial Administration gave them exclusively to zoos of scientific importance. The Congo has, so to speak, a "monopoly" on okapis, although the possibility of isolated specimens existing in the adjoining northern Ubangi forest of Equatorial Africa cannot entirely be ruled out. The distribution of this species is approximately as large as the area of Switzerland. No one is able to estimate the actual number of animals living there. They may well occur more frequently than was anticipated some decades ago. The number of tracks and skins which were found with the natives lead to this conclusion. A hunting official from the Ituri forest told me that there was one pair per square kilometer. This would add up to a total of several tens of thousands of specimens.

There are only a few whites who have seen an okapi in the wild. The habitual use by these forest creatures of the same paths had made their capture possible. Pits are dug on their paths. Such a pit is barely deeper than two meters, it tapers inward towards the bottom, and has very smooth walls. It is covered with a dense cover of thin branches with dead leaves on them. When the okapi capturers once showed me such a pit from a very close distance, I could not see anything; the dead leaves blended so well into the surroundings. Antelopes or even small duikers would not have trouble jumping out of these holes, but the okapis do not seem to be able to jump. During Belgian colonial times, groups which were to capture okapis used to have about 200 such pits on a distance of about 60 kilometers. Each of the pits had to be checked daily which kept twenty-five assistants busy. When an okapi had fallen into the trap, the discoverer first removed the branches and carefully covered the opening so that the animal would calm down. Then he went to get the other people. They quickly put up a two meter high fence woven from cut branches and lianas around the hole, so that the animal had no chance to escape. Then a troop of twenty laborers arrived who immediately started to build huts to accommodate them for the following weeks, since they would be busy here with the animal for quite some time. Close to the hole a round area of about thirty meters in diameter is enclosed with another woven fence. This fence is densely covered with green twigs. Then they made a runway between the pit and this pen which is also camouflaged with fresh greens. An African then crawled very cautiously next to the pit and began to chip down the rim of the pit, filling it with earth under the okapi's front legs, thus making a ramp which would enable the animal, sooner or later, to climb out. It would be captured when it went through the runway and into the pen, although it still was in its natural environment; the fence looked like a dense, green plant cover. The captors' work, however, continued. Only a few meters from this pen, connected to it by another gangway, they built and camouflaged another enclosure. The animal could then be driven from one corral into the other

How okapis are captured

without ever seeing humans, and one of the pens could be cleaned thoroughly and freshly covered with greens. In case the okapi received bruises or injuries from the capture, the wounds were dabbed with a swab of cotton on the end of a long pole. No caustics could be used because the okapi can reach each part of his body with his long, blue-black tongue.

Acclimatization at the capture station

While the new prisoner slowly became accustomed to people, a long, narrow gangway was built, which led over a distance of more than one kilometer to a road which could be reached by truck. There the runway ended on another ramp of the same level as a truck's platform. The truck backed up to it and a transport box was pushed to the end of the ramp. The box was also camouflaged. The okapi was not driven into the box, but rather was allowed to walk into it by itself. Once in the narrow corridor, where it could not turn around, it kept on going until it reached a dead end, which was the transport box. Then a sliding gate was let down behind him. The animal was then brought to another station for further training where it left the transport box in the same manner over a runway into a pen where it was to live for some time. So far it had not even been touched by a human hand.

Beginning in 1946, the excellent game authorities from the Belgian Congo in the area of the upper Uele managed the controlled capture of okapis in the Ituri forest. While formerly only young okapis were sporadically brought in by pygmies, now a perfect organization for the capture and the model taming compound of Epulu originated under the capable supervision of Mr. de Medina. When I visited there in 1954 to pick up the first of the Frankfurt okapis, fifteen okapis were living in the station pens. After the founding of the Congo Republic and the following internal turmoils, the last eight of the captured okapis were brought to Kinsangani and the station was abandoned. The Minister of Agriculture of the Congo Republic told me that the buildings and the equipment of Epulu had been destroyed during the last phase of the Civil War. We were pleased to learn in 1967 that rebuilding of the station had been begun and that the first two captured okapis had been sent to the Kinshasa Zoo.

Research on their habits in the wild

Very little is known so far about the life of the forest-type giraffe in the wild, despite the fact that many okapis had been sighted there. In contrast to the steppe, it is hardly possible to perform uninterrupted observations of animals in the tropical rain forests. In spite of the lush vegetation, the dense woods appear to be empty to the traveller or explorer. All the active life takes place high above ground level in the leaves of the giant prime forest tree tops. The few larger mammals, like the Red River hog, the bongo, and the okapi, rarely get within sight of a human. They can hear the approach of a man long in advance and they disappear in the shadow of the woods. The brown color and the pattern of stripes camouflage the okapi so well that it blends into its environment at a distance of twenty-five meters. The ground is covered with brownish, rotting leaves. Only in a place where a fallen tree has created a clearing can impenetrable brush grow. The larger clearings and the bush regions near the water are the actual habitat (biotop) of the okapi. In regions with a compact roof of green leaves, only low plant growth is possible and so the okapi lacks food.

In studies on its life, we were able to find the most information about its feeding habits. The places where it had eaten branches and its excrements could be analyzed in its tracks. The okapi feeds predominantly on the young shoots of the prime forest bushes. The long tongue grips the branches, pulls them close, and then the leaves are stripped off. De Medina has collected all of the feed plants of the okapi. About thirty species of thirteen plant families could be classified. It is amazing that among the food plants are very many *Euphorbia*. Okapis are also said to eat a kind of forest grass. The analysis of the excrements showed that the okapi also eats charcoal from trees burnt by lightning and that its need for mineral salts is satisfied by a sulphurous, slightly saline red clay which is to be found in the vicinity of water.

Okapis eat leaves

The habitat and the natural food which may be available are the most important considerations when keeping them in a zoo. In contrast to the accommodation of steppe-type animals, the okapis need a quiet pen with lots of shade. They have brooms of brushwood to slip through where they may scrub themselves for their grooming in the Frankfurt Zoo. In all leaf-eating animals, adaptation to another diet is difficult. Only during the summer months can we daily offer them bundles of freshly cut branches. In the winter, apples, bananas, cabbage, and onions are substituted for the greens. Although they are used to a diet of fresh leaves, the okapis have come to accept clover, lucerne hay, and air-dried branches.

After the okapis are exposed to the psychic and physical stress of transportation, they are exposed to still another hazard during the period of adjustment to the new environment and the adaptation to different food. The parasites become abundant. On the trip they may become infested with trypanosomes, which are dangerous blood parasites. Even worse is the infection by intestinal and liver worms. Most okapis which survived transportation to a zoo or the first year in the zoo eventually died of worms. All intestinal worms deposit a large quantity of eggs daily. In the wild, chances are rather slim that an okapi will eat his excrements so that it will develop in its body into a reproductive parasite. However, in the transport box the animal is constantly exposed to its excrements and, even later in the zoo, it can hardly be avoided. So the animal is frequently endangered by worm's eggs in captivity. From the eggs more parasites develop which then lay eggs, and this leads to an increased concentration and infestation of the host. In freshly imported okapis, more than four thousand eggs of different species of worms were counted in one gram of excrement. This means that the animal daily disposes of more than five million worms' eggs with its feces. More than thirty species of worms are known to live in the okapi. Many, but unfortunately not all of them, can be treated with medication in a zoo. The hookworm *Monodontella* is especially hazardous and difficult to treat. From its eggs, which are eliminated with the excrement, hatch larvae in the moisture of soil with sufficient warmth which, after approximately five days when they are less than one millimeter long, are able to penetrate the okapi's skin. On its way through the animal, the larvae keeps growing and settles when fully grown in the bile duct of the liver. In case of a heavy infestation, masses of *Monodontella* may completely block the bile ducts, which will cause the death of many imported okapis, as Mrs. Gijzen found out. Scrupulous cleanliness and the removal of all excrements are among the most important rules for keeping animals in a zoo. Since the hazardous larvae are able to survive in the slightest cracks in the ground, the usual stall cleaning is not good enough. The larvae are sensitive to dryness; therefore, a powerful heating system was installed in the floor of the indoor okapi pens of the Frankfurt Zoo. Every day after the floor has been washed, this was turned on to remove any moisture. Only then may the okapis return to their room.

Okapis are probably loners. Only in the mating season does the female, who may be accompanied by her latest young, associate with a male. It is improbable that the pair find each other by loud calling. It is more likely that they are guided by their well developed sense of smell. During the mating season in the zoo, the female utters an audible sound which is best described as "a soft, slight cough." Mating and copulation, which F. Walther studied thoroughly in the Frankfurt Zoo, is like the behavior of many species of antelopes. The two sniffing and licking animals circle around each other. Shortly after this, the male displays his superiority by placing his neck in an erect position, throwing his head upwards, and kicking forward with one front leg. Finally, the male drives in from behind and mounts. Soon after copulating the animals separate. In free ranging animals, according to de Medina, the pregnant females retreat into the dense forest to give birth. The young animal, which remains hidden in the forest for several days, is visited by the mother only after it has called. We know from zoo observations that the newborn are anything but mute. They are able to "cough" slightly, as the females do, they can bleat like a calf, and they may utter a whistling sound. When the mother and child are separated, voice contact becomes more frequent. During the first days, the mother also calls frequently.

After the okapis had adjusted to zoological gardens, the most important goal was their reproduction. Some pregnant females had already reached the capture station of Epulu, and had given birth to their young in captivity before reaching Stanleyville. The first birth to take place in captivity was the female "Lorraine," born on April 19, 1941 in Stanleyville, which unfortunately could not be raised after her mother died. It was a long time until the first birth occurred in Europe. The first young zoo okapi was born in 1954 in Antwerp, where the father had lived since 1948 and the mother since 1950. Unfortunately, neither he nor two other newborn could be raised. It was not until 1956 in Paris that the first okapi was conceived, born, and raised in a zoo. Meanwhile, several pairs from Epulu had been given to zoos in Europe and America, and births followed rapidly in Chicago, New York, Rotterdam, Basel, and Frankfurt.

The first okapi born in Germany was the young male "Kiwu" on September 9, 1960 in the Frankfurt zoo. The birth began at 2:00 A.M. and was completed after seventy minutes. The mother took care of her young immediately. Young okapis are born with a narrow black bristled mane from the neck to the back which disappears later. Furthermore, the white coat hair is much longer and smoother than the dark hair, and it hangs in fringes over the dark brown coat. The young Frankfurt okapi drank from his mother for the first time two hours after birth, and then it walked around continuously for almost two hours. It took the Basel female about two hours to give birth. During this time, she made constant vocal contact with the male in the pen next to hers. Parturition took place when the mother was standing, and the young fell into the straw. After less than half an hour, it was standing on its feet. The first nursing lasted more than thirty minutes. When barely six weeks old, young okapis begin to suckle on hay and other kinds of food. After nine months, they are able to feed on their own. The horns, found only on male okapis, in contrast to the giraffe, begin to grow no sooner than the third year. One exception was one of the Frankfurt okapis, the male "Josef," born on June 22, 1966, whose horns began to form soon after his first birthday.

Okapi females defend their young. I could watch quite distinctly in Epulu, where four females with three young were kept together in a pen, how excited the mothers may get. De Medina imitated lowing, the young's call of distress. Immediately all the adult females dashed towards us, threatening with clattering blows of both front legs at the ground and then kicking with their hind legs so that we hastily had to leave the pen. A young's call of distress may also cause the females to attack each other. Then the head is lowered under the opponent's neck and flung upward. The attacker turns slightly to the side shortly before striking the neck; hence the animals do not really touch each other. Okapis lay down when making a submissive posture. A specific imprinting of the young to its mother apparently does not exist. In Epulu I observed a young nursing at two females without being warded off. Older female okapis are said to really adopt another's young.

The first successful reproductions in captivity

Fig. 9-9.
An okapi mother, who believes her young is endangered, threatens another female with her lowered head.

Subfamily: steppe
giraffe

The giraffe, by
G. Grzimek

Distinguishing
characteristics

The second contemporary subfamily of giraffes are the STEPPE-
GIRAFFES (Giraffinae) which has only one species, the GIRAFFE (*Giraffa
camelopardalis*; Color plate, p. 262). The HRL is 3–4 m, the TL is 90–110
cm, the BH is 2.70–3.30 m, the crown height is 4.50–5.80 m, and the
weight is 500–750 kg. The animal has a very long neck and a steeply
sloping back. Both sexes have two to five horns, the tips of which are
rounded and covered with skin. The ears are narrow, pointed, rather
short, and the eyes are very large. The gestation period is from 420–468
days, and the nursing period lasts ten months. The young first take
solid food at three weeks, and first ruminate at four months. Sexual
maturity occurs at three years, and longevity is 20–30 years. There are
eight subspecies: 1. NUBIAN GIRAFFE (*Giraffa camelopardalis camelopardalis*)
has been largely exterminated and is now found only in Southeast
Nubia and West and South Ethiopia. 2. KORDOFAN GIRAFFE (*Giraffa camelopardalis antiquorum*). 3. CHAD GIRAFFE (*Giraffa camelopardalis peralta*).
4. RETICULATED GIRAFFE (*Giraffa camelopardalis reticulata*; Color plate, p.
262). 5. UGANDA GIRAFFE (*Giraffa camelopardalis rothschildi*). 6. MASAI
GIRAFFE (*Giraffa camelopardalis tippelskirchi*, Color plate, p. 261 and p.
262). 7. ANGOLA GIRAFFE (*Giraffa camelopardalis angolensis*; Color plate, p.
262), of which only a few remain. 8. CAPE GIRAFFE (*Giraffa camelopardalis
giraffa*), which has been killed off in the South.

For some time the different subspecies were described as distinct
species. This theory has been abandoned because all of them crossbreed and have fertile offspring. Also, in the same herd of giraffe,
individuals of very light or very dark color may occur. Pictures of
giraffes without any patterns were found in Egyptian tombs. Since
right next to them there were those with spots, this could not possibly
be due to chance. Albino animals, which occur in every species from
insects to primates, also occur in giraffes, although so far nothing is
known about an albino giraffe. Goodwin in 1938 in Kenya filmed a
male giraffe which was an almost pure white. The animal, however,
had dark eyes, so he could not be an albino. He was with a female
of normal coloration. In 1952, two Europeans found another white
male giraffe, who was with a rather light-colored, but normal, giraffe.
Hediger describes a single white animal from the Garamba Park
(Congo). It had dark, probably blue, eyes, a dark tasseled tail, and a
bone colored mane. Another white male, who was regularly seen for
more than ten years in a rather restricted area of North Tanganyika,
was never found in the company of conspecifics. He was safe from
native poachers because they related the animal's color with the
nearby peak of Mount Kilimanjaro and with a superstitious fear of
revenge from its ghosts.

Giraffes live in Africa south of the Sahara to Capetown, wherever
man has not destroyed them. When they stand on the open steppe in
small groups, occasionally in herds of more than fifty animals, or for-

Fig. 9-10.
Giraffe *(Giraffa camelo-
pardalis).* In the areas
marked with crosses the
giraffe has been
destroyed.

merly even by the hundreds, they cannot be overlooked. They are the tallest animals on earth. When they are between dead trees, they remain easily concealed from the observer because the spotted pattern on their neck resembles the bark of the tree trunks with the shadows of branches on it. Like most animals of the steppe, giraffes live in groups and graze together with ostriches, zebras, and antelopes. While this may increase the mutual safety of such a group, it probably is not a genuine group (see Ch. 14). Giraffes have good vision, although exact experiments on their sense of sight have not been made. Observations have shown that they are able to recognize conspecifics over a distance of one kilometer. Naturally, the eye performance of diurnal animals is influenced by the capability of seeing colors. Experiments by Backhaus in the modern giraffe house of the Frankfurt Zoo showed that the pigments red, orange, yellow-green, purple, green, and blue, which are clearly distinguished from different shades of gray, are probably recognized as colors. The animal groupings with giraffes as "watch towers" and "observation posts" get along excellently together, and the smaller animals stay among the giraffes.

A giraffe has few enemies. They are able to defend themselves with terrible blows with their front legs, even against lions. They can smash the skull of a large cat which is about to attack a young giraffe. Despite these defense capabilities, lions occasionally do attack giraffes. The lion will approach a single animal. A game warden in the Etosha Pan once watched a lion stalk a single giraffe. As soon as the giraffe saw him, she took off. But the lion caught up with a few leaps, jumped on her neck, clung to her with his claws, and probably bit through the vertebrae. Anyway, the large animal soon staggered and collapsed. In another case, the attacking lion did not get close enough before the giraffe became aware of him. Therefore, the chase lasted longer. Since a lion is not an enduring runner, he was probably somewhat exhausted before he started to leap on her back. He landed on her hindquarters, slipped down, and received the full blow of the giraffe's two hind legs on the side of his body. When the injured predator had not recovered after several hours, a game warden shot him. The chest cavity was caved in and almost all of the ribs were fractured. The death of another male giraffe, who while browsing in a treetop annoyed a leopard, should almost be considered an accident. The leopard jumped and damaged the giraffe's neck so badly that it soon died.

Backhaus's studies in the Garamba Park (Congo) proved that giraffes have a rank order, although most of the time they live together so peacefully that it is hard to decide which is higher in rank and which is lower. Only a superior one may bar another's passage. The higher ranking animals also carry their heads and chins higher. An inferior pulls his head downward slightly when the "boss" is passing by. A giraffe tries to impress or threaten a conspecific, including the keeper

Fig. 9-11.
Fighting male giraffes usually stand beside each other. By swinging their necks, they strike their skulls against the other's head, neck, chest, or side of the neck. Sometimes they also stand reversed-parallel so that each of the opponents boxes with the other's hindquarters.

in the zoo, by raising its head. No young male giraffe may show "flehmen" (lip curl) when a superior is present. This behavior is best known from horses: the head is slightly raised, the mouth is opened, and the upper lip is turned up. According to von Knapp, during the "flehmen" air is brought to Jacobson's organ, a cavity lined with olfactory mucus membranes and connected with the oral cavity and the nasal cavity. With this procedure, a male giraffe may test a female's readiness to mate.

If a giraffe is serious about warding off a conspecific, he walks straight up to him, bends his neck slightly forward, and dips his head. Thus he threatens with his short horns. Below the two frontal horns the northern subspecies have a third one, which is rather a hump, in the center of the head. Some giraffes even have two more bony knobs covered with skin. In all of these horns, the skin reaches almost to the tip and in the males the hair is rubbed off. These bare, blunt horntips act together with the large heavy head, as the main weapons in male encounters. In these fights, they never kick with their feet. Such a fight between rivals is an exciting spectacle even though the action seems to be in slow motion (see fig. 9-11).

These rank order fights between two individuals may last for a quarter of an hour or even longer. Often other males and females are standing nearby, although such fights do occur without a female audience. Sometimes a fighting male may press his opponent against a tree, or the two will circle around a tree trunk. If one of them finally gives in, he walks a few steps away from the victor, who will follow him only a short way with his head raised. While deer, antelopes, and many cats of prey drive the defeated rival out of their territory and keep him out of it, giraffe males generally live peacefully beside each other after they have determined who is the strongest. They may rub their necks together or graze peacefully immediately after the fight. Serious injuries or even fatalities rarely occur in such fights. Innes did observe in 1958 a male who became unconscious during such a fight and could not get up again for twenty minutes. A blow must have hit one of the large blood vessels at the base of the neck.

The Egyptians, who knew how to tame many game animals, kept giraffes as early as the eighteenth dynasty, which was about 1500 B.C. Then it was believed that giraffes were a crossbreed between a female camel and a male panther, according to the description of the Arabian geographer Ibn-el-Faqih from the eleventh century. The Romans in the era of Caesar also knew the giraffe. An aquarelle from 1559 shows a giraffe in the menagerie of Suleyman II in Constantinople. Today almost every child has seen a giraffe. But imagine the excitement when, for the first time, such a strange creature came to the Emperor in Vienna as a present from the Viceroy of Egypt, Mohammed Ali Pascha, in 1828. Shortly before the Viceroy had sent two giraffes to Paris and

London, but this was the first one to come to Germany. The animal travelled by sailboat from Egypt to Venice. From there it walked, as did all large animals for menageries at this time, via Hungary to Austria.

Single giraffes have been kept in zoos for more than twenty-eight years before they died of old age. In the second half of the last century, giraffes were very fashionable in all the zoos of the world. The first giraffe in America reached New York in 1873. The animal trader Hagenbeck imported a total of thirty-five animals in 1876, and another German trader brought another twenty-six to Europe.

Fig. 9-12.
This is how giraffes gallop.

A person cannot reach these huge animals on foot, but it is possible to catch up with them on horseback after 500–600 meters. When giraffes dash off, their ambling walk is clearly visible. Giraffes may gallop, although not often and only over a short distance. A giraffe may comfortably lift both front legs simultaneously when it throws its long neck backward in order to shift the heavy weight to the back. How does such an animal manage to cope with barbed wire fences when its feet seem to adhere to the ground? In Transvaal the giraffes at first simply walked through them and dragged the wires behind them. But after a few years of experience with these restrictions in their living space, they learned to jump them, much to the surprise of the farmers. Finally, even barbed wire fences of 1.85 meters did not present serious difficulties for them. They would then throw their heads and necks backward and jump it with the front legs first. It does not seem to bother them when their hind legs touch the upper wires in jumping. A difficult obstacle for giraffes is water. Nevertheless, Turner observed two giraffes in the Serengeti in broad daylight walking across a small lake. The water probably was one and a half meters deep. In the South Sudan, Scherpner saw three giraffes cross an even larger tributary of the Nile. According to the statements of his native companions, these animals must have swam. It was especially strange to see their necks stretched far forward; only the upper third of the neck and the head were above the water's surface.

When capturing giraffes, it is best to try to surprise a herd crossing open territory. Formerly, the selected animal was pursued on horseback. Now a leather loop on a long pole is slipped over the animal's head from a light, fast truck. The loop is prepared so that it will not pull tight and strangle the giraffe. None of the animals may be chased for too long because they would die of a heart attack. As soon as the speed of the captured animal has been slowed down, the captor's assistants jump down to give a hand. Furthermore, the animal is injected as soon as possible with a tranquilizer. As soon as the animal has settled down, it becomes surprisingly calm and even takes food. Only half-grown animals are captured, since full-grown ones would require very large transport boxes, thus causing too much trouble on the way.

The adaptation from natural food to the zoo diet begins in the capture station. A giraffe in the wild will rarely eat grass because it is so difficult for it to reach the ground with its long front legs. While drinking, which it may forego for several days, the animal has to spread the legs far apart or even "kneel" down. Giraffes are adapted for tearing foliage and buds from trees. With their long tongue, they may easily reach high branches and even get the last leaves from the tops of the thorniest acacia trees. In a zoo, a giraffe receives about four kilograms of oats per day, and first quality meadow or lucerne hay, as much as it likes. In addition, it is fed apples, radishes, onions, sometimes bananas and, above all, carrots. Furthermore, it receives fresh branches of deciduous trees in the summer. In the winter, green seedlings of barley and bundles of air-dried acacia branches are substituted. Naturally, its food must be hung very high. In the Frankfurt Zoo, the hayloft and forage rooms are on the second floor. The keeper may place the food right in the trough through a door on the wall on this level, which is just the right height for these giants. The hay baskets are also placed on this level.

Relatively early in the history of giraffe keeping, they could be bred. In the London Zoo in 1836, three males and a female arrived from the Sudan and developed very well. The female gave birth in 1839 to the first young to be born in a zoo. Since then, many giraffes have been born in zoos. However, the birth of a young giraffe is still a special event for all zoo people. In 1957, we observed such a birth in the Frankfurt Zoo. Labor pains came every ten to fifteen minutes. We discussed the problem of how the long stilt legs and the neck might be packed and folded in the short womb, which had not increased in size. At 11:45 A.M. out came one front leg, with its pointed hoof soft and swollen so that it could do no harm. Then, a little later, the second hoof followed. Five minutes before 12:00 the little head with the small horns flapped forward and down, coming out with a slight jerk. Restlessly the mother walked to and fro. Before reaching the wall, she would throw her huge head five meters upwards and back with a jerk and without ever really touching the wall. The long neck of the young kept gliding into this world, sometimes retreating a little, then coming out further. The young did not breathe yet, although it was alive. Its long, blue-black tongue was moving and licking its nose and mouth. Another silent jerk, which we thought we heard, and the shoulders and the hind part of the body slid out. In giraffes the female gives birth standing, so the young has to drop down for more than two meters. In spite of this, our young giraffe did not land on its head, but turned around during the fall. Gravity caused it to lay sidewards on its back. With a slight trembling at 12:14, the chest began to move. For the first time air streamed into the lungs which were still small and folded. Soon the wet, shaggy, little animal on the ground raised his neck,

Fig. 9-13.
To drink, the giraffe has to spread his front legs far apart or flex them so he can reach the water or the ground with his mouth.

tottering and uncertain. It swung to and fro and repeatedly touched the ground. The mane stood erect, but the clusters of hair stuck together like little matches. The mother lowered her head from more than five meters and licked her offspring on the head and body. Soon it made the first attempt to stand on its legs. But the meter-long legs would not respond. They slipped away, and the baby kept falling back. Finally, at 4:00 P.M., the keeper came to its aid. He put his foot behind the young's hind leg so that it had support, and then the little fellow staggered and tottered on its four long legs. At two meters, he was already taller than we were. A young giraffe in the Basel Zoo, who was only 1.20 meters at birth, grew to this size after only six months.

Clumsily, the young giraffe went towards its mother and between her legs. Over and over the mother stepped uncomfortably aside. But finally the young managed to press its head and neck under her belly between the mother's front legs and to reach the udder. At 8:45 P.M. it drank for the first time with a loud smacking, and the mother tolerated it for three minutes. She licked its back. Raising it was bound to be successful now. The young was healthy, it walked around, and the mother had accepted it.

Not every birth proceeds so smoothly. Even in the wild a mother and child may die in birth. Owen found a dead giraffe in the Serengeti where the head and front legs of the young were already out of the mother's body. There was no sign that predators were involved. Little is known about giraffe births in the wild. An observation from the Krüger National Park suggests that perhaps it takes place regularly in the center of the protecting herd. Nine giraffes stood in a circle around the female who gave birth. After twenty-five minutes the young was able to get up and to walk around its mother. All members of the herd who stood nearby touched the young with their noses. In the zoo conspecifics do not interfere with a birth either. Zoo giraffes may have their first young at five years, while the second one sometimes follows as early as sixteen months later. In Cincinnati a female giraffe had nine young until she was twenty-five, and in the Berlin Zoo a group of giraffes, which lived from 1932 until 1941, had seven offspring.

Only because of giraffes in zoos have we begun to learn more about the habits of the tallest creatures on earth. For a long time there were disputes over whether giraffes would lie down during the night to sleep because it would take them too long to get up in a case of emergency. The exceptional neck and leg length are responsible for an unusual resting position. Immelmann and Gebbing studied the sleeping habits of the giraffes in the Frankfurt Zoo. In many large terrestrial mammals, the duration of sleep at night is rather short, which seems to make sense biologically since a sleeping animal lacks protection. Adult giraffes in a zoo seldom lay down during the day, although young animals, especially at noon, will lay down several times for ten minutes

▷
Reclining masai giraffe (*Giraffa camelopardalis tippelskirchi*) in the east African savannah.
▷▷
Giraffes: 1. Giraffa (*Giraffa camelopardalis*), subspecies a) Reticulated giraffe (*Giraffa camelopardalis reticulata*), b) Angola giraffe (*Giraffa camelopardalis angolensis*), c) Masai giraffe (*Giraffa camelopardalis tippelskirchi*). 2. Okapi (*Okapia johnstoni*).
▷▷▷
Pronghorns: pronghorn (*Antilocapra americana*).

1a

1b

1c

♂

♀

2

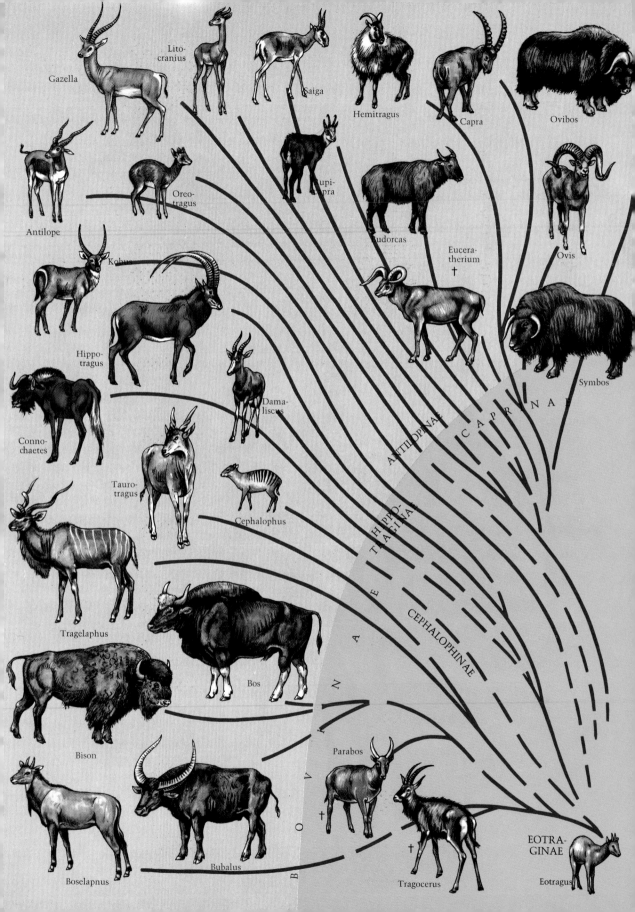

Gazella

Lito-
cranius

Saiga

Hemitragus

Capra

Ovibos

Antilope

Oreo-
tragus

Rupi-
capra

Budorcas

Eucera-
therium †

Ovis

Kobus

Euceratherium †

Symbos

Hippo-
tragus

Dama-
liscus

Conno-
chaetes

Tauro-
tragus

Cephalophus

Tragelaphus

Bos

Bison

Parabos †

Boselapnus

Bubalus

Tragocerus †

Eotragus

EOTRA-
GINAE

ANTILOPINAE

CAPRINAE

HIPPO-
TRAGINAE

CEPHALOPHINAE

BOVINAE

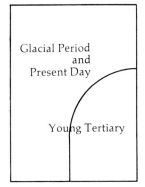

The phylogeny of the horned ungulates:
1. Eotraginae: *Eotragus.*
2. Bovinae: *Tragocerus, Boselaphus* (nilgai), *Parabos, Bubalus* (arna), *Bison* (bison), *Bos* (gaur), *Tragelaphus* (greater kudu), *Taurotragus* (eland).
3. Duikers (Cephalophinae): *Cephalophus* (banded duiker). 4. Hippotraginae: *Damaliscus* (blesbok), *Connochaetes* (white-tailed gnu), *Hippotragus* (sable antelope), *Kobus* (waterbuck). 5. Antilopinae: *Antelope* (blackbuck), *Oreotragus* (klipspringer), *Gazella* (Grant's gazelle), *Litocranius* (gerenuk).
6. Caprinae: *Euceratherium, Saiga* (saiga), *Rupicapra* (chamois), *Budorcas* (takin), *Hemitragus* (tahr), *Capra* (ibex), *Ovis* (wild sheep), *Symbos, Ovibos* (musk ox). This list is by Erich Thenius. A classification of the horned ungulates by Fritz Walther, which is slightly different, is in Chapter 12.

at most and without sleeping. The adults also have a siesta at noon where they stand close together, inactive and ruminating. They lay down for the first time in the evening after dusk when everything is quiet. Then they pull their two front legs and one of the hind legs under their body, while the second hind leg is stretched away to the side. The neck remains erect. The animals doze with the eyes alternately opened and closed. They remain on the alert, and their ears move. Occasionally, they may stand while they are dozing. The reclining giraffes get up approximately every two hours to defecate and to urinate, and then they will eat a little. After one hour, they will lie down again. With several interruptions, the total time a giraffe spends in lying down amounts to seven or nine hours. Within this period, however, the giraffe only attains a genuine deep sleep for several minutes each time. Then the formerly erect neck is bent backward like a handle, the chin touches the ground behind the tarsal joint of the stretched hind leg, and the lower jaw rests on the shank. The longest recorded duration of deep sleep with the head supported in an adult giraffe was twelve minutes; the shortest, when the animal woke up naturally, was one minute. The different phases of deep sleep never occurred in close sequence but were distributed rather evenly through the whole night. The total duration of the deep sleep phase is amazingly short. In adult animals it is rarely longer than twenty minutes. The Frankfurt young giraffe, however, slept soundly for about seventy minutes at the age of four months. While this deep sleeping in zoo giraffes could be observed only between sundown and early morning, I saw in the Serengeti in broad daylight an adult male deeply asleep in the middle of a herd.

Another question has puzzled physiologists for a long time. In an upright giraffe, there is a remarkably tall column of blood. When it stretches its head steeply upward in order to reach the tree tops, this column may well come to seven meters. This corresponds to a pressure of 500 mm mercury. In order to surmount the difference between the heart and the brain, the giraffe's heart has to attain a pumping pressure of 300 mm mercury. When a giraffe lowers its head to the ground and then raises it again, there would be a great difference of pressure in the brain. This would cause unconsciousness in humans. The blood returning in the neck veins should dash down to the heart with great speed. How does the giraffe's body cope with these problems? Professor Goetz from Capetown has studied this question. The former opinion was that perhaps the giraffe's blood was especially viscuous. However, it is hardly different from human blood, although it has, like a camel or a llama, twice as many blood corpuscles than we do. The giraffe's heart, which weighs more than eleven kilograms, transports approximately sixty liters per minute.

Goetz, anaesthetizing a place on the neck of a captured giraffe, introduced (through an incision) a rubber sounding-lead of 2.5 meters into the large heart artery in the center of the bloodstream. He placed some radioactive cobalt in the tip of this long antenna so that it could be located with a Geiger counter. Thus, it was possible to bring this probe in the large artery up to the base of the brain. There the blood pressure was recorded electrically. In the erect neck it was 200 mm. When the animal lowered its head to the ground, contrary to all expectations, it did not rise but dropped to 175 mm. Probably the solution to this problem lies in the valves in the large neck artery. They are able to interrupt the blood flow so that high excess pressure is generated in this artery, even if in the connecting veins, which also may be closed with valves, a negative pressure is prevalent at this time. So the huge neck artery, which actually would only have to bring the blood back to the heart, acts temporarily as a collecting vessel which compensates for the pressure in the brain.

Bernhard Grzimek
Erich Thenius

Fig. 9-14.
In the deep sleep phase, which lasts for only a few minutes, the giraffe bends its neck in a backward circle and rests its head.

10 Pronghorns

Until a few decades ago all those ruminants which had horn-sheathed weapons on their foreheads and were neither sheep, goats, nor oxen were classified as "antelopes." For this reason, the strange pronghorns, which once populated the prairies of the North American West in great numbers, were classed with this collective group.

Phylogeny, by Erich
Thenius

Now the pronghorn (*Antilocapra americana*) is seen as the last survivor of a once very multiform family of ruminants which was restricted exclusively to North America. In the Upper Tertiary strata of North America, many fossils were found which belong to the relatives of the pronghorn. They are predominantly the MERYCODONTES (*Merycodus*, etc.; Color plate, p. 32), who were even-toed ungulates with branched horns like deer antlers and high-crowned molars. In 1922, Winge and Hilzheimer still classified them with the deer, but later studies by Matthew, Furlong, and especially by Frick in 1937 showed that any similarity with the deer is only superficial and that the merycodontes belong to the pronghorns. The forehead construction of the branched merycodonte horns is a compound of bone material which is deposited at regular intervals. It cannot be compared with the basal bone construction of the deer's antlers because it does not "coalesce" with the "bar" but can easily be separated from it. The horns were covered with skin which was shed and renewed regularly. The actual bone plugs were not shed.

In addition to the merycodontes, other fossils of North American genera, such as the *Ilingoceras*, and the *Neotragoceras* belong to the pronghorns. The Pliocene forms of *Proantilocapra platycornea* and *Capromeryx* (with several species) are intermediate between the merycondontes and the contemporary pronghorn.

The pronghorn, by
B. Grzimek
Distinguishing
characteristics

The PRONGHORN (⊕ *Antilocapra americana*; Color plate, p. 263) is deer-sized with slender legs and a short tail. The HRL is about 100–130 cm, the TL is 17–20 cm, and the BH is 88–105 cm. The long head has a straight profile, a small muzzle, very large eyes (about the size of a horse's!), and medium long ears with narrow tips which are inclined towards each other. On the forehead is a pair of diverging horns. The bone plugs are small, sharp edged, unforked, and without cavities. The forked horn sheaths are renewed annually. The ♀♀ horns are weaker or missing. The pointed, narrow hooves are stronger in the front than in the rear; there are no pseudo-claws. The dense, stiff, brittle bristles are undulated, and the fleece is sparse. The white hair patch, which is longer on the rear (seven to ten centimeters), can be erected and spread as a "signal." The animal has well developed scent glands, including the postmandibular or cheek patch gland (five centimeters below the ear), the croup gland (odd-numbered, above the end of the sacrum), the tail gland (at both sides of the tail base), the hock gland (outward behind the hock), and the interdigital gland (on all four feet). The females have four nipples. The heart is very large—the pronghorn is a runner. There is a gall bladder. There are 32 teeth: $\frac{0 \cdot 0 \cdot 3 \cdot 3}{3 \cdot 1 \cdot 3 \cdot 3}$. The molars are small with high crowns.

Fig. 10-1.
Pronghorn (*Antilocapra americana*). The pronghorn was almost destroyed, but it now occurs in some areas of its former distribution, which are marked on this map, rather frequently.

When the white settlers of the last century moved to the Western United States in long treks of covered wagons, they found enormous numbers of pronghorns, which lived there in larger or smaller groups, between the Missouri River and the Rocky Mountains. Around 1800, there must have been more than forty million of them. They were as numerous as the famous herds of buffalo which were then slaughtered for the fun of it. On the rail line from Denver to Cheyenne in the winter of 1868/1869, daily wagon loads of dead pronghorns were brought to town. Along the rail line lived three to four million head. Three to four of the animals were sold for 25 cents.

More than 40 million pronghorns

Only now are people aware that these vast herds of buffalo and pronghorns would have been the ecologically best utilization of certain soils for human food. Of course, after killing these game animals, meat was provided by cattle farming, but cattle never yield the same quantity per hectare. Also game herds did not destroy the soil while now, after only one hundred years of cattle farming, wide parts of the United States have turned into dust bowls. Right now people are about to repeat these same mistakes in East and Central Africa with the support of Foreign Aid Programs. Only there the soil will be transformed into desert even faster under the tropical sun.

They were almost destroyed

The pronghorn is the only animal which annually sheds and renews the horn sheaths of its permanent horns. While it keeps only the bone plugs on the head, the horn sheath over them detaches every year, falls to the ground, and is consumed by mice, ground squirrels, and rabbits. Before the old sheath has been shed, new horn material, covered at first with a furry protective layer, is growing at the base of the horns. The process lasts four months until the new sheath is hard.

When a pronghorn perceives a wolf, a coyote, or something unusual in his environment, he spreads the long, still, white hair on the rear flatly apart so that they flash like two giant chrysanthemums and

Fig. 10-2.
This is how Ernest Thompson Seton presented the blinking and scent signal system of the pronghorns.

actually reflect the light. Other pronghorns can see this signal over a distance of several kilometers, and they immediately repeat it. The alarm spreads with lightening speed. Almost simultaneously, the scent glands at the base of these hairs secrete a strong scent which even humans with their rudimentary sense of smell can perceive over a distance of 100 meters and more.

These "blink-stink-antelopes" are the fastest of mammals on the American continent. They are able to run at a speed of eighty kilometers per hour over a distance of about one and a half kilometers. This means that no wolf, no coyote, and no greyhound can catch up with them. Only thoroughbreds may be fast over a long run. Nevertheless, it was possible to kill off the pronghorns by the millions over a short period because at first their strong curiosity turned out to be disastrous. The old settlers told stories about how all a person had to do was to lie down on his back and to kick about with his legs in the air, or to wave some red rag on a stick in order to lure the animals within rifle range. Meanwhile, they have gotten to know man better. Formerly, it would happen that they accompanied the wagon trains at a distance of only 100 meters, or that they would unsuspectingly dash through a camp at night.

Pronghorns are probably the best wild animals to cope with barbed wire. They do not jump the fences, but crawl under or dash straight between two strands that are thirty-five centimeters apart. In areas with good pronghorn populations the farmers are supposed to keep the lowest wire of the fence a minimum distance from the ground in order not to restrict the free movement of the pronghorns.

Once in a while, they delight in running for some time beside a car until they finally speed up, pass it, cross the road in front of it, and then slow down on the other side almost as if they had satisfactorially proven their abilities.

In exactly 100 years, man has caused these strange and pretty American animals to diminish from 40 million to 19,000 head. It seemed as if they might ultimately disappear. But around the turn of the century, mainly due to the publications and books of Ernest Thompson Seton, the famous naturalist and author, the American people became aware of the disgrace on their history through the sadistic slaughtering of the huge buffalo herds. Therefore, efforts were made to protect endangered species of animals, and national parks were founded. Thus, the number of pronghorns increased gradually. In 1924 there were 40,000 of them, and now the population is up to 400,000. In order to prevent farmers from killing pronghorns on their land as "pests," they receive premiums in some states. In Wyoming every hunter has to pay five dollars to the farmer who owns the land on which he shoots a pronghorn; in Texas, it may cost 30 to 100 dollars.

In late summer and in spring, some of the male pronghorns become territorial. According to F. Walther's observations, they mark single plants with a secretion from their post-mandibular glands. This secretion is not visible, but it is easily noticed because of its strong musk scent. The male sequentially paws the ground, urinates, and defecates at different places, and this may also be a way of marking the area. He does not establish larger manure piles, like some of the other bovidae do. Before or during the rut, a small harem may stay for hours or days with a territorial male. Apparently the females do not remain for a long time with the same male, but rather change frequently from one to another. Walther found that during the mating ceremony the male circles around the female with accentuated steps, alternatingly presenting to her the black patch on the right and the left side of his cheeks. In the mounting position, he clings to her flanks with the front legs and he keeps his neck erect. For the time being, it is not known for sure whether copulation takes place exclusively within the territories, or whether non-territorial males may also take part in reproduction. In any case, non-territorial males also intensively drive at full gallop single females during the rut in mid-September. The females mostly have twins. Just like the European roe deer, they make them stay put during the day and come only briefly, predominantly during the night to nurse. The two young always lie separately, often 80 to 100 meters from each other, and they have almost no scent. Thus, if a predator finds one such young, at least the other remains undetected in most cases. The mother attacks coyotes, foxes, and even eagles with her front legs and usually chases them off.

The pronghorns were the original natives of America, older even than American Indians and most of the other animals on this continent, none of which is closely related to them. The ancestors and relatives of the pronghorns are found in America where they lived one or two million years ago (see p. 267). While their contemporaries from that time, the large giraffe-camels and the hornless deer, have long been extinct, the pronghorns have managed to survive. Thus, they were true Americans for millions of years, while the other large two-toed ungulates, the bison, deer, moose, and wapitis migrated to this continent at a later time. Of course the Indians, horses, and white Americans were much more recent immigrants.

No one knows why pronghorns reproduce so rarely in zoological gardens, where they often cannot be kept alive for more than a year. In their natural habitat west of the Mississippi, they do rather well in captivity. For example, Dr. Poglayen had a beautiful breeding group in Albuquerque, New Mexico zoo. Even in the zoos of the Eastern United States they do not last long. The zoos of Washington, New York, Chicago, and the Catskill Game Farm have kept some for seven or eight years which occasionally reproduced. Except for a few specimens, they never last long in Europe. Since no one likes to keep

Pronghorns in zoological gardens

animals in a zoo which are bound to die soon, they are only rare wards in European zoos. Occasionally another attempt is made with new food compounds or a different method of keeping them. Thus, after World War II pronghorns came to the zoos of Rome, Paris-Vincennes, and Whipsnade near London. In Whipsnade in 1964 they reproduced for the first time in Europe, although most of these animals did not live long.

The report of L. Dittrich, director of the Hannover Zoo, is a good example of the difficulties encountered in keeping and breeding pronghorns in European zoos. The Hannover Zoo, which is especially experienced in keeping rare ungulates, has experimented since 1961 with the adaptation of pronghorns. For a long time, these animals were close to death, suffering from serious diarrhea, infestation by parasites, and poor appetite. At times they were so weak that they had to be carried in the morning from their stable to the pasture. Nevertheless, the two surviving animals in June, 1966, became the first to reproduce the species in Germany.

A female in poor health gave birth to twins after heavy labor pains. Lothar Dittrich reports on their condition: "Like all ungulates, the two soon stood on their legs and tried, alternately or at the same time to nurse. They increased in weight 200 to 250 grams per day during the first week. In the beginning, the two young animals resembled each other to such an extent that we could not tell them apart, so we had to mark their coats. Three weeks later, when they began to eat from their mother's food, the male was clearly recognizable by his slightly sturdier physique. After birth the mother drank large quantities of a linseed preparation, and ate great amounts of other food. Now the mother reached an optimal physical condition. Nursing of the young, which certainly meant the production of several liters of milk per day, seemed to be a problem for her. Of course, there was great joy in the Hannover Zoo about the successful reproduction, after five years of continual efforts at keeping the pronghorns."

Bernhard Grzimek
Erich Thenius

11 Horned Ungulates

Evolutionary biology has taught us that the phyla of the animal kingdom ascend from the lower to the higher. The simpler and more primitive animals evolve in the course of the earth's history into forms which are more complicated in structure and more specifically adapted. Therefore, in the animal literature of past decades, the description of the animal kingdom usually ended with man and his relatives, the primates. The determining factor in this case was the development of the cerebrum which led to man. However, according to the latest findings, it is no longer possible to place the primates at the top of the ascending line in the animal kingdom. The primates, and man included, still have very many primitive characteristics, besides their special adaptations. They demonstrate many links with the insectivorous root of all the higher evolved mammals (compare Vol. X and XI). Therefore, the end of our review of the animal kingdom has to be, according to modern zoological classification (systematics), another group of animals which, in addition to primates, represent another evolutionary climax in the mammals. Physically this group is more highly evolved than the primates, and has apparently reached its prime and greatest multiformity now. They are the horned ungulates, the last and richest in species of the family of even-toed ungulates.

To the large family of the ungulates (Bovidae, formerly Cavicornia), besides those various forms which formerly had been classified as "antelopes," belong the goats, sheep, and chamois (related to the musk oxen and the wild oxen). Modern mammalogy has classified all these partly small, partly medium-sized, and partly very large ruminating, horned even-toed ungulates into a number of equivalent subfamilies. Thus, the old classifications of "antelopes," goats, sheep, and wild oxen have been completely abandoned. The so-called "antelopes," especially, do not have a systematic or phylogenetic unity, so when we use this term now we must be aware that it is only an unscientific name which is still commonly used.

Family: horned ungulates

Distinguishing
characteristics

The HORNED UNGULATES (Bovidae) range in size between a hare and a buffalo, with an L of 55–435 cm. They have various shapes: small and graceful, "sack-like," medium-sized and slender, tall and slim, or bulky. The head has a short or long facial structure and a muzzle with a large to small naked or hairy nose region. The eyes are often large and the small to large ears often end in a point. In almost all cases, there are two horns (with the exception of the four-horned antelope; see Ch. 12). Sometimes both sexes have horns and sometimes only the ♂♂ do. The horns, which vary from very short to long (2–175 cm), have very different, species-specific shapes. The neck is short to very long, and, in some species, has a dewlap. The back may be straight or higher at the withers or croup. The tail, from very short to long, has a cross section which is round or wide-oval in shape. The legs are pencil thin or sturdy, and short to very long. The narrow to broad hooves may be short to very long. Most animals have pseudo-claws, and some are very large. The hair, which can be short and smooth or dense and long, often changes with the seasons. Sometimes the hair has a sparse or woolly feel. The sexes are of the same or of different colors.

They are herbivorous animals, but may at times eat animal food as a supplement. Some species depend heavily on drinking water, and others are largely independent of it. There are usually 32 teeth: $\frac{0 \cdot 0 \cdot 3 \cdot 3}{3 \cdot 1 \cdot 3 \cdot 3}$. There is a gap between the frontal teeth in the lower jaw and premolars and molars. The intestinal system is the same as in all ruminants. Their habitats are most varied: some species live in the dense prime forest, others in the open steppe, some in arctic and polar regions, and many in the hottest places on earth.

The rut in the temperate and northern zones is strictly seasonal, but in the tropics and subtropics it occurs year-round with peak periods. The ♀♀ have two or four nipples, a gestation period of five to eleven months, and in some species there are two births per year. Usually one young is born at a time, rarely two; in very rare cases there are three. In a few species the females, starting with the second birth, may occasionally have twins. Pubescence is at one half to three and one half years. Longevity is about ten to thirty years.

How we classify the
horned ungulates

It has not been until this last decade that the behavior of some horned ungulates has been studied in more detail. The systematic classification of some of the horned ungulate groups is not completely clarified. Many different and sometimes contradictory opinions exist. It is still especially difficult to classify those horned ungulates which formerly had been called "antelopes." With the great geographical discoveries in the second half of the nineteenth century, so many new species were discovered that were obviously not closely related to each other, so that systematic analysis of this group became most problematic. Therefore, the "antelopes" were divided into a number of equivalent subfamilies. Later it turned out that some of the new subfamilies were more closely related than others. Then another compilation was made, but it still remained unclear which of the characteristics would justify specific assignment to a particular group. Similarities in structure and habits may develop not only because of common ancestry, but also because of similar functions. Unfortunately, the cases where this cannot be determined with any certainty, based on our present knowledge, are rather numerous in the horned ungulates.

Thus, to date, numerous classification systems of horned ungulates have been created. Each of them has advantages and disadvantages but none is really satisfactory. In this area, the experts are still debating. Of course, in this book we can present only one classification, and we cannot describe others as well. After these preliminary remarks, the reader should not be too confused to find in the papers of other authors that there are different arrangements or other scientific names for a single species. Following a suggestion of Fritz Walther's, we will classify the family of horned ungulates into 15 subfamilies with 53 genera, 99 species, and 866 subspecies. For the classification of phylogenetic relationships in our colored genealogical trees (Color plates, pp. 32, 264, and 470), we found it more practical to combine some of these subfamilies into larger groups.

The horned ungulates are a geologically young group of animals. The oldest fossils which definitely belong to this family come from the Lower Miocene of Eurasia. Small even-toed ungulates, which may have looked like the contemporary kantschils, are the primitive form of these oldest horned ungulates. Such a primitive, hornless even-toed ungulate (Artiodactyla) was the genus *Archaeomeryx* (Color plate, p. 134), which is classified with the relatives of the chevrotains (see Ch. 7). from the Upper Eocene of East Asia. It still had very large pseudo-claws and a complete set of front teeth in the intermaxillary jaw.

Phylogeny, by Erich Thenius

The *Eotragus* species of the Miocene gives us a good idea of what the first horned members of this family may have looked like. They were animals of barely roe deer size with small, stretched or slightly inward-bent horns, a moderately long skull, and slender limbs (Color plate, p. 264). In appearance and habits, they may have been closest to the contemporary duikers. From the *Eotragus* species of the Middle Tertiary evolved the TRAGOCERINAE (subfamily Tragocerinae) whose closest living relatives may be the INDIAN NILGAI ANTELOPE *(Boselaphus tragocamelus),* and the FOUR-HORNED ANTELOPE *(Tetracerus quadricornis),* whose different genera *(Protragocerus, Miotragocerus* and *Tragocerus,* etc.) had horns of various shapes. *Pachyportax,* a relative of the nilgai antelope from the Upper Tertiary, is close to the original form of the wild oxen (subfamily Bovinae or tribe Bovini).

Due to fossil finds, the prehistory of the single subfamilies and related groups of the horned ungulates is rather well known. Their phylogenetic relationships, as well as some extinct forms, are given on pages 32, 264, and 470. Among the oldest fossils of wild oxen, the genus *Parabos* from the Lower Pliocene of Eurasia (about ten million years ago) deserves special mention. In the glacial period lived some conspicuous species of bison, among them the forest *Bison schoetensacki* and the steppe *Bison priscus.* Some of them were much larger than the contemporary American and European bison, and some forms had huge, expanded horns.

Contemporary horned ungulates, by Fritz Walther

The evolution of the forehead weapons of the horned ungulates began in the Miocene, approximately twenty-five million years ago. During this geologically short period, horn shapes of prodigious variety developed. There are the finger-long little daggers of a duiker, the meter-long spears of the oryx, the soft undulation of the gazelle's horn, the magnificent spirals of the kudu's horns, the hook-like curved horns of the chamois, and the massive horns of a wild sheep ram. In some species, the horns stand almost vertically on the skull, in others they arch backwards or curve forward, and in still others they expand to the sides. These shapes are so varied and typical that, in many cases, it is easy for the specialist to immediately determine the genus, the species, and often the subspecies. Horns are plugs of bone material which are covered with a horn sheath. In the ontogeny of a single individual, they grow from proturberances of the bone structure (horn-bone, os cornu) and are covered by skin. The horn-bones develop as autonomous bones in the lower layers of the skin (mesoderm) of the forehead. Shortly after birth they connect with the frontal bone which fuses them. Thus the plug-shape is formed. The external skin covering (ectoderm) which covers the bony center as a hollow cone changes into a horn beginning at the base. This horn sheath does not need to be dropped because it stretches with increasing growth corresponding to the length of the bone plug.

The multiformity of horn shapes

In some species the "anlage" (rudiment) of the horns is present at birth; in others, nothing is seen at this time. Some of the young animals have visible little horns at six months. Bulges and ridges on the sheaths are frequently found in horned ungulates, but in the adult animal they are not an indicator of age. However, species in a northern habitat may have regular "annual rings," which are notches in the horn by which the age of the animal in question may be determined. They are caused by variations in growth based on seasonal changes in metabolism and food. Many of the horns are screwed or wound. Two "types" of windings may be distinguished: 1. If the right horn is twisted to the right around the longitudinal axis from the top to the base, and the left one to the left, this is called normal or homonym winding. 2. However, if the right horn is twisted to the left and the left one to the right, this is called perverted or heteronym winding.

What are horns made for?

The question arises as to the biological meaning and purpose of these conspicuous structures. Some scientists think that the horns are simply decorative formations, while others feel that they may play a part as a kind of valve for metabolism or as a thermostat for temperature regulation. There may be some truth to these opinions, but one cannot help but agree with the oldest and most widespread explanation that the horns are basically and predominantly weapons. Of course, they are not defense weapons against predators, as was formerly taken for granted. In a considerable number of species, the females either do not have any or have only weaker horns than the males. But since females and young animals are exposed to the attacks of predators at least as much as males are, such a "prejudice" against females would not make sense. Formerly, it was often thought that the well-armed males would defend their females and offspring against predators, but this is, with few exceptions, just a tale.

How little the males actually defend the females and their young I could observe extensively in the Thomson gazelles on the Serengeti plains. In this species, the females have only small, often reduced horns, while the forehead weapons of the males are rather handsome. Their young are frequently hunted by jackals. During such an event, the mother and often another female try to ward off the predator by running between them and the young, feigning attacks, and trying to distract them and to wear them out. Sometimes they are successful and sometimes not. The number of times when a young is killed in spite of all the mother's efforts is considerable. But I have never observed or heard of a jackal being killed or even seriously wounded. In such a situation, a male may suddenly chase furiously after the young, the mother, and the jackal, not to come to their aid, but to stop the female. The male Thomson gazelles have territories where they stay while the females pass through. Then each male tries to retain females within his territory for as long as possible. If the mother pursues the jackal, the male completely misjudges the situation. He is only aware that a female is leaving his territory and so he tries to prevent her from doing this. Defense of the females and the young is not within his behavior repertoire; he is not able to recognize such a situation.

No weapons against enemies

Mutual defense or defense of an individual's own life against predators is rarely found in horned ungulates. To our present knowledge, this occurs only in musk oxen and in mountain goats regularly and with any success. Occasionally buffalo and some large antelope may fight back, but with obviously moderate chances. In the Serengeti, for example, Cape buffalo (not only young animals but adults as well) which have been killed by lions can often be found. Considering how many species of ungulates there are and how often they are chased and eaten by all kinds of predators, the cases where an attacker has been warded off successfully are only exceptions. Predominantly, the horned ungulates are chased and killed by their natural predators without any resistance. Neither the horn nor the hoof are for defense; flight, in most of the horned ungulates, is the best weapon in the "struggle for life." Therefore, it is rather improbable that warding off predators was a crucial factor in the evolution of horns.

The best weapon for defense against predators would be a strong, smooth spear, not too long and not too short. Only seen from a distance do some horns even come close to this "ideal" type. The obviously more highly evolved horn shapes, which often deviate considerably from this shape, are much less suited for defense against predators than are some of the primitive short daggers. Thus, if the horns were essentially weapons for defense, we would come to the absurd conclusion that their evolution led to a deterioration. In addition, the rings, bulges, and ridges would remain unintelligible.

Weapons against conspecifics

But all these absurdities appear in a different light as soon as we conceive of the horns not as tools for warding off enemies but as weapons for fights between conspecifics of the same sex. Suddenly, it makes sense that only the males are armed, or at least more heavily armed than the females, for it is they who, at least in the mating season, have to fight for territories, their harem, or at least some space free of rivals. Furthermore, in most species, males are more aggressive than females. Horns in females are found predominantly in those species where the individuals live together in large herds or groups. The more gregarious a species is, the sooner it may be expected that the females will "get into each other's hair." This is another example that shows that the horns are predominantly tools for encounters between conspecifics.

However, in most cases fights among conspecifics of the same sex are mostly ritual fights in horned ungulates. Their aim is not to harm or to kill but to demonstrate superiority and to make the inferior one give in or flee. Therefore, only a few species stab with their horns, as was formerly taken for granted; instead, they beat each other with the base of their horns, strike with the longer axis, or link the horns together so that the opponents are bound together, thus using all their strength for jerking and pushing. In all these forms of "ritualized" fighting the widening at the base of the horns, the bends and windings of the horns, the rings, bulges, and ridges which we find in all types of horns indeed have a meaning and a purpose.

Once the horns were there and had evolved accordingly, they could naturally be used for other purposes as well, for example, for scratching parts of the body or occasionally in desperate defense situations against a predator. As we know, Caesar fought with his slate pencil against his murderers for lack of something better. In spite of this, slate pencils are not manufactured as weapons for warding off enemies.

The evolution of horns clearly shows the great importance of ritualized fighting in the lives of horned ungulates. In some gazelles, the males have at least one or two, but often ten, twenty, or more, fights per day. With slight exaggeration, the horned ungulates might be called "the fighters among animals." If the horns were really dangerous weapons, these many fights would soon lead to the extinction of the species. Nature has prevented this from happening not only through the above mentioned evolution of the forehead weapons but also through certain changes (ritualization) in the fighting behavior itself.

Fig. 11–1.
The hornless female nilgai uninhibitedly butts another female's flanks.

The simplest and most primitive method of fighting was certainly beating and pushing against the opponent's body. This is indeed what the hornless females of some species do, although their beating is not dangerous; at worst it may cause sore spots. Also, the mountain goat, which obviously remained primitive in its fighting behavior with its relatively short horns, behaves similarly. The rivals stand reversed-parallel beside each other while each tries to push his horns into the other's flank. In Valerius Geist's opinion, this orientation of the opponents towards each other can probably be considered to be the original fighting position in the horned ungulates. In their fights, bad injuries may occur which seem to be close to the limits of just what a species can "afford" to do, although their buttocks are protected by thicker skin. When the horns increased in size and volume, it "became urgent" to change their shape and use. The simplest solution was to direct the attack to the least delicate part of the body, which means the horns. For this purpose, it became necessary for the opponents to change their orientation towards each other. The original lateral position in fights changed to a head-to-head position.

Fig. 11–2.
The male elands fight head to head with locked horns.

When animals fight head to head with rivals, they are able to intercept the attack and to direct it toward their own horns. In some species, the head has become the "fighting center" to such an extent that they normally direct attacks only toward this point. They would not even "think" of attacking the opponent from the flank, even if they could do so.

Years ago, when I kept a group of dorcas gazelles, there were two males of the same age, one of which was clearly stronger than the other. When the stronger one intended to fight, the weaker one often went into a "submissive posture." He turned his rear towards the opponent and lay down with his head stretched forward. This behavior was found in many horned ungulates and means just the opposite of a threat, where a horned ungulate presents his weapons, and of a display when he tries to appear large and broad. As soon as the animal showed the "submissive posture," he inhibited the attacker and largely stopped the attack. In the case of my dorcas gazelles, the superior male would stand for a while behind the submissive one. However, he had to discharge his drive somehow and there was no other opponent present, since a normal male gazelle would never attack a female. Therefore, he usually walked around the lying animal until he stood

Fig. 11–3.
The submissive posture (left) and the horn presentation with an erect head in dorcas gazelles.

head to head in front of him and then he pushed against his horns, even though with decreased fervor.

Such examples, and there are many more, show to what extent attack has been ritualized to a frontal attack. Naturally this ritualization has not reached perfection in all species, and even within the same species, there may be exceptions. Occasionally a human boxer will hit below the belt, which is not according to the rules, but compared with the total number of fights, these are exceptions to the rules in boxers as well as in horned ungulates. It seems justified to speak of a "fighting ethic" in horned ungulates, which, of course, has an innate basis.

A bullfighter who steps aside from the bull's attacks is anything but a hero. Rather, he treats the animal abominably by taking advantage of the natural fighting ethic of the bull. When a bullfighter kneels down in front of the bull, which is considered a courageous, high-spirited gesture, this behavior is analogous to the animal's submissive posture. In an animal with normal instincts, this will stop or at least reduce aggression. The bullfighter, taking advantage of this fact, attacks again, which is against the natural rules of the bull, in a manner and with methods which the instinct-bound animal cannot anticipate. Therefore, the victory of the human sadist over the bull with its intact instincts is a macabre spectacle, which a person with normal feelings who knows about the behavior of horned ungulates, cannot help but regard with abhorrence and disgust.

Although a stab or a push with the horns was the simplest and probably the most primitive fighting method, it is not frequently found among today's highly evolved horned ungulates. The most common fighting techniques are ramming, fencing, and pushing with the forehead. In those species where ramming is predominant, the base of the horns is enlarged or widened. The opponents, who dash towards each other either on all four or only two legs, clash together with their horns. Thanks to the massive base of the horns, which protects the skull like a shield, they can do this with great vigor. In fencing, the counterparts do not meet at the base of the horns but approximately in the center of the longitudinal axis. Therefore, their horns are long like spears, sometimes undulated or bent backward like a sabre. In these fights, they actually exchange blows, and each blow of the opponent is warded off adequately. The encounter resembles a sporting duel.

In fights where the foreheads are pushed together, the opponents lower their horns and push them into each other's horns so that the foreheads touch. Here each animal places the other's head between its horns, like a horse standing between the shafts of a cart. Other species hook together and entwine their horns so that some space is left between the heads. In these cases, the horns serve the purpose of binding the opponents to each other. This doubtlessly is the climax of ritualization, a competitive fight to measure strength without

Fig. 11-4.
Fighting Thomson's gazelles stand on their four legs.

Fig. 11-5.
Ibexes rise on their hind legs for fighting.

Fig. 11-6.
Sable antelopes kneel down on their carpal joints.

bloodshed. In species which have become highly specialized in this method of fighting, for example, the males of the greater kudu, the opponents may entwine their horns so that they cannot come apart again, and both animals will die miserably. Thus, there exists in horned ungulates both too little and too much ritualization with the corresponding transformation of the weapons.

Closely related to wrestling is another fighting method in which the opponent grasps the other's horns from the side and shakes them to the right and to the left, thus forcing an individual rhythm of motion on the opponent and "convincing" him of his own superiority. There are many more fighting methods and a species may use more than one. Sometimes a fight will begin with an exchange of blows, then turn into a pressing of the foreheads, and finally end with wrestling.

Formerly, it was believed that all such fights were about the females, but this is not always the case. Even when there are no females around, hard and vigorous fights start over territories. Of equal importance are the fights among members of the same sex about rank order within the groups. Fights also occur which coordinate the activities of a group or herd. If some members of a group are resting while others have the impulse to migrate, the restless ones will pester the resting ones until finally everyone is on the move. And last, but not least, healthy, strong males will also fight just for the fun of it; such playful fights are an everyday occurrence in gregarious species. It is not always easy to distinguish them from serious fights, since they may be very vigorous and high-spirited.

Almost certainly the hornless ancestors of the horned ungulates also fought with each other, only not with horns. They probably had a complete fighting behavior repertoire to which new methods were added along with the increasing evolution of the horns. This leads to the extremely interesting question of what happened to the phylogenetically older fighting methods and what they may have been like. We may obtain an approximate idea by comparing similarly built animals which are hornless. Such comparable forms are some llama species about whose fighting methods we are well informed thanks to the studies of Gauthier-Pilters. The variety of fighting techniques in these "unarmed" animals is amazing. The opponents jump towards each other, trying to throw each other down with the weight of their bodies. At the same time, they viciously bite and only their thick fleece prevents serious injuries. With their mouths closed, they deal out heavy blows with their muzzles. Raising the muscular neck, they press frontally against the opponent, trying to force him down or to glide underneath him to lift him up or to apply a bite from below. They jostle each other laterally, kick with their hind legs, and beat with their front legs.

Many of these "old" fighting methods are still found to a much

Fig. 11-7.
Male nilgais neck-
fighting.

Fig. 11-8.
A ritualized neck fight in
the mating ceremony of
the greater kudu.

Fig. 11-9.
The foreleg kick
(Laufschlag) was ori-
ginally a fighting
behavior (compare fig.
6-8). In dibatag and
many other horned un-
gulates, it became a
courtship behavior as a
prelude to mating.

greater extent in the unarmed females of the horned ungulates than in the well-armed males or in species where both sexes have large horns. Some of the "old" fighting methods, for example, biting, have actually been lost by the evolution of the horns. The better the forehead weapons are, the less likely it is that the animals will bite. The hornless females of kudu, sitatunga, and nilgai bite vigorously in fights, while the horned males of these species rarely or never bite. This even applies to the development of the individual animal. A young male sitatunga, whose growing up I observed very closely, would occasionally bite in a fight when he did not yet have horns or when they were still very small. But after the horns grew to their full size the snapping ceased. When hornless female antelopes attack by pushing the forehead, they frequently snap *in vacuo*, with their mouths lowered downward into empty space. The "modern" push with the forehead, so to speak, still activates the primitive bite, another indication of how closely these two phenotypically different fighting methods must be linked in the central nervous system.

The greatest surprise with respect to "old" fighting methods comes from the study of the mating actions of males. While such "fighting relics" have disappeared either completely or almost from the fighting repertoire of the males, they often reappear in many species in courtship behavior. This can be traced beautifully in the neck-fight. Fighting guanacos, which are hornless animals, direct their bites towards the opponent's front legs in order to throw him down. When the two opponents do the same thing, their necks naturally cross. Then one will try to press the other down, while the other pushes up from below. In this case, the neck fight is seemingly an almost incidental by-product of the biting fight. In the nilgai antelope where the females are hornless and the males have relatively short horns, both sexes often engage in the neck-fighting ritual but the males no longer bite.

The adult males of the genus *Tragelaphus* are well armed with spiral horns and, according to our present knowledge, do not fight with their necks. Only their hornless females and young do so occasionally. Here the neck fight is the most conspicuous courtship behavior of the males. The first phases of the mating ceremony are, for example, in the greater kudu a regular neck fight. The male pushes his neck over the female's, but he does it in a strongly ritualized manner, since he presses the female's neck down only slightly or not at all. The action has become more or less "symbolic." In the last phase of the mating ceremony, the neck fight becomes more sexually oriented. The male pushes the female's croup from behind with his neck. In the eland antelope, where both sexes have large horns, we find only this strongly sexual touching phase and a mimically exaggerated neck and head stretching where the neck muscles are strained. The regular neck fight, neck over neck, occurs only in the young of this species.

The same is true for other "old" fighting methods, for example, the foreleg kick (Laufschlag) where the male kicks with the front leg, often in an exaggerated manner and ritualized in slow motion, against or between the female's hind legs. Obviously, the foreleg kick and neck fights have the same function. In the same situation the kudu has neck fights but does not do the foreleg kick, while the gerenuk only kicks with the leg and shows no sign of the neck fight. In other species such as the water buck, both are done in succession or simultaneously.

Greater kudu.

Generally, we may say that many of the phylogenetically primitive fighting methods reappear as courtship behavior in the mating behavior of the horned ungulates. But were they really "transposed" there? Probably not. In the horses and camels, the preliminary phases of mating behavior are still a more or less ritualized fight. It is possible that this also was so in the unarmed ancestors of the horned ungulates. There, too, "old" unspecialized methods of fighting behavior may have been part of the mating behavior from the beginning. But while these "old" methods of true fighting became more and more unnecessary with the increasing perfection of the horns, they could well have remained in the mating behavior. While males still use them to demonstrate their dominance over the females, the horns, their major weapon for actual fighting, do not play any part of it. On the contrary, they are actually hidden from the female in some of the courtship behavior, especially when the animal stretches his head and neck forward and when he raises his nose forward and upward. Although the males behave rather aggressively during mating behavior, they "assure" the females at the same time that they do not "intend" to use their horns. Thus, while fighting and mating behavior are unmistakably different, the "fighting nature" of the mating ceremony has actually been maintained.

White-tailed gnu.

This is of great importance. In the mating ceremony of these animals, the female has to walk or stand in front of the male. This corresponds to the behavior of a defeated animal in a fight situation. During copulation, the female is literally in a "submissive" posture. Furthermore, most of the horned ungulates are "distance animals," as Hediger called them. The partners keep a certain minimum distance (individual distance) from each other or else there will be conflict. In copulation, the male necessarily has to break through the female's individual distance barrier. This requires some aggression on the part of the male so that the female will see him as dominant and will therefore tolerate his approach without serious defense. On the other hand, he must not behave so wildly that she runs away or there could be no mating. The "old" fighting methods, which have become transformed into harmless "modern" horn fights, suit these rather complicated requirements perfectly. Besides, there are still other "nonaggression guarantees," appeasement gestures, and affectionate

Oryx

Fig. 11-10.
The copulation positions of the different horned ungulates vary greatly.

behavior on the male's part and provocative and warding-off gestures by the females, which may alternate or be interwoven in various ways. These truly contradictory tendencies in both partners reach a rather fine state of equilibrium by various and very distinct mating ceremonies, which are the prerequisite for copulation.

Driving forms the basic structure of the mating ceremony. There the female strolls along in a more or less ritualized flight and the male follows, often over long distances, however mostly in winding, curving or circular paths. This appears different depending on the species or group of species. Even in copulation there are surprising differences in the male's posture which may even be significant in the zoological classification of the species. Sheep, goats, and some antelope males embrace the female's flanks with their front legs and rest the chest on her croup, and keep their neck freely erect. Hartebeests in the same situation rest their noses vertically on the female's withers from above. The Tragelaphinae males lay their necks and heads far over the female's back. However, the male gazelles raise steeply, do not embrace the female's flanks and do not support their chest. They are able to copulate while walking.

A strange phenomenon in the mating ceremony of horned and many other ungulates is "flehmen" (lip-curl). A male sniffs or licks at the fresh urine of a female, raises his head, opens his mouth slightly, and pulls his upper lip upward. Often he will turn his head aside and remain motionless for a moment, as if he were in a trance. Finally he licks his lips. All horned ungulates have a narrow duct lined with mucous membrance, the so-called Jacobson's organ, in the upper jaw between the oral and the nasal cavity. In "flehmen," probably the scent and gustatory substances reach this organ and the male consequently learns about the female's condition in the estrous cycle. Afterwards he either loses interest in the female or he intensifies his driving. Although "flehmen" can be stimulated in an experiment with other scents the female's urine seems to be the releasing factor under normal conditions. The females also show "flehmen," but not as frequently as the males.

The mating succeeds only when the pair is not disturbed during driving. Several males cannot chase the same female because they would get in each other's way and begin to fight. This, too, is arranged differently in different species. In some cases, the males of a group establish a rank-order long before the mating season begins; hence it has been decided which is the "alpha animal." The "authority" of the highest ranking animal is such that none of the other males would dare to get in his way during the mating ceremony. He is the only one to have the right to copulate, and the other males may stay nearby or even in the herd without causing any disturbance. In other cases, a male separates himself from the others with a group of females and does

Fig. 11-11.
The male Thomson's gazelle has sniffed the female's urine and shows the lip-curl (flehmen).

Animals of the Indian Dry Forest

Two-toed ungulates: 1. Gaur (*Bos gaurus*, see Ch. 13). 4. Arna (*Bubalus arnee*, see Ch. 13). 5. Axis deer (*Axis, axis*, see Ch. 8). 6. Indian sambar (*Cervus unicolor*, see Ch. 8). 8. Nilgai (*Boselaphus tragocamelus*, see Ch. 12). 9. South Indian wild boar (*Sus scrofa cristatus*, compare Ch. 4). ☐ Proboscidians: Indian elephant (*Elephas maximus*, see Vol. XII). ☐ Predators: 11. Tiger (*Panthera tigris tigris*, see Vol. XII). ☐ Rodents: 7. Palm squirrel (*Funambulus palmarum*, see Vol. X). 10. White-tailed porcupine (*Hystrix leucura*, see Vol. XI). ☐ Monkeys: Entellus langur (*Presbytis entellus*, see Vol. X). ☐ Birds: 12. Woodpecker (*Brachyternus benghalensis*; see Vol. IX). 13. *Buceros bicornis* (see Vol. VIII). 14. *Molpastes cafer* (see Vol. IX). 15. *Megalaima haemacephala* (see Vol. IX). 16. *Acrocephalus stentoreus* (see Vol. IX). 17. *Aethiopsar fuscus* (see Vol. IX). 18. Wild jungle fowl (*Gallus gallus*, see Vol. VII). 19. Peacock (*Pavo cristatus*, see Vol. VII). ☐ Reptiles: 20. *Calotes jubatus* (see Vol. VI). ☐ Insects: 21. *Erasmia sanguiflua* (see Vol. II). 22. *Papilio clythia* (swallowtail family, see Vol. II). 23. Weevil (N.A.) (*Protocericus colossus*, see Vol. II).

not tolerate any other male nearby, although such "harem groups" with their "pasha-males" seem to be much rarer than it was formerly believed. However, another custom is prevalent in the horned ungulate males. They occupy, often long before the mating season begins, distinct territories and fight or ward off every male who comes in to this "prohibited" area. This is advantageous in several respects. The occupant of the territory does not need to get upset at every conspecific of the same sex around him, but only at those that enter his territory. Furthermore, the neighbors as well as the "bachelors" who are not territorial come to know his territorial boundaries and avoid them. This saves energy. Since the territories are often relatively small, quite a few males can have their territories in a limited habitat and thus take part in reproduction. Such a "field" of territorial males breaks up the large female herds into small groups. Each of these small groups then remains for a shorter or longer period with a territorial male. It is easier for him to find females in the proper phase of estrus from such a small group than it would be in a large herd.

All kinds of territoriality are found among the horned ungulates. Some species may be territorial for years or even for life, while others establish their territories only for a season, which may last anywhere from a few hours, to days, or months. Within the same species or even subspecies, populations of a certain area may be distinctly territorial while in another area they will occupy territories only occasionally. These territories also vary greatly in size. The Uganda kobs have territories of only fifty meters in diameter, while the territories of the approximately equal-sized Grant's gazelle may be two kilometers long. There are all gradations in between. In some species, only the males are territorial, while in others apparently both the male and the female are. In some the same group of females will stay for weeks and months in the same male's territory, although in others the female herds pass at best daily through their territories and stay for a while with the males. Sometimes all the females of a specific population take part in visiting the male's territories and, in other cases, apparently only the females in heat do so.

Some species mark their territories with urine, feces, or secretions from special skin glands which are applied to grass, bushes, and twigs. These scent marks help the occupant of the territory to find his way around the area. Furthermore, they indicate to conspecifics that the place is occupied, even if the owner has left it for a short while. However, the fact that a horned ungulate marks does not necessarily mean that he is also territorial. The animals may also mark their migration routes or special points of interest for orienting in an area with scent material. When an individual animal frequently marks a limited area, he identifies himself as the owner of that territory, not by the markings per se but by the high number of individual markings and

Fig. 11-12.
The male blackbuck "marks" by applying some of the scented secretion from his preorbital gland on a twig.

the manner in which they are distributed. This way the animals may emphasize the boundaries or the markings may be so arranged so that they are denser in the center. In some cases, they even combine the two methods. They mark the boundaries with fecal piles and apply their gland secretions increasingly near the center of the territory. In some species, males and females mark, while in others only males do. Finally, there are species which do not mark at all with gland secretions. They open their glands only in situations of great excitement, in fights, or in mating, when they cover themselves with a "cloud of scent," which may give greater security to the individual animal and intimidate rivals.

Territories are often mistaken for home ranges where single animals or certain herds stay for either a short or a long time. These home ranges are usually much larger than are the territories. The animals divide them, like territories, into resting areas, watering places, grazing areas and defecation places. They may occasionally mark them, but they do not establish boundaries where they would behave differently within as opposed to outside of them. While the animals in the home ranges generally live in rather peaceful coexistence, in their territory the fighting drive is increased and the occupant attacks intruding conspecifics of the same sex. Roughly speaking, the territory is a "defended" area, while there is no fighting about the home ranges.

Gregarious species, which live in large numbers, are especially inclined to become territorial during the mating season. The males then leave the herd and the female groups are mostly smaller in the male's territories than outside of them. Territoriality may be viewed as a method of avoiding the disadvantages of mass formations while still being part of a large aggregation. It may also help to prevent an overgrazing of certain habitats, which is of great importance in arid steppe regions which have little water. As a side effect of territoriality, the herds scatter and distribute themselves better in favorable months, during and after the rainy season. Thus, the forage plants are fully utilized and the feeding areas do not become overly utilized.

When the food supply decreases in the dry season, only a radical measure, migration, which is the exact opposite of territoriality, can help. While territoriality is distinguished by an inclination towards separation and absolute readiness for fighting and mating, migration is notable in that sexuality and aggression reach their lowest level as the need for association increases. Males and females come together in compact herds, often in large mixed associations. Between these two poles, the lives of many steppe animals run their course. Their annual migrations insure that during droughts the vegetation will not be totally destroyed. Therefore, it is easy to understand the disastrous effects which result from replacing gazelles and antelopes with domestic animals in just such arid regions. Neither cattle nor other domestic animals have such a "mechanism of distribution" nor do they ever migrate on their own impulse. They remain compact masses in an area until the shepherd drives them farther on, which usually occurs after the last blades of grass are consumed.

In the horned ungulates of the forests and mountains, the varied environment prevents the large aggregations. Inevitably there is always something between individuals, a bush, a tree, or a rock, which impedes coherence and establishment of larger herds. For this reason, the animals there naturally live further from each other. In much more uniform steppe country an animal itself becomes a "landmark," a reference point in the landscape which actually invites approach and association. Therefore, the smaller antelopes in the dense forest live singly in "permanent territories," in pairs, or in very small groups. The wild sheep and wild goats in the mountains do not seem to be territorial at all.

Except for periods of migration and mating, the formation of groups of the same sex is the rule in many horned ungulates. Usually within a species, the female herds are larger than those of the males, the "bachelors" herds. In such groups, some rank-order also exists, but it diminishes with the increasing size of the groups or herds. Often there is an "age hierarchy," where each adult male in good health is necessarily superior to a younger adult male, just as the young adult is to a subadult male, and the subadult to an adolescent male. Within these age groups, no strict rank order is evident but one individual may control one or two others. The same applies to female herds, although in some species, the mothers have distinctly closer relations with their older, subadult, or adult daughters even after the birth of another young. In captured white-tailed gnu, such a bond was still evident after two years.

There are two kinds of relations between mother and young in horned ungulates, the "stay put" (Walther) or "lay out" (Ewer) type and the "follower" type. In the "followers," the young animal remains in close contact with its mother during the first days and weeks, follows her like a shadow and often rests nestled against her. In the "stay put" type, the young lays down on its own in a hidden place and remains there when the mother goes considerably farther away from it and returns only at the next nursing period. Then she usually does not go directly to the young's hiding place but calls from a distance. The young hurries to her, is nursed and licked, remains for a while up on its legs, jumps around her, and finally walks away on its own to lay down again in the high grass, behind a bush, or in another adequate place. Naturally, a "follower" may also rest some distance from the mother, just as occasionally a young "stay put" young may walk behind its mother. Furthermore, there are species with a greatly shortened "stay put" period. Interim solutions and gradations are also

Fig. 11-13.
The newborn chamois is a "follower." Immediately following birth, it walks behind its mother.

Fig. 11-14.
Young gazelles are the "stay-put" type (Grant's gazelle).

possible and exist, but in species which have been studied closely in this aspect, one or the other "type" was prevalent.

In birds one spoke originally of "altricial animals" (those who remain in the nest) and "precocial animals" (those who follow the mother shortly after hatching). These terms have been modified for mammals. The "lay out" type were considered those young that are born in a rather undeveloped state at birth, while those young which can use their senses and limbs shortly after birth are called the "follower." According to this definition, all the horned ungulate young would be of the "follower type." However, this does not exactly agree with reality, because the term "follower" is not quite adequate for young who, although fully developed at birth, do not walk around but "stay put" for most of the day in their hiding places. In this case, the distinction between the "stay put" or "lay out" and the "follower" types is much more correct and absolutely necessary. Formerly, it was believed that only the human baby was such an "altrical precocial animal," hence man held a special position in the animal kingdom in this respect. The evidence from the horned ungulates has shown that this presumption is rather doubtful.

Upon closer observation, more differences between "stay puts" and the "followers" have become evident which are rather significant. If a young gnu, a typical follower, is not standing on its legs with 10–20 minutes after birth, I would seriously expect something to be wrong with the animal. In a young of the "stay put" type, however, I would not expect to see it standing this soon and I would not worry too much even after forty-five minutes. After one hour, however, it would be about time to get up. Once a little "follower" has struggled onto its legs, it immediately will stand correctly on its little legs while, as a rule, the "stay put" type young walk by touching the ground with the pastern joint during the first days. In both types, the young are licked dry by their mothers right after birth. Later, however, only the mothers of the "stay put" type lick their young while they are nursing them or thereafter in the ano-genital region. During the first days, they will also eat the feces and drink the urine of the young. This maternal behavior is obviously related to the "staying put" of the young. The best possible protection for such a little fellow, besides its vanishing from sight, is its complete odorlessness. It already has all the scent glands, but they probably do not function yet. Thus, the only things which might disperse odor and thus endanger the young are their urine and feces. Therefore, the mother stimulates urination and defecation far away from the hiding place by massaging with her tongue. She furthermore takes each particle from the young's coat which might later produce odors. Afterwards, the young is as odorless as is possible by mechanical means.

To nurse, the "stay put" young responds to the calling mother and

Fig. 11-15.
The young blackbuck "stays put" near a tree trunk.

Fig. 11-16.
The mother approaches and the young runs to her.

Fig. 11-17.
The mother tests the young's scent.

Fig. 11-18.
The young is nursed and soon after it returns to its place.

later goes back to its hiding place by itself. The mother usually does not set foot on this place and with good reason. Many horned ungulates have scent glands between their hoofs which impregnate their tracks with a special odor which facilitates the finding of conspecifics. Naturally, this "perfume" is also conspicuous to predators which have a good sense of smell. Therefore, the mother's tracks had better not lead to the young's hiding place because then jackals and other predators might easily discover it. As mentioned before, the young's scent glands are not yet functioning. In some species, even the mother herself is not able to follow her young's track. She must remember the hiding place, and she is rather helpless if she does not find the young there at the next nursing period. In such cases, she will often search for miles for him, but only visually. I have often seen predators which did not detect the "lying out" young with their sense of smell. Hyenas and jackals passed only a few meters' distance from the "lying out" young gazelles. When their path led by chance directly to the young, these would finally jump up two or three meters from them and run away, and a wild chase began. Hyenas, jackals, and baboons frequently attack young gazelles when they are with their mothers and on their legs during the short nursing period.

Thus, the young gazelles of the "stay put" type who, anthropomorphically speaking, seem poor and abandoned creatures, are in no worse a situation than the "followers." For example, it often happens that a young gnu following its mother is seized by a hyena. Although both the "stay put" and "follow" method are dissimilar, they are of equal biological value. Both have advantages and disadvantages which are approximately balanced. Interestingly enough, "following" is found more frequently in the alpine and the truest steppe animals, while "staying put" occurs in forest and bush dwellers. Therefore, these kinds of young behaviors seem to be not much an adaptation to avoid predators but rather an accommodation to other environmental conditions.

In addition to the species-specific characteristics, certain external and internal stimuli also determine "following" as well as "staying put." When a young has had enough to eat, it feels like laying down and staying put, while hunger may finally bring the most determined "stay put" young on its legs and increase its readiness to follow. The external sign stimuli for "staying put" are a "depression in the ground and something vertical" which may be a tree, a bush, a rock, a termite heap (in a zoo the walls of the stall, the fence, or even the legs of a person) and, in some cases, "some kind of a roof," e.g., the canopy of a tree. The releasers for following are more varied and complex. The young perceives mainly visually the mother's movements as the object to be followed. The mother is closest to the young. She is larger, heavier, faster, and she walks in front of it or passes it. Therefore, the

What is innate in horned ungulates?

young animal reacts to proximity, mass, size, and quick movements. Planned and chance experiments, however, have shown that it is also possible for other objects as well to release the reactions of the young. Thus it can happen that the young in exceptional cases while it is being nursed, follows another fast-moving, large object, e.g., a car. It is unable to recognize its mother individually until after it is several days old.

The same is true with the olfactory, tactile, and sense of smell. Many young horned ungulates seem to react innately to sound. Usually, the first thing they hear is the contact call from the mother. After a short while, they react only to such calls and, sometime later, exclusively to the individual calls of their mother. However, when they are bottle-raised by humans, they can also become "imprinted" on a whistle, the horn of a car, or a human word. Olfactory imprinting to the mother or a "substitute mother" occurs even more rapidly. In any case, young horned ungulates are able to learn during the first days and even hours of their lives. Once a young white-tailed gnu less than two hours old fell down a steep slope. From this one bad experience, it immediately learned how to behave the next time at a steep slope. My young bleshok "Mausi" behaved in a very similar manner when she stumbled down a stairs on the first day of her life.

What the young
animal must learn

Therefore, when compared with human infants, the young horned ungulates have a considerably greater learning potential. At a later age, they do not remain at this high level or even become "stupid". Actually they grow up into their own respective worlds, which is so unlike our human. Hence it is quite difficult to understand them. Usually we are well able to understand and interpret the actions of the primates who are closely related to us, although it becomes more difficult when we come to dogs and, what is even worse, to the horned ungulates. To understand them requires intensive work and long, thorough studies of their behavior. The intelligence levels of the various kinds of mammals, which man determines, do not in most cases correspond to their actual intelligence but rather to the standards we apply to ourselves. At best, they may give evidence to what extent there are or are not similarities between that species' psychological and physical state and ours.

Birth in horned ungulates takes place within one to two hours. The mothers then stand or lie down, either more or less on one side or alternating between the two. Sometimes the birth begins while the mother is lying down and ends when she is standing. Immediately after birth, the mother turns to the newborn and, in a licking-eating motion, removes the fetal membranes. Then she licks the young's whole body and especially the head. This is apparently important to stimulate standing. Sometimes, it looks as if the mother were pulling the young up while licking at its head. The placenta, which comes out one to four

The birth of horned
ungulates

hours after birth, is eaten by the mother. These true herbivores thus become temporarily carnivorous, as Hediger puts it. Some species, like the gazelles, eat the placenta piece by piece as it comes out of the vagina so that practically nothing touches the ground. Others, such as the eland antelopes, bring the placenta out in one piece and then gulp it down. And still others, the gnus, might do the same but in the wild they could only get a few bites because vultures immediately converge on the placenta and eat it.

Apparently, all horned ungulate mothers eat important substances in the placenta. Thus, it is only superstition when farmers prevent their cows, sheep, and goats from eating it because it "would be harmful to the animals." In contrast, there are observations from zoological gardens of some mothers who did not eat the placenta and would then not accept their young. In any case, this should be a cue for zoo people that something might be wrong, especially in "stay put" animals where it is naturally much harder to determine whether or not the mother has accepted the young than it is in the "followers." According to the observations of Dieter Backhaus, it is important that a maturing female watch a birth before she have her first young. Not all young females in captivity have this opportunity, and this may be an explanation of why primiparous females sometimes do not accept their young.

Infantile behavior in adult animals

Some typical behavior patterns of young appear in an altered form at a later age. The "staying put" changes in the older animal into a "lying down in fear." During the first days after birth young animals that are frightened while they are walking around, will drop flatly to the ground and remain down. The behavior of adult animals is the same when they adopt a submissive posture. Following does not disappear completely at an older age. A moving conspecific stimulates an adult animal to follow him, especially when it is in a mood to wander or to migrate. This may result in the single files of migrating wild herds whose members partly drive each other and partly "pull" the others along.

The different ways in which animals groom their bodies, their skin, and their coats were described by Hediger with the term "comfort behavior." Horned ungulates groom themselves by nibbling with their lips, licking with their tongue, gnawing with their teeth or with the mouth open, scraping with the incisors in the lower jaw. They also scratch themselves with their hind hooves and horns. Except for minor species-specific differences, they all behave more or less the same. Except for the wild oxen, which are somewhat less capable in this regard, a horned ungulate is able to reach practically all parts of its body by at least one of these means by itself and to rub on them. Horned ungulates also often lick, nibble, and gnaw each other. Some species do so more than others. Very strikingly in social grooming, they prefer to groom the front part of the other, the head, the neck, and the shoulder region. The mother, in contrast, takes care of the young's whole body when licking it dry, legs included. During the mating ceremony, the males preferably lick the females' genital region or the croup. Except during mating and care of young, comfort behavior in horned ungulates is not directed at all to such places on their partners which are especially dirty or which are hard for the animal to reach by itself. Since in such a relationship those animals groom each other who are "on good terms" with each other, we may assume that hygiene is not as important to the animals as the gentle-friendly touch. When horned ungulates groom and scratch themselves during combats, this, like grazing in such situations, is a kind of conflict behavior.

Other forms of comfort behavior are restricted to only a few species. Wild oxen like to roll on the ground and goats like to rub themselves on objects. However, most of the "antelopes" do neither one. Pawing is a peculiar behavior. In wild oxen, gnus, and other species, it serves to prepare a place to roll or wallow in. In addition, in this and in other species, it occurs in searching for food, before lying down, before urinating and defecating, and as a threatening gesture before fighting. Goats frequently paw before they lay down, but almost never before they urinate and defecate. Gazelles do just the opposite. The males and females of some species paw about equally frequently, while in others this action is predominantly or exclusively up to the males. Tragelaphinae paw only rarely. As the horn is the most important "tool" for encounters among conspecifics, so the hoof is the most important "tool" for coping with the environment. The walk of the horned ungulates may be divided into two types: the cross-walk and the amble. During the one phase of the cross-walk, the front and the hind legs on one side of the body come close together, while the legs on the other side are farthest apart from each other. In the amble, the front and hind legs on the same side reach forward synchronomously, while those of the other side reach backward. Slow motion pictures show that a pure amble is very rare. Usually the hind leg is somewhat ahead of the front leg on the same side of the body. Furthermore, even true amblers may cross-walk under certain conditions, for example, in a very slow forward motion or in climbing a steep slope. The amble may occur vice versa in animals of the cross-walking type. Despite these transitions and exceptions, generally we can clearly distinguish between the amble and the cross-walk in horned ungulates, it is species-specific. Amblers are mainly steppe animals and closely related species which live in the brush and fringe forest along rivers. The cross-walk is found in forest and alpine species. Since the more primitive duikers and dwarf antelopes are exclusively cross-walkers, we may consider the cross-walk to be the older form of locomotion.

All species cross-walk while trotting. The gallop, which looks different in the horned ungulates, may be defined as the "jumping

Fig. 11-19.
The markhor in the cross-walk.

Fig. 11-20.
The dibatag is an ambler.

gallop" and the "rocking horse gallop." In the "jumping gallop," the animal begins with the front legs, stretches the body, and comes down again with the front legs, one after the other. Meanwhile, the hind legs leave the ground and almost simultaneously reach forward, sometimes passing the outside of its front legs. The hind legs then return to the ground and the animal begins the next leap with the front legs. While the body is stretching and contracting, all four legs are off the ground. This is called the "floating phase." The floating phase in stretching may, especially in small and medium-sized forms, change over into a long jump, where the animals keep the front legs stretched forward almost horizontally and the hind legs angled closely towards the body. In small species and forest animals, the jumps in a gallop consist of a series of single, high jumps which are clearly interrupted as one jump follows the other. In contrast, steppe animals keep their bodies close to the ground and their legs work underneath like a well lubricated engine, so that the single jumps merge into one another and the animals dash along like arrows. There is no "floating phase" during stretching. During the floating phase when the body is contracted, the animal makes a rocking movement in the air, bringing the hind legs forward with it and putting them back on the ground. The rocking movement is especially obvious in the hartebeest. Some species perform both types of gallop.

The jumps in horned ungulates are of the same kind. In addition to a "jump over something" which is found in forest and alpine animals, there is the jump onto something which only alpine animals have. Steppe animals have a special kind of jumping, the "stotting," where the animal springs upward from the ankle joints with all four legs at the same time, bringing the legs forward and landing on the ground with them almost simultaneously. The animals do not intend to surmount obstacles with this jump. Even horned ungulates in which "stotting" is especially prevalent do not like to jump over ditches or other horizontal obstacles, as well as bushes, fences, or other vertical obstacles. They will not jump at all or only in cases of extreme emergency. "Stotting," which serves the purpose of moving forward, is often an alarm signal which seems to be coupled to the increasing or decreasing excitement of the animal. Therefore, these jumps often occur in running games or at the beginning or the end of a flight, but not when the enemy is close and closely pursuing the animal. If we take "stotting" to be an alarm sign, it is quite comprehensible why young animals are more inclined to use it than are adults. In this manner a young may attract its mother's attention to the fact that it has left its hiding place and it will "call," if possible, for help. The question remains unanswered as to why an adult gazelle will flee from hyenas and wild dogs predominantly in a stot, while they do this much less frequently when chased by cheetahs and lions.

Animals do not have verbal communications as humans, but they are able to communicate with many different expressions—by touch, scents, sounds, postures, and gestures. Some part of olfactory communication in horned ungulates is by the above mentioned scent gland secretions, although there is little known about just what information is passed on and how it influences the recipient. We know a little more about vocalizations. We can distinguish mainly the calls of alarm and excitement, of fear and pain, contact calls, as well as threat, mating, or driving calls. In some species, these calls are not made in the oral or throat cavity but they are nasal sounds which may be described as a kind of snoring or snorting. Sometimes (e.g., in gazelles) the frontal nasal cartilages are transformed into resonance chambers. Such vibrating calls may sound different depending on the distance from which they are heard. Many sounds of horned ungulates are so soft that a person has to stand nearby in order to hear them. In their softness, these sounds really do not seem to be appropriate to the large or medium sized animals that produce them. For quite some time, I thought that a certain call was a bird's twitter, until I finally found out that it was the driving call of the bushbuck. The ibex, springbuck, dik-dik, and other species utter piercing alarm whistles through their nose.

Today, we are best informed about visual expressions. Although horned ungulates have facial expressions, they do not seem as important to them as in primates or even in man. Ear and tail movements also do not seem to be as important for intraspecific communication as one may expect. Instead "pantomime" predominates in horned ungulates. It begins with an orientation of the whole body towards the partner, laterally from the front, frontally, oblique, or at a slant from the rear, etc. The meaning depends on the orientation of the "recipient" to the "sender." For example, when one member of a herd turns its hindquarters towards another member, it makes a difference whether or not the other is oriented towards him with the head or its hindquarters. In the first case, it could be an invitation to follow but, in the second, it expresses complete indifference and, therefore, constitutes an "offer of peace" in partners who are not on the best of terms. It is surprising how much horned ungulates are able to express by these simple means of physical orientation towards one another. There are more gestures: the animals stretch upward, decrease in size, or bend their back; they stretch the neck out horizontally, deep down, or steeply upward; they pull the head towards the body or stretch it forward etc. An emphasized forward stretching of the head and neck may be an invitation to a neck fight and thus mean a threat, if the neck fight occurs in the species in question. If it occurs only in the species' mating ceremony, this posture means courting. In species which make no use of the neck in this way, the same gesture may perhaps express inferiority or submissiveness.

Fig. 11-21.
In white-tailed gnu, stretching the neck forward is the beginning of a submissive gesture.

Fig. 11-22.
Submissive lying down in front of a higher-ranking animal is the complete submissive gesture in white-tailed gnu.

Fig. 11-23.
The lowered horns are a threat posture in a male eland.

Fig. 11-24.
The threatening male Grant's gazelle "weaves," rhythmically beating his horns against plants.

It is important for the animals that the expressive gestures are neither mistaken nor overlooked. Naturally quick movements are striking, but they are of short duration. Therefore, the horned ungulates frequently have repeated, rhythmical expressive gestures. Movements in an exaggerated slow motion may also attract the others' attention. Therefore, a slowed down and emphasized walk is widespread as a means of expression. Furthermore, there are actions, especially attack movements, which may be "symbolic." They are performed at some distance from the partner and redirected into the air or to another object, a bush, a tree, or the ground. Humans act similarly in situations when they bang their fists on the table instead of hitting the other person. Sometimes this "symbolic aggression" is directed away from the partner while at other times it is directed at him. Of course, the latter is much more challenging. In any case, threatening actions, the gestures derived from them and those very similar to them play an extremely important part in the life of horned ungulates, not only in actual fights but also in mating behavior and as a means of transmitting moods within a herd. An unbelievable number of gradations is possible. These may roughly be distinguished as:

a) Threatening in a narrow sense, which expresses an immediate readiness for fighting ("I shall fight!"). The weapons are presented to the opponent in a posture from which the animal may attack immediately. b) Demand for space ("This place is mine"). It is emphasized by actions which have some reference to space, for example, marking. c) Display behavior ("I am the greatest"), which expresses a claim of superiority without a direct intention of fighting. Therefore, the animals do not display by presenting their weapons but by demonstrating them by size, they predominantly stretch themselves upward and/or stand laterally to the other.

Fig. 11-25. Mutual display (male Grant's gazelles).

All these actions, depending on the condition or mood of the opponent, may work in different or even almost opposite ways. It will intimidate a weaker conspecific or one who is not ready to attack and cause him to retreat. But it acts as a challenge to an opponent of equal strength or a stronger one who is ready to attack. Then the opponents increase each other's fighting mood by threats and counter-threats until they finally fight. There is no question that many fights are avoided by the threatening gestures and related forms of behavior. The "differences of opinion" are settled on this level of expression which is well suited for conserving strength. Therefore, horned ungulates attack each other only in the rarest instances without any preliminaries. Usually they try to settle a quarrel first with threats, territorial or other, or displays.

Fig. 11-26. The male nilgai stretches his neck forward in a threatening display.

Most horned ungulate species are under heavy pressure in the struggle for life. What someone once said about the rabbit applies equally well to small and medium-sized species: "Each and every one wants to eat it." Therefore, one might presume that these animals live in continuous fear and readiness for flight, and that they had developed special methods to avoid predators. But this is not the case, at least not to the extent to which it might be expected. Naturally these animals are always on the alert, but they do not lead a life of terror and constant restlessness at all. They do not necessarily flee as soon as they are aware of the presence of a predator in the vicinity. They flee only after the predator has approached to within a certain distance (Hediger's flight distance). This flight distance differs according to predator species and the situation. In some cases, it is amazingly close. Hyenas, for example, not only attack young but also adult gazelles and gnus. Nevertheless, the Thomson gazelles in the Serengeti Plains allow hyenas, in most cases, to approach within 30-60 meters before they retreat. Their flight distance from lions is 100-150 meters. They often flee from cheetahs already at 200-400 meters, and from wild dogs at one kilometer. Since cheetahs and wild dogs are the most specialized gazelle hunters, "bad experience" may well play a part.

Sometimes gazelles may flee from hyenas when they are still 500 meters and farther away from them. An old hunter's tale relates that animals of prey could determine by looking at the predators whether they were full or hungry. Actually, the behavior of the predator either releases or promotes flight. Thus, the flight distance is shorter towards a lying hyena than from a standing one; it is larger from a walking hyena, and greatest yet from a galloping one. Furthermore, it makes a difference whether the predator is only passing by or coming straight towards the herds, whether it was visible for a long time or suddenly appeared, whether it was only one animal or several, and, in the latter case, whether they were in hunting formation or not. Certainly horned ungulates also learn by trial and error, but often sign stimuli seem to be of greater importance although they may lead to mistakes. For example, a wart hog which suddenly appears and dashes towards the herd may cause a mass flight even though wart hogs usually do not endanger adult gazelles and, at other times, may even graze among them.

If a predator has started to chase a herd, the most important task for the individual animal is to discover as quickly as possible: "Does it mean me?" The simplest method for the fleeing animal to be sure is to cross directly in front of the predator's path in a quick dash. If the predator follows then, all efforts are made to escape and if he does not, the problem is over. Obviously this "test" is rather widespread. The animals often show this behavior towards passing cars, which sometimes places quite a strain on the driver's nerves in areas with much game.

Compared to the various forms of fighting, mating, and expressive behavior, flight behaviors are rather simple. At first, the animal dashes

▷
The spiral horns of the greater kudu (*Tragelaphus strepsiceros*) are not effective in warding off enemies. But in rival fights among conspecifics, this horn shape is meaningful.

▷▷
Delamere's bushbuck (*Tragelaphus scriptus delamerei*), a subspecies of the bushbuck in Kenya. These pretty antelopes, which are the size of a white-tailed deer, live in the densest plant cover and are difficult to observe.

Flight signals

Mass flight protects the individual

Spiral-horned Antelopes

1. Nyala *(Tragelaphus angasi)*
2. Lesser kudu *(Tragelaphus imberbis)*
3. Greater kudu *(Tragelaphus strepsiceros)*
4. Mountain nyala *(Tragelaphus buxtoni)*
5. Bongo *(Taurotragus euryceros)*
6. Sitatunga *(Tragelaphus spekei)*
7. Bushbuck *(Tragelaphus scriptus)*
8. Eland *(Taurotragus oryx oryx)*
9. Giant eland *(Taurotragus oryx derbianus)*
10. Nilgai *(Boselaphus tragocamelus)*
11. Four-horned antelope *(Tetracerus quadricornis)*

Wild Oxen

12. Tamarou *(Bubalus arnee mindorensis)*
13. Anoa *(Bubalus depressicornis)*

Hartebeests

14. Topi *(Damaliscus lunatus topi)*
15. Hunter's sassaby *(Damaliscus lunatus hunteri)*
16. Lelwel *(Alcelaphus buselaphus lelwel)*
17. West African hartebeest *(Alcelaphus buselaphus major)*

More subspecies of hartebeest and sassaby are portrayed on page 407.

Horned ungulates as domestic animals

off. If it is not then immediately pursued, it finally stands still and looks back. Forest and brush animals like to "hide" in such situations behind bushes and trees. Dwarf forms often lie low. When severely pursued, gazelles double like hares and swamp antelopes take to the water. However, this exhausts the possible means of escape. Large numbers and the general turmoil in the beginning is of advantage to herd animals. By confusing the predators, it makes it difficult to orient the attack towards one individual animal. They snap in all directions and chase after several animals before they often find themselves alone without having caught a prey. However, if they succeed right from the beginning in focusing upon one single animal, it will be caught in at least seventy-five per cent of the cases. No hunting method of predators is perfect, and neither is any flight method of the prey; just for this reason the two may coexist. Many species make up for their losses with high birth rates. Therefore, it is more important that each female become pregnant again as soon as possible after a relatively short gestation in order for these complicated mechanisms to avoid and to ward off their enemies and evolve. Of crucial importance for this purpose are the social organizations and the perfection of the mating ceremony. Both are highly evolved in many horned ungulates, and in this lies their biological strength. They reproduce, so to speak, in a "race with death."

When observing the varied species and often large herds of African steppe game animals who are always threatened by various predators, one can hardly escape the impression that the predators' hunts of horned ungulates, in many cases, mean a little more to them than a thunderstorm. It is possible to take precautions and, when something happens, to do something about it, but basically one is powerless. Lightening strikes, leaving one dead and sparing the others. Peace returns and life goes on.

Man has taken some of his most important domestic animals from the horned ungulates—cattle, sheep, and goats. In ancient Egypt as in some other early cultures, domestication had been tried with many species. Besides deer, horned ungulates have always been man's preferred game animals. Therefore, a hunter occasionally caught an animal alive or tried to keep and raise a young. Men probably discovered rather soon that it was easier to have animals around which were tame or half-tame so that in case of need, they could be slaughtered; this was far easier than stalking wild game. Obviously not all species were equally suited for domestication, and by trial and error man eliminated most of them. However, this does not mean that there are no other species that might be bred and used as well. With the Europeanization of much of the world, the white man has distributed his domestic animals all over the earth, crowded out the endemic domestic animals, and consequently neglected the utilization of native game animals. This process has often been disadvantageous for vast regions and the game living there.

Those wild species which are candidates for domestication must not be food specialists, since an animal which has highly specific food requirements, e.g., for specific plants, is more difficult to keep. Since even the most fertile horned ungulate does not have as many young as rabbits or other small mammals, raising them for meat, leather, and wool was worthwhile only in medium-sized to large species. All pigmy forms, therefore, were eliminated right from the beginning. Furthermore, the animals had to get along well among themselves and, to a certain degree at least, be suited for contact with people. Therefore, man could successfully domesticate only gregarious animals; it would not have worked with loners. Furthermore, for obvious reasons, territorial species had to be excluded, and those with highly complex mating ceremonies, and which were especially aggressive. In addition, species whose young are either followers or at least have only a short "lay out" period are easier to handle than those whose young "stay put."

Certainly man later bred domesticated animals to suit his purposes, but the above mentioned characteristics probably were in the primitive forms from the beginning. Such wild species still exist today and interestingly enough, they are either the immediate ancestors or close relatives of our domesticated animals. Therefore, if cattle behavior seems to be limited if not downright boring compared with the behavior of other wild ungulates, this has not necessarily been a direct consequence of domestication. Man probably has managed to get only such species "under control" which had certain natural suitable predispositions in this respect. Whether a "loss of instincts," has taken place in our domestic animals, as it is sometimes said, cannot be satisfactorally answered as yet. Therefore, too little is known about the domestic animal's behavior, but there is no question that there are some important and extremely interesting problems to be discovered about man's own "self-domestication."

The history of the horned ungulates is a part of our own economic history through the millennia. It begins with the Stone Age hunter who stalked bison and ibex and leads to modern cattle farming where the breeding stock is artificially inseminated and the milk is taken from the cows with electric milking machines. For millennia, men as hunters of game and as herdsmen of domestic animals in Africa, Asia, and North America have lived exclusively or substantially on horned ungulates. Some of these people, at least, were aware of this fact. In a Kiowa Indian tale, the Great Spirit told them at the time of their creation: "Here are the buffalo. They shall be your food and clothing. But if one day you see them perish from the face of the earth, then know that the end of the Kiowa too is near." Never has a "spirit"

Horned ungulates as prey

spoken more truthfully to his people. The destruction of the immense buffalo herds by white hunters also removed the last of the prairie Indians. The history of the settling of the American west by the white man has mainly been the history of cattle which, driven by cowboys, came to the country in endless herds. Even now the economic existence of whole nations, for example, Australia and South America, depends on the sheep and cattle which man has introduced there. For some of the young African nations whose vast desert and steppe regions are not suited for cattle farming, it is now a fateful issue to see whether they will be able to sensibly use their natural game resources, especially the many antelopes.

Horned ungulates in cult and art

In the cultural life of mankind to date, the horned ungulates have also played an extremely important part. In India the zebus were and still are considered to be sacred. The ancient Egyptians worshipped apis bulls as a godhead. Zeus abducted Europe in the form of a bull, he-goats pulled Donar's wagon, and the natives of Israel danced around the golden calf. The Christian religion, too, could not find a better symbol for the guiltless suffering of the Saviour than "the lamb which bears the sins of mankind." In the history of each of the great religious founders came a period when "he went into the desert and lived among animals," and everyone familiar with the animal geography knows that this predominantly meant horned ungulates. In art the horned ungulates have repeatedly served as models from the earliest Stone Age cave drawings to contemporary masterpieces. Only now does science begin to realize what the study of horned ungulates may have to offer.

Human cruelty

While the subject of "horned ungulates and man" leads one to the peaks of the cultural life, it also reveals the greatest of human errors. There is hardly a meanness, cruelty, and perversion which has not been committed on horned ungulates by man. Poachers in Africa still shoot their poisoned arrows into the herds and then trot patiently behind them until they find the animals which have perished miserably. In some places, Africans and Europeans have poisoned water holes, which are scarce enough in the dry season, and later they either collected the carcasses of the animals who had died in agony or just let them rot. White hunters, interested only in a "trophy," sometimes have to fire five or more shots at the "tough" African antelopes until such an animal finally comes down. Then a native is often ordered to apply the coup de grace to the victim with a panga or a spear. It has even happened that a white "sportsman" strictly requested that this be done very slowly. Buffalo are still being poached with wire slings and then killed by rocks thrown at them; gazelles are caught in nets and then slain with clubs, etc.

Traps, as they are still used in some places, could have come from a medieval torture chamber. They consist of a ring of sticks with pointed tips pointing inwards which are bound together and placed over a small pit. When an unfortunate animal steps into them, the tips of all these sticks penetrate its leg and hold it tightly. Attached to this ring by means of a cord is a club which is spun around by the movements of the desperate animal in the attempt to get free and which smashes his legs. Thus, the animal is disabled, but still alive. The flesh cannot spoil and the trapper is in no great hurry to get it. Other "well constructed" traps are of the same "quality."

In the Spanish version of the bullfight, the death of a living creature is a source of public amusement. In the slaughterhouses, in kosher butchering, and in the noble sport of hunting, there are scenes which occur, in spite of all humanitarian efforts, which should cause any sensitive person to vomit. Such cruelties and absurdities are not committed only by primitive people who "do not know what they are doing," but also by members of civilized nations and persons of the highest standing.

Schillings, the German explorer of Africa, describes in detail how he killed single female Grant's gazelles. He himself had discovered that such females usually had young only a few hours or days old in a hiding place nearby and which then were consigned to a miserable death. One hardly knows what to say about the behavior of such an "animal lover" and "conservationist." The English veterinarian Chorley, whose name, as Bernhard Grzimek once said, should be remembered along with desecraters of temples and destroyers of cultures, recommended the total extermination of game animals to fight sleeping sickness and Nagana's disease. The English colonial administration accepted Chorley's suggestion, and for decades while implementing the so-called "Tsetse-Control Program," all game animals in vast areas were killed. Meanwhile, we know how frightfully wrong Chorley had been. Glover and other scientists found out that many of the eradicated species of horned ungulates, e.g., waterbuck and gazelles, are never or only rarely stung by the disease-carrying tsetse flies. Therefore, only in very rare cases are they the disease transmitters. Since small mammals like rats or mice may be infected in large numbers by the dangerous diseases, killing the game never could prevent the spreading of diseases. One of the most important findings was that the larvae of the disease-carrying insects do especially well in shady soil. The grown insect, too, needs plenty of shade in order to stay alive. It dies from continuous exposure to sunshine. Shade in the wild comes mainly from the leaves of bushes and trees. The game animals feed on them, thus keeping them short, while cattle, which had been introduced to the areas after the mass killing of the game, do not eat them or at least not to the same extent. Thus, the mass eradication of African game animals has in effect not reduced the conditions essential for the life of tsetse flies and thus the diseases which they carry, but it has

Mr. Chorley's extermination of horned ungulates

actually favored and promoted them. Although these facts are generally known today, this kind of killing is still continued in some areas, and the gentlemen in charge of the tsetse control program only gave another name to their enterprise, since the old one has become "unpopular."

Means to protect horned ungulates

After the many outrages which man has perpetrated on the horned ungulates, now sensible men and governments are trying to preserve the surviving species in national parks. These "islands of peace," in many cases, were established by the colonial authorities. When the countries in question became independent during the last years, animal lovers all over the world were deeply concerned about the further existence of some of these parks. Fortunately, most of these apprehensions have not come true. Many of the young African nations, who had many other problems to solve after declaring their independence, made the most meritorious efforts to preserve their national parks. One of the determining factors, doubtlessly, has been the increasing interest by tourists, who come from countries scarce in wildlife to enjoy the delightful view of the large animal herds. Thus, the animals brought foreign exchange to the Africans, and the African population developed a certain pride in their multiform fauna which is so unique in the world. May this development continue for the benefit of animals and man.

Erich Thenius
Fritz Walther

12   Duikers, Dwarf Antelopes, and Tragelaphinae

Central Asia was probably where the horned ungulates originated. The various species migrated from there into their present areas of distribution. Because of the fossil discoveries in the Siwalik Mountains in North India and in Pikermi near Athens we know that in the Upper Tertiary period a fauna existed at the foot of the Himalayas and in Greece, which largely can be compared with the present one in Africa. Among them are the immediate ancestors of our contemporary species. Originally the horned ungulates probably lived in a forest and brush habitat. During the course of their distribution, they also came to the steppes, swamps, mountains, deserts, and the tundra.

The subfamily of the DUIKERS (Cephalophinae) most clearly resembles the small forest-type horned ungulates of the Tertiary period. They range in size from a hare to a deer. The HRL is 55–145 cm, the TL is 7–17 cm, and the BH is 30–85 cm. The animal has an arched back, slender legs, and a relatively plump body. The facial part of the skull is not much longer than the cranial part, and the head has a humped nose which is bare. The tail has a small tassel. The short horns, which are no longer than the ears, are straight or slightly bent downward or upward. In the ♀♀, they are weaker or missing. The forehead has a full tuft which partially or completely hides the horns. The preorbital glands are slits in the cheeks. Most animals also have interdigital glands. There is no color difference between the ♂♂ and the ♀♀, although the ♂♂ are somewhat larger and heavier than the ♀♀. The females have four nipples. They eat vegetarian food in addition to protein. The habitat is the forest or the savannah. There are two genera (*Sylvicapra* and *Cephalophus*) with fifteen species and seventy-one subspecies.

The GRAY DUIKERS (genus *Sylvicapra*; Color plate, p. 401) live outside the rain forest and the desert in scattered forests, in brush and in bush steppe. The HRL is 85–115 cm, the TL is 8–18 cm, and the BH is 45–70 cm. The ears are long and narrow and the straight horns are short and

Subfamily: duikers

The gray duiker

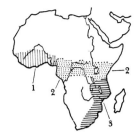

Fig. 12-1.
1. Maxwell's duiker *(Cephalophus maxwelli).* 2. Harvey's duiker *(Cephalophus harveyi).* 3. Red duiker *(Cephalophus natalensis).*

The genus
Cephalophus

Fig. 12-2.
1. Black duiker *(Cephalophus niger).* 2. Blue duiker *(Cephalophus monticola).*

slender. There is a dark stripe from the nose to the lower part of the forehead. Gestation lasts four months and usually only one young is born. There is one species, GRAY DUIKER *(Sylvicapra grimmia),* and many subspecies.

The gray duiker avoids the dense forest. In its wide distributional range it actually avoids the prime forests of Central Africa. When these animals roam about the bush steppe, they continuously flick their tails. In flight, in which they zigzag with occasional high jumps, they move the tail up and down so that the white underside flashes. They are seen predominantly at night, although in quiet areas they may well come out during day, especially in cloudly weather. They live singly or in pairs. They spend the day resting, hidden in the high grass or under shady bushes. It is said that they are not as dependent on water as many other bush animals.

Captive gray duikers usually tame quickly, although the males may then become rather aggressive, as frequently happens in horned ungulates after they have lost their natural shyness because of daily contact with their keepers and because they have no other way of discharging their aggression. All medium-sized or small predators are a threat to the gray duikers. They are usually too small a prey only for the lion. Birds of prey and large snakes also attack them frequently.

In contrast to the gray duiker, the other DUIKERS (genus *Cephalophus;* Color plate, p. 401) predominantly inhabit the tropical rain forests. The HRL is 55–145 cm, the TL is 7–17.5 cm, and the BH is 30–85 cm. The short ears are round and the short horns are bent backwards at an angle. The gestation period is four months and usually one young is born. There are fourteen species with forty-seven subspecies: 1. MAXWELL'S DUIKER *(Cephalophus maxwelli).* 2. BLACK DUIKER *(Cephalophus niger).* 3. BANDED DUIKER *(Cephalophus zebra),* with its black vertical stripes which conspicuously mark its body. 4. OGILBY'S DUIKER *(Cephalophus ogilbyi).* 5. JENTINK'S DUIKER (◊ *Cephalophus jentinki),* a large species with an HRL of approximately 135 cm, a BH of 75–85 cm, and a coat color which is reminiscent of the Malayan tapir. 6. RED-FLANKED DUIKER *(Cephalophus rufilatus).* 7. BAY DUIKER *(Cephalophus dorsalis).* 8. GIANT DUIKER *(Cephalophus sylvicultor),* with an HRL of 115–145 cm and a BH of 65–85 cm. 9. GABOON DUIKER *(Cephalophus leucogaster).* 10. BLACK-FRONTED DUIKER *(Cephalophus nigrifrons).* 11. HARVEY'S DUIKER *(Cephalophus harveyi).* 12. BLUE DUIKER *(Cephalophus monticola),* the smallest species, with a BH of only 30–40 cm. 13. ABBOT'S DUIKER *(Cephalophus spadix).* 14. RED DUIKER *(Cephalophus natalensis).*

The areas of distribution for individual species of duikers are sometimes rather small. They follow the belt of the large rain forests, and the number of species decreases from west to east and to the south, just as the rain forests also become more scarce towards the east and the southeast of the continent. Unfortunately, there is little known so far about the habits of the duikers. The red duiker is said to climb slanted tree trunks like a goat and to browse on the foliage. Harvey's duikers seems to be strongly dependent on water. The giant duiker uses nest-like caves under fallen tree trunks as a resting place. While the head scars of male duikers indicate that the males have fights with each other, no one has ever described such a fight in detail.

Several informants have discovered that duikers eat termites in addition to their herbivorous diet and even occasionally kill birds, both in the wild and in zoos. This was checked more thoroughly with a bay duiker in the Zurich Zoo. According to Kurt, they had become aware of the animal's desire for meat because occasionally pigeons would be missing, and only the duiker could be logically suspected. When they gave him a freshly killed pigeon, he ate it immediately. Then he was offered live birds several times. As soon as the duiker saw the bird, he took wind, his head stretched forward, nostrils expanded, and ears close to his head pointed towards the bird. He approached the bird with his head lowered, and as soon as the bird tried to escape, the duiker chased it. He killed small chicks instantly with a bite in the chest or abdomen. He killed pigeons that were about to fly off or still on the ground by blows with his forelegs. Then he grabbed the victim with his mouth and carried it to his resting place. He would eat the whole chicks. However, he bit off the head of a pigeon with his molars and ate it. Then he sniffed the dead body, bit off the legs and wings, and left them. Then he took the headless rump with his mouth, sucked blood and intestines out of it, and ate all or part of it, biting pieces off with his molars from the front to the rear. This behavior is even more unusual since there are no characteristics of the duiker's teeth which would indicate a meat diet.

So far the most extensive observations on the behavior of duikers in the wild and in captivity have been made by Aeschlimann. The Maxwell's duikers, which he studied, have a territory with distinct resting places and other areas of special importance to the animals. These places are connected by paths where the secretions from the cheek glands are applied. Furthermore, males and females smear each other's faces with these scent marks. When more than one pair is kept together in captivity, it is clearly shown that this mutual marking takes place most frequently between the male and one specific female or, in very rare cases, only between two females. The animals feed mainly on grain and fruit. As long as they have fresh leaves, they need only a little water. Their alarm call is a kind of whistling.

Frädrich saw that the young banded duikers of the Frankfurt Zoo were also marked. They are the "stay put" type which know their mother individually from the third day on. In the zoo, they came to their mother for nursing; they were not called by her. The banded

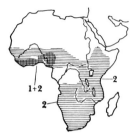

Fig. 12-3.
1. Ogilby's duiker (Cephalophus ogilbyi). 2. Grey duiker (Sylvicapra grimmia).

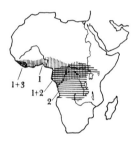

Fig. 12-4.
1. Bay duiker (Cephalophus dorsalis). 2. Black-fronted duiker (Cephalophus nigrifrons). 3. Banded duiker (Cephalophus zebra).

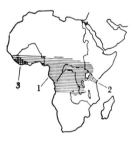

Fig. 12-5.
1. Giant duiker (Cephalophus sylvicultor). 2. Abbot's duiker (Cephalophus spadix). 3. Jentink's duiker (Cephalophus jentinki).

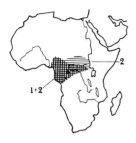

Fig. 12-6.
1. Gaboon duiker *(Cephalophus leucogaster)*. 2. Red-flanked duiker *(Cephalophus rufilatus)*.

Subfamily: dwarf antelopes

The royal antelope and suni

Fig. 12-7.
1. Royal antelope *(Nesotragus pygmaeus)*. 2. Bate's dwarf antelope *(Nesotragus batesi)*. 3. Suni *(Nesotragus moschatus)*.

duikers bite each other in fights. In the mating behavior of gray duikers, blue duikers, and black-fronted duikers, I observed the "foreleg-kick" as a courtship behavior (see fig. 11-9).

Even though, as far as we know, keeping duikers does not present insurmountable difficulties, they are not frequently displayed in zoological gardens. Probably their "exhibition value" is too small for the majority of the visitors. Some species, however, are actually zoological rarities. Only a very few specimens of Jentink's duikers have been shown since their first discovery in 1892. In several cases, duikers have reproduced in zoos. The Frankfurt Zoo, for example, has had success with the rare banded duikers and the large giant duikers. In captivity, duikers become more than ten years old.

The smallest horned ungulates are in the subfamily of DWARF ANTELOPES (Neotraginae). The animals range in size between a hare and a deer, they have an arched or straight back, ears medium to long, and a short to medium-long tail. The horns, which are as small or as large as the ears, are found almost exclusively in the ♂♂. Some species have a tuff on their head, others a mere hint of one. The ♀♀ are often somewhat larger than the ♂♂. The females usually have four, rarely only two nipples. They live in the bush steppe, the steppe, or the semi-desert.

This subfamily can be divided into five tribes: DWARF ANTELOPES proper (Neotragini). 2. DIK-DIKS (Madoquini). 3. BEIRA-ANTELOPES (Dorcatragini). 4. KLIPSPRINGER (Oreotragini). 5. STEINBOK (Raphicerini).

Strictly speaking, the genus ROYAL ANTELOPE *(Neotragus)* with one species and the genus SUNI *(Nesotragus)* with two species and seven subspecies belong to the dwarf antelopes (Neotraginae). The HRL is 50–62 cm, the TL is 5–13 cm, and the BH is 25–38 cm. These hare-sized animals have a graceful shape and round back with a croup which is higher than the shoulders. The head is rather large relative to the body, with a naked nose region and large eyes. The horns, which are present only in ♂♂, are slender, pointed, and directed at a slant towards the back. The neck is short. The medium-long legs are graceful and lack pseudo-claws, which are replaced by a naked spot of skin. There is no tuft on the head.

With BH of only 25 cm, the ROYAL ANTELOPE *(Neotragus pygmaeus;* Color plate, p. 428) is the smallest present-day horned ungulate. The name "royal antelope" comes from a native name. In Africa this little animal is often called the "King of hares." A little larger (BH 30–38 cm) are BATE'S DWARF ANTELOPE *(Nesotragus batesi)* and SUNI (◊*Nesotragus moschatus;* Color plate, p. 428). The suni's preorbital glands are said to have a musk scent. They predominantly inhabit forests and dense brush; in other areas, they are rarely seen. At dawn and during the night, they are quite agile, but in the daytime they come out only in the early morning and the late evening. When disturbed they take off, circle around trees and bushes, and soon disappear. They are known to make weak barking and piercing whistling sounds.

To the DIK-DIKS (Madoquini) belong the species *Madoqua* and *Rhynchotragus*. They are slightly above hare size, with a graceful shape, slender legs, and a round back which is higher at the croup than at the shoulder. The delicate head is graceful and short with a forehead tuft. The trunk-like nose is movable. The ears are medium long, the eyes are large, and horns exist only in the ♂♂. The narrow, long, and pointed hooves have very small pseudo-claws, which are still visible. The females have four nipples. There are five species: 1. SALT'S DIK-DIK *(Madoqua saltiana)*. 2. RED BELLY DIK-DIK *(Madoqua phillipsi)*. 3. SWAYNE'S DIK-DIK *(Madoqua swaynei)*. 4. GUENTHER'S DIK-DIK *(Rhynchotragus guentheri)*. 5. KIRK'S DIK-DIK *(Rhynchotragus kirki;* Color plate, p. 426).

In spite of their small size, the dik-diks unfortunately are heavily persecuted by man. With respect to man they have only a very short flight distance and, therefore, it is easy to kill them with a club which every native in their habitat carries with him. The small skins, which come on the market labelled as "gazelle's skin" are made mainly into gloves. Each pair of gloves requires at least two animals; hence it is easy to imagine how long it will take until these dwarf antelopes are extinct in some areas. The threat of extinction of such a small species is much greater than in larger animals, which because of their size are more conspicuous, and one becomes more readily aware of their decreasing numbers, which results in protective measures.

In the dik-diks of the genus *Rhynchotragus*, the nose is extended into more of a trunk than in other dik-diks. They are able to turn this "trunk" in all directions. Both sexes have a tuft of reddish hair on the forehead which may stand erect when the animal is excited. Probably all dik-diks live in pairs in territories which, according to my observations on Kirk's dik-dik in the Serengeti, are occupied for years or even during the entire life of the pair. At the boundaries and in other important places, there are many fecal piles of thirty or more centimeters in diameter which are connected by trails from gland secretions. Only the males mark with their preorbital glands. Pawing, urinating, defecating, and marking in male dik-diks have become so linked in a sequence that right next to each pile of feces a strongly marked grass or twig can be found. The females use these piles of feces too, but without pawing, although the male sometimes actually "demands" her to do so. Then he paws her feces apart and places his own excrements on top of them.

After a gestation of about six months, the dik-dik female has one young which "stays put" for a long time, but otherwise grows quickly. It is fully grown at twelve months, but may be sexually mature at six months. The groups of three individuals which are occasionally seen consist of the parents with an almost or fully grown young. Later the

Dik-diks and rhynchotragus

Fig. 12-8.
1. Red-belly dik-dik *(Madoqua phillipsi)*.
2. Swayne's dik-dik *(Madoqua swaynei)*.
3. Grysbok *(Nototragus melanotis)*.

Fig. 12-9.
1. Guenther's dik-dik *(Rhynchotragus guentheri)*.
2. Kirk's dik-dik *(Rhynchotragus kirki)*. 3. Salt's dik-dik *(Madoqua saltiana)*.

young are turned out of the parents' territory. The young males are chased off by their father. In female young we have no information about how they are removed. Because of the rapid maturation and early pubescence, the male young may stay with their parents for a relatively long time. Naturally, one day the old male will consider the young as a rival. When the young male, probably in a childlike manner, approaches the mother the old male attacks him. The young male drops down into the submissive posture, thus stopping the attack. This may happen fifty times or more during one night. Even though these continuous attacks certainly are not pleasant for the young male, he may well stay in the parents' territory for several more weeks. One day it will be too much for him, and he "moves out." In most cases, he will not go far but will settle in "no man's land" between the territories of his parents and their neighbors.

Apparently, young females behave similarly. Single dik-diks on the move will usually be young ones. It is possible that such individuals will form pairs, or that they will move into a territory in the vicinity, where a partner of their sex is missing. This would confirm the old hunters' experience that a killed partner of a dik-dik is replaced by the next day.

The territories seem to be of different size depending on the population density. Next to my house in Banagi (Serengeti National Park), there was one approximately 500 meters long and 200 meters wide. But in other areas, I saw several pairs of dik-diks so close together that their territories could have hardly been more than 50 meters in diameter. The resting places, especially those in the daytime, are often located at the edge of the territories. Even though dik-diks may be seen in broad daylight, their main activities take place in the morning and late afternoon. On nights with bright moonlight, they are often up and around uninterruptedly from evening until three in the morning, when they rest until shortly before daybreak. They apparently have longer resting periods between activity phases on dark nights. Their alarm call is a rather loud nasal whistling. In flight, they jump off like duikers with several big leaps and soon stand still behind trees and bushes when they are no longer pursued.

The dik-diks need brush with plenty of plant cover for their habitat. In the Serengeti, they live mainly in the fringe forest of the Korongos, where there are creeks with water only in the rainy season, or in the brush around groups of boulders. In other areas, it is said that they are not very dependent on water. Because of their predominantly nocturnal life and their habitat in dense brush, a thorough study of dik-diks is not always simple and it is sometimes rather difficult. However, it is a most delightful experience to watch the activities of these interesting and lovely dwarfs among the antelopes by the magic light of an African moon. Because of their insignificant appearance, dik-diks are only seldom shown in zoos, although they do well there and have reproduced several times, for example in the Hannover Zoo.

The only species of the BEIRA-ANTELOPES (Dorcatragini) is the BEIRA (◊*Dorcatragus megalotis;* Color plate, p. 426). The HRL is 80–90 cm, the TL is 10–12 cm, and the BH is 55–60 cm. The rather short head has a wide forehead, pointed muzzle, a slightly humped, pointed nose, large eyes, and long, oval ears which are somewhat pointed and conspicuously large and wide. The horns, found only in ♂♂, are parallel, far apart, and rise steeply. The neck is rather short and the round rump is higher at the croup than at the shoulders. The dense coat has long hair, as a deer's. The female has two nipples.

Beira antelopes

With their large ears, the beira almost looks like a miniature kudu. Living in dry hill and mountain country with bushes, usually in pairs or small groups, they are not abundant anywhere. Mornings and afternoons, they roam about grazing, while at noon they rest. Details on the habits of this rare, small antelope are not known.

The KLIPSPRINGERS (Oreotragini) consist of only one species, the KLIPSPRINGER (*Oreotragus oreotragus:* Color plate, p. 426). The HRL is 75–115 cm, the TL is 8–13 cm, and the BH is 50–60 cm. The body is compact, rather long-legged, with a round back which is higher at the croup than at the shoulders. The rather small head is short with a wide forehead. Their eyes, preorbital glands, and ears are all rather large. The horns rise steeply and parallel to each other. The legs are sturdy, and the long hooves are pressed together on the sides. The animal stands on the worn tips of the hooves which can be spread far apart. The coat is thick, dense, and hard. The females have four nipples. There are eleven subspecies, and in the one in the Masai country the ♀♀ also have horns.

Klipspringers

As we know, all hoofed animals walk "on the tips of their toes." Among these the klipspringer is the only one to touch the ground only with the very tips of its vertically rising hooves. Translated into human terms, this would mean that a dancer would not, as usual, trip on the tips of her toes, but on the tips of her toe nails. When the klipspringer is standing still, it brings its front and hind legs together in one spot, which is rather unique in horned ungulates. In addition, all horned ungulates when lying down, except the klipspringer, are unable to stretch both front legs forward and parallel to each other. This occurs only in goats which are not closely related to it.

The hoof tips in klipspringers are flat at the base and harder at the edge than in the center. Therefore, they wear down in the center and the sharp rim extends beyond the horn material of the center. This produces the same effect as a nail on a mountaineer's boot. Probably it also creates possibly a certain sucking action. Thus, the klipspringers have two safeguards against slipping. Their body structure already indicates that they are not animals of the plains but that they are highly suited for living in a rocky habitat.

In contrast to many other alpine animals, klipspringers live on all kinds of rock. They are not restricted to larger, continuous mountain ranges, but are also frequently found on isolated boulders outside the mountain regions. Their coat is rather unusual, too. It has a peculiarly speckled "salt and pepper" pattern of an almost olive shade and it is rugged and brittle as in no other antelope. When the animals shake their coats, it sounds like the distant rattling of a porcupine, according to Rösl Kirchshofer. To catch a klipspringer in a zoo it is necessary to grab him securely and carefully, or else one ends up with a bunch of hair and no animal. When the females bite each other, they tear out clusters of hair. The klipspringers love to nibble the secretion from each other's face, which in both sexes exudes from the preorbital glands and dries up quickly. Both males and females also mark with it. In a large zoo pen, I found only one large pile of excrements used by all the animals where they placed their feces. Obviously this was not a territorial marking but more of a "restroom."

Wild klipspringers are found mostly in pairs, although occasionally there are singles. A pair, which I knew individually, I met almost daily in the morning on the same boulder. There the two stood "guard" and warmed themselves in the sun. I saw them in this place during the rainy season as well as some weeks later. When the country dried out, they disappeared from this area as did all the other klipspringers. With the beginning of the next rainy season, the klipspringers were back. On the day of their arrival, six animals were together. This was the largest number of klipspringers I ever saw in one group. A few days later they had formed pairs again; the boulder mentioned above was occupied by a pair, possibly the same one as before. Klipspringers generally are known to remain in one place. Under unfavorable circumstances, however, they seem to give up their territories temporarily and to migrate; they return when conditions have improved.

The klipspringer's alarm call is a loud whistling through the nose. Sometimes they jump in rocking horse fashion in the same place. In flight they usually run downhill at first; the animal thus leaves the "lookout" where it had stood. Fearful and distressed klipspringers make roaring sounds with their mouths open. During copulation, the males hum briefly and softly. While fighting, they butt hard with horns and forehead, leaving the front part of their body back and down, and dashing forward from this position. However, I have seen only one serious fight so far between two males. It looked as if each one tried to reach from the front under the other and then to push against his chest.

A rather strange scene once took place in a zoo when three klip-

Fig. 12-10.
1. Beira *(Dorcatragus megalotis)*. 2. Steinbok *(Raphicerus campestris)*.

Fig. 12-11.
Klipspringer *(Oreotragus oreotragus)*.

springers were to be caught to be transferred into another pen. The animals were in a small, completely empty stall, and the two keepers entered it ready to catch them. Klipspringers usually are rather tame in captivity, so they did not show any signs of being alarmed when the men came in. When the keepers were about to grab one, it escaped, ran to one of its fellows, held its head under the other animal's body, and then it could be seized without further trouble. The next one did the same and the last one, for lack of anything better, stuck its head into a corner of the stall. It actually looked as if each one "believed" it had disappeared as soon as it hid its head. The "ostrich's policy" seems to occur in certain situations in several species of animals.

The last tribe of the dwarf antelopes are the STEINBOKS (Raphicerini). They range in size between that of klipspringers and deer. Only ♂♂ have horns. There is no tuft, but occasionally there is a hint of one. The nose region is naked. The ears are medium-long to long. The slender hooves are pointed at the tip with small or missing pseudo-claws. The tail is short. The females have four nipples. There are three genera (grysbok, steinbok, and oribi) with three species and twenty-four subspecies.

The GRYSBOK (Nototragus melanotis) is named for the white hair scattered on its coat. Its HRL is 63–75 cm, the TL is 5–8 cm, and the BH is 45–55 cm. It lives in groups in the grasslands and belongs to the few species of horned ungulates which have not been exterminated in the Boer Cape Province, probably because of their small size and concealed way of life.

Grysbok, steinbok, and oribi

A little larger is the STEINBOK (Raphicerus campestris; Color plate, p. 426). The HRL is 70–90 cm, the TL is 5–10 cm, and the BH is 45–60 cm. There are no pseudo-claws. The popular name comes from the habit of the Boers who named all the creatures they saw when they first settled in South Africa with names of those European species with which they were familiar. We find such confusing names in antelopes, too. Unfortunately, many of them were translated into German. This animal has nothing to do with the ibex (Steinbock) of the European alps.

A startled steinbok begins its escape with great speed, as do many of the small antelopes. Sometimes it will also make high jumps. After a short while, it stands still, lowers the head to the ground, and lays down in a den, high grass, or behind a bush. As far as the human observer is concerned it has "disappeared from the earth." It is even said that hotly pursued steinboks occasionally flee into aardvark holes or other caves. In the Serengeti area, I saw steinboks only occasionally. Apparently males and females live singly, probably in territories. Only during the mating season does a male visit a female so that then pairs are seen. In courtship, the male kicks with the front leg towards the female's hind legs, as dik-diks and klipspringers do. In contrast to

Fig. 12-12.
Oribi (Ourebia ourebi).

dik-diks, both male and female paw after defecating, so that sometimes the feces are more or less covered with soil. The mating and breeding seasons vary depending on the area and latitude. Gestation is said to be seven months.

Still taller, but very graceful and long-legged is the ORIBI (*Ourebia ourebi:* Color plate, p. 426). The HRL is 92–110 cm, the TL is 6–10.5 cm, and the BH is 50–67 cm. The relatively small head has a straight profile. The short tail has a naked underside and a pointed tassel. The pseudo-claws are small, short, and broad. The uniformly dense coat is almost without any fleece. A gestation period is approximately seven months, and one young is born. Longevity is up to fourteen years.

Like their relatives, the oribis live singly or in pairs, although sometimes there are small groups of up to six animals. They do not occur in dry areas as do steinboks, but they inhabit areas deeper in the brush forest. Only males mark with their well developed preorbital glands. They establish a pattern of several manure piles which probably indicate the boundaries of their territories. Both sexes have a round, naked spot of skin below the base of the ear without any signs of glands, and nothing is known about its significance. Their alarm call is a whistle. In flight, they make great leaps with the front legs stiffly forward-down and the hind legs more closely pulled to the body even more frequently than other dwarf antelopes. When they jump a sequence of such leaps it is like the "stotting" of the gazelle (compare Ch. 15).

Subfamily: Tragelaphinae

While the smallest horned ungulates are found among duikers and dwarf antelopes, the largest are found in the Tragelaphinae (subfamily Tragelaphinae). The animals range in size between that of deer and oxen. They are slender, long-legged, some straight-backed and others are higher at the withers or croup. The tail is moderately long to long. The facial part of the skull is long to very long with a large bare nose region. There are great differences in color between the sexes. The females have two or four nipples. There are three tribes (spiral-horned antelopes, nilgais, and four-horned antelopes) with four genera, ten species and forty-five subspecies.

Distinguishing characteristics

The SPIRAL-HORNED ANTELOPES (tribe Tragelaphini) have spiral or lyrate horns, which in the genus *Tragelaphus* are found only in males and in both sexes in *Taurotragus*. The HRL is 105–345 cm, the TL is 18–100 cm, and the BH is 65–180 cm. The body form ranges from small and graceful to large and bulky. The head is moderately large in relation to the body. The large, wide ears are slightly pointed. The neck has a partial dewlap. The long and slender legs have large and strong pseudo-claws. The medium-long tail has bushy or short hair and a tassel. The females have four nipples and a gestation of seven to nine months. One young is born, rarely two. Longevity is ten to twenty-five years.

The males of the spiral-horned antelopes (genus *Tragelaphus*) have more or less spiralled horns. The ♀♀ are hornless. The ♂♂ are often considerably larger and darker than the ♀♀, with much stronger necks which are enhanced by longer hair or imposing throat manes. The neck mane is more or less distinctive. In the ♂♂, there is also a mane on the back which continues to the base of the tail. There are six species:

1. The GREATER KUDU (*Tragelaphus strepsiceros*; Color plates, pp. 299, 302). The HRL is 195–245 cm, and the BH is 120–150 cm. The magnificent spiral horns reach 168 cm in length. 2. The LESSER KUDU (*Tragelaphus imberbis*; Color plate, p. 302). The HRL is 110–140 and the BH is 90–105 cm. The horns are not as far apart at the tips. They may be as long as 90 cm. There are more stripes, they are closer together, and are more pronounced. There is no throat mane. 3. The MOUNTAIN NYALA (*Tragelaphus buxtoni*; Color plate, p. 302). The HRL is 190–260 cm, the BH is 90–135 cm, and the horn length is up to 110 cm. 4. NYALA (*Tragelaphus angasi*; Color plate, p. 302). The HRL is 135–155 cm, the BH is 80–115 cm, and the horns measure up to 80 cm. There is a very remarkable color difference between the ♂♂ and the ♀♀. 5. The SITATUNGA (*Tragelaphus spekei*; Color plate, p. 302). The HRL is 115–170 cm, the BH is 75–125 cm, and the horns reach a length of 92 cm. The hindquarters are much enlarged, and the croup is higher than the withers. The hooves, which are up to 10 cm long, can be spread far apart (adaptation to swamp and aquatic habitats). 6. The BUSHBUCK (*Tragelaphus scriptus*; Color plate, p. 302). The HRL is 105–150 cm, the BH is 65–100 cm, and the horns are up to 55 cm long. The body is like a deer's, and has a croup higher than the withers. The horns are weaker and lyrate.

The greater kudu is the most handsome specimen of the genus and probably one of the most beautiful of antelopes. In its wide area of distribution (see map, fig. 12-13), it still lives in rocky hill and mountain country with scattered or dense brush and in brush covered plains. It needs waterholes with easy access. When water and food become scarce during periods of severe drought, they roam over great distances. Otherwise, they are rather stationary. Whether they are territorial in the strict sense has not been clearly determined until now. Since the great cattle plague at the end of the last century, they belong to the rare game in East Africa, but are still abundant in Rhodesia and Southwest Africa. In Southwest Africa, the artificial wells and irrigations works have made the greater kudu into an actual follower of civilization. Since they are excellent jumpers which can easily take obstacles of two and a half meters high, the "kudu-proof" fences of the farms are not high enough.

Usually a group of kudus consists of six to twenty animals, or sometimes even thirty to forty. There are mostly females with their young and younger males. Strong males, which apparently join the herds only during mating season, form groups of their own or live singly at other

Fig. 12-13.
1. Greater kudu (*Tragelaphus strepsiceros*).
2. Lesser kudu (*Tragelaphus imberbis*). 3. Nyala (*Tragelaphus angasi*).
4. Mountain nyala (*Tragelaphus buxtoni*).

The greater kudu

Fig. 12-14.
1. Bushbuck (*Tragelaphus scriptus*). 2. Sitatunga (*Tragelaphus spekei*).

times. In a "bachelor herd," males of all age groups, beginning with yearlings, may live together. The kudus, which feed predominantly on foliage, eat at night or in the morning and the evening. During the day they rest in the dense brush. In the morning and in the evening they drink water, and where it is lacking, they feed on juicy roots, tubers, and bulbs. The gestation period is seven to eight months. In some regions, as in Rhodesia, birth occurs throughout the year, while in other parts of Africa there seems to be some correlation between breeding times and rainy seasons.

In addition to men, lions, leopards, and wild dogs also pursue the greater kudu. Cheetahs may become dangerous to the females and the young. Depending on the situation, startled kudus "hide" from their pursuers, skillfully taking advantage of any cover, or they take off with tremendous jumps. Then the tail is curled up so that the tip almost touches the base, and the white underside becomes visible. Males in flight in the dense brush are said to lay their horns back on the shoulders. When they are walking on the alert, they make strange "butting" movements with their heads. Their alarm call is a bark, similar to the roe deer's, but much louder and in males lower.

The large, spiral, and sideward-sweeping horns of the males are especially suited for "wrestling" with the horns locked into each other (see Ch. 11). In zoological gardens, I frequently observed how kudu bulls pushed their horns backwards over the shoulder against trees, fences, and other substitute objects. It is possible that this also may be part of the fighting ceremony, although then the animals would have to stand parallel to each other. I have a picture of kudus with their horns locked together in such a way that they must have utilized this approach. Although fighting bushes and trees and ploughing the ground with the horns has been observed so far only in zoological gardens, it occurs so frequently that it is hard to believe it is only a redirected activity as a result of captivity. Probably, it also occurs in the wild where it may well serve the purpose of marking the area.

Otherwise, true and conspicuous actions of markings are not known in the kudu and its relatives. Sometimes the males brush with their cheeks, exactly where the cheek glands in duikers are located, along trees and the females' bodies. There are skin glands in this body region which certainly could be used for scent marking. In fights, the females not only bite but also box with their hornless foreheads towards the opponent's flank. Furthermore, they butt with their mouths closed. If a female is not yet ready for mating, she may use these fighting methods in warding off bulls. It may even happen that a female actually "beats up" a young male, although naturally she would not dare to do this to an old bull. In these cases, it becomes clear that kudu bulls are exceptionally inhibited in attacking females. The males do not counter these attacks but tolerate them without any defense on their

Fig. 12-15.
In the mating ceremony the greater kudu displays, standing erect and laterally in front of the female.

Fig. 12-16.
The lateral display of the greater kudu towards an opponent.

Fig. 12-17.
The male catching up from the side in the mating ceremony of the greater kudu.

part. They stand lateral and highly erect in the display position. However, it is just this "immobility" that finally allows the bull to get his way with the female. When she stops her attacks, he may begin to drive her. Old bulls only need assume the lateral display posture and the female will take off, and driving begins.

A kudu bull in the zoo who considered me a rival displayed laterally to me, but in another, obviously more serious form. He also lowered his head, and the hind legs were pulled far under his body so that his back was arched. When I circled around him, he would follow, constantly presenting the broadside of his body. Interestingly enough, he never did attack from this position. He always attacked frontally.

A kudu bull begins the mating ceremony by stepping laterally into the female's path with his head highly erect. Then a ritualized neck fight from the side or front may also occur. When the bull has thus caused the female to move, he follows her, uttering short, soft and strangely "whining" calls. He then bends and stretches the neck forward and with all his muscles strained it forms a flat U. In this posture, he approaches her withers which he brushes briefly from the side backwards with his neck over hers. He retreats again, and the game is repeated until the female slows down. Then the bull walks straight behind her and slides his head and neck straight along her back until finally, she stands still and he can mount.

The greater kudu young are classic examples of the "stay put" type (see fig. 11–14). As soon as they are a little older, however, they may quite vigorously demand nursing. They follow their mother eagerly, approaching from the side and circling around her close to her front legs. If the mother does not want to run down her young, she must stand still and the young instantly dash to the udder to drink. When still older, the young may even become aggressive toward the mother if she does not immediately respond to their demand for milk. They push their mouths into her flank, box with their foreheads, jump at her, and even try to neck fight with her. Usually the young drink for about five to ten minutes. Licking of the anal region of the young, which it flaps over with its tail, is of great importance in the kudu.

In kudus, besides the running and fighting games, we also find the so-called "intention games," which might also be called "threat games" because many forms of threat behavior are shown. The kudus then run and encircle each other with the front legs spread apart and with a deeply lowered chest, like dogs about to invite another dog or a person to play or to go for a walk. The meaning of these "intention games" is an invitation to join running or fighting games. These playful fights do not differ in principle from serious fights, which means that there are no different behavior patterns. Whether the fight is serious or not is shown by the fact that the behavior can quickly change into another form. The animals lick each other in between, perform high or long jumps, or suddenly run away with neither animal being defeated. All this does not happen in serious fight. In running games each animal usually "practices" by himself, but this is so stimulating to the others that very often several animals run around, Kudus also like to play with all kinds of objects. Bulls, for example, throw broken branches into the air with their horns and then catch them again; females throw twigs with their mouths. As is true of all animals, the young and subadults are much more inclined to play than older individuals.

Zoological gardens like to keep greater kudus because they impress visitors with their beauty and appearance. In zoos, where they reproduce regularly, kudu females are pregnant for seven to eight months. Longevity in captivity exceeds fifteen years.

The lesser kudu

In contrast to the greater kudu, the lesser kudu has a much more restricted distribution and inhabits predominantly dry areas with thornbrush in the plains and hills of East Africa. Otherwise, it is a true copy of his larger cousin in appearance, behavior, and habits. They live in small groups of up to six animals, which in most cases consist of a male with several females and their offspring. Herds of males and single animals also occur. The fighting behavior is similar to the greatest kudu's. However, I observed that the lesser kudu more frequently employs the "horn pressing" which occurs in moderately serious fights where the opponents confront each other with their heads deeply lowered downward. After first touching each other's noses, they roll their foreheads along the other's until the horns touch. Now a trial of strength takes place as each opponent tries to ram his horns against the other's and force the rival's horns back on his neck.

In the mating ceremony of the lesser kudu, the neck fight is not as important as it is for the greater kudu. Instead, they have a truly delightful "demonstration of superiority." The male stands with his broadside to the female. Also standing erect, she wards him off now and again with symbolic butts with her mouth. Then the two try to outdo each other, although the male always remains "victor" simply because of his larger size. Finally, her nose points perpendicularly to the sky, and he nibbles at her chin in appeasement. As in the greater kudu, young males during the lateral display are often attacked by females which butt with their forehead, and they have the same inhibition to attack in this situation as their large cousins. In driving, the lesser kudu male literally "adheres" to the hindquarters or shoulders of the female, leaning his head closely against her body and bending his neck at least as much as the greater kudu. He also may circle in front of the female, thus stopping her the same way the young forces its mother into nursing. He often reaches with his nose and forehead under the female's abdomen, much like a young kudu. This is probably a ritualized fighting behavior because this process also occurs in the fights of the *Tragelaphus* females.

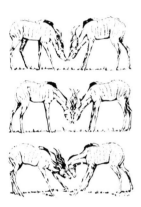

Fig. 12-18.
"Horn-pressing" in the lesser kudu: ritualized nose-touching (above), rolling foreheads against each other (center), and "horn-pressing" (below).

Fig. 12-19.
A lesser kudu pair resting in a close "star formation".

When two or more members of a group rest at the same time, their positions are anything but coincidental. When the animals expect some disturbance but do not know from which direction it will come, they usually lie in a "star formation." Their hindquarters are either very close together or actually touching, but the heads and the frontal parts of their bodies point like the rays of a star in different directions. However, when they expect the danger to come from a specific direction, all animals lie in such a way that their heads point in that direction. In addition, there is an "orientation towards a valuable object." The animals lie down so that they may keep an eye on the "precious" member, for example, the favorite female or the young. In the resting position, lesser kudus make sure that they have the cover of trees or bushes at their back. This behavior is more conspicuous in them than in many other horned ungulates.

Fig. 12-20.
In the mating ceremony of the lesser kudu, the male displays his superiority by raising his head high.

Observing this charming species is a true aesthetic delight for every animal lover. Nevertheless, lesser kudus unfortunately are less frequently seen in zoos than their large cousins. They have already reproduced in several zoological gardens, and the gestation period is approximately seven months. Lesser kudus reach an age of twelve to fifteen years.

The MOUNTAIN NYALA has the smallest distribution of all *Tragelaphus* species. They inhabit an area of only about 150 square kilometers in Southern Ethiopian hill country at an altitude of 2500 meters and higher, where they live above the belt of woodlands in a habitat of mountain heather regions *(Erica arborea)*. This rare species was not discovered and scientifically described until 1910. At first the mountain nyala had been classified as closely related to the South African nyala-antelope, but today many zoologists consider it a very close relative of the kudu. Even though mountain nyalas have been kept in zoological gardens, e.g., in Berlin, and have also reproduced there, so far no scientist has made a more extensive study of them. Therefore, no details are known about their habits and behavior.

The mountain nyala

The true nyalas or lowland-nyalas live only in the South African plains or mountainous regions, predominantly in places with sufficient water, in brush, steppe, and fringe forests along rivers. Nyalas form herds of six to forty animals. As in other Tragelaphinae, there are female groups with only one strong male, groups of males or females only, and occasionally groups of single old males. After a gestation period a little over seven months, the female gives birth, usually in August or September, to one young.

The nyala

In the New York Bronx Zoo, a larger group of nyalas has been kept for quite some time in spacious outdoor enclosures, where Burckhardt has studied them thoroughly. On the way to the watering place, this group would walk in a single file. One specific female walked ahead. She had "conquered" this position peacefully. In the beginning, she

Fig. 12-21.
The male lesser kudu, in an upright display, unflinchingly tolerates forehead butts from the female.

The sitatunga

walked a little bit faster than the others. When she had reached her place she prevented all attempts by other females to take the lead by vigorous threats or butting with her mouth. When she ran, the others would also run. When she stood still, the rest of the group would stand still. The old bull, who was always the last one in the line, made the animals move. Obviously, only the first and the last positions in the line are of importance to nyalas. When the old male felt uncomfortable because of the observer's presence, he would drive the females back and then approach the person all by himself with butting of his head. In this case, while one specific female was the leader, the male played the part of the "sheep dog."

In kudus, bushbucks, and other species of *Tragelaphus* it struck me that even when the animals wanted something high up they only rarely stood up on their hind legs. For example, I saw sitatungas repeatedly reaching with stretched-out tongues for the leaves of a tree, but they could not reach them even though they were only a few centimeters away. They would raise one front leg, but they would never really stand erect. Burckhardt saw in nyalas that one single individual female was able to stand erect on the hind legs while reaching for leaves, but the other members of the the group were clearly unable to do this. In encounters between the nyala males, the display walk is especially well developed. The opponents approach each other frontally in slow motion and periodically present to each other their broadsides. The mating behavior of the nyala is intermediate between the greater kudu and the lesser kudu. Here we find, according to Burckhardt, ritualized neck fighting as well as the male reaching underneath the female. The display also involves lifting up and bending the head down in alternation with the bulls raising the hair on their backs and flapping their tails over one thigh. Some of the kudu's behavior obviously has become more ritualized and symbolic in the male nyala.

Of all *Tragelaphus* species, the SITATUNGA is best adapted for life in the swamp and water. The hooves can be spread far apart and carry the animal safely over muddy ground. On solid ground the sitatungas actually move rather clumsily and with difficulty. They are excellent swimmers, often spending the day resting in the swamp or in the water, and taking to the water in flight. When in danger, they may submerge out of sight so that only the nostrils are visible. This is rather unusual for a horned ungulate. Because of these habits the sitatunga inhabits only places with swamp forest, marshes, reed and papyrus jungle, or other very moist regions within its large area of distribution. Its preferred diet is aquatic plants, and the shoots of bamboo and papyrus.

Sitatungas do very well in zoos and they reproduce easily. They have reached an age of up to seventeen years in captivity. Even though their existence in the wild is at present not actually endangered, the sitatunga is among those species which ought to be absolutely protected.

They are not really abundant anywhere, and unfortunately, it is very easy to hunt them with the most primitive weapons. Natives have dogs which chase them out of their hiding places, and when the sitatungas flee into the water, they are killed from boats. On several islands in Lake Victoria, which had been uninhabited for a long time, the sitatungas had reproduced nicely. When the human settlers returned, they were completely destroyed within a short span of time.

Males fight hard, sometimes going down on their front legs, which are stretched forward so that their elbows almost touch the ground, and then they push forward and upwards towards the opponent. Females snap with their mouths, box with their foreheads, and use all kinds of neck fighting. In the mating ceremony, the driving male slides his neck and head along the female's back from behind and pushes her down rather vigorously, in contrast to the kudu. A high spirited female sitatunga counters with an almost endless variety of neck fighting, boxing, and snapping movements, most of which are "symbolic," which means that they are directed forward and not towards the male.

Even though the young sitatungas are typical "stay put" animals like their relatives, I observed a remarkable variation in the Frankfurt Zoo: The mother did not call the young but went straight up to the lying fawn, licked over its nose or looked at it directly. When she walked on, the young jumped on to its feet, hurried after her, and drank. Nursing usually lasts five to ten minutes, or sometimes, with brief interruptions, up to twenty minutes. Sitatunga females give birth to one or two young after a gestation period of seven to seven and a half months. In zoos they would nurse other female's young in addition to their own or separately. In a slightly over-populated zoo pen, it struck me that only the mother stayed away from her young, which was "staying put," while the other adults would lie right next to it. When a young was resting beside a female, this certainly was not its mother. When alarmed, the females make a sneezing sound, which is unusual since otherwise the alarm call of *Tragelaphus* males and females sounds like a bark.

The bushbuck

Like the sitatunga, the BUSHBUCK also is strongly enlarged in the rear; thus it is higher at the croup than at the withers. There are twenty-four subspecies, which may differ greatly in shade and banding, one of which is the HARNESSED ANTELOPE *(Tragelaphus scriptus scriptus)*. Bushbucks live in plains, alpine regions up to 4000 meters, prime forests, fringe forests along rivers, or tree-covered spots in the steppe plains. Since they are good swimmers, they have even managed to populate islands, such as those in Lake Victoria. Often bushbucks may be found in the immediate vicinity of human settlements. Even though bushbucks in the wild live predominantly singly or in pairs, it is possible to keep a group of them together in zoos without complications, as long as there is sufficient space for the animals to lie down to rest, always singly and with a large distance between the individuals. Thus, they are the most distinct "loners" of the genus. According to Verhayer, the males and the females have pear-shaped territories of one to three hectares. Where these territories meet at the "stem of the pear," there is a jointly used space where, according to Verheyen's observations, the animals meet in the early afternoon.

During the threat display, the male bushbuck erects the hair on his back and stretches his head and neck forward. He has the same posture in driving, only without the erect back mane. In the mating ceremony, there are no actual preludes. The male approaches the female with his head lowered forward and down. Sometimes she assumes the same posture for a moment and stands opposite him like his mirror image, but then she turns around and takes off. The male follows with his neck stretched forward and approaches her side, uttering calls made only when he is driving females. Often he brushes his cheek and the side of his neck along the female's hindquarters, as does the lesser kudu. Often he licks the female in the genital region, at the thigh, on the back, and especially at the base of the tail. The inhibition to attack a female seems to be less than in the kudus. At least it happens in zoos that the male will push his horns against the female's hindquarters, causing bloody injuries.

Eland antelopes

Distinguishing characteristics

The ELANDS (genus *Taurotragus*) mainly differ from the spiral horns because their females also have rather handsome horns. The tails of the adult animals are like a cow's tail. There are two species: 1. BONGO (*Taurotragus eurycerus;* Color plate, p. 302). The HRL is 170–250 cm, the TL is 45–65 cm, the BH is 110–125 cm, the weight reaches 220 kg, and the horns may extend up to 100 cm. The body is similar to the nyala, only a little more bulky. The lyrate horns are very thick and compact, long, and quite elegant. The tail tassel reaches down to the heels. The long, narrow hooves have pointed ends. The pseudo-claws on the front legs are long and pointed, while on the hind legs they are shorter and rounder. 2. ELAND (*Taurotragus oryx;* Color plate, p. 302). The HRL is 230–345 cm, the TL is 50–90 cm, the BH is 140–180 cm, the weight reaches 1000 kg, and the male horns extend up to 113 cm. Their size is like that of an oxen, but with a more slender rump. The animals have a straight back, hump-shaped withers, a dewlap between the throat and the chest, and horns which are slanted back and upward. In the ♂♂, the horns are very bulky with a thick base. The hooves are rather narrow and short, the pseudo-claws are large and strong, and the tail has a thick, medium-long tassel. There are five subspecies.

The bongo

The bongo's position in the zoological classification has changed several times. Until lately, it was considered as a genus of its own *(Boocercus)*, then it was classified with the spiral-horned ungulates *(Tragelaphus)*, and then again with the elands *(Taurotragus)*. Many experts consider this handsome horned ungulate with its reddish-brown coat and vivid white pattern to be one of the most beautiful antelopes. It is also one of the rarest and most expensive zoo animals. The only places outside of their African native habitat where the bongos have been displayed were in Rome, Antwerp, London, New York, and Cleveland. Today only the Cleveland Zoo still has bongos, while Antwerp has two crossbreeds from a sitatunga and a bongo. Recently the Zoological Gardens of Frankfurt and Basel purchased bongos in Africa which are expected to arrive soon. Director Reuther of the Cleveland Zoo found out that the bongo females come into heat approximately every three weeks. Otherwise, not many studies have been made on these zoo animals.

Fig. 12-22.
Bongo *(Taurotragus euryceros)*.

Little is known about life in the wild, and the reports which do exist are contradictory. For example, some authors state the bongos occur in Nigeria, yet others dispute it. Some informants say that they will not jump over obstacles, like fallen trees, fences, etc., but, if possible, crawl underneath them. On the other hand, a zoo bongo did jump a fence. Furthermore, it has been said that the bongo wallows in mud, similar to European red deer. Since none of its relatives, to our knowledge, do this, such behavior would be rather surprising. Apparently no one has actually seen a bongo doing this, but conclusions were drawn from places in the ground which looked like wallows.

Like the other Tragelaphinae, the bongo's preferred habitats are areas with dense brushwood. They need water, feed mainly on leaves, shoots, and fruit, and live singly, in pairs, or in small groups. Furthermore, they are said to pry bush and tree roots out of the ground with their horns. Injured bongos have occasionally attacked people. The young animals in captivity quickly become tame with their keeper. Up to the present time, the bongo has remained a legendary animal, not only to science but in its native countries as well. Attilio Gatti, the Italian explorer of Africa, has found that the natives, especially the Pygmies of the Congo forest, believe that the bongo has magic powers and strange abilities. They say that the bongo is able to hang from branches with its horns so that it could drop down on an unsuspecting hunter. They also claim that it would eat poisonous plants so that its meat would not be edible to humans, and that when it was pursued it would dive under water and stay there until the following dry season, losing its coat and feeding on fish.

Fig. 12-23.
Eland *(Taurotragus oryx)* with subspecies: 1. West Sudan and East Sudan giant eland *(Taurotragus oryx derbianus* and *Taurotragus oryx gigas)*, 2. other subspecies.

Gatti captured a young bongo in the 1930s. The little animal learned to walk around him and even to make a figure eight around him when he came with the milk bottle. However, Gatti misinterpreted this behavior. In order not to overfeed the little animal, he took the bottle away several times during the feeding and turned away. He believed that the bongo initiated this motion and concluded that the bongo had an especially quick power of comprehension. Of course, this kind of "animal behavior" looks different now that we know about the manner in which young kudus and sitatungas, and probably bongos as well, circle around their mothers to force them into nursing.

Even though the bongo is distributed in suitable forest regions from West Africa to Kenya and Uganda, it is the rarest large antelope and should be completely protected.

The eland

The eland antelope's horns are not lyrate but almost straight. The "screw characteristic" of the African Tragelaphinae horns, however, is evident here as well. The lower third, or sometimes half, of each horn is wound in spiral ridges around its longitudinal axis. Among the five subspecies of this large antelope are the WEST SUDAN and the EAST SUDAN GIANT ELAND (⚥ *Taurotragus oryx derbianus* and *Taurotragus oryx gigas*), whose 180 cm height is especially worth mentioning. They live more in the brush and forest than in the steppe where elands may also occur in the fringes of forests, but they are found more frequently in open spaces. The pronounced dewlap of the steppe elands begins near the throat, while in the giant elands it starts at the lower jaw. The adults of all subspecies have a forehead tuft which they impregnate with the moist soil or occasionally with their urine when they plough the ground with their horns. Then they brush this paste on tree trunks, bushes, or, in the zoo, on the stable walls as markings.

As all Tragelaphinae, elands always march in a crosswalk. When one is close enough, a strange, cracking noise becomes audible. With each step of the heavy animals, the two scales of their hooves are spread apart and when the leg is lifted they clack back together again. Even though elands are well able to gallop, they usually take off in a trot which they may, if necessary, continue for a long distance. Often some high and long jumps initiate the flight, and in open country one animal may even jump over another. In Serengeti the elands are the shyest game by far and have an extremely large flight distance from people. Allegedly, this is caused by the Masai's hunting of elands. Generally the Masais, as genuine pastoral tribes, do not hunt, except in a "demonstration of courage" with lions and rhinoceroses. But according to their faith, God created the oxen for the Masai, and so they also go after wild oxen, including the eland, which they consider to be an ox, in addition to the cape buffalo.

Elands as domestic animals

This similarity between eland and cattle in size and weight has inspired Europeans to attempt taming and domesticating these animals. Presently, some herds of tame elands are being kept in Rhodesia and South Africa. Before the turn of the century, a Boer farmer tried to breed some eland antelopes, which the animal trader Hagenbeck later bought from him. Interestingly enough, all these attempts ended in failure except for one case. This exception took place not in Africa, but in the southern Ukraine. There in 1896 a country squire, Friedrich Falz-Fein, had acquired some eland antelopes and turned them loose with several other species of game animals on his estate Askania Nowa in the southern Ukraine steppe. In spite of all political changes and wars, the descendents of this herd are still there. The eland antelopes are tended by mounted herdsmen, and the females are even milked regularly. Therefore, it would be possible to domesticate the eland which would be of greater economic value in Africa as a supplier of milk and meat than the countless cattle herds which are now turning the steppe into desert with their grazing.

When young elands in a zoo are forced into continuous contact with humans by keeping them in a stall or in small pens, all of a sudden an urge to escape may arise after the "stay put" phase. Then it makes no difference whether the mother and the other animals have been tame for a long while or not. This often causes many problems for the keepers. The young flee immediately and may run into fences and gates as soon as a person comes close to them. This acute readiness for flight lasts for several weeks. Later it decreases, and the once very shy animals may then become completely tame. In a zoo where the elands inhabited a very large pen, we had the same difficulties with each young that grew up outdoors. It is possible that this behavior has been an obstacle in attempts to domesticate elands. In the literature there is a report on a crossbreed between eland and cattle. However, this statement is rather controversial. Soviet scientists have tried to obtain such crosses for many years, even with artificial insemination, but not one of the experiments was ever successful.

Like kudus and their relatives, the elands may also display against rivals with their broadside, keeping the head up and slightly turned away from the opponent, and presenting the huge shoulder region to the rival. Attacking bulls slowly approach the opponent, gradually getting "in touch" with their horns. Only after they are in actual contact with horns do they push forward with all their might. Hence, with tame animals one can usually get out of range of the slowly developing attack. On the other hand, the bull approaches so calmly that people who are not familiar with his behavior do not realize his intention to attack until it is too late. In addition to fighting, elands also use their horns to break branches off of trees since they like to eat foliage.

The mating ceremony of the eland is very simple and actually consists only of driving. The bull approaches the center of the female's rump from the side in a very stretched-out posture or he slides his neck and head from behind on her back. In mounting, he rests his neck and head on the female's back. After a gestation period of eight and a half to nine months, the female usually has only one young. The births, which occur throughout the year, are more frequent at the beginning of the rainy season. In captivity, female elands may occasionally nurse other young, as does the sitatunga. The older young may also drink, sometimes even when they are fully grown, and the mother will not reject them. Even though the younger colt receives less milk as a result,

Fig. 12-24.
Male eland catching up with the female from the side in the mating ceremony.

I have never heard of any rearing problems attributed to this practice.

The huge forest elands now have a much smaller distribution than they had several decades ago. The West Sudanese subspecies *(Taurotragus oryx derbianus)* is found only in a small number in a few areas of West Africa. In contrast, the steppe eland still occurs rather frequently. They are more gregarious than the other *Tragelaphus* species. Besides smaller groups, they may join herds of 200 and more animals. In the Serengeti region, they predominantly associate with zebras, with which they migrate annually over considerable distances. They spend the rainy season on the steppe and the dry season in the brush. Adult elands are in danger mainly from lions. Against smaller predators which might be able to attack their calves, eland mothers are more aggressive than antelopes usually are. I once saw a small herd of females as a unit take on a cheetah and actually chase him away. Another time they chased a leopard.

The nilgai

The only species of the tribe Boselaphini is the NILGAI *(Boselaphus tragocamelus;* Color plate, p. 302). The HRL is 180–200 cm, the TL is 40–55 cm, the BH is 120–150 cm, and the weight reaches 200 kg. The ♂♂ are about one fifth larger and heavier than the ♀♀. The animals are long-legged, with a long head, a short neck, and a slanted back which is higher in the withers. The front part of the body is heavier and bulkier than the rear. The ♀♀ heads are long and narrow, as in female red deer. In the ♂♂, the heads are more compact, with a straight nose and a slightly rising forehead. The short, sturdy, rather steep horns are only found in the ♂♂. The legs are long and slender, the hooves are narrow with pointed tips, and the pseudo-claws are short, broad, and flat. The slightly bushy tail has a tip like a tassel, and reaches down to the heels. A gestation period is about eight months (247 days), and one or two young are born.

The name "nilgai" is said to come from a combination of an Indian and an English word: "nil" means "blue," "gai" came from "cow." The popular English name "blue bull" also suggests the male's blue-gray color. The nilgais and blackbucks belong to the best known antelopes of India, and are also most frequently found in zoos. They live in small groups in scattered brush or grass steppes which contain isolated areas with trees and bushes. Although they are said to be territorial, there are no details known about the way in which their territoriality is expressed. We only know that within their range, they have preferred paths, water holes, defecation and resting places. In contrast to the African Tragelaphinae, the nilgai often rise on their hind legs while browsing on the foliage. They are not dependent on having water in the vicinity, although nilgais drink at least every two or three days. Then they may "wash" themselves occasionally, wetting their rumps with the dripping muzzles.

In India their mating season begins at the end of March, so that the young are usually born in December. In European zoos, however, nilgais are born mainly in the summer. At the first birth, the female has one young, while after that twins are the rule. Like in all Tragelaphinae, the young "stay put"; in twins, it is clearly revealed that occasionally, at least in a zoo, the mother's legs are an invitation to lie down. One of the twins frequently lies near the mother's front legs and the other near the hind legs. When the mother walks off, the young toddle behind her. Thus, "staying put" becomes meaningless and the mother often has trouble until the young have learned to use this innate behavior in an appropriate manner.

The very "stretched-out posture" which all Tragelaphinae show in many variations, was first observed in the nilgai. The threatening male approaches the opponent in slow motion and presents his head, which is horizontally stretched forward, and the broadside of his body. While this behavior is in the African Tragelaphinae, except for the bushbuck, predominantly a courtship gesture of the mating ceremony, in the nilgai it is a threat against the rival. Here it may be considered an initial invitation for the neck fight which in nilgais occurs as a regular part of the fighting ceremony between the males. In addition, the bulls also fight with their horns. In every intense fight, they go down on the carpal joints (the apparent "knees" of the front legs). Severe fights occur also between females; the animals box with their hornless foreheads, jump at each other, engage in neck fights, and bite each other. When captured bulls have little opportunity to fight, they may redirect their aggression and turn it against people, which is extremely dangerous.

India is also the native habitat of the strange FOUR-HORNED ANTELOPE (*Tetracerus quadricornis;* Color plate, p. 302), the only member of the tribe Tetracerini. The HRL is 90–110 cm, the TL is 10–15 cm, the BH is 55–65 cm, and the weight reaches 15–25 kg. This animal is the size of a gazelle, and shaped like a deer, except it is higher in the croup. The slender head has a narrow muzzle. The horns, which exist only in the ♂♂, emerge rather steeply, slightly arched, and directed upward and inward. Often an additional pair of almost vertical horns grow close together on the anterior part of the forehead and they may occasionally drop off. The slender legs have narrow hooves with pointed tips and flat broad pseudo-claws. A gestation period of seven and a half to eight months produces one to three young.

The habitat of this unique "four-horned" antelope is found in regions made up of meadows, groups of trees, bushes, and water in the vicinity. The animals drink at least once a day. Four-horned antelopes, which live singly or in pairs, probably keep their territories for a long time. When disturbed they sneak off, duck down, or flee in short jumps. During the mating season from June until August, the males are very aggressive.

Fritz Walther

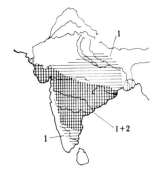

Fig. 12-25.
1. Nilgai *(Boselpahus tragocamelus).* 2. Four-horned antelope *(Tetracerus quadricornis).*

The four-horned antelope

13   The Wild and Domestic Oxen

Subfamily: Bovinae
by A. Wünschmann

Distinguishing
characteristics

The fourth subfamily of the large horned ungulate family (Bovidae) consists of the WILD OXEN (Bovinae). The HRL is 160–350 cm, the TL is 18–110 cm, the shoulder height is 60–280 cm, and the weight varies between 150–1350 kg. The animals are barrel shaped with a plump rump. The broad skull is only slightly reduced towards the muzzle. In more primitive forms the parietal bone is pushed less to the back of the head, while in the higher evolved forms it is pushed more to the back so that the axis of the skull appears to be broken. There are horns in both ♂♂ and ♀♀; in more highly evolved forms they sit on the upper edge of this break; in a crosscut the horns are round or triangular, more so in ♂♂. The mouth drools continuously and is bare of skin. In some species, there is a skin fold and/or a large muscle behind the neck and withers, which is supported by the extended spinal process of the dorsal vertebra. There are preorbital or interdigital glands. The short-haired coat clings smoothly, although rarely it may be woolly or extended like a mane. The tail has a more or less heavy tassel. The females have four nipples at the udder. There are 32 teeth: $\frac{0 \cdot 0 \cdot 3 \cdot 3}{1 \cdot 3 \cdot 3 \cdot 3}$. The four genera (Asiatic buffalos, African buffalos, true cattle, and bisons) have seven subgenera, nine species and 21 subspecies.

Phylogeny and
distribution

Compared to other mammals, the bovinae are phylogenetically a relatively young group. The oldest fossils come from the beginning of the Upper Lignite era. Now they are far past the peak of their evolution, and their wild forms inhabit only a small part of the range they had during their prime in the Glacial Period about 1,000,000 years ago. Asia is the original habitat of the wild oxen and most of the present species still live there. There and in Africa, the wild oxen could survive in the wild, although some species and subspecies are threatened with extinction. In contrast, the wild oxen of Europe and North America were almost completely destroyed in historical times, and the few survivors spend their lives under careful protection in conservation areas and zoological gardens. South America and Australia have no wild oxen, but today their descendants which have been domesticated by man, yet are partly feral, live there and all over the world.

Wild oxen live in a variety of habitats. Many species prefer the dense forest, others the open grass steppes, and some occur in the mountains with elevations up to 5000 and 6000 meters. But they all need water, especially since most of the species have a distinct tendency to wallow or to bathe. Being ruminants, the wild oxen are true herbivores whose diet, according to the habitat, consists of grass and herbs, foliage, buds, shoots, and the bark of trees and brush, swamp and aquatic plants or even moss and lichen.

Habitat and food

Among their sense organs the sense of smell is best developed although their sense of hearing is also good. Vision seems to be of lesser importance and seems to play only a small role in orientation, at least at a greater distance. Motionless objects are usually seen only at close range. According to the results of experiments by Bernhard Grzimek and his assistants in the Frankfurt Zoo, wild oxen, as well as other horned ungulates, have a relatively well developed color vision and are able to distinguish the primary colors of blue, red, green, and yellow from all shades of gray offered for comparison.

Sensory capacities

During the mating season, the bulls fight among each other. The gestation period of the wild oxen is, depending on the species, nine to 11 months. Twins are very rare. According to the findings of Naakt-geboren (1963) on domesticated and wild oxen, their calves are born in almost all cases, like in all horned ungulates, head first in a stretched out position and they are soon able to follow their mother and the herd.

Reproduction

With few exceptions, all wild oxen are gregarious animals with distinct social behavior which also appears in various forms in cattle. The basic unit of a cattle group is the small herd which consists of one to three dozen animals. The composition of the group alternates between mixed herds, groups of females and young, and exclusively male associations, according to the three main phases of the year, the mating season, the time when the young are born, and the scarce dry or winter months. At times several small groups join large herds. The adult bulls often separate from the herd and become loners, but during the mating season they again try to make contact with the herd.

Social behavior

The Swiss ethologist, R. Schloeth, from 1956 until 1961, spent many weeks on horseback among the semi-feral French fighting oxen at Camargue in the Rhone delta. In more than 2000 hours of observation, he was able to collect detailed observations on the behavior of this primitive race of oxen, which in many respects are close to the extinct aurochs. The social habits of the camargue oxen show many parallels to the behavior of wild oxen. Schloeth distinguishes between interactions in which smell, hearing, sight, and physical contact is differentially important. The scent control is a sniffing at the head, anal, and genital region, and it is mainly performed by bulls on cows, or in rarer cases between bulls or from a cow on a bull. The vocal expressions of oxen are quite varied. With lowing, grunting, or roaring, according to Schloeth, 11 different meanings may be expressed: threatening sounds when a rival approaches, alarm calls, fighting and chasing sounds, herding calls, calls of distress, calls of gathering, contact calls between cows and calves, mating calls, and sounds made during play.

Since oxen have only a poorly differentiated muscular system in the face, their facial expressions are limited, but this is compensated for with a variety of other expressions in which the position of the head and the horns in orientation to the neck and rump are significant. "The head is the center of the visual expression," as Schloeth puts it. For example, in lateral threatening the ox presents the broadside of his body to the opponent, the neck appears, depending on the intensity of the threat, more or less lowered or arched, and the horns are vertical to the ground. However, if the ox turns his head away, this expresses a certain appeasement. The position of head and neck also denotes the individual animal's rank within the herd. The higher ranking animals approach conspecifics of equal rank or lower rank before physically contacting them, frontally with a slightly lowered neck and the head distinctly stretched forward, while low ranking animals in the same situation approach higher ranking individuals with their neck deeply lowered.

Among the contact relationships are licking the shoulders, which may last quite some time and usually is performed by the lower ranking animal on the higher ranking one, and the habit of pushing with the forehead and horns against each other, mostly with peaceful intentions, but occasionally aggressively. Some of these behaviors are directed at a distinct partner; pawing the ground with the front legs, rubbing the sides of the head and neck on mounds or bushes which are used repeatedly for this purpose, and stabbing one horn into the ground. With all these movements, an ox expresses self-confidence or challenges conspecifics. McHugh, in 1958, described similar expressive actions in the American bison, such as rolling on the ground in situations of great excitement, for example, in mating season.

Individuals communicate by means of these expressions and the animals claim their place in the rank order or change it. This behavior occurs mainly during the mating season when the bulls constantly try to chase each other off. Fights occur more often when more peers live together. After a prolonged lateral threat display consisting of pawing, rubbing the head, and digging a horn into the ground, the actual horn fight develops with a sudden attack by one of the two opponents. By butting and pushing with their foreheads firmly against each other, the opponents try to get one another into a position which will enable them to strike with their horn into the other's body. The only escape is flight through a quick turnabout.

The reproductive behavior of the semi-wild Camargue oxen also resembles closely that of most wild oxen. In the beginning of the mating season, the herd merges closer together. The bulls constantly watch for each indication of sexual behavior in the herd and fight over the females. As soon as a cow shows signs of heat, all bulls try to stand as close by her as possible. The highest ranked bull always claims the cow most advanced in the estrous cycle. This "guarding" has also been described by McHugh in the bison. It may last for days until the female is ready to copulate. It is best described as a "marriage of days or hours" and it is the only individual relationship between the sexes. During "guarding", the bull tries to mount the cow and when, after several such attempts, she finally is ready for copulation, the coition follows with the short, powerful jump which is typical of all oxen.

On the relations between mother and young, Schloeth reports: "During the first two or three days, the cow stays close to the newborn calf. Later it joins peers and spends most of the time resting with them. During this phase, the young is nursed an average of six times within 24 hours, and the cow meets her calf only for nursing. Either she calls her calf or the young becomes restless and begins to call for the mother. At first a calf will often go to other cows, and it is only through their threatening movements that it becomes aware of its error. While nursing it is oriented in an acute angle to the axis of the mother's body, like all young horned ungulates. This orientation is innate as is the first search for the milk source after birth at any angle to the mother's body. Older calves may sometimes diverge from the usual suckling position, especially in cases where the nursing has been interrupted too soon. Then they follow the departing mother and try to reach the nipples through her hind legs so they can drink for a few more minutes."

At a young age, oxen are very playful, and the adults will occasionally join the games. There are regular "playgrounds" where the young frisk about high spiritedly, chase each other, have mock fights, run races, and sometimes even chase their own tail. Age mates form actual teams within the herd, although often the mothers are their playmates. When calves playfully mount one another or play "mother and child", where one calf pretends to nurse the other, they are already expressing sexual behavior.

The territorial behavior of the Camargue oxen is probably like that of most wild oxen. Neither herds nor single individuals have, according to Schloeth's observations, distinct areas which would be marked or defended. But certain areas are connected by frequently used paths; grazing grounds, water holes, resting places, and points which are preferred for grooming. These include rubbing trees and bushes which are damaged with the horns and the legs, and places which the animals prefer for pawing. They pierce the horns into the ground, or rub the head and neck. In the American and the European bison and the African buffalo similar forms of behavior have been described.

To the buffalo of Asia and Africa belong some of the phylogenetically most primitive representatives of the true oxen as well as some rather advanced forms. These are from medium to very large. The HRL is 180-300 cm, the TL is 18-110 cm, and the weight reaches 900 kg. They have a massive build. As they age they have only scarce hair growth. The horns are larger than in other wild oxen. There are two genera (Asiatic buffalo, and African buffalo) with three subgenera. All buffalo prefer the forest, savannah, or brush over the steppe which does not provide any cover, and they are enthusiastic about wallowing.

Asiatic and African buffalos probably diverged and developed into two separate races at an early period in their phylogeny. There were Water buffalo in the Glacial Period in North Africa and ancestors of the African buffalo once lived in Asia Minor (Palestine) producing during the Glacial Period giant forms with long horns (genus *Homoioceras*). However, the difference between the contemporary buffalos of Asia and Africa are large enough to classify them into two genera (*Bubalus* and *Syncerus*). From the Asiatic buffalo of the genus *Bubalus*, the Anoa is separated as a subgenus of its own *(Anoa)*.

Asiatic buffalo

The ANOA (*Bubalus depressicornis*; Color plate p. 302) is the smallest and oldest of all contemporary wild oxen. The HRL is approximately 160 cm, the shoulder height is 60-100 cm, and the weight is 150-300 kg. The rump is barrel-shaped. The legs are medium long and almost slender like an antelope's. The animals have a graceful form and a skull tapering off towards the muzzle, and rather short, straight, compressed horns which in cross section are flat triangular or elliptical. The maximum horn length is 39 cm. The calves have a thick, woolly, golden brown coat. There are two subspecies; The LOWLAND ANOA (⟡ *Bubalis depressicornis depressicornis*) in the lower elevations of Celebes, and the MOUNTAIN ANOA (⟡*Bubalis depressicornis fergusoni*) in the mountains of Central and Southern Celebes. According to Erna Mohr, the existence of a third subspecies (*Bubalus depressicornis quarlesi*) is possible.

Even though the anoa looks like a transition between an antelope and a wild ox, it gives, to quote Hilzheimer in the fourth edition of *Brehm's Animals of the World*, the impression "of a little ox, is lazy and reluctant to move like his relatives are, often stands for a long time in the same place, ruminating or eating, seeming to care little or not at all for his environment. His usual locomotion is a slow pace, but once in a while he may make a few awkward jumps, just like an ox."

The anoa's distribution is restricted to Celebes, where the two or three subspecies are now threatened with extinction. Fossils have also been found only on the island of Celebes and never on the continent. It has evolved as a true island form of small stature. All anoas inhabit swamp forests and jungles and, since they flee from civilization, the increase in farming in their habitat gradually forces them back. In 1937, the lowland anoas were still said to be relatively numerous in some areas of North Celebes, but presently they are only scattered throughout the remotest parts of the swamp forest. Before the introduction of firearms, the anoa was feared by the natives and only rarely hunted. Unfortunately, this has completely changed. The anoas are hunted ruthlessly everywhere for their tender, good-tasting meat, their horns, and their skins which some tribes use for dancing clothes. Therefore, danger of extinction is increasing, and the laws for their protection remain ineffective as long as they are not enforced.

Fig. 13-1.
1. Anoa *(Bubalus depressicornis)*
2. Water buffalo *(Bubalus arnee)*, distribution of the wild form.

When injured or cornered, the anoa attacks blindly, a fact which has been seen repeatedly in zoos. Therefore, anoas in captivity should always be kept in sufficiently spacious pens where they may withdraw when they feel pressed. It is especially recommended that the bulls are kept separately from the cows and the young. The anoa seems to live, in contrast to most of the other wild oxen, predominantly singly or in pairs, or in rarer cases, in small family groups. The gestation period lasts 275 to 315 days. Mating does not seem to be restricted to distinct periods. In zoos births occurred at any season with a slight rise in March. The natural diet of the anoa consists of grass, herbs, leaves and fruit, as well as swamp and aquatic plants. Like nearly all oxen, it needs water and moisture nearby for bathing and wallowing.

In zoological gardens, anoas have been kept successfully, in former years much more frequently than now, and in some cases they have also reproduced regularly. Some individuals have reached an age of up to 28 years. In the San Diego Zoo, California, where between 1942 and 1964 a total of 26 anoas were born, the zoologist, J. M. Dolan, kept a studbook of all anoas in captivity on behalf of the International Union for Conservation of Nature (IUCN) with the purpose of keeping exact control over the development of this rare wild ox.

The WATER BUFFALO of the subgenus *Bubalus* (Color plate p. 345) are, in contrast to the anoa, huge, bulky animals. The HRL is 250–300 cm, the TL is 60–100 cm, the shoulder height is 150–180 cm, and the weight reaches almost 1000 kg. The huge horns are crescent shaped, gently bent backward, and almost on one plane from which only the tips protrude slightly (record length 194 cm). The crosscut is triangular with heavy ridges across the flat upper side. The hair is short and sparse. The broad hoofs are long and can be spread far apart (adapted to the swamp habitat). There are no obvious differences in the size and the horns between the ♂♂ and the ♀♀. There is one species, the WATER BUFFALO *(Bubalus arnee)* with six subspecies: 1. ARNA *(Bubalus arnee arnee).* 2. ASSAM WATER BUFFALO *(Bubalus arnee fulvus),* a rare brown animal found only in protected areas. 3. CEYLON WATER BUFFALO *(Bubalus arnee migona),* pure and not crossed with domestic buffalos, and found only in the Yala Game Reserve. 4. *Bubalus arnee subspecies,* is perhaps not a genuine wild buffalo, but feral. 5. BORNEO WATER BUFFALO (*Bubalus arnee hosei*), is smaller, of pale gray color, found only in the valleys of the Miri and the Baram River in North Borneo. 6. TAMAROU (*Bubalus arnee mindorensis;* Color plate p. 302), the smallest subspecies, which has some characteristics similar to the anoa, found only on the Philippine Island of Mindoro, and it is the rarest wild ox on earth.

The population of wild water buffalos, from which the domesticated buffalos that are distributed all over the world originated, kept decreasing especially during the last decades. In prehistoric times, they lived from India eastward to China, westward over Mesopotamia to North Africa, and probably even in Europe. Until lately, a last herd which, in the opinion of some scientists, may have originated from the North African water buffalo of the last Glacial Period, was living in the Ischkeul swamp near Bizerta in Tunisia. All subspecies of the Asiatic wild buffalo have become very rare or are threatened by extinction in their last refuges. The large arna is found only in a few places in the Central Provinces of India and in Nepal, Assam, and Bengali. The population of the Assam water buffalo in the Kaziranga Preserve, which is famous for its great Indian rhinoceros, were estimated by Heinz-Georg Klös at 400 head, and by Wolfgang Ullrich at a maximum of 800 animals.

In spite of strict regulations for their protection in some areas, the water buffalo are hunted everywhere in an uncontrolled manner and are poached by the natives, often with poisoned arrows. The meat is eaten by some Indian castes, and the skins and the horns, which are made into various utensils, are in great demand commercially. An especially serious danger to the wild buffalo is the fact that they cross with tame or feral domestic buffalo. Not only does this decrease the number of pure wild buffalo, but also the wild animals become repeatedly infected with dangerous contagious diseases from the domestic buffalo. Thus, in Bengal and Indo-China, the water buffalo were decimated to a catastrophic extent by the cattle plague.

Even more than all the other wild oxen, the water buffalo depends on the vicinity of water. They live in the moist grass jungles and brushwood areas, in river valleys, and in lowland swamps near the edge of forests, and less in the dense forests and the brackish lagoons at the coast. There they eat grass, herbs, swamp, and aquatic plants. In 1963, Wolfgang Ullrich observed a herd of Assam water buffalo in the Kaziranga Game Reserve for four weeks. The animals used to stay for several days near a lake, which was located in the western part of their home range, and eat the grass and the herbs at the edge of the lake. "As early as seven o'clock in the morning, the arnas would lie in the wallows and stay there until afternoon, mostly until around five p.m. By beating their front legs, the animals who lie on their bellies splash fresh mud on their bodies. They also use their horns as shovels.

They dip the horns into the mud with sideward movements of the head and throw the mud on their backs. In addition, they roll in the mud." The home range observed by Ullrich was about two and a half kilometers long and 400 meters wide. At the other end of it was a grazing area, although the buffalo did not visit it as frequently as the lake. Wandering at night between the two places, they used the great Indian rhinoceros' paths. They also grazed predominantly at night.

Herds of water buffalo usually consist of ten to twenty animals, although concentrations of 100 and more buffalos have been observed. The herd observed by Ullrich consisted of 17 head: "...one old black bull, one adult, young, gray bull, seven adult and two subadult cows, four calves under one year of age, and two calves of about one and a half years." An especially strict rank order within the herd did not seem to exist, but the old bull who usually stood several hundred meters apart from the other animals was respected by all members of the group. In flight this bull kept the herd together and drove single animals back to the herd with sideward blows of his horns even when they were only a few meters away from the others. On their trips, the herd of buffalo usually kept to an invariable marching order. Some of the older cows formed the "advanced guard"; a group of three consisting of one high ranking old cow, one young cow, and a young bull were the "rear guard". In case of danger, they covered the retreat of the herd and, if necessary, attacked. The calves were always kept in the center, also during flights the herd dashed off in a bunched-up group. They always fled towards the high grass jungle and never into the water or to the wallows. In all cases, however, the animals fled at first only several meters, then they formed almost a half-circle, stood still, and looked back at their pursuers—a behavior which is also known from the African buffalo. When threatening, they stamped the ground with their hoofs in typical manner of buffalos.

In water buffalos, like in all wild oxen, old bulls separate from the herd and become loners. As a rule, they do so on their own, although often decrepit or sick animals are chased away by the younger bulls. The mahouts, the drivers of Indian elephants, especially fear these cross and aggressive loners and try to avoid them with their riding elephants, if possible. At the boundary of the Kaziranga Game Reserve, it often happens that old arna bulls take the lead of a domestic buffalo herd and mate with the cows.

According to Ullrich's observations, the water buffalos are almost equal to the great Indian rhinoceros in the biological rank order. The two often graze next to each other in the Kaziranga Game Reserve and use the same wallows without bothering one another. When a rhino comes too close, the buffalo withdraws "quietly, slowly, and without any sign of excitement". In contrast, elephants seem to avoid the arnas, and riding elephants are said to have been attacked and repeatedly

Social life in herds of water buffalo

Old loners

injured by the buffalos. Barasinghas, sambar and hog deer, muntjacs and wild pigs also share the habitat of the water buffalo, but always stay out of their way. The arnas are frequently accompanied by cattle egrets and oxpeckers, which eat the insects buzzing around the herd as well as other small animals that they have stirred up. The long-time foe of the buffalo, besides man, is the tiger, but it usually attacks only young animals. Several reports say that the scent of the tiger excites the buffalos so much that they followed its tracks in closed formation into the dense jungle and, having tracked it down, would trample the large cat to pieces.

Reproduction

Mating and births in water buffalos, as in most tropical animals, are not seasonal. The whole year round calves of all sizes are seen among the herds. Only in the northern part of their distribution, does mating take place predominantly in the fall with births accordingly taking place in early summer. After a gestation period of 300 to 340 days, the cow gives birth to her reddish-brown to yellow-brown villous calf. The calves, which are nursed for six to nine months, reach pubescence by the end of their second year. Wild water buffalo may become 20 to 25 years old. They probably have never been kept in a zoological garden.

The tamarou

The TAMAROU (Bubalus arnee mindorensis) has undergone, in the seclusion of the Philippine Island of Mindoro, a similar special evolution as the Anoa has on the island of Celebes. It, too, is small like the island forms of many species. Measuring 100 to 120 centimeters at the withers, it is only slightly larger than the anoa with which it also has in common the white patches at its head, lower side of neck, and limbs. However, the basic color of its short-haired, rather dense coat is a gray black or sometimes a dark brown, and thus is similar to the water buffalo's. The dimensions of their skulls and the cross sections of their sturdy, plump horns also resemble the water buffalo's. Since the island of Mindoro in the Glacial Period was still connected to the continent, these buffalo have split from the water buffalo stem much later than the anoa which, on the island of Celebes, had been separated from its continental relatives during the lower Pliocene. Fossils gave proof that the tamarou once lived on Luzon, the main island of the Philippines.

Presently the tamarou is extremely endangered. According to a publication of the Philippine zoologist C. G. Manuel in 1957, there are an estimated 19 herds of tamarou at Mindoro with only 244 animals which, however, are distributed over all parts of the island so that they hardly ever meet. This even increases the likelihood of total destruction. Even though strict laws for their protection have been passed and generous preserves have been established, enforcement is practically impossible. A conservation area of approximately 8000 hectares has existed for the tamarou since 1961 near Mount Iglit (1000 m) in the southern corner of Mindoro, but it can be checked only occasionally by the single game warden on the island. In 1964 the

Americans, Lee and Martha Talbot on behalf of the IUCN, investigated the tamarou. The two scientists learned that armed farmers and cattle ranchers on Mindoro arrange regular "hunts" for the animals. One tamarou killed in such a roundup had 167 bullet wounds. Hediger also received the impression on his visit to Manila in 1965 "that this species is mainly endangered by poaching because Filipinos approached me who offered to assist me either in making a guaranteed kill or in catching a tamarou." Most of the natives have great fear of the tamarou, since it is considered, like the anoa, to be extremely aggressive and dangerous. Some tribes of the primitive original natives of Mindoro hunt it with spears and pits dug on its paths wherever they see a chance to do so. The destruction of the forests and contagious cattle diseases also represent a serious danger to the last tamarous. This subspecies was especially stricken in 1930 by the cattle plague raging among domestic cattle.

Originally the tamarou preferred the dense prime forest which had earlier covered the whole island. In mountain forests, it is said to occur at elevations up to 1800 meters. The tamarous of the Mount Iglit Game Reserve live on grassland near the forest. During the rainy season from June until October, some grass species grow so high that the animals find cover in them. When the grass becomes long and hard in the rainy season, the buffalo feed mainly on young bamboo shoots. "The animals' preferred habitat," report Lee and Martha Talbot, "probably would be an area at the edge of the forest where safe cover, open pasture, and water to drink and wallow in are close together."

Similar to the anoa, the tamarou also lives singly or, in rarer cases, in small family groups. But this may be an effect of severe hunting as well as the transition to a nocturnal life. The farmers at Mount Iglit remembered that the tamarous, in the beginning of the area's cultivation, still used to graze during the day and were rather tame. With increasing persecution, they became more concealed and aggressive. Finally, after about five years, they had completely changed to a nocturnal life. Hardly anyone has ever managed to observe tamarous in the wild; therefore, the scarce information on their behavior is rather contradictory. For example, C. G. Manuel claims that this wild ox would neither bathe nor wallow, that it would absolutely shun water, and in the rain it would immediately flee under dense plant cover. Hediger found this statement confirmed in the female tamarou of the Manilla Zoo. This animal never used the pool in its pen and preferred to lie in the dry sand. In contrast, Walker mentions in his extensive publication on mammals (1964) that the tamarou, like all bovidae, likes the water and wallows frequently. Lee and Martha Talbot found several creek wallows in a typical tamarou habitat at Mount Iglit. They stayed for several days on the island, but saw only one tamarou.

Births seem to take place at the beginning of the dry season when

▷

1. Yak (Bos mutus), a) and b) horned and hornless domestic forms (Bos mutus grunniens), c) Wild yak (Bos mutus mutus)
2. Gayal (Bos gaurus frontalis)
3. Kouprey (Bos sauveli)

▷▷

Left: Borneo banteng (Bos javanicus lowi, bull in front, cow in rear) Right: Gaur (Bos gaurus, bull)

▷▷▷

Left: European bison (Bison bonasus) Right: American bison (Bison bison, bull in front, cows in rear)

▷▷▷▷

Left: African buffalo in strict sense (Syncerus caffer caffer, bull in front) Right: Forest buffalo (Syncerus caffer nanus, with the cow in front and the bull in rear. The Forest buffalo in this picture are externally closer to the African buffalo in strict sense)

▷▷▷▷▷

Left: Zebus (races of Bos primigenius taurus, in front is the dwarf zebu, Khillarizebu in the middle and Watusi cattle in the rear) Right: Water buffalo (Bubalus arnee, in front is the Arna, white domestic buffalo in the middle, European domestic buffalo in rear)

1a♂

1b♀

1b♂

1c♂

2

3

𝓕·67

Domestic cattle *(Bos primigenius taurus)*
1. Spanish fighting ox
2. Dahome dwarf cattle
3. Scottish highland cattle
4. Black and white lowland cattle
5. Lower Bavarian piebald cattle

The domestic buffalo or carabao

food conditions are best. In the subsequent mating season, the animals are in an optimal state of nutrition. Natives told C. G. Manuel about a strange mother and young behavior. In case of danger, the tamarou cow would reach under her calf with her horns, place it on her back, and then flee as quickly as possible.

So far the tamarou cow in the Manila Zoo is the only animal in captivity. However, the first step in an attempt to save this rarest wild oxen from final destruction would have to be the founding of breeding groups in several renowned zoological gardens.

It is possible that the rice growers in North India or Indo-China tamed the water buffalo the first time in the third millennium B.C. The domestic buffalo *(⚥ Bubalus arnee bubalis)* soon became of great economic significance and together with the zebu constitutes the most important tropical domestic animal. Altogether, an estimated total of about 75 million tame water buffalo live on the earth, 50 million of them in India and Pakistan alone and 20 more million in other parts of East Asia and Southeast Asia. Buffalo have also been introduced to Japan and Hawaii, and lately to Central and South America. In 1825 domestic water buffalo came to North Australia where they became feral and now live by the thousands in the brush steppes. In Asia Minor and North Africa the wild water buffalo lived in prehistoric times, but they were never domesticated. The Indian domestic buffalo, which came to these regions with the Muslims, became, especially in Egypt, the most important domestic cattle and spread from there as far as East Africa and the islands of Zanzibar and Mauritius. On Madagascar live several herds of feral buffalo. The water buffalo came to Italy perhaps as early as the Lombardic era during the mass migrations. However, it was probably much later that the Arabs brought them from Egypt to Sicily and South Italy. In Southeast Europe their appearance strikingly coincides with the Turk's domination in the Middle Ages.

In India where the domestic buffalo lives next to his wild ancestors, the two rarely differ. In the body and horn size the domestic buffalo falls only a little short of the arna. Domestication is still continuing now in some parts of India. Again and again wild bulls impregnate the tame cows, in most cases with the consent of the herds' owners, who often catch wild buffalos to crossbreed them with their domestic cattle to freshen up the blood.

Formation of breeds in the domestic buffalo

In the other parts of the domestic water buffalo's range, however, numerous breeds have developed which differ considerably not only from their wild ancestors but also among themselves in size, weight, color, and horn shapes. In Indonesia, plain white or black, black and white, red and black, brown and white and irregularly dappled animals occur in addition to the natural color. In most breeds of the domestic buffalo, the horns are much shorter and directed more backward than in the wild buffalo. Even hornless animals occur.

Like its wild ancestor, the domestic water buffalo depends considerably on water. Therefore, it is of greatest economic value as a work animal in the tropical and subtropical swamp regions where it is superior to other kinds of cattle because of its limited requirements, its resistance to diseases, its strength, intelligence, and acuteness of its senses which exceed other cattle. The domestic buffalo has become identified with rice cultivation since it is especially well suited for tilling the flooded rice fields. Petzold describes the various uses of buffalos in Vietnam. Not only do they pull ploughs, they also draw the heavy two-wheeled carriages, haul the bulky bundles of wood from the bamboo forests, power the cane-crushers' wheel, and are even being used for riding. The animals, which are worked from early in the morning until late in the evening, rest, graze, or take a bath only during the hottest time of the day around noon. "Then they immediately go to the nearest puddle and submerge their whole body so only the head with the broad, crescent-shaped horns shows above the water."

How the domestic buffalo is used

Another valuable characteristic of the domestic buffalo is their good nature and the ease with which they can be tamed. These huge animals can be controlled without difficulty by six to eight year old children, if they know them. In Hediger's opinion, olfactory stimuli play an important part in the taming. In areas where buffalos are accustomed only to the scent of the natives, the animals may panic when a white person suddenly appears. In addition to the domestic buffalo's rate of work, the milk production ranks the highest. In Italy, domestic buffalos, which were kept in stables, yielded an average of 1971 liters in one lactation period with a total of 172 milkings. The fat content usually is eight per cent, which is three and a half to four per cent higher than in the other cattle breeds. Furthermore, buffalo's milk has about five per cent more dry substances and contains one per cent more protein than cow's milk. The meat supplied by the buffalo is of importance mainly in India where the zebu is considered sacred and may not be butchered. The skin and horns are valuable materials for the production of useful items.

With increasing mechanization of agriculture, the water buffalo becomes gradually less important as a working animal in Europe, for example, in the Balkans and the Ukraine. In Italy the number of buffalos has also considerably decreased since the Pontine Marshes and the Maremma of Tuscany were drained. While the stock of domestic buffalo in Italy in 1947 was still an estimated 12,000 animals, it presently is only 2000 to 3000 head.

Domestic buffalo are frequently displayed and regularly bred in zoological gardens, although some of the zoos prefer the hardier, cold resistant European forms which do not need the warm stables in winter as do the Indian water buffalo. The record longevity in a zoo was 29 years.

African buffalo

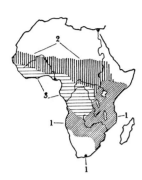

Fig. 13-2.
African buffalo *(Syncerus caffer)*, subspecies:
1. African buffalo in the strict sense *(Syncerus caffer caffer)*
2. Sudan buffalo *(Syncerus caffer brachyceros)*
3. Forest buffalo *(Syncerus caffer nanus)*

Distinguishing characteristics

Their habits

In Africa the water buffalo are represented by the African or Cape buffalo and the dwarf or forest buffalo, both of the genus *Syncerus*; (Color plates pp. 344, 352-3) who combine, like the Asiatic buffalo, primitive and advanced characteristics. Between the extremes of the giant steppe buffalo, who are barely smaller than the Indian arna, and the small dwarf or forest buffalo, all transitionary forms occur. Otto Antonius explained in 1930 that the multitude of transitionary forms, which occurred in the crossbreeds in the overlapping regions between the prime forest and the steppe, may be attributed to the increasing dessication of the African continent. Forest buffalo and steppe buffalo are an extremely impressive example of Hilzheimer's prime forest-steppe-rule. In closely related animals, the forest dweller is always the smaller and more primitive type, and the steppe form is the larger, more highly evolved type. In the invariable environment of the tropical rain forest, the species are "preserved," while the open landscape with its abrupt transitions accelerates selection and hence evolution.

The forest buffalo followed the receding woodlands, while the Cape buffalo moved up with the expanding savannahs. Thus, in wide regions a mixed breed of the two forms developed. In the Queen-Elizabeth-National Park (West Uganda), for example, black massive Cape buffalo with huge horns, small short-haired forest buffalo, and all transitionary forms live in the same herds. Therefore, the modern classification places the African buffalo in one species, the AFRICAN BUFFALO *(Syncerus caffer)*, with three subspecies: 1. CAPE BUFFALO *(Syncerus caffer caffer)*. The HRL is about 260 cm, the withers are 170 cm, and the weight is 800 kg (allegedly even up to 1200 kg). The ♂♂ in most cases are distinctly heavier than the ♀♀. The horns, which in a crosscut are flat and semicircular, first curve downwards and upwards so that the tips point to the center. In old ♂♂, the horns have a span of more than one meter. The sparse hair which in old animals may be missing at parts of the rump, is dark brown to black in adults and in calves dark red to black brown. 2. The DWARF OR FOREST BUFFALO *(Syncerus caffer nanus)* is small to medium sized with withers of 100-130 cm high. The short horns, which end in points, are directed backwards and curve upward and to the outside. The bases of the horns are far apart and not helmet-like over the forehead. The dense coat is a blazing red to a reddish brown with long fringes of a pale yellow inside and at the rim of the ears. 3. The GRASS OR SUDAN BUFFALO *(Syncerus caffer brachyceros)*, is medium sized, rather long-legged and has withers 120-140 cm high. The horns are longer than in forest buffalo, but they are not curved downward as far as in Cape buffalo. At the base they are moderately to greatly enlarged. The western species have flatter horns with ridges across the upper side (similar to the Asiatic Buffalo). They are reddish-brown to black in color.

Buffalo are found in any African landscape from the flooded deltas of the West African rivers and the evergreen rain forests of the Central African Basin to the semidry grass steppes in the east and south of this continent. They climb up to elevations of more than 3000 meters. Mountaineers even found a buffalo skeleton at about 5300 meters. on the east slope of Mawenzi, one of the peaks of Mount Kilimanjaro. Cape Buffalo adjust as hardly any other large mammal of Africa can, but they need water to drink and to wallow in. Now, human settlements limit their habitat. The cape buffalo likes an area where there are open pastures as well as covering jungles, forests or reeds, swamps, and water. Only at dawn do the buffalo leave their protective cover and march for several kilometers to their grazing grounds. At daybreak they withdraw again into the forest or dense brush. They feed mainly on grass and herbs, but not too much on foliage and branches. The dwarf buffalo, an animal of the forest, lives, according to another prime forest-steppe-rule, only in small groups, in pairs, or even singly. The buffalo of the steppe, in contrast, are herd animals. The size of a herd is usually 30–60 individuals, but it depends on the given food supply, the number of water holes, and the changes due to the mating and calving season. During the dry months, several herds may form a herd consisting of several hundred, or formerly even several thousand animals. There are herds which consist of cows and calves only, and yet others of exclusively very old and very young bulls.

According to the findings of R. Verheyen in 1954, a herd of buffalo has no actual leader. The lead is taken at any time by the animal who is most familiar with the area which the animals are passing through. Often different cows take turns, but an old bull, whose range the herd is about to pass through, may lead them for a short while. According to Verheyen, the single African buffalo bulls represent a normal state, thus they are not sick, injured, or chased away from the herd by stronger rivals. Adult bulls are inclined to separate from the herd and to move alone, each setting up its own permanent range which must include access to water, a wallow, good grazing ground, and cover. These territories usually are located along the old established paths of the large herds, so that these bulls never lose contact with the larger herds. Usually, a single bull does not leave his territory unless he lacks food, is driven by the urge to mate, or threatened by danger; otherwise he will not seek out conspecifics. Only on their daily walk to the common mud hole do single bulls of an area join in with small groups or mingle for a short while with a larger herd. The younger bulls without a permanent territory roam restlessly about in groups. Sometimes very old bulls associate with them. In any case, the herd is the social unit where each member finds protection, security, and refuge in case of danger.

The old cows are cautious and vigilant, especially when they have calves. Even the slightest uneasiness is transmitted immediately to the

▷

In the Alps cattle farming has influenced the landscape to a great extent. Most of the alpine meadow region is "cultivated landscape" which soon would be covered with woodland again if it were not for the domestic cattle kept there.

▷▷

Two strong African buffalo bulls in the thornbrush.

▷▷▷

Cattle egrets (Bubulcus ibis) catching insects stirred up by the buffalo.

rest of the herd. With highly erect heads, the nostrils directed against the wind to smell and the horns placed back on the neck, the animals check the source of danger with all their senses. The bulls are in the front and the cows and calves are behind them ready for immediate flight. In such situations, the unusual habits of the bulls prove to be biologically meaningful. Apart from the main group, they present the advance and the rear guard and cover the retreat of the cows and the young animals.

How dangerous is the African buffalo?

Much has been written about the dangerous African buffalo. Indeed, more big game hunters have lost their lives to buffalos than to any other African game. In most of these cases, the accidents were caused by the fact that the hunters did not know the behavior of an injured Cape buffalo well enough. Such an animal usually returns in a semi-circle to its own tracks and waits motionless in the cover for its pursuer. Since such a cornered, injured animal is unable to flee, attack is the only alternative. By nature the Cape buffalo is no more "vicious" than any other animal. The peacefully grazing herds of buffalo in the national parks and game reserves of East Africa hardly step out of the way of tourists' cars and seem like the cattle in the pastures of the tourists' home countries. The lack of danger to them has reduced their flight distance to a minimum, and their amazing curiosity often causes them to neglect any precaution.

Buffalos help each other

Like many other gregarious animals, buffalos do not only find protection in association with their conspecifics, but they also try to help each other in certain conditions. Verheyen reports on the following episode: "One day I shot from a very close distance an adult bull who had attracted my attention with his self-confident behavior. A few steps away from him were two other bulls who immediately turned against me. Even after two more quick shots which knocked down the injured animal, his companions withdrew to barely ten meters. Suddenly, the dying animal uttered his metallic sounding death cry. One of the other buffalo became so upset that he seemed to forget about my presence and came to help the unfortunate animal. With an unfathomable strength, he tried to pull up the dying animal, reaching with his horns under the body and trying to lift him with all his power. The injured animal turned to and fro, and his helper, upset at the roaring and furious at the frustration of his efforts, threatened me so that I hastily retreated. Two hours later the two bulls still stood guard next to their dead companion." Other observers have also reported that Cape buffalo have supported injured conspecifics and stayed with them. In the Krüger-National Park, four strong bulls came to the aid of a cow who had strayed from the herd and had been attacked by three lionesses; they chased off the cats.

The herds of African cattle have turned vast areas into desert with overgrazing.

The mating season varies depending on the region and the climate. At the beginning of mating, the single bulls try to make contact with the herds and fight insistently for superiority and possession of the cows. At first the rivals try to impress one another by threat displays. With their heads held erect they prance towards each other with stilted steps, paw the ground with their hoofs, and display their huge weapons with a threatening shaking of the head. Breathing hard they stand still when they are approximately ten meters from each other. All of a sudden, they lower their horns and rush towards each other. With sounds that can be heard over a long distance, their enormous heads clash together. Repeatedly, the opponents rush against one another. Then these colossi weighing almost a ton crash into each other, pushing and shoving each other forwards and back with their mighty horn bosses until one of them breaks off the fight in quick turnabout and leaves the place. The dangerous tips of the horns are not used in these "tournament fights". Therefore, such a fight, in most cases, ends without bloodshed. However, it happens quite frequently that the tips of the horns become hooked together and break off. In the Queen-Elizabeth-National Park in Uganda I saw a buffalo bull whose one horn, together with the wide base, and a part of the skull had been torn off.

Mating and fighting behavior

The calves are born after a gestation period of 10–11 months (300 to 330 days) at the end of the dry season, when fresh greens are growing everywhere and the cows find the best food conditions. The young, which are nursed for at least a half year, are able to reproduce by the age of one and a half to two and a half years.

Birth and raising of the young

The Cape buffalos hardly interact with other species of animals. Elephants and hippopotamuses range higher in the biological rank order, and the buffalos avoid them. John Owen observed in the Mara-Game-Reserve in Kenya a "friendship" between a Cape buffalo and a rhinoceros: "An old Cape buffalo bull was constantly in the company of a female rhinoceros and her young. The buffalo became aware of our vehicle much earlier than the female rhinoceros and quietly escorted her away. Finally, the three disappeared together." Even more frequently than other large animals, buffalos are visited by tick birds (oxpeckers) which rid them of skin parasites, the blood sucking ticks of the genus *Ixodes*. The buffalos also try to get rid of their irksome tormentors by wallowing for hours and rubbing at termite heaps and tree trunks. Except for man, the buffalo is mainly pursued by lions, which usually dare to attack only sick, aged, or young animals. A healthy adult Cape buffalo is able to defend himself effectively. Often lions have been chased off, injured, or even killed by buffalos. Old, experienced bulls sometimes even attack lions and chase them from the vicinity of the herd. The leopard endangers at most only newborn calves.

Behavior towards other animal species

However, the most severe threat to the Cape buffalos all over Africa is the cattle plague, a terrible contagious disease which spread to South Africa by the end of the nineteenth century and which, from the end

Danger from cattle plague

of 1890 to the beginning of 1891, in East Africa within a few months had killed millions of cattle and about 90 percent of the buffalo population. Another siege of the disease from 1896 until 1897 almost completely destroyed the rest and brought many other species of horned ungulates close to extinction. Some of the bongo, the greater kudu, and the roan antelope in East Africa have not yet recovered from it, but the buffalo population increased well within 20 years and is no longer endangered. However, these catastrophies showed clearly that the Cape buffalo are almost as susceptible to the cattle plague as the domestic cattle. It is even feared that they are of some importance in spreading the disease as a host animal. Therefore, the danger for African hoofed animals in general, and for the Cape buffalo in particular, is not over yet.

While all subspecies of the African buffalo crossbreed and have fertile offspring, so far no such results were obtained between the African buffalo and other wild oxen or domestic cattle. All attempts to turn the Cape buffalo into a "useful" animal by crossbreeding it with cattle have been without success. Neither has the taming of this huge wild ox been possible so far. As a useful animal, the Cape buffalos, as well as the eland antelope, would probably be of greater economic value to Africa than the many cattle herds which are now turning wide sections of this continent into deserts by grazing, because both species utilize their food plants more efficiently and with a higher yield of meat than any domestic cattle.

In captivity Cape and forest buffalo may become rather aggressive. Cape buffalo are kept in many zoological gardens; in the Berlin Zoo, they have reproduced regularly for years. The record longevity was more than 26 years. The small forest buffalos, however, are rather rare in zoos and their population is decreasing steadily in the wild. In Germany, they are only in the zoos at Berlin and Dresden.

True oxen (genus *Bos*)

The most varied group of oxen by far are the TRUE CATTLE in the narrow sense (genus *Bos*). All have the oval or round crosscut of the horns in common, and the flexion of the skull also appears to be stronger than in the more primitive buffalo. The primitive form of cattle-in-strict-sense was split off from the ancestors of the buffalo as early as the Middle Miocene, about 25 million years ago. We distinguish, depending on the classification, three or four subgenera: Koupreys, aurochs, and yaks.

The gaur

The KOUPREYS (Subgenus *Bibos*) are striking because of their especially high back and head. On the front part of the back there is a ridge of muscle supported by the extended spinal processes of the third to 11th dorsal vertebra. The lower parts of its legs are like "boots", a white or a kind of yellow which contrasts with the darker upper part. There are two species (gaur and banteng), both of which have become domesticated, and six subspecies.

The largest of all contemporary wild oxen is the GAUR (*Bos gaurus,* Color plate p. 342). The HRL is 260–330 cm, the TL is about 85 cm, the height at the shoulders is 165–213 cm, and the weight is 700–1000 kg. The ♂♂ are about one fourth larger than the ♀♀. The horns, in the ♂♂ especially, are strong and at the base are slightly curved downward, then bent backward and up with the tips pointing inward (length of horns 60–115 cm, distance between tips of horns up to 120 cm). In old ♂♂ the horns often are considerably shorter from wearing out by use. There is a high bulging forehead ridge between the horns. The skull is stretched, with a long muzzle which appears to be indented in the upper part, and a nose bent forward. In adult ♂♂ the dewlap consists of two parts at the chin and throat and there is a huge hump over the withers. There are three subspecies: 1. the INDIAN GAUR (⚬ *Bos gaurus gaurus*) is dark brown to black with a high ridge on the forehead. 2. ⚬ *Bos gaurus readei* is black and the ridge between the horns is less distinct. 3. the MALAYAN GAUR OR SELADANG (⚬ *Bos gaurus hubbacki*) has no ridge on the forehead. All the subspecies are endangered and in some areas extinct.

The preferred habitat of the gaurs are shady hill and mountain forests up to elevations of 2000 meters, with brush that is not too dense and is interrupted by clearings. They also inhabit the light bamboo jungles in the lowlands and open grass plains near forests. In spite of their bulky shape, they can move even in difficult terrain and in the mountains with the greatest ease, especially in flight. They need water for drinking and bathing, but they do not seem to wallow. For food they prefer fresh grass, tender bamboo shoots, a variety of herbs, and the young sprouts of bushes and brush. Since they avoid civilization and stay away from settlements, they do not do damage to the plantations—unlike the water buffalo. The natives fear the valiant oxen and hardly bother them. Therefore, their population is endangered less by poaching than by the increasing limitation of their habitat.

However, hunting the gaur, the "noble game among the oxen," as Oscar Kauffmann has called him, has always been considered a special challenge. Certainly this has contributed to his decimation. In many regions of India, the cattle are driven into the forests to graze and they infect gaur herds with hoof and mouth disease, cattle plague, and other contagious diseases against which they do not have any immunity, causing severe losses. A total hunting prohibition and an effective control of diseased cattle, however, have led in some areas to an increase of the almost extinct gaurs.

The gaur, too, lives gregariously, but the herds are smaller than those of the buffalo and consist mostly of eight to twelve or, in rare cases, more than 20 animals. One to two old bulls are usually with a herd, plus two to three young bulls, five to ten cows with their calves, and several subadults. Old bulls often live alone, and the younger bulls sometimes join in bachelor groups. Only during the driest season

Distinguishing characteristics

The gaur's habits

Fig. 13-3.
Gaur (*Bos gaurus*)

when the food is too scarce, do the single groups form large herds. Ursula and Wolfgang Ullrich have observed gaurs in the South Indian province of Madras in 1965 and reported on their experiences:

"The gaur herds we met in the jungle of Mudumalai consisted of nine to 26 animals. Calves were in all herds, where they graze together and remain during the migrations. They seem to form 'kindergartens,' and the mothers of these calves stay in the vicinity. Thus, the defenseless calves are constantly surrounded by cows who are ready to defend them, and when the herd moves on, the 'kindergarten' of the gaur calves remains in the center of the herd. It is difficult to recognize a leader. Often old cows walk ahead of the herd or in flight bring up the rear, looking back repeatedly. The old bulls, who are conspicuous with their shiny black coats, do not seem to take part in this cautious behavior. When a young bull or a cow utters an alarm call, they are the only ones who do not throw up their heads to look for the possible enemy; they usually continue grazing. The alarm call is a light snorting which immediately alerts the other members of the herd. When the source of the danger is discovered, the animal who had alarmed the others directs his head towards the enemy, thus pointing out the danger. This is accompanied by a grumbling mooing with the mouth open. Immediately the herd looks in this direction, ready for flight or for defense. The most alert and suspicious animals are the mothers of the young calves."

Charles S. Ogilvie, the long-time Chief Game Warden of Malaya, observed in Malay gaurs that usually one of the cows stands guard with her head up, while the other animals are grazing. In a moment of danger, however, the bull is in charge of the protection of the herd, and he faces the enemy in front of the herd and covers its withdrawal. The gaur is the only wild ox who does not attack frontally but approaches the opponent broadside, lowering its huge head and hind-quarters, and striking from the side with its horns. Therefore, the horns of the old bulls are worn on one side or even broken off. According to George B. Schaller, the lateral threat is a display behavior. The opponent is supposed to be intimidated by this presentation of the profile which is especially impressive because of the high hump on the withers. In encounters concerning rank order among gaur bulls, this behavior is usually a substitute for a fight.

Daily activities

According to Ursula and Wolfgang Ullrich, the gaur's daily activities begin at about eight in the morning "when they slowly rise from their beds, as their resting places in the jungle are called, graze on the dry grass of the light forest, and move into the dense thornbush (*Lantana camara*) where they disappear at about 9:30. There they feed, reaching with their long tongues for young sprouts in the brush. When the oppressive heat of the mid-day paralyzes the jungle, they lie down in the thornbrush to sleep, rest, and ruminate. At about 3:30 they con-

tinue to feed, and when the sun sets behind the mountains they walk
back to the light forest which probably offers more security from predators in the darkness of the night because there are more clearings.
On their way to their nightly resting places, they eat the sun-dried
grass which grows on the ground of the scattered jungle." On days
with cloudy skies or rain, the animals remain active all day long. When
the rainy season begins, the gaurs move from the moist forest valleys
into dry regions at higher elevations where they escape the millions
of flies and mosquitoes.

The mating season in India is from November until March. According to Schaller's observations from another area, the peak is in March
and April. Then the loner bulls rejoin the herds and fight with the
herd's bulls over the cows. Sometimes during the mating season they
may stay several hundred meters away from the herd, as Kauffmann
reported in 1911. The mating call of the gaur bull is different from the
sounds of all other oxen and may be heard in the jungle, according
to Schaller, over a distance of one and a half kilometers. It attracts not
only the cows but also other bulls. Oscar Kauffmann describes this
sound from the wild and compares it to the roaring of red deer: "This
mating call consists of an organ-like sound and begins, unpleasantly,
with a quavering and dragging e-e-e, which gradually changes to a-
oo-uu and ends in a tremendous crescendo. During the following days,
I often listened for hours to these mating calls which lasted eight to
ten seconds each, and which the bulls would also utter during the
evening and at night."

The calves are born after a gestation of 270–280 days. In southern
India this occurs mostly in August and September, but according to
the Ullrichs, it may happen in November and December when the
rainy season is over. The births of gaurs in the Berlin Zoo were
distributed over the whole year. Twins are rare, although Ogilvie, in
Malaya, observed the Seladang herd in which one cow had several
times given birth to and raised twins. Usually the young are nursed
for nine months and become pubescent in their second or third year.
Like many other gregarious ungulates, the expectant mother separates
from the herd for a few days and returns when the young calf is strong
enough to follow the others. Without the security of the herd, the
mothers are especially cautious and suspicious. A mother may attack
blindly whenever she feels that she or her newborn calf is endangered.

The gaurs live together peacefully with other animals who share
their habitat. They may even form a loose association, for example,
with sambar deer or the wild pigs whose continuous vigilance is of
advantage to them. The gaurs are far less afraid of tigers than of man.
Even though they are constantly on the alert for tigers, they are aware
of the herd's power and would rather fight than flee. The American
zoologist George B. Schaller, who studied tiger behavior for one year

Reproduction

**Behavior towards
other animal species**

in the Indian Kanha-National Park, writes on this subject: "In most cases, the tigers get only an average of about half of the young born annually. The tigers will attack fully grown gaurs only in rare cases, since they are obviously impressed by the strength of the gaurs' horns. There are authentic reports of tigers which have been pierced by gaurs and buffalos." The gaurs usually do not avoid the elephants. Kauffmann reported that both species visit the same mud holes at night, while Ogilvie saw gaurs which mostly left their grazing grounds and their salt licks when elephants "stuck their long noses into them." Ogilvie calls the red jungle-fowl the "gaurs' best friends." They walk fearlessly in front of the grazing wild oxen's feet to catch the swarming insects. Vice versa the gaurs take advantage of the constant alertness of the fowl which "seem to hear the inaudible and to see the invisible." In the Seladang herd, Ogilvie observed in 1955 that one cow had a gaping wound at the base of the horn, which a young red jungle-fowl cock cleaned of dead tissue and maggots several times a day for more than two weeks. The cow apparently liked this treatment of her injury. When the little cock was near, she would lay her head flat on the ground and tip the sick horn invitingly.

The gaurs are among the rarest animals in zoological gardens, and their successful reproduction, as it has occurred in the Berlin Zoo for many years, is even more exceptional. Presently, there are only about two dozen gaurs in nine different zoos of the world. The Berlin Zoo was assigned by the IUCN in cooperation with the IUDZG to keep the studbook records. The oldest age a gaur reached in captivity was 24 years.

The gayal

The domestic form of the gaur is the GAYAL (*Bos gaurus frontalis*; Color plate p. 341). Smaller than the gaur, the HRL is 270–280 cm, the height at the shoulders is 140–160 cm. The ♂♂ are larger and heavier than the ♀♀. The legs are shorter, and the lump on the back is weaker and almost level, giving the rump more of a barrel shape and making it lower than in the gaur. The skull is distinctly shorter with a much wider and flatter forehead than in the gaur. The shorter horns are plumper, conical, and without bends, and they protrude from the sides of the forehead. The neck is thick and muscular, and the double dewlap at the chin and throat is well developed.

Many characteristics of the gayal are typical of domestication. Besides the normal brown-black to black animals, there are also piebalds and white ones. For a long time, it was not even clear whether the gayal originated from the gaur, or whether the two were different species of wild oxen, or whether the gayal had resulted from a cross between the gaur and domestic cattle with banteng blood, or zebus. Thorough anatomical studies of the skull, especially the ones by Bohlken in 1958, settled the question about the origin. Back crosses with the wild gaur bulls still occur frequently and are made purposely by breeders to improve the race. In some areas, crossbreeds between gayal and zebu are welcome domestic animals with economically valuable characteristics.

Many external characteristics of the gayal are similar to the gaur. They also have something in common in behavior and habits. Feral gayal herds live in the same habitat as the gaurs and are said to move equally skillfully in the mountains. Their nature has been considerably changed by domestication and they are of a much quieter temperament than their wild cousins. Vienna Zoo director, Otto Antonius, saw in 1932 that the gayal, like the gaur, does not attack frontally but laterally. The gayal bull of the Schönbrunn Zoo on different occasions uttered the call typical of gaurs. However, this sound seemed to have lost its original significance in this case. A comparison of such behavior may be of great importance in questions of origin. In this case, it was also further proof that the gayal is the domesticated descendant of the gaur.

Obviously, gayals have never become domesticated animals in the strict sense. Their herds live in a semi-tame state near jungle villages and come to settlements only in the evening to lick salt. The capture of feral gayals is done with salt as bait and with tame gayal cows, who are driven into the forests to attract their wild conspecifics and to keep them in the vicinity of the village. The gayal has little economic value. Only in some regions of Northern India are the animals used for field work. In part they are important as meat supply, since their milk is rarely drunk. To many tribes of Northeast India and Burma, they serve mainly as sacrificial animals. The Nagas use them as a kind of "currency" in exchange for trading goods and objects of value. They buy women and pay penalties with gayals. Only in places where the gayals are crossed with cattle do they show their characteristics of domestication and occur in different colors.

Much smaller and lighter in weight than the gaur is the BANTENG (*Bos javanicus;* Color plate p. 342). The HRL is 180–200 cm, the BH is 130–170 cm, and the weight is 500–900 kg. These animals are built more gracefully than the gaur with longer and more slender legs. The hump on the back is lower, more level, and stretched out longer. The ♂♂ are one fourth larger and heavier than the ♀♀. The bulls usually differ conspicuously from cows in color and horn shape. The hair is short, dense, and smooth. There are white "boots" and a very striking white patch on the rear of the hindquarters. The smooth horns are round, pointed, flat, and enlarged at the base. in the ♂♂, they curve slightly downward at the sides and then bend flatly, or they curve in a crescent upward so that the tips are directed inward. In old ♂♂, there is callous hairless skin on the forehead, similar to the cape buffalo's "helmet."

There are three subspecies: 1. JAVA-BANTENG *(Bos javanicus javanicus),* which is greatly endangered with only 400 animals left in some Javan protected areas. 2. BORNEO-BANTENG *(Bos javanicus lowi),* which is greatly

The banteng

Fig. 13-4.
1. Banteng *(Bos javanicus)*
2. Kouprey *(Bos sauveli)*

endangered, and in some regions already extinct. 3. BURMA-BANTENG *(Bos javanicus birmanicus)*, which is a lighter color than the two island forms. It has stronger horns and is also greatly endangered everywhere.

For good reasons, the banteng is said to be the most beautiful of all wild oxen. The single forms may be colored in various shades—adult bulls are mostly dark to black-brown, sometimes even blue-black, and the cows usually are of a gorgeous red-brown. In 1952 Appelman also found red-brown bulls and blue-black cows in East Java. The Swedish writer Bengt Berg, in 1955, killed pale brown banteng bulls in Burma. Living in Thailand are bantengs whose brown coat parts have fine white spots. Sometimes the patches on the hindquarters and the "boots" are not light, but are a darker color or completely missing. In some areas, there are no differences in the color of bulls and cows.

This magnificent wild ox has decreased in all parts of its range in spite of various protective measures. In Java this decimation may be blamed on changing the original environment into densely populated settlements. On the island of Borneo, the Dajaks hunt the valiant wild oxen not only for their meat, but also for their heads. Since head hunting of people is now prohibited, they take the banteng's head as a substitute. Fossils show that bantengs formerly had lived on Sumatra, but disappeared at an unknown time. On the island of Bali the wild bantengs fell victim to increasing cultivation and land clearing even in historical times. The banteng flees from cultivation and avoids the vicinity of human settlements and plantations. It inhabits the untouched woodlands from the coast up to elevations around 2000 meters. Here it prefers swamp forests with brush, bamboo jungles, light forest, and grasslands with scattered woodlands. Like the gaur, it is fond of young, juicy grass, herbs, leaves, blossoms, and the sprouts of trees and brush, and young bamboo shoots.

The herds mostly consist of two or three younger bulls with two to 30 cows, calves, and subadults. The old bulls usually separate from the associations of their conspecifics and visit the cows only in the mating season. The daily activities of the bantengs also resemble those of the gaurs. From early in the morning until about nine, the animals graze in the open meadows and then retreat into the shade and the shelter of the denser forest to ruminate. In the late afternoon, they become active again. They are said to be active sometimes at noon or at night also. In Udjung Kulon, the largest protected area of Java, the breeding season is predominantly from July until August. In other regions, the calves are also born at the beginning of the dry season. Gestation lasts 270 to 280 days. The calves are a full yellow-brown color with a distinct black streak on the back. They are nursed up to the ninth month and become pubescent in their second or third year.

Unfortunately, there are only few observations of the banteng in the wild. F. Millet in 1930 in India reported the coexistence of bantengs and gaurs. He saw a banteng cow, whose herd apparently had been destroyed by disease, join a herd of gaurs. He also found the skeletons of gaur and banteng herds, which doubtlessly had fallen victims to a contagious disease, beside each other. F. J. Appelman reports that on Java, in addition to the tiger, the adjags *(Cuon javanicus)*, which belong to the red dogs and hunt in packs, are the most dangerous natural predators of the bantengs. These wild dogs also occur in the habitats of the two other subspecies of banteng.

Presently bantengs are kept in about 23 zoological gardens of the world, although they probably are not all wild bantengs. Pure wild bantengs reproduce regularly, for example, in the zoos of Berlin, Frankfurt, Rotterdam, and in some of the American zoos. In the thriving herd of the Berlin Zoo, 24 purebred Javan bantengs were born and raised between 1957 and 1967.

The banteng has been domesticated on the island of Bali as it has been on Java. There are no correct records about the time of this domestication, but it may have been long before our chronological records. On Bali, wild bantengs were captured and tamed in the Middle Ages, and the same was done on Java in the 18th century. The domestic banteng, the Bali cattle, are distinctly smaller than the original wild form. The bull's horns do not have the significant inward curve of the wild banteng's, the skull is smaller as if it had remained in a juvenile state, the withers are flatter, and the external dimorphism of the sexes is less. Instead, the muscles, especially those of the hindquarters, are more strongly developed. The animals grow more rapidly and become pubescent earlier. The true Bali cattle differ relatively little from the wild banteng, according to W. Ch. P. Meijer, because the Balinese have never attempted the systematic selection of distinct characteristics for their domestic animal. Nevertheless, there are piebalds, white, black, and yellow shades, different kinds of spots, and other variations, in addition to the wild color. On the island of Bali, the domestic banteng probably was always isolated, but on surrounding islands it was crossbred with other cattle.

The Bali cattle

Today, crossbreeds of banteng and zebu are distributed all over the island of Indonesia and in wide areas of Northern India. From crosses between Bali cattle and Madura cattle came a very long-legged race which, according to Hediger, is even used for races on the island of Madura. The economic value of purebred Bali cattle and their crossbreeds is considerable. They are mainly used as draft and work animals and also as mounts. They produce milk and are slaughtered. Released or escaped Bali cattle become feral in a short time and many of them live, according to Meijer, on the wide savannahs in the south of Celebes.

One of the large animals, which was only recently discovered, is the KOUPREY (◊ *Bos sauveli;* Color plate p. 341). The HRL is about 220 cm, the

The kouprey

BH is 190 cm, and the weight is 900 kg. The ♀♀ are approximately one fourth smaller and lighter than ♂♂. The hump from the withers to the back center is like in the gaur and banteng. The body is slender and long-legged. There is a very well developed dewlap from the throat to the mid-chest, which in old bulls is sometimes so pendulous that it drags through the grass. The lower parts of the legs, like in the gaur and the banteng, have white "boots". The long slender horns are similar to the wild yak's. They first slant backward to the sides, then upward to the front and finally grow with the tips inward. The crosscut of the horn base is oval and has bulging rings in old animals. The horn length in ♂♂ is about 80 cm while in the ♀♀ it is about half of that.

The kouprey ranges in size between the gaur and the banteng. Its zoological classification has not yet been cleared up. Professor Achille Urbain, the former director of the Paris Vincennes-Zoo, found the horns of a kouprey in the home of veterinarian Dr. Sauvel in North Cambodia for the first time. In 1937 he described the new species from a male calf of the same region which lived in his zoo until 1940, and he named it in honor of Dr. Sauvel.

Is the kouprey a wild ox?

The question of whether the kouprey is a genuine form of wild oxen or just a feral domestic ox cannot be answered definitely yet. In 1961 H. Bohlken made a thorough comparative study of the skulls and found "that the kouprey represents a mixture of *Bibos* (mainly banteng) and *Bos* (mainly zebu) in its external, as well as in the skull characteristics. Thus, the kouprey is either a genuine wild ox situated between the banteng and gaur on the one hand and the aurochs on the other, or a crossbreed between the banteng and zebu which became totally feral many generations ago. The zoologist Braestrup in 1960 even considered the kouprey as a surviving form of the aurochs which had remained primitive. Haltenorth and Bohlken agree that the kouprey should be considered a genuine wild species, so long as there is no clear proof of its descent from feral cattle. They classify this animal as a subgenus of its own *(Novibos)* between the subgenera *Bibos* and *Bos.* Wharton assumes that the kouprey may have become domesticated temporarily during the Khmer culture, 400 to 800 years ago, and then became feral after the downfall of this culture.

A peculiarity of the male kouprey is a strange corona of horn fringes right below the horn tips. They are the remnants of the juvenile horn sheath which is pierced by the permanent horn which is growing underneath in the animal's fourth year. This process occurs in all horned ungulates, but in all other cases, the splinters are ground off so quickly when the animal ploughs with his horns into the ground, that only a flat, gradually disappearing ridge marks the line between the old and the new horn. Due to the complicated curve of the kouprey's horns, it cannot manage to push its tips into the ground, so the corona of fringes remains for an unusually long time. According to Erna Mohr in 1965, the horn is usually stronger at the inner than at the outer side. The distribution of the kouprey is, according to Wharton, restricted to a continuous area along both sides of the Mekong River in Northeast Cambodia, in the far south of Laos, and in the most western part of Vietnam. Wharton estimated in 1957 the number of koupreys in this area to be 650 to 850 animals. However, it is feared that the political turmoils of the last years have also seriously damaged the rest of this population.

Like the banteng, the kouprey prefers to inhabit rather open landscape, light savannahs and meadows with woodlands, and scattered jungle and forests with much brush which provides protection and shade. Salt licks and drinking places with water even in the dry season are of vital importance. It feeds on the same food as the banteng. Therefore, herds of kouprey and of banteng often live in Cambodia in a loose association. They graze and wander together, especially after the mating season. However, groups of these two species stay away from each other and do not mix. According to Sauvel, a kouprey herd originally consists of one bull, eight to ten cows, and some subadults. The bulls walk behind the herd which is led by an old cow. After the mating season in the beginning of June, the herds split up into groups of cows and of bachelors. As in all oxen, old bulls often become loners. The mating season is in April and May, and the calves are born after a gestation of approximately nine months in December and January, or sometimes even in February. Cows about to give birth separate and rejoin the herd with their calves after about one month. They nurse the calves for a half year. The daily activities of the koupreys are similar to those of other wild oxen. According to Wharton's observations over the years, they do not wallow. They are said to be even shyer than bantengs and gaurs.

Its habits

Unfortunately, this rare and especially interesting ox is not presently kept in any zoological garden. For further research about its descent, and to save him from the dangerous situation in his habitat, all efforts should be made to capture some koupreys alive and to secure them in zoological gardens, where the species would stand the best chance to escape total destruction.

No other event in early history was of such comparably far reaching significance for the development of human culture as the domestication of oxen. The DOMESTIC CATTLE *(Bos primigenius taurus)* made possible the step from the primitive hoeing of the Upper Stone Age to highly developed agriculture and, thus, it became the very basis of the Asiatic-European culture. Therefore, the question of the descent of cattle has interested breeders, zoologists, archeologists, and historians, and has led to various theories. However, the modern zoological research on domestication presently confidently considers one wild form, the AUROCHS *(Bos primigenius primigenius;* Color plates pp. 163-4),

The aurochs

to be the one ancestor of all domestic cattle. The aurochs and its descendants, the various races of domestic cattle, are the subgenus *Bos* among cattle-in-strict-sense.

Fossils from the Tertiary Siwalik stratum of North India prove that the origin of this wild cattle has to be in India. From there it gradually spread to the north. In Germany the first evidence of the aurochs comes from the Riss Ice Age in Eurasia during the Pleistocene (about 250,000 years ago). Its distribution finally covered all of the temperate zone of the old world of Europe (except for Ireland and Central and North Scandinavia) over North Africa, Asia Minor, and North India eastward to the coast of the Chinese Sea. The aurochs of the Glacial Period, for example, *Bos primigenius namadicus* of India and the *Bos primigenius hahni* of Egypt, considerably exceeded in size the post glacial forms of Europe. Probably unfavorable food conditions have caused the gradual reduction of the aurochs in Europe. Despite this fact, the aurochs became extinct, particularly in its Asiatic range, by the early historical times. In Central and Western Europe, the last aurochs died off between 1200 and 1400 A.D., after a period of extensive forest-clearing from the ninth to the eleventh century had begun their demise. Around 1400 the aurochs still occurred in the area which now is East Prussia. In 1927 Leithner reported, "In the account books of the Teutonic Order between 1200 and 1400, aurochs are often mentioned, which indicates that they had been preserved by this order for hunting purposes. There are even some notes which say that the knights of the order captured aurochs and presented them as valuable gifts to the sovereigns of Europe."

The aurochs survived longest in Poland, especially in the then dukedom of Masowia, and in Lithuania, where vast woodlands with hardly any settlements still gave them refuge. By the end of the 16th century, the last of the wild aurochs in the forest of Jaktorow, 55 kilometers west southwest of Warsaw, were put under the special protection of the sovereign. They were kept in a game preserve, tended by game wardens, and in winter received hay as an additional food. These last aurochs had become half domesticated. About their sad end, Otto Antonius reports that, according to the existing records, in 1564 eight loner bulls were still living, in addition to 22 old cows, three young bulls and five calves. In 1599, 24 animals were still said to have been there, but in 1602 there were only four. In 1620 only one single cow was left, and she died in 1627. Thus, this species was finally extinct.

What did the aurochs look like? We are able to obtain quite an exact picture of them through the many discoveries of bones and skeletons, descriptions in old travel books, the tales of several peoples, and a multitude of pictures. The people of the Lower Stone Age drew almost lifelike pictures of aurochs in the famous prehistoric wall paintings in the cave of Lascaux in Southern France. Probably the best repre-

Fig. 13-5.
In this region the aurochs (*Bos primigenius primigenius*) still existed in historic times.

sentation is the so-called "Augsburg Aurochs Picture", which an unknown artist from the 16th century had painted on wood. The English zoologist Hamilton Smith discovered it in an Augsburg antique shop and published it in 1827 in a black and white reproduction.

From this, we may picture the AUROCHS (*Bos primigenius primigenius*) as follows: the HRL was up to 310 cm, the BH in the ♂♂ was 175–185 cm, the weight was 800–1000 kg. The ♂♂ were about one fourth larger and heavier than the ♀♀. The animals had a light build with rather long legs. The frontal part of the body with its deep chest was more voluminous than the hind part. The back was straight with only a slightly declining croup. The southeastern forms had a small hump, a muscular neck, and either no dewlap or only a small one. The slender head was long and had a straight profile and large forehead. The horns were up to 80 cm long, pointed, sturdy, and the crosscut was circular. There are different forms; most grow out to the sides, then turn upward and forward, with the tips directed slightly inward. The horns are white with black tips, although in Egypt they were probably plain black. The coat is short and smooth. In the winter it is dense and almost curly. The ♂♂ of Northern Europe were black-brown with a light streak on the back, a white circle around the chin and muzzle, bright curly locks on the forehead, and lighter underside and inner sides of the legs. In Southern Europe they were brown or gray-brown, and in Africa, red-brown with a white or pale yellow saddle. The ♀♀ were plain brown, but darker in winter, while the calves were red-brown.

Thus, externally the sexes were quite different in the aurochs. The bulls were much larger and heavier than the cows. This sexual dimorphism even misled the zoologists for some time. They thought that in Europe there was a smaller subspecies which had been described as dwarf-aurochs (*Bos primigenius minutus*). It was not until 1927 that V. Leithner proved that these alleged dwarf-aurochs were the cows of *Bos primigenius primigenius*.

Our knowledge about the biology of the aurochs is sparse. Originally, its preferred habitats probably were light forests, valleys, and water-meadows, the edge of woodlands with open pastures, and grazing grounds with scattered trees from the plains up to lower mountain elevations. In Africa the aurochs also lived in the steppe, and in the Pyrenees they went up as far as the timber line. The continuous swampy woodlands of Poland and Lithuania were the last refuge for the European aurochs which were pursued everywhere else. Like many other wild oxen, they also lived in smaller herds consisting of one bull and several cows with their calves. They were probably more active in daytime than at night. The adult bulls, especially, may have been rather agile and aggressive, since auroch hunts were supposed to be dangerous. During the mating season in August and September, the bulls had severe rival fights. The calves were born after a gestation

Aurochs *(Bos primigenius)*, ♂♀

Celtic domestic cattle, ♂♀ according to fossils found at Manching near Ingolstadt.

One of the present high class breeds, ♂♀

30 60 90 120 150 180 210 240 270 300

Fig. 13-6.
From the huge Aurochs rather small races of cattle have been derived through the process of domestication. In addition, the considerable size differences between the sexes in Aurochs was reduced. Improved high class breeds are much larger, even though their bulls do not nearly reach the size of the aurochs bull.

period of about nine months in May and June. In spring and summer, aurochs fed on grass, herbs, branches, buds and leaves; in the fall they gorged themselves on acorns; during the winter months they ate the dry leaves in the forests. Except for man, the wolves were their only predator, but they could only overpower calves or sick animals.

The domestication of the aurochs

As to the place and the time of domestication of the aurochs, various opinions have been expressed. It is certain that cattle is one of the oldest, perhaps the oldest, domesticated animal. In 1906 Duerst found in Turkestan remainders of tamed cattle which he dated as belonging to the period between 8000 and 7500 B.C. However, this age determination has been questioned. Zebus and other highly developed races of cattle were in Mesopotamia and India as early as 4000 B.C. Today, it is assumed, based on the research of Reed in 1960 and 1961, that the wild ox had been domesticated in the sixth millenium B.C. Old documents show that, at first, the cattle in Mesopotamia, Egypt, Persia, and India had only been objects of worship. Hahn (1911) relates the taming of the wild ox to lunar cults of the people from the Hoeing Era of the Upper Stone Age. The moon with its regularly changing phases was the symbol of fertility, and the crescent-shaped horns of the ox became the hallmark of the lunar goddess, who, at specific times, demanded consecrated animals as a sacrifice.

In the beginning, oxen from the wild were still captured for this purpose, but soon people changed to keeping the captured animals in pens. Thus, they eventually formed the first herds of cattle in the holy groves of the goddess. Since only certain organs of the animal were burnt during the ceremony, the participants ate the meat, and many peoples may have come to consider the cattle a useful supplier of meat. The use of sacred animals to pull a plough was at first also a ritual act. But as the plough lost its sacred value, cattle became more of a work animal. The utilization of the milk came thousands of years later.

Hilzheimer raised the question as to why only the aurochs and not also the European bison was domesticated even though both species lived side by side throughout all of Eurasia. Friedrich Falz-Fein in Askania Nowa (Ukrania) proved that it was possible to tame the European bison. Therefore, domestic cattle may have originated only in such areas where the aurochs and not the European bison lived; in the southeastern Mediterranean, South Asia, or, according to other authors, in the south of Spain. Lusatia has also been named as the place where the aurochs were domesticated. The most probable explanation, however, is that the aurochs were domesticated at several places and at different times under various conditions, and that from there the multitude of the present races of cattle developed.

Domestic cattle breeds

The many breeds of cattle throughout the world (Color plates p. 345, p. 346) have been classified systematically as well as phylogenetically by many different methods. Such breeds which are closest to the aurochs, have been called PRIMIGENIUS BREEDS. Other forms, the BRA-CHYCEROS or SHORTHORNS, formerly were considered to be derived from another wild form or even from the banteng. There are many more types or groups of types. These classifications may be rather informative for the determination of the origin and the relationships, but they do not consider the very important and conspicuous differences between breeds in body shapes and its economic utilization. We shall single out only some types of cattle which are of special importance.

In Europe there are some primitive races of cattle which still conspicuously resemble the aurochs in certain external characteristics, such as body structure, the shape of the horns, coloration, and in part, temperament. They are the ones which have been least transformed by man through breeding, and they are of lesser economic value than the select breeds. The plain black CAMARGUE OXEN and the less uniform, but often more auroch-like colored SPANISH FIGHTING OXEN have retained the wildness, the aggressiveness, and the quick movements of their ancestors, which makes them highly suitable for the bull fights of Latin nations. Details which describe the meanness and cruelty unworthy of man in this popular "amusement" are described by the ethologist Fritz Walther in the chapter on the horned ungulates (Ch. 11). The fighting oxen are medium-sized, relatively light animals which live the year round in an almost feral state in the pasture. The primitive COR-SICAN COUNTRY CATTLE, which have been bred somewhat more for meat and milk, resemble the aurochs especially in color. The mostly silver or dark gray STEPPE CATTLE, which are kept from Hungary through Southern Russia to Central Asia, have become enduring draft animals with limited capacity for fattening and milk production. Their horns resemble those of the aurochs, but they often become considerably longer. The same applies to the red-brown, long-haired Scottish highland cattle. The milky white English PARK CATTLE, which have been kept for 750 years on the estates of some British landed proprietors, for example in Chillingham and Chartley, in a semi-feral state, is like the aurochs in build and behavior to such an extent that formerly they were thought to be its last direct descendents. They probably are the descendents of oxen which had been kept in sacred groves.

In the zoos of Berlin and Munich, the brothers Lutz and Heinz Heck began, about 40 years ago, a breeding program with the goal to breed back to the wild ancestoral form by systematic crossbreeding of primitive cattle breeds which had retained characteristics of the aurochs. The Heck brothers reasoned that single characters of the wild form, which were eliminated with increasing domestication due to so called mutational losses, might have been preserved in some cattle breeds. Therefore, it should be possible through crossbreeding to recombine these characteristics in one animal. Such a crossbred product would theoretically resemble the wild form more than the parent animals.

Table 13-1. Synthesis of the Milk of

	Water %	Protein %
Reindeer	63.3	10.3
Buffalo	80.4	4.6
Cow	87.0	3.5
Goat	87.9	3.5
Sheep	82.0	5.8

The camargue breed

Spanish fighting breed

Corsican Country breed

Steppe breed

Scottish highland breed

English park breed

"Back-breeding" of the aurochs

Some Domestic
Animals

Fat %	Lactose %	Ashes %
22.5	2.5	1.4
9.3	4.7	0.8
3.9	4.9	0.7
4.3	4.3	0.9
6.5	4.8	0.0

Such experiments were made in Berlin with Camargue oxen, Spanish fighting oxen, Corsican cattle, and English park cattle, in Munich with Hungarian steppe cattle, Scottish highland cattle, gray and brown alpine races, black and white cattle from Frisia, and Corsican cattle. With this, the Heck brothers felt they had indeed succeeded in re-creating a "modern auroch" which eventually bred true with respect to the wild-form characteristics.

These crossbreeds became the subject of heated arguments. Otto Koehler showed in 1952 that in the ox with his 60 chromosomes an astronomical number of combinations of characteristics is possible. According to the laws of inheritance, a back-breeding of specimens with pure genetic make-up from such a small number of animals in so few generations would be impossible. Furthermore, an expert on domestication, W. Herre, stated in 1953 that the "re-created aurochs" did not resemble the picture of the extinct aurochs. They did not nearly attain its size and its remarkable sexual dimorphism. Nevertheless, the experiments of the Heck brothers did breed from primitive breeds of domestic cattle "model animals" which resembled the wild aurochs if only in color and shape of horn. Thus, they gave us at least an approximate picture of the aurochs. It should be emphasized that a once totally extinct species never can be really recreated by man. The noteworthy experiments of the Heck brothers cannot and should not raise any hopes for the "resurrection" of extinct animal species.

Present-day cattle breeds

The external forms of domestic cattle depend upon what they are mainly utilized for; whether predominantly for milk, meat, or work. Furthermore, the form is influenced by different conditions of climate, soil, and food in the breeding areas.

The Central European cattle types are divided into the HIGHLAND BREEDS of Southern Germany, Austria, Switzerland, and into the German sub-alpine mountains, and into the North German LOWLAND BREEDS. The highland breeds are more or less equally bred for milk, meat, and work. They are medium-sized animals of medium weight, with good fattening qualities and performance and an average yield of 3400 liters of milk per year with a fat content of approximately four per cent. Since record yields in any one direction usually excludes high performance in the others, these highland breeds that are used for milk, meat and work usually produce only average yields. However, this turns out to be the most economic for many small farmers with often scarce pastures. To the highland breeds belong: the SIMMENTAL PIEBALD CATTLE of West Switzerland, whose breed has been spread over wide parts of South and Central Germany; the lighter and more deli-cately built GRAY-BROWN HIGHLAND CATTLE from East Switzerland, which now is raised to a great extent also in the Allgäu Mountains in South-west Germany; the plain yellow and red Central German highland cattle; and several other undemanding, smaller, piebald breeds in

Simmental piebald
cattle

Gray-brown
highland cattle

limited breeding areas. The PINZGAU CATTLE, a race with an annual yield of 3200 liters of milk, is from the Salzburg-Austria region and is frequently kept in the Alps. They have a typical marking of a large white patch on the back, which increases in width from the withers to the croup, furthermore, they have white undersides, chest, and thighs. In the Alps cattle herds are kept in stalls only during the winter. In early May, they are driven to the alpine meadows, which are located in the mountains above the farms and which have a summer pasture with an abundance of flowers and herbs. There they range freely until September, tended only by the alpine cowherd who, in his hut, produces cheese and butter out of the milk. Before they start out for the mountains, which usually is a festive event, the cows often have determined fights among themselves about rank order within the herd. The established rank order then remains unchanged for the rest of the summer. In the Swiss Canton of Valais, these cow fights are a popular spectacle for all the valley's inhabitants. Every year the owners bring their best lead cows to the village of Conthey near Sitten, where the animals fight bitter pushing and wrestling bouts for the title of "queen" before a large audience of spectators.

Pinzgau cattle

The North German LOWLAND BREEDS are bred for beef and for their milk, and of all the German cattle, they make up the bulk in numbers and are of the greatest economic value. On the average, they become heavier than most of the highland breeds and with 3900 liters per year they bring a much higher yield of milk. Record yields can even reach around 11,000 liters, although the fat content of 3.8 per cent is slightly lower than in the highland cattle.

North German lowland cattle

The black and red piebald plains cattle are bred mainly in the coastal regions of the North Sea and the Baltic Sea with their year round pastures and moist cool sea climate. The black piebald races, especially, have penetrated far to the interior. The plain red, delicately built, medium sized ANGLER CATTLE from the Angel peninsula near Flensburg is a German cattle breed which has been bred especially for high milk production. The angler cows with an average weight of only 450 kilograms produce up to 4000 liters of milk per year, which is about nine times their body weight. For approximately 120 years in Schleswig, the heavy English Shorthorns were bred mainly for beef, but they also contributed to the improvement of most lowland races and some of the highland forms.

Angler shorthorn

The ABERDEEN-ANGUS was imported from Scotland into the United States in 1873. The Angus are moderately low-set cattle with a deep side, a straight top line, a wide back and loin, a long rump, and a deep, bulging quarter. The Angus is more cylindrical than most breeds of beef cattle. It is one of the smaller breeds, with cows fitted for the show ring weighing around 635 kg, and fitted bulls weighing 900 kg or more. The color of the Angus may be red or black, with Red Angus being

Domestic American Cattle, by Leslie Laidlaw

Angus (Beef Cattle)

registered in a different association. White markings are undesirable. The head has a prominent eye and a dished face. Angus are polled (naturally hornless) cattle. They are an early-maturing breed. The cows are good mothers and produce more milk than most of the other beef breeds. Angus have a nervous temperament and need good management. While they lack the size and scale desired by many modern breeders, they do produce carcasses of very high quality. Angus place consistently higher in carcass competition and hold their own in live competition. The Angus are somewhat less suitable as range cattle than some of the beef breeds, and are more popular in the midwestern corn belt. However, they are less susceptible to "pinkeye" than cattle with white around the eye.

Shorthorn (Beef and Dual Purpose)

SHORTHORN CATTLE were developed in England, but American cattlemen were not interested in the breed until it was improved in Scotland, becoming the Scottish Shorthorn. Both milking and beef type Shorthorns were imported into the United States in 1783. Beef shorthorns are one of the heaviest of the British beef breeds, with bulls fitted for the show ring weighing up to 1000 kg, and fitted cows weighing 775 kg. The shorthorn is more rectangular than other beef breeds. Shorthorns may be horned or polled (naturally hornless). These cattle vary in color from a deep red to a solid white, with roan or spotted cattle being acceptable. Shorthorns are the heaviest milking of the beef brands, which can be a handicap under range conditions. They also lack the grazing and rustling ability of other breeds, and are therefore more suitable for farm herds.

Milking shorthorns are of the same color patterns as the beef type, but are longer and more angular. Fitted cows weigh between 635 kg and 820 kg, and fitted bulls generally equal or exceed two thousand pounds. They have a more nervous disposition than the beef type. Milking shorthorns are suitable for farms where diversification is desired. They produce less milk than strictly dairy-type cows, but produce calves which can be fed out as beef. They withstand adverse weather better and require less housing than dairy cattle. Milking shorthorns produce milk with approximately four percent butterfat.

Hereford (Beef)

HEREFORD CATTLE were first imported from England in 1817. These cattle are tall and heavily muscled. The Hereford has a striking color pattern with a red body and white markings. The face, breast, flank, switch, and crest are white, with white also present in varying degrees on the underside. The legs below the knees and hocks are also white. The color pattern of the Hereford will not appear in crossbreeding unless the cross contains less than one-sixteenth of the other breed. Herefords may be horned or polled (naturally hornless). They mature earlier than shorthorns, but are not as early maturing as Angus. The Hereford has a rugged constitution enabling it to withstand heat, drought, and winter exposure. They have a heavier hide with a thicker, curlier coat of hair than many of the beef breeds. Herefords are excellent grazers and their rustling ability is appreciated by range cattlemen. Hereford carcasses lack the quality of Angus carcasses, but are readily bought by packers because of their high cutability (percent of lean to fat). Hereford cattle are susceptible to pinkeye when exposed to strong sunlight. The Hereford cow also produces less milk than many of the other beef breeds, but the calves are rugged enough to withstand range conditions. Hereford cattle also show efficient feed conversion under feedlot conditions.

The CHAROLAIS breed originated in France, but entered the United States in 1930 from Mexico. Charolais are one of the larger beef breeds with mature cows weighing from 570 kg to over 900 kg, and mature bulls weighing from 900 to over 1250 kg, depending upon conditions. Charolais are white or creamy white. The hair coat is short in summer, but is longer and thicker in winter. Most are horned, but some are polled (naturally hornless). No special effort is being made to breed for the polled characteristic, since most cattle are dehorned shortly after birth. The calves grow well when nursing their dams, and show good rates of feed conversion in the feedlot. They are useful under range conditions, being good grazers, and are easily able to walk long distances. They do well in hot weather, and can withstand reasonable cold. The bulls of the breed are frequently used in upgrading commercial herds lacking size and ruggedness. Charolais produces a carcass with little external fat. The breed is sometimes faulted for being low in number of calves weaned per number of cows bred. They lack the uniformity of breeds longer established in the United States, but their size and growing ability are increasing their popularity.

Charolais (Beef)

The HAYS CONVERTER cattle have been developed within the last seventeen years in Canada. A Holstein bull was bred to Hereford cows. A Hereford bull was used on these cattle and the blood of brown Swiss cattle was introduced. The Hays Converter is now recognized as a breed in Canada, the United States, and Russia. Mature cows weigh between 730 and 820 kg, and mature bulls weigh approximately 1250 kg. The cattle may be horned or polled. Hays Converters may be either red with a white face, or black with a white face. There are no fertility or calving problems in the breed. The disposition of Hays Converters is as good as that of the other beef breeds. The cattle forage well and show the best feed conversion rates of any of the breeds of cattle in Canada. The carcass lacks the high quality of the Angus but is high in lean meat and low in fat.

Hays Converter (Beef)

MURRAY GREY CATTLE were developed in an area along The Murray River in Australia. The Australian Government prohibits the exportation of live animals, but does permit the exportation of semen from top quality bulls. The breed was developed from crosses of Angus and shorthorn. The color varies between silver grey and dark grey. The

Murray Grey (Beef)

cattle are polled (naturally hornless). Mature cows in breeding condition weigh approximately 660 kg and bulls weigh around 900 kg. The Murray grey cows produce approximately thirty percent more milk than Angus cows. The calves are small at birth and there are few calving problems. The calves grow rapidly and soon compensate for their low birth weight. Murray greys are excellent grazers and are able to withstand severe winter conditions. Their ability to do well under feedlot conditions has not yet been determined. The Murray grey has a more placid temperament than the Angus. Carcasses are of high quality. Semen has been imported into the United States since June, 1969, and is being used in crossbreeding experiments.

The American Brahman (Beef)

The AMERICAN BRAHMAN was developed in the United States by blending several of the Brahman varieties of India. Brahman cattle may be light to dark grey, white, red, roan, and brown. These cattle are characterized by a hump over the top of the shoulder and the neck. The ears are large and drooping, and there are extra folds of skin under the neck and throat. These cattle have well developed sweat pores and perspire freely. An oily secretion is produced from the sweat glands. The horns of the Brahman usually curve upward with a rearward tilt, but may curve down and back. Cows in breeding condition weigh approximately 550 kg and bulls weigh 820 kg. Cattle fitted for the show ring may weigh 180 kg more. Brahman cattle are more able to tolerate high temperatures and humidity than other breeds of cattle. They are also resistant to insects. The cows are good mothers and provide a very adequate milk flow. There is an extremely low incidence of cancer eye in this breed, even though they frequently graze in the sun during the hottest portion of the day. The Brahman gains rapidly for grazing conditions, but lacks the beef conformation of several of the other breeds. Brahman cattle are also noted for their wild dispositions on the range, but may become docile with proper handling. This breed is used extensively in crossbreeding programs in the southern areas of the United States.

Santa Gertrudis (Beef)

SANTA GERTRUDIS cattle were developed in Texas. Crossbreeding of shorthorn and Brahman cattle began in 1918. The Santa Gertrudis cattle of today carry approximately five-eighths shorthorn blood and three-eighths Brahman blood. These cattle are a solid, deep red color. The hide is loose with some folds along the dewlap and underline. The hair is short and straight. Santa Gertrudis have somewhat large, drooping ears. The mature steers and cows of this breed average about two hundred pounds more than Hereford and shorthorn cattle of the same age. Santa Gertrudis cattle combine the ability of the Brahman to withstand hot humid weather with the beef conformation and milder disposition of the shorthorn.

Other breeds of cattle developed from the Brahman are the Brangus and the Charbray.

BRANGUS cattle carry five-eighths Angus blood and three-eighths Brahman blood. They are black and polled (naturally hornless). The crossing of the breeds was started in 1932 and the breed association was formed in 1949. Brangus cattle combine the ability of the Brahman to withstand Southern conditions with the high carcass quality of the Angus.

Brangus (Beef)

CHARBRAY cattle were developed between 1936 and 1942 by crossing the Brahman with Charolais. The breed association was founded in 1949. Charbray cattle contain at least one-eighth and no more than one-fourth Brahman breeding. The cattle are a creamy white at maturity. There is little evidence of the Brahman hump, but the dewlap is heavier than that of the British breeds. The cattle are horned. Mature cows weigh between 775 and 1000 kg while mature bulls weigh between 1250 and 1450 kg. The cows are good milkers and the calves grow rapidly. Charbray cattle do well in hot, humid areas and are expanding into other areas where the climate is less severe.

Charbray (Beef)

The first registered RED POLL cattle to become a part of the modern breed were imported from England in 1873. Medium or deep rich red is the preferred color of the breed, and the very light and very dark reds are discriminated against. White is permitted in the switch and on a small area forward of the naval. The head is neat and well balanced with a pronounced poll. No scurf or horny growths are tolerated. The red poll, like the other dual-purpose breed, the milking shorthorn, is longer legged than the beef breeds, and lacks the depth and spring of rear rib common in dairy cattle. They are more active than most beef breeds, and more placid than dairy breeds. Cows weigh between 550 and 775 kg, and bulls in breeding condition weigh approximately 865 kg. Red poll cows produce less milk than strictly dairy breeds, but produce more acceptable carcasses. The color of the milk produced is between the extreme yellow milk produced by some breeds and the white milk produced by others. The average butterfat is slightly over four percent. The cows have a large udder, but it frequently lacks desirable shape and balance.

Red Poll (Dual Purpose)

The first permanent HOLSTEIN herd founded in the United States was established in 1852, although cattle had been imported from England prior to 1625. Holstein cattle vary in the amount of black and white present, but large black spots on a white background are preferred. No roan or solid-colored animals may be registered. Red and white cattle are eligible for registration in a separate breed association. The Holstein is the largest dairy breed, with mature cows weighing 680 kg and bulls fitted for the show ring weighing 1000 to 1100 kg. Cows weighing over 900 kg and bulls weighing up to 1350 kg are common. Holstein cattle are the predominant dairy breed in all but the southern states. Holsteins have an enormous capacity and do well in good pastures or on grain feeding. Veal calves are larger than calves of the other

Holstein-Fresian (Dairy)

breeds at the same age. The tendency of the cows to put on flesh when not producing milk increases their salvage value when slaughtered. Holstein milk is lowest in butterfat, averaging 3.68 percent and is whiter than the milk of other breeds. Holsteins are late maturing, reaching full production at five to six years of age. They can withstand hot weather if adequate shade is provided. The cows of the breed are quite docile, but the bulls are nervous and untrustworthy.

Jersey (Dairy)

The first registered JERSEYS were imported from Jersey, England, in 1850. Jerseys range in color from light grey to almost black, and may have white markings. Jersey cows are somewhat more nervous than other breeds, but are usually docile and easy to manage. The Jersey is one of the most refined breeds, especially about the head and shoulders. Jersey cows range in weight from 365 to 550 kg. Bulls are less refined and range in weight from 365 to 550 kg. These cattle are good pasture animals, and can take more heat than many other breeds. Jerseys are also good feeders, making efficient use of grain and concentrate. Jerseys produce milk containing approximately 5.3 percent butterfat, but they do not produce the quantities of milk desired by many commercial breeders. Jerseys are early maturing and may be put into production at an earlier age than any of the other breeds. They are also long-lived. The lack in size of Jerseys is a handicap when cows are finished as producers and sent to market. Dairymen using a great deal of bulky roughage do not like the Jersey as much as some of the other breeds. However, they do well under poor grazing conditions.

Guernsey (Dairy)

GUERNSEY cattle first entered the United States in 1830 or 1831. They were imported from Guernsey Island, one of the Channel Islands of England. The Guernsey is a medium-sized, well-proportioned breed with clearly defined areas of fawn and white, and a white switch. Guernseys show a marked secretion of yellowish pigment about the ears and in the switch. These cattle are longer in the face than Jerseys. They have well-tapered horns that incline slightly forward. Guernseys are intermediate in dairy temperament, constitution, and early maturity. Guernsey cows range in weight from 365 to 635 kg and bulls weigh between 570 to 1025 kg, with a weight of 770 kg preferred. These cattle rank first or second in numbers registered in at least forty-two states. The bulls are widely used in upgrading commercial herds. Guernseys are efficient producers of milk and butterfat, and do well under adverse conditions, either hot or cold. Guernseys rank between Jerseys and Holsteins in grazing ability, exceeding the Holsteins in grazing ability, but not doing as well as the Jersey. They exceed the Jersey in ability to utilize roughage, but cannot equal the Holstein in this respect. Guernsey milk is of a golden color, containing four to six percent butterfat. Guernsey cattle often may be faulted for the occurrence of weak backs. The udders often lack length of fore attachment and height of rear attachment, and therefore tend to be somewhat round.

The first BROWN SWISS were imported from Switzerland in 1869. Their color is a shade varying between a light silvery brown and a deep dark brown. There is usually a lighter shade down the back and around the poll and muzzle. The hooves and switch are black. The horns are white with black tips, and turn forward, then up and out. The Brown Swiss are a large breed, with cows weighing 550 to 730 kg and bulls weighing from 750 to 1050 kg. The cows have docile and quiet dispositions, and the bulls are easier to handle than those of most other dairy breeds. Brown Swiss do well in rough sections of the country where considerable rough feed is available, and require less care and attention than some of the other dairy breeds. The cattle are hardy eaters of harvested feeds and have fair grazing ability. Their massiveness and ruggedness make them ideal for average farm conditions. Many cattle of the breed lack the refinement and quality of the other dairy breeds. They are slow maturing and lack udder size. The butterfat averages about four percent.

Brown Swiss (Dairy)

AYRSHIRE cattle were imported from Scotland. The first herd established in the United States was begun in 1860. These cattle are one of the most stylish and distinctive breeds of dairy cattle. The color may be brown, mahogany, or any shade of red, with well-defined areas of white. Black or brindle-colored markings are objectionable. More natural fleshing is present in the Ayrshire than is seen in the Jersey or the Guernsey, thereby increasing the salvage value of the cows when they are taken out of production. The Ayrshire is a medium-sized breed with cows weighing between 465 to 635 kg and bulls in breeding condition weighing approximately 820 kg. The cows have large, well-attached, well-balanced udders. The horns are one of the most distinctive characteristics, turning outward and upward in a very characteristic manner. Occasionally naturally hornless cattle are produced, and these may be bred for the future, since the horns must often be removed to prevent serious injury in fights between cows. These cattle are unsurpassed in foraging and rustling ability under adverse feeding or climatic conditions. They make good use of grain and therefore need less to stay in good condition. They produce an excellent flow of milk and average four percent butterfat. Ayrshire have more excitable dispositions than desired. Some cows are not persistent in their milk flow in a lactation. There are also some Ayrshire cows with small udders and teats.

Ayrshire (Dairy)

Like all domestic animals, the reproductive functions in cattle are overdeveloped and unlike wild oxen, are no longer a part of a special behavior ritual. Under normal conditions, bulls in many breeds become sexually mature at nine to twelve months. Depending on the early or late pubescence of the breed, the kind of food, and the use of the breed, however, the bulls are first used for breeding no sooner than at one and a half to two years, and the cows at one and three quarters to two and a half years, after the animals are fully grown. In the highly specialized breeds in the temperate zones, the cows come into heat every three weeks throughout the whole year for an average period of 30 hours. It is shorter in the cold season and longest in the spring. Presently, artificial insemination is economically advantageous and is increasingly important in cattle breeding. The gestation period is 240 to 320 days, with an average of about 285 days. The birth of twins, which is determined by inheritance, occurs in domestic cattle at a ratio of two in 100. Strangely enough, the female in fraternal twins usually has reduced genitals and is sterile. Such females externally resemble the castrated males and are used in the European alpine countries as draft animals.

In other countries, breeding of high quality cattle is carried out similarly as in Germany. The HORNLESS CATTLE, which are found mainly in North Europe, in Great Britain, and other places, have been developed at different times and in several places by selection of mutations and since improved by further breeding. They are not a separate breed.

Zebus

Of special interest are the ZEBUS (Color plate p. 345) which are distributed in many breeds in Asia, Africa, and more recently, in South America. Next to the water buffalo, they are the most important domestic animals of the tropics. With their conspicuous hump on the withers, they are slightly reminiscent of bantengs and guars. The Dutch zoologist E. J. Slijper found in 1951 that the zebu's hump is neither a compound of muscles supported by the spinal process of the vertebra nor a fat storage comparable to the camel's hump, but that it consists of the greatly enlarged rhomboid muscle. Since this muscle has virtually no function, it could develop into such a "luxury structure" which, in Slijper's opinion, is merely significant as a secondary sex characteristic. Other characteristics of the zebu are the slender long-legged stature, the long stretched, narrow skull, the large drooping ears, and the well developed dewlap. The most frequent color is gray, but there are all transitory shades between black and white and red and red-piebald types. The shapes and positions of the horns differ greatly in the various breeds. The handsome gudzerat-zebus have especially long horns which are curved upward in a crescent shape. Dwarf zebus are bred predominantly on the island of Ceylon.

While Keller in 1909 still claimed that the Indian zebu was simply a domesticated banteng, we now know that the zebus also originated from the aurochs, probably from the early extinct Indian form of *Bos primigenius namadicus*. However, in the Malayan Archipelago and in Northern India, zebus are frequently crossbred with Bali cattle and their offspring are fertile.

Use of the zebu

The utilization of the zebu is basically the same as of other domestic cattle, but their milk and meat production performances do not equal those of the European races. Due to their great agility and manageability they are excellent draft animals, especially in front of the plough and the cart. They are even used as mounts.

In India the zebu is sacred. According to Hindu religion, the divine soul of a human being comes back in every cow. To kill such a cow would be to murder the gods, a mortal sin which is still punishable by death in the religious Nepal. Therefore, a zebu which has outlived its use is not taken care of but is simply left to its fate. All India is crowded with tens of thousands of such ownerless zebus which are close to starvation. They stray over roads and fields, and are chased away by everybody, but are not killed by anyone. They block traffic in the cities, eat garbage and paper, and finally die miserably. Only a few of these "divine cows" are fortunate enough to end their lives in one of the cow asylums established by well-to-do, religious Hindus. For the permanently undernourished 400-millions of Indian people, this religious luxury, in our eyes, contributes to the economic catastrophe. Approximately 160 million zebus live in India, but one of these cows yields barely 180 liters of milk per year, which is less than one fifteenth of a German dairy cow. At least 60 per cent of these economically useless cattle, if slaughtered, would be potentially available as a source of protein for the starving population, and the rest might be used to breed healthy high yielding cattle. This would basically improve the food situation, but at this time plans of this kind are religiously and politically not acceptable solutions in India.

The sacred cows

The zebus had come from Asia to Africa by the time of the early Egyptians, when they were crossbred with the primigenius oxen and developed into large numbers of transitory forms. Many African tribes still carry on with cattle breeding for purposes associated with religious cults without consideration of their economic value. This original form of keeping cattle is still performed by the nilotic pastoral tribe of the Watusi in the Central African Lake Region. They breed cattle with exorbitantly long horns by crossing the Old Egyptian Longhorn cattle with zebus. Meat production is of only little importance; although besides the milk, fresh blood is drunk which is taken from the neck vein by blood-letting. A choice breed of WATUSI CATTLE (Color plate p. 345) are the INYAMBOS which Hediger saw in Ruanda: "These magnificent animals, with huge lyrate horns which at the base may attain a size of a half meter, are still cult animals which are cared for and carefully tended. Each of the animals has two or three caretakers whose function is almost of a sacerdotal nature. Under no circumstances would an inyambo ever be slaughtered, milked, or bled. The only gain are their feces, which are used as fuel, and perhaps the urine, which is utilized for personal hygiene." The Masai, the Samburu, and other pastoral nomads of East Africa have similarly uneconomical

The African humped-back cattle

Watusi cattle

forms of cattle breeding. Their giant herds of zebu-blooded cattle have become dangerous competitors for food to the game animals on the scarce pastures.

The yak

Distinguishing characteristics

A rather close relative of the aurochs is the YAK (sub-genus *Poëphagus*). In the ♂♂, the HRL is up to 325 cm, the BH is over 200 cm, and the weight is 1000 kg. The ♀♀ are only third of their weight. The back has 14 pairs of ribs instead of the 13 in all other oxen. The withers are formed by an extended spinal process of the vertebra, and since the seventh cervical vertebra is much higher, the line gently slopes to the croup. The skull is bulky, broad and well proportioned because of the considerable length of frontal and nasal bone (in large ♂♂ 70–80 cm). The horns are beautifully curved and similar to those of the aurochs. The size at the base is 50 cm, the length extends up to 95 cm, and the tips are up to 90 cm apart. In ♀♀ the horns are much weaker and shaped more irregularly. The rather long legs have strongly enlarged hoofs (adaptation to life in the swamp) and psuedo-claws which serve as supports when the animals climb in the high mountains. The tail is covered with long hair from the base, and ends in a tassel. The coat on top of the head, withers, and back is rather short but densely matted, while on the shoulders and sides of rump it is longer like a mane. The head hair is curly. The muzzle is sometimes lined with white hair, and the naked nose region is reduced to a narrow band on the upper lip. There is only one species, the YAK (⊹ *Bos mutus*; Color plate p. 341).

The wild yak

The wild yak inhabits the North Tibetan desert steppes, which have no trees and bushes, and are located at the lower elevations around 5,000 meters, interspersed with marshes, swamps, and lakes. The steppes are covered with steep slopes of clay and slate in the higher regions up to 5,200 meters. There the dense coat protects it against snowstorms and ice cold temperatures. In June it sheds the winter coat in large patches. The strikingly bushy tail tassel is sought after as trophy and is used by the Tibetans as a fly whisk. The yak's sense of smell is excellent, although its eyesight is only mediocre. Zoologist Ernst Schäfer, who made three successful expeditions in Tibet between 1931 and 1939, was able, with the necessary precautions, to approach to within a distance of about eighty meters, a group of five adult wild yak bulls, until the leader became aware of him.

The yaks belong to those animals which flee from civilization. Consequently, easily satisfied, this animal retreats from man into the most arid regions, although the scarce plant growth then forces it to migrate continuously from one pasture to another. In the humid and hot months of August and September, they stay in the highest elevations that have permanent snow, and during the rest of the year they graze on the sparce herbs, swamp grass, and lichen in the valleys and plateaus. They need water, but they eat snow instead of drinking only when they have to. They graze predominantly in the morning and in

Fig. 13-7.
Yak *(Bos mutus)*, distribution of the wild form

the evening. Ernst Schäfer contributed detailed reports on the habits and behavior of these interesting wild oxen: "The wild yak is an extremely lazy animal. If it finds 'good' pasture, it may not move away for days, but will alternatingly graze and lie down on the cold ground to ruminate, or it may take a refreshing bath in Naka lakes or steppe creeks even in very cold weather. When severe blizzards, which occur regularly until July, roar over the plateau, it turns its broad bushy tail into the storm and remains motionless for hours. The old bulls are grim, fierce fellows who usually turn into dangerous enemies of man when they are disturbed during their lazy comfort, especially when they have been shot. They will attack, head up, with great speed, with their tail raised like a banner which is whipped to and fro on the back."

Fig. 13-8.
Playfully fighting Yaks

The famous Russian explorer Przewalski saw herds of wild yaks with thousands of animals at the end of the 19th century in North Tibet. "At all times except for mating season, the old bulls live singly or in small groups of three or five. Young adult bulls are often found in herds of their own of 10 to 12 individuals, with one or two old bulls among them. Cows, adolescent bulls, and calves, however, join huge herds which may consist of several hundred, or even thousand, animals. In such large numbers, it is difficult for them to find enough food. This way, the calves have the best protection from attacks by wolves." Schäfer, too, reports that except for man the Tibetian wolf is the only enemy of the wild yak. In winter large packs of wolves are said to approach even loner bulls and sometimes bring them down. In high altitudes where neither man nor wolf go, yaks usually attain the record age of about 25 years when they finally die of old age.

The mating season begins in September and lasts for one whole month. The old bulls join the cows for some weeks and have bitter rival fights among themselves during which each animal tries to push his horn into the opponent's flank. While such encounters seldom end with the death of one of the animals, serious injuries occur often. However, the wounds heal quickly in the sterile air at these altitudes. Only during the mating season do wild yaks make strange grunting sounds. The rest of the time he is not very vocal. Hence, Przewalski gave it the scientific name *mutus,* the "mute". Cows give birth to their calves in June after a gestation period of nine months. However, they give birth only every second year because the young are dependent on them for one year. The yak is considered to be fully grown only at six or eight years old.

The distribution of the wild yak during the Glacial Period extended to Northeast Siberia. Eighty years ago it still reached south to the headwaters of the Hwangho and Yalung Rivers in the Chinese Province of Tsinghai. But since then its distribution has been gradually narrowed down by the well-armed, rapacious highland nomads from Tibet and the Mongolian pastoral tribes from the north. Today, wild yaks occur only in North Tibet, from the Karakorum in the West eastward along the south slope of the Altyn-Tahg over the Kuenlun Mountains to the Nanshan Mountains. In 1964 Bannikow estimated these scattered remainders at 3,000 to 8,000 animals; other information is far more pessimistic. Protective measures are difficult to enforce in these pathless alpine regions, so that for this species, too, the danger of being destroyed seems inevitable. A last resort would be to found breeding groups in zoological gardens. Presently, only a few wild yaks live in the zoo at Peking, and a single male animal is said to live in the zoo at Rostow on the river Don.

The domesticated yak

The yak had probably been first domesticated in the first millenium B.C. by the Tibetians of the western Kuenlun Mountains or of East Pamir. The DOMESTIC YAK *(Bos mutus grunniens)* is much smaller than its wild ancestor. Besides those with the wild black and pure white color, there are brown, yellow, reddish, gray, and piebald domestic yaks. Their coats resemble the wild ones, although in most cases they have an even longer stomach mane. Their horns are weaker; even hornless animals are not too rare, at a ratio of 1:100. Now the domestic yak has a much wider distribution than the wild original form. It occurs in the west approximately as far as Buchara, in the south to Bhutan and Nepal, where the animals are kept in pasture farming, and in the east and north up to the interior of Mongolia and to the mountain chains south of Lake Baikal. Because the domestic yak makes frequent grunting sounds, in contrast to the wild yak, he also is called the "grunting ox". The estrous is irregular and domestic yak females usually have calves every year. Wild yak bulls seem to have a strong dislike for their tame descendants. There are earlier reports that they would attack caravans and gore tame yak oxen that were only half their size to death.

Since the domestic yak requires little food and is insensitive to cold temperatures, it is the best suited domestic animal in Asia at elevations above 2000 meters. It easily carries loads of 150 kilograms over the steepest mountain paths, and it is even used as a mount. The milk production is somewhat lower at about 400 liters per year, but the milk has a high nutritious value with a fat content of seven to eight per cent. In Tibet a milk powder is prepared (besides butter and cheese) by a special process of coagulation and it is used for provisions. The meat tastes good, although only the old animals are slaughtered. The yaks are sheared once a year and the yield of about three kilograms of coarse wool is spun into yarn and then made into blankets, tent covers, bags, and ropes. Garments are made from the soft wool of young animals. The yak's manure is often the only fuel in the highlands of Tibet which have no trees or bush.

In the frontier areas of Tibet, yaks are often crossbred with Chinese cattle. Such crossbreeds are in great demand, since they are especially strong and good natured animals, for ploughing and use in caravans. According to the reports from Scheifler (1963) about an experimental farm in the Altai of the Soviet Union, it was possible to obtain, by crossbreeding yaks and domestic cattle, an animal which combines the valuable characteristics of the two animals—a good meat and milk production. Crosses between yak and zebu also give more milk and better meat and are excellent work animals for carrying loads at moderate altitudes. The yak even has been crossbred with the gayal. The male offspring of all breeds crossed with yaks are said to be sterile.

The third and last group of wild oxen are the AMERICAN and EUROPEAN BISON (genus *Bison*), whose fate has been disastrously influenced by man during the last hundred years. American and European Bison, which produce fertile offspring, resemble each other so closely that it is debatable whether the two are true separate species or merely subspecies. Fossils have shown that they are the last survivors of a once multiform group of huge wild oxen which at the end of the Glacial Period were distributed from Europe over all temperate Asia to North America and from Alaska to Mexico.

The European bison

The largest mammal of the American continent is the BISON (*Bison bison;* Color plate p. 343). The HRL in the ♂♂ is up to 300 cm, the BH is 190 cm, and the weight is up to 1000 kg. The ♀♀ are smaller and lighter by one fourth or one third. The frontal part of the body is overdeveloped, giving the bison its characteristic profile. The broad skull is more massive and is carried lower than in European Bison. The chest cavity is larger and the pelvis is one fourth smaller than in the European Bison. The spinal process of the last cervical vertebra and the first ten of the dorsal vertebra are greatly extended so that the withers are especially pronounced. The short sturdy horns are ridged at the base by growth. The horns come out to the sides of the skull, curve backwards in a simple curve, and slant upward. The tips are rather blunt and directed slightly inward. Due to unusual hair growth, the frontal part of the body is even more emphasized. The hair on the head, neck, shoulders, withers, and front legs is up to 50 cm long. On the forehead between the horns, it forms a thick hood which may flap over and hang down to the muzzle. The hair at the chin, beard, and dewlap have grown into a dangling mane, while on the front legs there are forearm cuffs which may reach as far as the pastern-joints. Occasionally there are white, gray, and piebald bison. There are two subspecies: 1. PLAINS BISON *(Bison bison bison).* 2. WOOD BISON (⊹ *Bison bison athabascae),* which is more similar to the European bison. It is larger, the rump and legs are longer, the hooves are wider, the horns are almost twice as long, and denser coat is darker.

The American bison

Distinguishing characteristics

In English usage, the bison is erroneously called the "buffalo." The equally incorrect term "Indian Buffalo" occurs even in travel and adventure books. The Plains Indians of North America depended on

The American bison is not a "buffalo"

Fig. 13-9.
Bison (*Bison bison*),
shaded: former distri-
bution; black: present
distribution.

Massacre of the
bison

bison hunting until the conquest and colonization of the West by the whites. They even considered the skin of white bison to be sacred and worshipped it in their hunting cults as a fetish. A. B. Szalay reported in 1932 on the great significance of the bison in the North American Indians' life: "All actions of the Indians, all their habits, concepts, conditions of life, views, their whole life was connected in the closest sense with the bison. To the dying Indian the shaman would say the following words: 'You came from the buffalos on earth, now you go home to the animals, to your ancestors, and to the four spirits. May your way be gentle.' There are no peoples in the world's history which has ever been so intertwined with any animal to such an extent as the Indians with their bison."

Around 1700 in North America, from Alaska along the east slopes of the Rocky Mountains southward to Northeast Mexico and across the continent almost to the Atlantic, approximately 60,000,000 bison were living, which roamed over the grasslands in almost endless herds. The Indians' bow and arrow hunting did not diminish these giant herds and neither did the severe winters, droughts, prairie fires, and other natural catastrophies to which many of the animals fell victim every year. In the fall the bison migrated 350 to 650 kilometers south to spend the winter on better grazing grounds, and in spring they returned to the north. These "buffalo paths," stamped through the centuries, were used by the first white settlers on their way west.

With that began one of the most shameful epochs in the history of man's conquest and one of the greatest animal tragedies. As long as settlers only tried to keep the migrating bison herds from their cultivated land which they had struggled so hard to wrest from the wilderness, the destruction of the bison population remained within limits. About 1830 the methodical destruction of the bison was begun. This deprived the Indians, who were fighting for their hunting grounds, of their fundamental conditions of life. The construction of the Union Pacific Railroad, the first large railway across the continent, which was begun in 1865, initiated the final phase of this cruel mass destruction. Thousands and thousands of the unsuspecting wild oxen were killed to supply the railroad workers with fresh meat. At this time, William F. Cody, called "Buffalo Bill," who had been hired for this purpose by the Kansas Pacific Railroad gained rather dubious fame when he killed 4,280 bison within 18 months. The most loathsome was the popular passenger's "sport" of shooting bison from the train, which the railroad companies arranged some years later. There the travellers were allowed to kill as many bison as they pleased from the train windows. From the dead animals only the tongue was cut off as a delicacy and the carcasses were left to rot in the piles along the railroad tracks.

After a short while, there were no bison left in a wide area on both sides of the railways. The bison had been divided by this mass murder into a northern and a southern population. Methodical destruction of the southern population was begun in 1871 and completed in 1875. Well organized groups of shooters butchered around 2,500,000 animals per year and used them for commercial profit. After the construction of the Northern Pacific Railroad, the northern herds were destroyed in the same manner between 1880 and 1884. In barely two decades, one of the largest concentrations of game animals on earth had been almost completely destroyed. The political goal for the massacre had been reached. The Indians who had survived the decades of debilitating fights were starved out and could no longer offer resistance. In 1889 Dr. William T. Hornaday estimated the number of remaining wild bison in the U.S.A., including one small herd of 200 animals which lived protected in Yellowstone National Park, at 835 head.

Fortunately, there were some men who were interested in protecting animals of the wild. Under the supervision of Dr. Hornaday, on December 8, 1905, they met in the lion's house of the New York Zoo, founded the American Bison Association, and saw to its financial support. This association under Dr. Hornaday succeeded literally at the last minute to awaken the public conscience. In 1907 the American government established a bison preserve in the Wichita Game Park in Oklahoma which was populated with a herd of 15 bison from the New York Zoo. Later on, other preserves were established in Montana, Nebraska, and Dakota; they also were populated with animals bred in captivity. The bison of the Yellowstone National Park, too, increased in number.

Rescue at the last moment

At the same time, efforts were made in Canada to protect the bison. In 1907 the Canadian government purchased the then largest existing herd of 709 animals from a private owner and brought them to Bison Park in Wainwright, Alberta, which was created just for this purpose. There the number of bison increased to 5,000 head by 1920. For the few surviving wood bison, the government founded in 1915 the large Wood Buffalo Park between Great Slave Lake and Lake Athabasca. Unfortunately, in the following years thousands of surplus plains bison, which spread tuberculosis, crossbred with the wood bison, and brought this subspecies close to extinction, were also brought there from the Wainwright Bison Park. Fortunately, in 1960 in the remotest corner of the Wood Buffalo Park, a last herd with approximately 200 pure Wood Bison were discovered so that there is a chance that perhaps the continued existence of this subspecies may be assured. The total number of all bison presently living in the protected areas of Canada and the United States is again up to approximately 30,000 head. Thanks to the energy of some far-seeing men, the destruction of this magnificent creature, which once had inhabited the earth in larger numbers than any other wild ox, was prevented at the last moment.

Habits of the
American bison

Before the naturalists of the last century had an opportunity to know the bison, it was destroyed before their very eyes. Therefore, all our knowledge on bison habits and behavior comes from recent observations. During the summer, plains bison feed predominantly on the grass and the herbs of the steppe, and the wood bison feeds on leaves, shoots, and the bark of trees and bushes. In winter they have to settle for lichen, moss, and dry grass which they dig out of up to one meter of snow with sideways movements of their broad skulls. At least once a day, the herds visit the water holes to drink. Only when thick ice covers the lakes and the rivers will they eat snow. According to the research of McHugh in 1958, among the bison's vocal expressions are different kinds of hollow grunting sounds which can be heard when the herds are on the move. During the mating season, the bulls utter a throaty rumbling roar which can be heard over a distance of five to eight kilometers when there is no wind. When the bulls are in a fighting mood, several of them often join in the roaring "concerts". Snorting and sneezing vocalizations occur during play and in sexual excitement. McHugh observed that bison are extremely playful up to the age of two years, although the older animals may occasionally play. Fighting and active games occur most frequently, and there is playful mounting. Closely related with play in bison, and often its beginning, is the distinct urge to explore the unknown. McHugh reports on this:

"Curiosity is highly developed in the bison. They thoroughly examined new or strange objects, for example, my observation platform, dropped Wapiti antlers, a skeleton, horses, a porcupine, prairie dogs, etc. Young animals were more curious than adults. The members of the herd came to inspect newborn calves, to lick, and to sniff at them. The older calves were especially interested in the new little calves." An exploring animal approaches an unknown object with its tail stiffly stretched out horizontally or vertically erect. This is a significant sign of excitement in bison and it also occurs on other occasions. This strange kind of curiosity may well have been one of the causes for the rapid destruction of the bison. The body of a freshly killed conspecific and the scent of blood seem to be irresistably attractive to other members of the herd. They sniff at the carcass and push it with their horns with increasing excitement, as if they are trying to make the dead animal get up. Then a hunter may easily kill from his hiding place one animal after the other. It must not have been difficult to kill within one hour a whole herd of 60 to 70 animals.

How dangerous is
the American bison?

Nevertheless, the bison is not a harmless animal and will charge quickly when it is cornered and able to recognize its enemy. With its highly developed sense of smell, bisons are able to scent sources of danger over a distance of one or two kilometers. According to the

In the Kalahari

Animals of prey: 1. African wild dog (*Lycaon pictus*, see Vol. XII). Artiodactyla: 2. and 3. South African Oryx (*Oryx gazella gazella*, see Ch. 14). 4. Springbuck (*Antidorcas marsupialis*, see Ch. 15). The forehead weapons of the ruminants are usually used in intra-specific encounters (see Ch. 11). Only exceptionally are they also used for defense against enemies. In oryx it may happen that they defend themselves with their long horn spears. African wild dogs are the most dreaded beasts of prey on the African steppes. No prey can easily escape from them because they chase with unremitting endurance. They have an exhausted oryx at bay (2), and desperate defense is not at all effective. While he throws one of the attackers into the air with his horns, the other dogs seize him from the rear and from the side in order to throw him down and to tear him to pieces. The oryx group dashes off in panicky flight (3). A group of springbuck also takes off with high jumps.

Body care

observations of Garretson in 1938, they are able to scent water over seven or eight kilometers. Their sense of hearing is also good and, according to McHugh, their vision has often been underestimated. Grooming plays an important part in the bison's daily activities. They scrub their heads, necks, and sides of their bodies on trees, branches, or tree trunks which, from the abundant use over the years, have no more bark and appears to be polished. They also like to roll in loose sand and dust, and these places are used repeatedly. Prairie dog colonies often seem to be especially inviting. The bulls in the mating season and when they are in a mood for fighting frequently discharge much of their excitement by rolling in the dust on the ground.

There does not seem to be an actual day-night activity cycle in bison. Even though the animals graze early in the morning and at dawn and ruminate around mid-day, they also graze during the day and sometimes at night. In spite of their plump shapes, the bison move with amazing ease and endurance. On their migrations, they cross the prairies on their trampled paths at a fast pace, which may change into a rapid trot, and finally reach record speeds of 40 to 50 kilometers per hour in a gallop. Usually an experienced, old cow leads the herd migrations, but several animals take turns so that there is no actual leader. McHugh distinguishes between bull herds with two to twelve members and cow herds with an average of 23 animals, including calves and some subadult bulls.

Reproduction

During the mating season from May until September, groups of bulls and cows form large herds. The rank order fights of the bulls often end with serious injuries, or sometimes even with the death of one opponent. The victor associates with the courted cow for several hours or some days, and the association ends after successful copulation. Later the large herds split up again into smaller groups, and old loner bulls often roam among them. The cows give birth to their red-brown calves after nine months gestation. They sometimes separate from the herd and go to a protected place for calving but often the births take place within the herd. The young are nursed for one year and become pubescent in their second or third year.

Longevity in bison, like in most wild oxen, is 20 to 25 years. Crosses between bison and domestic cattle are possible, but their young offspring are usually sterile. In zoological gardens, plains bison are frequently seen, usually in large pens, which at least reflect a little of the past splendor of the life of these large wild oxen of the North American prairies. They reproduce regularly in captivity. A breeding group of purebred wood bison, however, presently exists only at the Alberta Game Farm in Canada.

The European bison by H. G. Klös

The EUROPEAN BISON (♦ *Bison bonasus*; Color plate p. 343) is equally impressive. The HRL is 310–350 cm, the TL is 50–60 cm, the BH up to 200 cm, and the weight goes up to 1000 kg. The conspicuously high withers are formed by the greatly extended spinal process of the dorsal vertebra. The head, which is usually carried low, is below the top of the withers. The horns grow from the frontal bone first to the sides, then they turn upward, and incline forward, so the tips are turned inward (record horn length according to Ward is 50.8 cm). The hindquarters have rather weak muscles. The dense coat of dark to black bristles is longer at the frontal part of the body, especially at the chin, neck, and forearms. The beard grows at an angle into the throat mane, and the forehead hair flaps forward. The tail hair increases in length from the base to the tip, but it does not have a true tassel. Around the muzzle and the eyes the hair is short and smooth. In shedding, the old fleece comes off in patches. There are two subspecies: 1. EUROPEAN BISON or WISENT *(Bison bonasus bonasus)*. 2. CAUCASIAN BISON *(Bison bonasus caucasicus)* which has been destroyed, although the bloodline still exists in some breeding groups in zoological gardens.

Distinguishing characteristics

From the two forms of wild oxen formerly living in the European woodlands, the aurochs were extinct by 1627; the European bison, however, was saved after a hard fight to preserve the species. From the Glacial Period, we have its picture from many beautiful cave drawings. Oppian described the bison in 200 B.C. with the following words: "The bison are horrible animals, similar to the oxen, that live in Thrace. They have manes like lions, pointed, curved horns with which they throw people and wild animals into the air. Their tongue is very rough, like a file, so that they may tear the skin by licking." Another document by Gesner (1516–1565) describes it as being "ugly, terrible, with much hair, a thick and long mane like a horse, bearded, and altogether wild and misshapen."

How the European bison survived

However, those who were fortunate enough to see, in one of the large bison preserves or even in the National Park of Bielowecza, a full grown European bison coming out of the shady forest to a clearing will have to admit that this "ugly" animal is the most majestic and most powerful of European game animals.

The European bison is the closest relative of the North American bison. Both species originated from *Bison sivalensis*, whose fossils have been found in North India. A group of them migrated over the Himalayas which were then still low, and crossed the then existing landbridge north east into North America and evolved into the American bison; another group migrated westward and became the European bison. Each of these two groups developed into two different lines: a steppe and a forest type. The steppe bison became extinct during the Glacial Period. The forest type bison, probably during the last Glacial Period, developed into an alpine and a plains form, of which the alpine form, the Caucasian bison, became extinct in 1927. However, the bloodline of this bison exists in varying degrees in the breeding groups at zoological gardens.

Fig. 13-10.
Former distribution of the European bison *(Bison bonasus)*. The black triangle marks the prime forest of Bielowecza, the only place where the European bison is presently free-ranging again.

Fig. 13-11.
Plains bison *(Bison bonasus bonasus)*

Fig. 13-12.
Caucasus bison *(Bison bonasus caucasicus)*

The studbook of the
European bison

Once, the habitat of the European bison reached across Europe and probably as far as Siberia. It was forced back not so much by severe hunting as by steadily expanding human settlements and the increasing clearance of the woodlands. The number of European bison decreased steadily with land cultivation in Europe. In the beginning of the nineteenth century, the remaining plains bison had retreated into the forests of Bielowecza. There, in the heartland of Poland southeast of Bialystok, 300 to 500 animals led a hidden existance. The number of this herd varied constantly, but on the whole, the number of this herd steadily decreased, partly due to poaching and partly because of the overcrowding of the area. The turmoils of World War I brought utter ruin, and the final shot came in 1921 from a poacher. Erna Mohr writes as follows: "The last free European bison in the Bielowecza forest was killed on February 9, 1921, by a former forester from this area by the name of Bartlomeus Szpakowicz: may his name always be remembered like that of Herostratos!" (A Greek who set fire to the temple of Artemis at Ephesus so as to become famous. Ed.)

This February 9, 1921, seemed to be the fateful day for the largest European game animals. Fortunately, before the herd ceased to exist, some of the bison, a total of 56 animals (27 males and 29 females), among them five male calves and five female calves, had been given to zoos and private game reserves. With these few zoo and preserve bison, a small group of optimistic bison breeders and keepers began the almost hopeless struggle to preserve the species. Under the leadership of the former director of the Frankfurt Zoo, Dr. Kurt Priemel, the International Association for the Preservation of the European Bison was founded in August, 1923 at the Berlin Zoo. Its aim was to systematically breed the last bison, to regularly exchange experiences within and beyond the boundaries of the country, and to make a detailed study of these animals. The still existing European bison were listed in a studbook. Goerd von der Groeben accepted the position of keeper of the studbook, and in December, 1924, he made a report on the existing stock of European bison which was anything but hopeful. According to this report of the 66 animals, six were very old and could not be bred any more, and 22 more were still too young to be bred at the time.

After years of preparatory work, in which Dr. Erna Mohr from 1927 on participated, the first studbook on the European bison was finally published on April 15, 1932. Not only did it contain all purebred European bison living at this time, but also all of their traceable ancestors. Each purebred bison had a number and a genealogical name whose first letters indicated where the animal had been bred. For example, the meaning of Pl = Pless, Po = Poland (Plains Bison), Pu = Puszcza Bielowecza (animals with the prevailing bloodline of the Caucasian bison), Ar = Amsterdam, Be = Berlin, F = Frankfurt,

Ko = Copenhagen, Sp = Springe, etc. Four breeding lines were distinguished: 1. The extinct Bielowecza-line. 2. The Pless-line, also consisting of purebred plains bison. 3. The zoo-line with crossbreeds from the Bielowecza and the Pless-line, and purebred plains bison. 4. The Caucasus-line whose animals have blood of the alpine bison.

Dr. Erna Mohr, Hamburg, has been keeping the studbook of the European bison for decades. She did much to reestablish the breeding program of the European bison. After the setback due to World War II, Dr. Jan Zabinski of Warsaw and Dr. Erna Mohr restored the studbook. Presently, the data are examined every two years and published in Warsaw. Formerly, the zoo breeds of Amsterdam, Berlin, Budapest, Vienna-Schoenbrunn, and the game preserves of Boitzenburg and Springe (Hanover) were of importance. Today, the main breeds of the European bison are in Poland, the Soviet Union, Germany, and a number of other European and overseas zoos and game preserves.

Erna Mohr reports on the bloodlines of the presently existing European Bison: "All former and present breeds of the European bison are traced back directly or indirectly to Bielowecza and Pless (= Pszczyna). The most important branch of them is the one in the Berlin Zoo, which has influenced many of the other breeds mainly with the output of surplus bulls. However, single female animals have also been given to other breeds, and some of them went as far as New York. Because of their skill in keeping and breeding bison, most of the Berlin bison developed into huge and beautiful animals with high backs." From 1881 until 1966, a total of 57 European bison have been bred and raised in the Berlin Zoo.

The animals do not need heated accommodations. They live in a log hut, and the doors are open during the night so that they may enter or leave at will. However, most of the time they are outdoors, and in the winter they will even allow snow to cover them. It is important, depending on the size of the group, to have one or several stalls to separate animals advanced in pregnancy or cows with young calves. The pen in the Berlin Zoo, which is 900 square meters, has proved to be large enough for a group of one bull and six cows. The sandy Brandenburg soil is very favorable for the bison's need of taking sand and dust baths.

Appropriate food plays an important part in raising strong and healthy animals. The following list is representative of the food given the Berlin bison: A full grown bison bull receives daily about 2.5 kg energy food, 12 kilograms of good quality hay, and as many beets as he will eat. In the summer, beets are replaced by sliced turnips, and fresh greens are fed instead of hay. Fresh willow branches proved to be a good additional food, although the animals do not show signs of deficiency when this supplemental food is not available. Carefully balanced and rich nutrition is especially important shortly before

Where do the present European bison come from?

Keeping them in zoological gardens

shedding and for pregnant cows. Each animal of the Berlin group has its own trough in the stall to guarantee a uniform distribution of the food.

Reproduction and
longevity

A European bison cow is sexually mature at two years, and a young bull is capable of coition at two years. The youngest proven father is "Wouwerman" who produced his first offspring at the age of 15 months. A European bison bull is full grown in approximately eight years. For about five years, he is in his prime, and then slowly the aging process begins. However, cows may still have calves when they are old. For example, "Beatrice" had young when she was 21 years old, and "Planta" at 22. So far the authentic record age for a European bison was 27 years, while the cow "Beatrice" that came from Berlin died in Amsterdam at the age of 26 years.

The European bison cows are pregnant for nine months (260–270 days). About three or four days before the birth of the calf, we separate the pregnant cow from the herd so that she may give birth without being disturbed, and we return the cow and calf to the herd no sooner than three or four weeks later. The calves who weigh, at birth, about 40 kilograms are nursed for a half year or longer, but they will begin to eat from the mother's food after three weeks. During this time it is important that the calf has a loophole in the fence so that it may get to a trough of its own. In contrast to American bison, the European bison's cow pushes her calf away from the trough, and without the above mentioned assistance, the calf would not get enough food.

The European bison
exists on open range
again

After the continued existence of the European bison seemed to be guaranteed through successful breeding in zoos and game preserves, a small herd was turned loose in 1956 in the prime forest of Bielowecza. In 1963, it consisted of 57 animals of which 34 had been born in the wild. The animals are in good condition. In the winter they receive a supplemental feeding of eight kilograms of hay per head per day, and Krysiak has determined that two European bison may live on an area of 1000 hectares without seriously damaging the forest. In this free ranging herd, it was finally possible to study the social behavior. It turned out that during the mating season the herds split up into small groups of eight to ten animals which are usually led by the oldest cows. In winter and spring, the groups rejoin to form a herd, although the old bulls stay away from it and form a separate group.

With the release of one herd at Bielowecza in Poland and another in the adjoining Soviet part of the prime forest, one of the great goals of the International Association for the Preservation of the European bison has been reached. The huge wild ox exists again in the wild in Europe. However, as bitter experiences have taught us, it is better not to have all animals of a valuable species in one place. There are always chances of loss due to contagious diseases or warfare, and then it is better to have the animals in many different locations. Therefore, in

will be of importance for the continuing existence of the European
the future the zoological gardens and game preserves all over the world
bison more than for any other species.

In 1967 the total number of European bison in the world was 860
head, more than 15 times the number in 1923, in spite of war and the
destruction of many flourishing breeding stations. This is the kind of
success which may be wished for all endangered species.

Heinz-Georg Klös
Arnfried Wünschmann

14 Hartebeests, Roan and Sable Antelopes, and Waterbucks

This group of animals includes such different animals as harte-beests, blesboks and gnus, sable and oryx antelopes, and water and reedbucks. The American zoologist George Gaylord Simpson in 1945 included them all into one systematic unit. But since they do vary greatly from each other externally, we shall follow in this chapter the older division into three subfamilies: 1. HARTEBEESTS (Alcelaphinae). 2. ROAN and SABLE ANTELOPES (Hippotraginae). 3. REEDBUCKS (Reduncinae).

Subfamily: Hartebeests

The hartebeests (subfamily Alcelaphinae) are approximately the size of red deer. The horns in the ♂♂ and the ♀♀ are medium, large, and S-shaped or angled backward like hooks. The basal axis of the cranium is 30-45 degrees from the facial part of the skull. There are preorbital and interdigital glands. There are no or only slight color differences between the sexes. The females have two nipples. The two tribes (hartebeests and gnus) have three genera, five species, and 32 subspecies.

Distinguishing characteristics

To the TRUE HARTEBEESTS (Alcelaphini) belong the hartebeests in a narrower sense (genus *Alcelaphus)* and the SASSABIES (genus *Damaliscus*). Until today there have been great differences of opinion about how many species of hartebeests there were and which should be consid-ered a species and which a subspecies. Therefore, we have decided to agree with the opinion of Th. Haltenorth who suggests a general lumping into species. Thus, the reader may find that forms which are described in this volume as subspecies are called true species in other publications.

Hartebeest

Distinguishing characteristics

According to this modern classification there is only one species in the genus *Alcelaphus*, the HARTEBEESTS (*Alcelaphus buselaphus*; Color plates p. 302, p. 407). The HRL is 175-245 cm, the TL is 45-70 cm, the BH is 110-150 cm, and the weight is 120-225 kg. The animal has a large head, long legs and a sloping back, which declines from the much higher withers to the croup. The narrow and long head has a straight or slightly arched profile at the eye region with a narrow muzzle. The horns are shorter and thinner in the ♀♀ than in the ♂♂. The shapes vary in the subspecies. The slender legs have hoofs with pointed tips, and short and pseudo-claws. The thin, round, medium-long tail has a crest which ends in a long, thin tassel. There are 15 subspecies, many of which differ conspicuously in shape, coloration, and shape of horns.

The best known of these subspecies are: 1. NORTH AFRICAN HARTE-BEESTS *(Alcelaphus buselaphus buselaphus)*, which formerly lived in North Africa, Egypt, North Ethiopia, and Palestine, but has been extinct in these areas for at least 50 years. 2. WEST AFRICAN HARTEBEEST *(Alcelaphus buselaphus major)*. 3. LELWEL *(Alcelaphus buselaphus lelwel)*. 4. TORA (♀ *Alcelaphus buselaphus tora)*. 5. KONGONI *(Alcelaphus buselaphus cokii)*. 6. LICHTENSTEIN'S HARTEBEEST *(Alcelaphus buselaphus lichtensteini)*, is smaller than other hartebeests. 7. KAAMA *(Alcelaphus buselaphus caama)* exists sporadically in South Africa in some protected areas.

The Boer name "hartebeest" means "tough ox." The Boers found out that Hartebeests were rather enduring runners and could not be chased down on horseback. Furthermore, they would have to be shot many times if the first shot was not immediately fatal. But this "being tough" did not help most of the subspecies. The hartebeests have been largely decimated and in wide areas even exterminated with modern firearms. The North African subspecies, the "game of the desert," had once been kept in great numbers in a semi-tame state in Ancient Egypt. It has disappeared from most of its habitat since the end of the last and the beginning of this century and may now be extinct. Some animals were said to have been seen in 1943 in the Southern Rio de Oro, but there has been no report of them since. The same would have happened to the South African subspecies if they had not been saved on some farms and game reserves.

With their long, narrow heads and steeply sloping backlines, the hartebeests at first glance appear to be strangely misshapen and less elegant than other antelopes. But this rather unfavorable impression disappears immediately when the animals are in motion. Like all members of the subfamily, they use the amble walk, and in a fast run they change to an odd "rocking horse gallop." The most impressive locomotion of the hartebeests is their springy, bouncing trot where they carry their head close to the neck like a bridled horse. Their habitat is the open steppe and the light brush, where they feed mainly on grass. Usually they drink every day, but they can get along for several days without water. They do not roll, but they like to "kneel" down and to rub mud on their horns and foreheads with which they then smear on their flanks. Only in some places do they form larger herds; usually they live in groups of five to 30 animals.

Lately we have learned some details on the habits and behavior of hartebeests through the studies of Backhaus on the lelwel hartebeest and of Gosling on the kongoni, which I had also observed. The groups

What does "hartebeest" mean?

Fig. 14-1.

Hartebeest *(Alcelaphus buselaphus)*

1. present distribution;
2. extinct.

Duikers (see Ch. 12.

1. Grey duiker *(Sylvicapra grimmia)*
2. Black-fronted duiker *(Cephalophus nigrifrons)*
3. Black duiker *(Cephalophus niger)*
4. Banded duiker *(Cephalophus zebra)*
5. Bay duiker *(Cephalophus dorsalis)*
6. Blue duiker *(Cephalophus monticola)*
7. Giant duiker *(Cephalophus sylvicultor)*

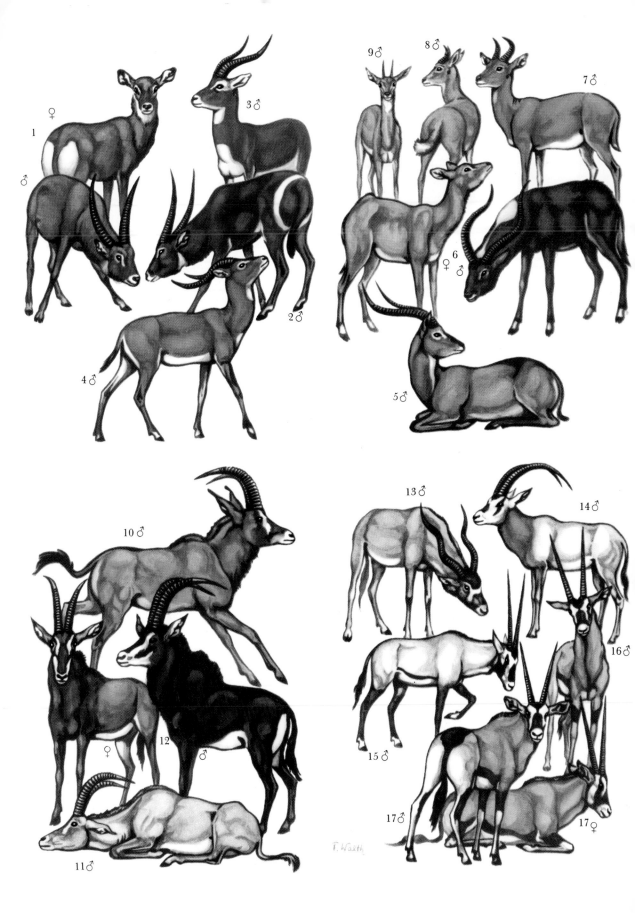

Waterbucks

1. Defassa waterbuck *(Kobus ellipsiprymnus defassa)*
2. Common waterbuck *(Kobus ellipsiprymnus ellipsiprymnus)*
3. White-eared kob *(Adenota kob leucotis)*
4. Uganda kob *(Adenota kob thomasi)*
5. Lechwe waterbuck *(Hydrotragus leche)*
6. Mrs. Gray's or Nile lechwe *(Onototragus megaceros)*
7. Bohor reedbuck *(Redunca redunca)*
8. Mountain reedbuck *(Redunca fulvorufula)*

Grey rhebucks

9. Grey rhebuck *(Pelea capreolus)*

Roan and sable antelopes

10. Roan antelope *(Hippotragus equinus)*
11. Blue antelope *(Hippotragus equinus leucophaeus)*, destroyed
12. Sable antelope *(Hippotragus niger)*

Oryx antelopes

13. Addax *(Addax nasomaculatus)*
14. Scimitar-horned oryx *(Oryx gazella dammah)*
15. Arabian oryx *(Oryx gazella leucoryx)*
16. East African oryx *(Oryx gazella beisa)*
17. South African oryx *(Oryx gazella gazella)*

Sassabies

Distinguishing characteristics

consist of associations of females or males. Some of the adult males leave the "bachelor" groups and occupy territories. The size of the territories depends on the terrain and the density of the population, whether the males stand farther apart from each other or closer together, and whether a specific group of females remains in a male's area for some time or just for a few hours. The males mark with their preorbital glands and have defecation places within the territories. Their urine is not important for marking.

In fighting, hartebeests, like the other species of the subfamily, go down on their carpal joints. Often the opponents will cross their heads while they are still standing, so that they are cheek to cheek. They rub and nibble at the cheek and neck before they lock horns. Pawing of the ground plays an important part as a threat gesture, although it also occurs before they plough the ground with their horns and before they smear mud on their shoulders. Besides horn-fighting, neck-fighting may occasionally occur. More often than in most horned ungulates, fights of the hartebeests cause serious injuries on the neck, chest, and shoulder. Males in pursuit often butt fleeing opponents on the rump with their horns. Besides ritualized fighting behavior, true damaging fights also occur.

Hartebeests belong to the few horned ungulates in whose mating behavior we observe no driving. The male approaches the female with his tail erect, the neck and head stretched forward, and the ears in a sideward position and slightly pressed downward. He then often steps in front of her, turns and walks a few steps. If she follows, this seems to signify a kind of "agreement," The two then stand in specific position to each other, one behind the other, or at right or acute angles to each other, or reversed parallel. This could be considered kind of driving in place. Then the male finally gets into a reversed-parallel position to the female and sniffs at her genital region. If she is in the right phase of estrous, she will lift her tail even higher, he will turn around so that he stands at right angles to her with his head at her anal region. Usually, he mounts from this position. These preludes may well be omitted or considerably reduced, especially if the same male has previously copulated with the same female. The young are born after a gestation period of approximately eight months. In East Africa most births take place at the beginning of the so-called "Small Rains." Today hartebeests are seen in zoological gardens less frequently than formerly. While their longevity in the wild is 12 to 15 years, zoo hartebeests have become almost 19 years old.

Modern classification distinguishes two species of the sassabies (genus *Damaliscus*). 1. SASSABY *(Damaliscus lunatus;* Color plates p. 302, p. 407, pp. 414-5). The HRL is 150-205 cm, the TL is 40-60 cm, the BH is 110-130 cm, and the weight is 110-150 kg. The horn length reaches 72 cm. Compared to the following species, they are tall and bulky.

There are nine subspecies, among them the KORRIGUM *(Damaliscus lunatus korrigum),* the SOUTH AFRICAN SASSABY *(Damaliscus lunatus lunatus),* the TOPI *(Damaliscus lunatus topi),* and HUNTER'S ANTELOPE *(Damaliscus lunatus hunteri),* which many authors classify as a separate species because of the missing dark facial patches and because of the wide horns. 2. BONTEBOK or BLESBOK *(Damaliscus dorcas;* Color plates pp. 61-2, p. 407). The HRL is 140-160 cm, the TL is 20-45 cm, the BH is 85-110 cm, and the weight is 80-100 kg. The horn length may reach 47 cm. Compared to the former species, they are smaller and more graceful. The animals have distinctly contrasting white marks on the front on their faces, on their legs, chests, abdomen, and anal region. There are two subspecies, the BLESBOK *(Damaliscus dorcas philippsi)* and the BONTEBOK (⊹ *Damaliscus dorcas dorcas).*

The Bonteboks were brought close to extinction in their South African habitat by the Boers. There only a few of them are left in narrowly restricted, very small, protected areas of the Cape Province. However, the blesbok population seems to have recovered somewhat during the last decades and, therefore, has again a somewhat wider distribution. Previously such numbers of them were said to have existed that "the country was red from them." According to some reports, blesbok males defend their own territories and establish huge manure piles in them. Strangely enough, they often lie down right on top of them. Males and females mark with their preorbital glands predominantly single long and hard stems. In several European and American zoos, breeding groups of this beautiful antelope are kept so that this species is probably no longer endangered.

Some years ago I raised a young blesbok in the Zurich Zoo. The little animal turned out to be an extreme "follower" type; it could literally not be left alone for one minute or else it would bang against walls, doors, and fences without regard for its own safety. I had no other choice than to keep it in my room. Right from birth, "Mausi," as it was called, responded to sounds and came running to me as soon as I tried to imitate the mother's call note. At first, it would go to everyone who called, but later it learned to distinguish my voice and responded only to it. After 24 hours, it recognized my scent and took its bottle only from me. Other people could feed it only when they wrapped one of my shirts around the bottle. The following response of the animal was released by movements of legs and in the dark by the sound of the steps.

The radiantly white marking on forehead and nose, for which the blesbok is named is not seen in the young. In contrast, this region is even darker than the rest of the body. Young blesboks and bonteboks are much lighter than the adults; often they resemble young hartebeests to such an extent that it is hard to tell them apart.

In their range, their habitat and food requirements and in their de-

Bonteboks and blesboks

Fig. 14-2.
1. Sassaby *(Damaliscus lunatus),* 2. former distribution of bontebok and blesbok *(Damaliscus dorcas)*

Topi

pendence on water, the sassabies are very much like hartebeests, although sassabies are found in even drier regions or, on the other hand, in flood plains and in dense brush. Presently, we know most about the East African subspecies, the TOPI. Like the kongoni, the males and females go down on the carpal joints and dig in the moist soil with their horns, so that afterwards thick lumps of dirt stick to the horns. However, they do not smear this on their bodies. Even more than the hartebeests, they like to stand for hours on higher ground, especially on termite mounds. This was formerly interpreted rather simply as "standing guard," although now we are more cautious about this interpretation. Often the animals cool off their legs this way in a breeze close to the ground; they also display in this manner, marking their territory in a very visible way.

In the Serengeti the mating season is from December until March, during and after the "Small Rains." After a gestation period of seven or eight months the young are born. The births usually coincide with the beginning of the rainy season or after scattered showers. During the mating season, the male may be recognized within a herd from a distance by the way he holds his head. Only he carries his head almost continuously stretched forward and up. The females keep it more normal on the approximate level with the rump and with the nose directed forward and downward. Males which keep their heads erect this way are always the rulers of an area. The others live in bachelor groups.

Topis also mark with their preorbital glands and establish several defecation places in their territory. Furthermore, the males are in the habit of galloping along the boundaries. At least they do this when there are no females with them, or right after the arrival of the females. All of a sudden such a male will take off, run to the boundary, and gallop around his whole territory. Three territories, which I knew in more detail, were between 200–400 meters in diameter. The females are not territorial, but they may remain for several days with the same male. In other cases, they just move on. As long as the females are moving within his territory, the male often takes charge of the group which walks in a single file.

A special peculiarity of the topis is to jerk their heads while they are running and especially when they begin their march. Other hartebeests and sassabies also do this. Originally this was a threat gesture, but here it is used as a means of coordinating the activities of the group. Translated into the human language, it would mean "let's go!" When territorial males are by themselves, they often meet at the boundaries where they perform a rather strange scene. They stand head to head, side to side, or reverse parallel to each other. Then they scratch or groom their shoulders, graze, paw the ground, defecate, go down on their carpal joints, and dig in the ground with their horns. Sometimes they do the same thing at the same time, and sometimes they alternate their activities. When they stand opposite each other head on, turning the head to the side plays an important part. Both animals may, at the same time, turn their noses into the same or opposite directions. Sometimes this is all that happens, at other times, the matter becomes more serious. The attacks begin with an animal throwing its head upward, jumping with the front legs in the direction of the opponent, and performing "symbolic" butting and warding-off movements. After the two have once again demonstratively "looked away" from one another, they throw themselves down on their carpal joints and tangle each other's horns. With "well acquainted" neighbors such a fight lasts but a few seconds; then both opponents jump back on their feet, "look away" from one another and move—sometimes grazing—into different directions. In this manner the males establish the boundaries and which are later "confirmed."

When females move into the territory, the male walks toward them in slow motion with the tail stretched out almost horizontally, the head forward and up, and the ears pointing to the sides and down like in zebus. The front legs are bent sharply and he lifts them very slowly as high as possible. He stalks around his "intended" female and finally stands beside her with his hindquarters closer to her than his front-quarters in a so-called "star formation." If at this point the female takes off, the male just looks after her; however, if at another time the females are about to leave the territory, he goes after them in a wild chase. If the female stays with the male, his attitude changes. He now carries his head like a bridled horse, turns around to her, bends his hind legs and kicks with a front leg in the direction of her hind legs; copulation then immediately follows.

The young are of the "follower" type, although they lie down more often and for longer periods than the adults. Hence, topis "invented," so to speak, a compromise solution between "staying-put" and following. Several females with young of approximately the same age form a group. The young rest together in a "kindergarten" and one of the mothers stays with them as a "governess," while the other females walk off to graze. In case of a disturbance, the mothers return immediately to their young and each one takes care of her own, walking away with it after having briefly nursed it. The young walks beside or behind the mother, next to her hind leg, or sometimes beside her front legs. It is possible that the large blue-black patches on the shoulders, front legs, thighs, and hind legs of the adult animal facilitate orientation for the young.

Topis produce quite a variety of sounds. Besides the snorting alarm call, there is the growling call of the male, which he utters predominantly during mating behavior or when he is herding females. Furthermore, there is the low, somewhat grumbling contact call of the

Fig. 14-3.
The topi male prances around the female when courting her.

▷
Hartebeests in a narrow sense

1. Hartebeest (*Alcelaphus buselaphus*), subspecies a) Lichtenstein's hartebeest (*Alcelaphus buselaphus lichtensteini*), b) Kongoni (*Alcelaphus buselaphus cokii*), c) Kaama (*Alcelaphus buselaphus caama*), d) Tora (*Alcelaphus buselaphus tora*).
2. Sassaby (*Damaliscus lunatus*), subspecies a) Sassaby (*Damaliscus lunatus lunatus*), b) Korrigum (*Damaliscus lunatus korrigum*)
3. Bonte or blesbok (*Damaliscus dorcas*), subspecies a) Bontebok (*Damaliscus dorcas dorcas*), b) Blesbok (*Damaliscus dorcas philippsi*). Pictures of other subspecies of hartebeest and sassaby are shown on page 302.

▷▷
Gnus

1. Brindled gnu (*Connochaetes taurinus*), subspecies a) Southern brindled gnu (*Connochaetes taurinus taurinus*), b) Eastern brindled gnu (*Connochaetes taurinus albojubatus*), c) Johnston's brindled gnu (*Connochaetes taurinus johnstoni*).
2. White-tailed gnu (*Connochaetes gnou*)

1a♂

1b♂

1c♂

♂

♀

2

F. Walth.

mother and the soft "answer" of the young. The young also utter an almost squeaking sound during their running and jumping games. Even though topis are not at all rare in their East African native habitat, they are not often seen in zoological gardens.

Gnus

Distinguishing characteristics

Closely related to the hartebeests are GNUS (tribe Connochaetini) which are about the size of red deer. The smooth horns of both ♂♂ and ♀♀ grow sharply bent or are hook-like. The coat is short and smooth, and there is a mane on the forehead, the upper and lower side of the neck, chest, or throat. The large gnus have a short neck, slender legs, and a long tail. The muzzle is broad and the eyes are relatively small. There are two species, which were formerly classified into separate genera:

1. WHITE-TAILED GNU (*Connochaetes gnou;* Color plate p. 408) is smaller. The HRL is 170–220 cm, the TL is 80–100 cm, the BH is up to 120 cm, and the weight is approximately 180 kg. The back is straight. The mouth region has many long white hairs, and the back of the nose, neck, and withers have an erect short hair mane, while the chest mane is longer.

2. BRINDLED GNU (*Connochaetes taurinus;* Color plates p. 408, p. 416) is larger. The HRL is 175–240 cm, the TL is 70–100 cm, the BH is 115–145 cm, and the weight is 145–270 kg. The back slopes from the withers to the croup. The coat and mane colors differ depending on the subspecies. There are five subspecies, among them the EASTERN BRINDLED GNU (*Connochaetes taurinus albojubatus*) and the SOUTHERN BRINDLED GNU or BLUE GNU (*Connochaetes taurinus taurinus*).

The white-tailed gnu

The white-tailed gnu is one of the oddest and most conspicuous horned ungulates. Its habitat has always been limited to South Africa, where it has been exterminated in the wild by the boers. However, a few head were preserved on some large farms in a semi-wild state. From there they were brought into several protected areas where they have reproduced well. Thus, relatively many white-tailed gnus have come into zoological gardens of Europe and North America where they heard the piercing trumpet-like "hoeitt" call of the male white-tailed gnu. This gnu, with its "high-spirited-exaggerated" movements and the erect manes on the nose, neck, throat and chest, looks like a veritable monster. Single males in captivity may occasionally become aggressive towards their keepers; otherwise the animal is absolutely harmless.

The brindled gnu

Our knowledge of the brindled gnu has been greatly increased during the last years due to the observation of Estes in the Ngorongoro Crater and of Watson in the Serengeti. Quite surprisingly, it was found that the habits of the Serengeti-gnus are rather different from those of the crater-gnus, even though they belong to the same subspecies and their habitats are close to each other. The gnu herds in the Ngorongoro Crater roam about, depending on the condition of the pasture and the rainy and dry seasons. But this cannot be compared with the immense migrations of the Serengeti-gnus. Twice a year the giant herds cross the national park, so that they are in the Southeast during the rainy season and in the Northwest, outside the park boundaries, in the dry season.

Bernhard and Michael Grzimek had discovered that the herds were outside the protected area for a considerable part of the year. Unfortunately, because of incorrect surveys, the boundaries of the national park had been established before this became known and it is hardly possible to change the course of the migrations. As Bernhard and Michael Grzimek found out, the gnus follow short grass and other specific forage plants on their migrations. They leave areas where such plants are scarce or depleted by grazing. Right now, this does not matter in the Southeast, since the Masai live there. The pastoral Masai tribe has always gotten along well with the gnus and probably will continue to do so for some time. But in the Northwest, the herds travel into densely populated farmland where everyone is against them, and where they are killed by the thousands.

The gnu population in the Ngorongoro Crater is easier to observe since it is more stationary. There the animals form herds consisting of females, bachelor herds, and territorial males. Estes was able to show that single males remained in their territories for weeks, months, and even for years. Each territory is slightly less than 100 meters in diameter. In contrast to this, the serengeti-gnus occupy territories only in the mating season. Right out of the migrating herds, some males stop and become territorial, haggle about boundaries, and try to keep some of the females with them. These territories may exist for only a few hours, and then the males return to the migrating herds and move on with them. The longest period that I observed a territorial male in the Serengeti was five days.

The East African gnus have rather well defined mating and breeding seasons. Nevertheless, the breeding season may vary somewhat annually, which probably is linked to the beginning of the "Small Rains." Most of the births in the Serengeti occur in February and March, and in the Ngorongoro Crater somewhat earlier. Most young are born within a span of only four to six weeks. Strangely enough, the mating season lasts much longer. It begins during the long rains in April, runs at its peak until June, and may even last as long as July.

The gnus also prefer to fight on their "knees." According to Estes' observations, the encounters of territorial males at the boundaries look very similar to those of the topis (p. 405 ff.). Each of the males has a defecation place in his territory where he also deposits the secretions from his preorbital glands. He smears the secretion into the grass but does not mark single stems as distinctly as many other species do. Gnus often roll extensively, always after having pawed; they also paw before lying down. Young gnus belong to the clearest "followers" in the horned ungulates. According to the latest counts in the Serengeti

Fig. 14-4.
1. Brindled gnu (Connochaetes taurinus).
2. White-tailed gnu (Connochaetes gnou).

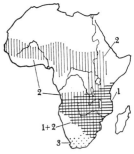

Fig. 14-5.
1. Sable antelope (Hippotragus niger)
2. Present distribution of the roan antelope (Hippotragus equinus).
3. Former distribution of the exterminated blue antelope (Hippotragus equinus leucophaeus).

region, more than 300,000 gnus still live there. It is a breathtaking view to see these tremendous herds pass by in endless single files or when their grazing herds cover the steppe to the horizon. I hope it will be possible to preserve, at least in the Serengeti, this reminder of an era when the earth was still unspoiled.

It has always been said that gnus and other African steppe animals are not only gregarious among themselves, but they also associate with animals of other species. But this happens much less frequently than was formerly thought (comp. Color plate p. 416). Books about Africa tell of the advantages that associations between different species would have, such as warning each other in cases of danger. This, too, is only half the truth. Alarm signals are widely understood by most of the savannah ungulates, and this has nothing whatsoever to do with whether or not they associate with other species. For example, the alarm whistle of the reedbuck is responded to by many species of game animals, even though reedbucks are strictly solitary animals. Furthermore, any animal which in a fast, running flight attracts the others' attention may well release flight in them. However, this can not be defined as a mutually advantageous association. Relationships in animals are often more due to chance than man, with his urge to look for purpose and meaning, would like to admit. Game species which harmoniously graze together in the wild actually do not have any close relationships. This can be shown in captivity, when they turn out to be absolutely incompatible when they have to live together permanently due to the lack of space in a zoo.

Subfamily: Roan and sable antelopes

Distinguishing characteristics

The most conspicuous characteristic of the ROAN and SABLE ANTELOPES (subfamily Hippotraginae) are their long horns which are equally well developed in both sexes. They range in size from that of deer to horse. The horns have a spear form, and are loosely screwed around the longitudinal axis or are curved like a scimitar. The backline is higher at the withers, and some have a neck mane. They have preorbital or interdigital glands. The length of the cranium is approximately one third of the facial part of the skull. There is no or only a slight difference in color between the sexes. The females have four nipples. There are two tribes (Hippotragini and Orygini) with three genera, five species, and 18 subspecies.

Hippotraginae

Because of their crescent-shaped horns which are bent upward and back, the members of the hippotragini tribe, genus *Hippotragus*, are conspicuous. The HRL is 188–267 cm, the TL is 37–76 cm, the BH is 100–160 cm, and the weight is 150–300 kg. The ♂♂ are about one fifth to one fourth larger and heavier than the ♀♀. The animals have slender legs, a bulky neck, a large head, and a slightly declining backline. The gestation period is 270–280 days, and one young is born. In captivity they live almost 17 years. There are two species: 1. ROAN ANTELOPE (*Hippotragus equinus*; Color plate p. 402) in which the ♂♂ and ♀♀ are of the same color. The underside of the neck has longer hair. The horn length is 50–95 cm. The molars are larger and broader. There are seven subspecies, among them the BLUE ANTELOPE (*Hippotragus equinus lencophaeus;* Color plates pp. 61–2 and p. 402). 2. SABLE ANTELOPE (*Hippotragus niger;* Color plate p. 402). The ♂♂ and ♀♀ have different colors. The horns are 70–173 cm. There are three subspecies, among them the GIANT SABLE ANTELOPE (⊹ *Hippotragus niger variani*).

Both species of sable antelopes are animals of the brush forest, of meadows with scattered trees, and the fringe forests along rivers. Although they may live without water for two or three days, in the dry season they drink, if possible, two to three times per day, especially at noon. They are aggressive animals. Except for man, only lions seem to be dangerous to the adults.

The roan antelope lives in small groups of three to 15 individuals, consisting of an old male with females and their young. Only seldom were herds larger than 40 to 60 head observed. Younger males are said to live in separate groups. The roan is stationary, but in the Serengeti, where they occur only in very small numbers, the ranges of the individual groups are very large. One distinct group there had been under observation for several years near Banagi, and it always contained about 10 to 15 animals. By preference, this group lived in two different areas, one during the rainy season and the other during the dry season. The two areas were about 20 kilometers from each other. When the group was about to change areas, the old male acted as "scout." He showed up at the other place by himself, then he disappeared again and returned a few days later with the females and the young. At the end of their stay, he was the first to leave the area, and the females were by themselves for several days until they, too, finally moved on. Unfortunately, I did not observe whether the male came back to "collect" the females this time. During this observation, a second male, who was clearly younger but otherwise fully grown, was with this group.

Except for a walk to the water hole at noon, the animals were active only in the morning until about nine and in the afternoon from five o'clock on. Although there were many bushes and trees in the area, they fed exclusively on grasses and herbs on the ground. When they were not disturbed, they spent the remainder of the day standing or lying a closely circumscribed area. They were probably active at night. When roan antelopes flee, they first utter a snorting sound and then, carrying the head horizontally, they dash off one after the other. With their strong, massive bodies, the black and white markings of their face, the mighty horns, and the long ears, they are a magnificent sight in the sun-drenched bush. Since their ears with the tufts at the tips are almost of the same length and size as the horns, the first Europeans who saw roan antelopes from a long distance thought that these

Roan antelope

▷

Defassa-waterbuck *(Kobus ellipsiprymnus defassa.* In these deer-sized "antelopes," only the males have horns.

▷▷

Topis *(Damaliscus lunatus topis,* on the steppe at dusk.

animals had four horns. The young also have very long ears, but the face marking is not yet developed. In the wild the roan antelopes get along well with giraffes, zebras, impalas, and other game animals. They are usually accompanied by tick birds.

Reproduction is said not to be seasonally limited. In the mating ceremony, according to Backhaus, mating occurs with circling and a ritualized kick with the front leg as a courtship gesture. Roan antelopes go down on their "knees" when they fight. Although they are otherwise rather shy, they are, as the closely related sable antelopes, among the relatively few horned ungulates which, when pursued, stand their ground and defend themselves with their horns.

The southernmost subspecies, the blue antelope, was probably the first antelope to be destroyed by white settlers. Surrounded by abundant game in the Cape Province, the Boers apparently were not able to imagine that their continuous and completely uncontrolled killing would some day inevitably lead to the destruction of the fauna. The Boers also had a misguided biblical belief according to which each animal was created only for the ruthless exploitation of man. Now the South Africans are trying to atone for the sins of their ancestors by contributing to the conservation of animals and by establishing national parks, for example, the famous Krüger Park. However, extinct forms can never be resurrected.

Sable antelope

The sable antelope, which has a body height of about 140 centimeters is somewhat smaller than the roan. It appears to be "lighter" and more "elegant" than the roan, and it is a very handsome animal indeed. In contrast to the roan, it carries its head close to the neck during flight, which is reminiscent of the deportment of a bridled horse. Its herds, which often are a little larger than the roan's, keep together in flight. Otherwise, the two species seem to be rather similar in their habits. For example, the sable antelope also goes down on its "knees" for fighting. Guggisberg once watched a display encounter between two males. The two stood for quite some time reversed and parallel to each other with their tails slightly raised, heads carried high, and the chin pulled towards the neck before they went down into the fighting position.

Brindled gnus and steppe zebras at the water hole. This often-described "association" is rather questionable. Both species inhabit the same biotope, have similar habits, and do not avoid each other. Therefore, they are often found together. But this certainly does not constitute a true association, where one seeks and remains in the other's vicinity.

Generally, sable antelopes in the wild get along rather well with other species of game animals. However, there are reports that they occasionally have attacked kudus. In zoological gardens, sable antelopes sometimes turn out to be rather aggressive. When they dash forward to attack passing visitors, they may even damage their horns on the fence.

Lately, a case where a solitary male remained master of the field against a whole pride of lions was photographically recorded. The lions had encircled the animal, but did not quite dare to come closer because the antelope turned against each of the approaching lions and attacked it. Finally, the male rushed at full speed towards a lioness who fled from him. Thus he broke through the encirclement and escaped. The "critical distance," which is the distance from the enemy when an animal will not flee but will defend itself, seems to be rather large in sable antelopes.

As the name indicates, the sable antelope's coat is of a dark color. Near Mombasa, where the northernmost sable antelopes are found today, only the old males are black and the females and younger males are red-brown. In other subspecies, the females and males are more or less of the same color. The giant sable antelope of Central Angola is especially famous among hunters for its enormous horns. Unfortunately, it belongs to the endangered species.

The ORYX (tribe Orygini) range in size from that of a reindeer to a horse. The horns in the ♂♂ and ♀♀ are long, spear- or bow-shaped or loosely screwed around the longitudinal axis. The coat is short and smooth or somewhat denser and rougher, and some have manes at the neck or throat region. The females have four nipples. The two genera have three species and seven subspecies.

Oryx

The ORYX (genus *Oryx*) has a HRL of 160–235 cm, a TL of 45–90 cm, a BH of 90–140 cm, and a weight of 100–210 kg. These deer-sized animals have medium to long and slender legs, a slender to compact neck, and a straight back which is higher in the withers. The eyes are relatively small, and the preorbital gland is in the process of reduction or it has already disappeared. The medium-long ears end in a point, sometimes with a brush at the tips. The round, thin tail ends in a long tassel, which begins rather high and reaches to or below the heel. There is one species, the ORYX (*Oryx gazella*; Color plate p. 402), whose horns are straight spears or slightly curved. There are eight subspecies, among them the EAST AFRICAN ORYX (*Oryx gazella beisa*), the SOUTH AFRICAN ORYX (*Oryx gazella gazella*), the ARABIAN ORYX (*Oryx gazella leucoryx*) and the SCIMITAR-HORNED ORYX (*Oryx gazella dammah*), whose horns are sabre-shaped, and bend downward. Some authors classify this as a separate species.

Distinguishing characteristics

The white oryx of Arabia and the scimitar-horned oryx of the Sahara are genuine desert animals which have been greatly reduced in numbers. Presently, the Arabian oryx is close to extinction. In both cases, the natives who hunted them ruthlessly from convoys of cars or even from planes are to blame. According to what we consider very optimistic estimates, at most 500 Arabian oryx antelopes are still living in the residual area which remains of their former distribution, which was from Palestine and Syria to the southern coast of Arabia. Meanwhile, the Sultan of Muscat and Oman has prohibited their hunting, but a true and effective enforcement in these desert regions is not possible. Stewart and some other Englishmen tried to capture several of these animals in order to establish a breeding group in a safe area.

Scimitar-horned oryx

In 1962 a male and two females came to Phoenix (Arizona) where the animals have a climate similar to that in their natural habitat. Later it was possible to transport some more Arabian oryx which had lived in captivity in England and Arabia. The animals have reproduced well, and this small herd has increased to 17 head. They are kept in a semi-wild state so that they might later be returned to the wild, provided that eventually men have become more sensible.

The times when the scimitar-horned oryx, the "game of the desert" inhabited all of the northern Sahara from West Africa to Egypt, belong to the past. Presently, scimitar-horned oryx are found only at the southern edge of their former distribution. Although their number has not yet decreased to the extent of the Arabian oryx's, they do belong to the seriously endangered antelopes. Unfortunately, all steppe and desert animals are much more endangered by the ruthless destruction of our modern technology than animals of the brush and forest regions. They have no chance to withdraw from the hunter's sight, especially when he uses an airplane. Scimitar-horned oryx may live for weeks and months without water. Depending on the seasonally varying feeding conditions, they will migrate long distances; for example, they migrate during June and July, at the time of the summer thunder storms in the West Sahara, northward into the interior of the Sahara; in the hottest season from March until May, however, they go southward to the Sudan. The animals are said to be active in the morning, in the evening, and especially during the night. Fortunately, some zoological gardens in the U.S.A. and Europe have breeding groups of this endangered species, and eventually this may be the only chance to save it for the future.

The South African oryx

The South African oryx is found in steppe and semi-desert regions. It was exterminated only in the Cape Province area, as were almost all of the large animals, but otherwise, it is not yet rare. In the older literature, it was also called "Passan," a name which obviously did not stick. The Boers called it, as was their custom, "gemsbock" (chamois), although the South African oryx has nothing in common with the European alpine game. In the Kalahari region, it likes to dig for tschamma, a fruit similar to a melon, as well as for roots, with its front legs. This fruit and its roots contain much water and apparently enable these animals to inhabit regions with little water. After a gestation period of about nine months, only one young is born, as is true in all oryx. The young frequently "stay put," according to Steinhardt's observations, next to termite heaps. As long as the young is still small, the mother allegedly leaves the herd and stays with her young one.

Of all the oryx subspecies, the South African oryx are the most vividly marked. Interestingly enough, oryx, as all roan and sable antelopes, have markings on their faces which have something in common with the gazelles. In the gazelles, the pattern of the cheek stripe begins at the preorbital glands; in the roan and sable antelopes and especially in the oryx, this pattern is even more developed and distinct. It looks as if a "structure," which originally developed phylogenetically from a scent organ, had been "transferred" into the eye region where it is now very conspicuous. Oryx antelopes no longer have functioning preorbital glands.

The behavior of the oryx also resembles that of the gazelles. In some respects, they might be called enlarged gazelles who evolved in a certain direction. For example, when adult oryx males defecate, they do so in the same deeply squatting posture of adult male gazelles. Furthermore, they paw the ground before defecation and in other situations as the gazelles do. However, urinating and defecating in the oryx are not combined as in male gazelles (see Ch. 15).

If a male drives a female which is not yet ready for copulation, she may withdraw from him by stepping beside him in the reversed-parallel position. She avoids his attempts to drive and to mount by circling around in a very limited space. This may happen in all horned ungulates and usually the male eventually ceases driving. In gazelles this behavior, although it is still an exception, occurs more frequently than in other horned ungulates. The male gazelle often will not give up his attempts. In oryx, and probably in roan and sable antelopes, there is, besides the courting foreleg kick (Laufschlag), a circling which is a fixed component of the mating ceremony. In a zoo pen where a male has only one or a few females, this behavior may become exaggerated. The male may drive a female even though she is not in heat, and the animals will circle around each other 30 times and more when the female seriously tries to withdraw. This is not entirely harmless because the oryx male has less of an inhibition to attack a female than the male kudu, and he is inclined to attack the female, vehemently driving in such "exaggerated" circling.

Other trends in the mating-circling of the oryx are obviously linked with the threat and display rituals. Among them is the highly erect posture of the male, the head-lowering of the female, and the reversed-parallel position of the two. All these are behaviors which are also found in certain specific situations. In these threat and fighting methods, the similarity with some gazelle species may be clearly recognized (see Ch. 15). No gazelle would go down on the "knees" in a fight, which the roan and sable antelopes do rather frequently, but in all of them an exchange of blows and "fencing" are especially frequent and typical, in addition to "wrestling" and "locking of horns." This "fencing" method occurs especially rarely in other horned ungulates. To sum up, forms of behavior which are equal to or similar to the gazelles' and which occur only now and then in other horned ungulates, are found with striking frequency in roan and sable antelopes. Therefore, roan and sable antelopes and gazelles seem to be

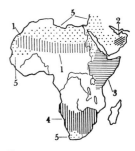

Fig. 14-6.
1-4. Present distribution of the oryx (Oryx gazella):
1. Scimitar-horned oryx (Oryx gazella dammah),
2. Arabian oryx (Oryx gazella leucoryx), 3. East African oryx (Oryx gazella beisa) and related subspecies, 4. South African oryx (Oryx gazella gazella) and related subspecies,
5. Areas where the oryx has been exterminated.

Fig. 14-7.
"Courting-circling" in the oryx

Fig. 14.8.
The foreleg kick (Laufschlag) in the mating behavior of the oryx.

Addax

Distinguishing
characteristics

Tame antelopes in
ancient Egypt

Fig. 14-9.
Male oryx "fencing"
during a running game.

Fig. 14-10.
Addax (*Addax nasoma-culatus*)

more closely related to each other than the usual systematic classification would indicate.

The second genus of oryx consists only of one species, the ADDAX (⊹*Addax nasomaculatus*; Color plate p. 402). The HRL is 150-170 cm, the TL is 25-35 cm, the BH is 95-115 cm, and the weight is 60-125 kg. Built like a reindeer, these animals have a short neck, long rump, medium-long sturdy legs, and a straight back with the withers slightly higher. The rather large head has a considerable tuft on the forehead, especially in old ♂♂. The eyes are small and the preorbital glands are not visible. The moderately thin horns, which slant backward and up, and slightly to the sides in large flat spirals, are almost round in a crosscut. They have low, flat hooves with flat plantars which are not sectioned into balls. The round, thin tail has a short tassel on the end. The winter coat is longer and rougher than the summer coat.

Like the scimitar-horned oryx and the North African hartebeest, the addax was kept semi-tame in large numbers in ancient Egypt. In pictures from this time, addax stood in stables like goats. They were fed out of troughs, led on a bridle, and were probably slaughtered for cult purposes. The keepers in the "Old Kingdom" has even developed a special procedure to protect themselves from the pointed horns of their wards. During growth in the young animals, they were gradually bent over, probably with special clamps. Since 1900 the addax has disappeared from Egypt, and even before that time, it had been exterminated in the north of Algeria, Tunisia, Libya, and Senegambia. Similar to the scimitar-horned oryx, it lives presently only in the southern parts of its former range and belongs to those species that are seriously endangered by passionate hunters.

The enlarged hooves and the well developed pseudo-claws probably prevent the animal from sinking into the sand. In small groups of only five to 15 animals, the addax roams across deserts and semi-deserts in a continual search for grazing grounds. Formerly, there were herds of hundreds and thousands of animals. They may go for weeks and months without water. Like camels, the walls of their stomachs are said to be transformed into alveolar reservoirs for liquids. Some authors disagree on this, but lately it has been categorically reasserted. The animals are active in the morning, evening and night. In order to protect themselves against strong winds and the sun's radiation, they dig holes with their front legs and lie down in them. Sometimes such resting places are located halfway beneath solitary rocks. Reproduction is not limited to special season. Allegedly, the addax has a gestation period of 10 to 12 months, which would be exceptionally long for a horned ungulate. A few days after birth, new copulations occur.

Although formerly addax antelopes were kept rather frequently in zoos, for example, in Berlin and some of the large American zoos, they are now among the rarities. Therefore, the breeding groups of the zoos of Brookfield and Hanover are of great importance. Each animal born there helps to preserve the species for future generations. In zoos they can become 18 years old.

The REED and WATERBUCKS (subfamily Reduncinae) are classified into two tribes (waterbucks and grey rhebucks) with six genera, eight species, and 43 subspecies. They range in size between the roe and the red deer. The horns, which only ♂♂ have, are from short to long, straight, and crescent-shaped or curved like an S. In some animals, there is a relatively small neck mane and a throat and cheek beard. Coloration of the sexes is identical at times, although at other times it varies distinctly. The females have four nipples. They live near water in savannahs and forests, in swamp regions, and in high grass-steppe or rocky hill country.

Subfamily: Reedbucks and waterbucks

Distinguishing characteristics

The WATERBUCK (*Kobus ellipsiprymnus;* Color plates p. 402, p. 413). The HRL is 180–220 cm, the TL is 22–45 cm, the BH is 100–130 cm, and the weight is 170–250 kg. The coat is strand-like and there is a neck mane; the tail reaches almost to the ground. The horns are simply curved and far apart at the base. Among the 13 subspecies are the DEFASSA WATERBUCK *(Kobus ellipsiprymnus defassa)* and the COMMON WATERBUCK *(Kobus ellipsiprymnus ellipsiprymnus),* both of which were formerly considered to be separate species.

The waterbucks are rather stationary and do not actually migrate, although the groups, in the Serengeti for example, move considerably, depending on the changes in the rainy and dry seasons. The territorial males occupy large areas, so the neighbors generally have no contact with each other. The female groups, which remain with these males for some time, associate in separate groups or in large herds, sometimes numbering 30 or more animals during the dry season. Although births occur the whole year round, they increase in number at the beginning of the rainy season, as is true for many animals of the African bush or steppe.

Waterbuck

After the male has determined a female's estrous condition by stimulating her to urinate and "flehmen" (lip-curl, see Ch. 11), he begins to drive her, resting his lower jaw and throat on her croup. At the same time, or independently of this, the male kicks his front leg in the direction of her hind legs. Sometimes, reversed and parallel standing may also occur. The female responds by carrying her head and neck stretched forward horizontally or deeply lowered forward and down. Concurrently, she snaps her mouth "into empty space" in front of her. The same posture and biting movements also occur when a female passes close to a strong male. Since adult males do not behave this way, the animal identifies herself as a female and "asks for permission to pass."

Before a fight, the opponents stand opposite each other with their heads deeply lowered. They fight forehead to forehead, pressing

▷
The gerenuk *(Litocranius walleri,* see Ch. 15) eats leaves. Often they rise on their hind legs for browsing.
▷▷
The Frankfurt Zoo has successfully bred gerenuks. This species was formerly considered especially hard to keep. Modern feeding methods now make it possible to provide for their specialized food requirements.

◁
Dwarf antelopes

1. Oribi (Ourebia ourebi)
2. Klipspringer (Oreotragus oreotragus)
3. Suni (Nesotragus moschatus)
4. Steinbok (Raphicerus campestris)
5. Royal antelope (Neotragus pygmaeus)
6. Kirk's dik-dik (Rhynchotragus kirki)
7. Beira (Dorcatragus megalotis)

Buffon's kob

◁
True antelopes

8. Springbuck (Antidorcas marsupialis)
9. Impala (Aepyceros melampus)
10. Blackbuck (Antilope cervicapra)
11. Slender-horned gazelle (Gazella leptoceros)
12. Pelzeln's gazelle (Gazella pelzelni)
13. Henglin's gazelle (Gazella tilonura)
14. Red-fronted gazelle (Gazella rufifrons)
15. Thomson's gazelle (Gazella thomsoni)
16. Speke's gazelle (Gazella spekei)
17. Mountain gazelle (Gazella gazella)
18. Dorcas gazelle (Gazella dorcas)
19. Soemmering's gazelle (Gazella soemmeringi)
20. Robert's (Grant's) gazelle (Gazella granti robertsi)
21. Grant's gazelle (Gazella granti granti)
22. Dama gazelle (Gazella dama)
23. Gerenuk (Litocranius walleri)
24. Dibatag (Ammodorcas clarkei)

against each other with their horns crossed. They may also stand like a team of horses beside each other and grapple with each other with hooked horns to test their strength. When the opponents are of unequal strength, the weaker changes to this method where the stronger one may not use his full body weight. Ullrich once observed two waterbucks starting a fight while they were standing in a creek. One jumped at the other and tried to press him below the surface. But this probably was an exception.

Disturbed and pursued waterbucks often take to the water. As an adaptation to frequent and lengthy stays in the water, their long, stranded hair seems to be "lubricated" by skin glands. When this hair is stroked, the fingers are extensively covered with a thin, brownish, tarlike grease that has a distinct, turpentine-like odor.

A close relative is the BUFFONS'S KOB (Adenota kob; Color plate p. 402). The HRL is 125–180 cm, the TL is 18–40 cm, the BH is 70–105 cm, and the weight is 50–120 kg. The hair is smooth, there is no mane, and the tail only reaches halfway to the heels. The rather short and thick horns, which are lyrate and curved in an S-shape, are close together at the base. There are 13 subspecies, among them the WHITE-EARED KOB (Adenota kob leucotis) and the UGANDA KOB (Adenota kob thomasi).

Buechner, Schloeth, and Leuthold have studied the Uganda-kob's territorial behavior very thoroughly. These animals form herds consisting of females, bachelors, and territorial males. Depending on the population density and local conditions, the territories are either sporadically scattered or concentrated in a very small space. Such "arenas" are in the hill country near creeks where the grass is short so the males are clearly visible. The males do not mark with excrement and gland secretions, but by displaying themselves and by frequent whistling. In the Semliki Game Reserve where Buechner did most of his research, one arena, which is approximately 200 meters in diameter, consists of 12 to 15 almost circular single territories, each of which is 20 to 60 meters in diameter. Some territories have joint boundaries, while others are separated by "neutral zones." Since the grass in the center of each territory is grazed off and trampled down, the boundaries are strikingly outlined by the higher grass.

Some males stay in such a territory for only one day, and others for several days, weeks, or even months. One to two times per day each male leaves his place to drink and to eat. If another male intends to conquer a territory, he rushes from the boundary to the center of an arena. He then seems to aim for a specific territory, perhaps one which he once had occupied before. Sometimes the attacker succeeds in driving the owner from his territory and occupying it himself, but more frequently the rightful owner remains master of his field. Often a male may dash to the center of the mating arena, where he is threatened by the owner and chased out again. The territorial males do not fight each other but defend their territories by displaying and threatening. The fighting stage is reached only when a courting male mistakenly comes too close or enters the territory of a neighbor. Usually the male will remain at the border when a female changes from his territory into another.

According to Buechner's observations, females in estrous are the ones to come into the territories. The male then walks towards her in his display walk with his neck erect and bent backwards; then he lip-curls, sniffs her urine, kicks with his front leg, and mounts. After copulation, he whistles, licks his penis, and sniffs extensively at her flanks. Such a "sequel" to the mating ceremony is not known in any other horned ungulate.

Even more dependent on water and swamp areas is the LECHWE WA-TERBUCK (*Hydrotragus leche*; Color plate p. 402). The HRL is 130–180 cm, the TL is 30–40 cm, the BH is 85–105 cm, and the weight is 60–120 kg. The coat hair is stranded and the tail reaches almost to the heels. The horns, which are thinner and longer than in the buffon's kob, are far apart between the tips. The long, narrow hooves have large, triangular and pointed pseudo-claws, which may be spread far apart. There are three subspecies.

The lechwe waterbucks always live at the edge of water, in the flooded and swampy marshes near rivers and lakes, in reed jungles or swampy meadows. They follow the water level; when it rises and wide areas are flooded, they disperse over the marshy ground, and at its crest they may even go onto the surrounding hard ground. When the water level drops again, they gather in low grounds and lagoons. They eat swamp grass and aquatic plants in the shallow water. Only when they rest do they go to dry banks, sand bars, or islands. In flight they take off with their heads and necks stretched foward; the gait at first is a trot, then it becomes a gallop and they leap high over obstacles. They predominantly take to the water when they are frightened, and they are enduring swimmers, although they start to swim only when their long, narrow hoofs can no longer reach the ground.

The peak of the mating season is from October until January. After a gestation period of seven months, the females give birth in a dry area to one young, and they nurse it for three to four months. The infant mortality is as high as 50 per cent and increases at the end of the first year, allegedly from box flies. Usually the lechwes live in groups which range from five to 50 animals, although much larger herds are not rare.

The most distinct color differences between the sexes, perhaps in any horned ungulates, are found in MRS. GRAY'S WATERBUCK (*Onototragus megaceros*; Color plate p. 402). The HRL is 135–165 cm, the TL is 45–50 cm, and the BH is 80–105 cm. The horns are similar to those of the lechwe waterbucks, but they are thinner, longer, and more steep, with tips that are very far apart. The old ♂♂ are dark on top with many white mark-

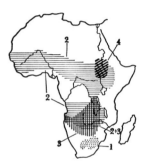

Fig. 14-11.
Waterbuck *(Kobus ellipsi-prymnus)*

Fig. 14-12.
1. Grey rhebuck *(Pelea capreolus)*. 2. Buffon's kob *(Adenota kob)*. 3. Lechwe waterbuck *(Hydrotragus leche)*. 4. Mrs. Gray's or Nile lechwe *(Onototragus megaceros)*.

ings; the females are much lighter with less distinct markings.

When I saw Mrs. Gray's waterbucks for the first time, I almost doubted that they were both from the same species. The adult males have large ears and a mahogany color with a pure white fist-sized patch behind their horns. From this patch a white band passes along the neck and turns into a broad white saddle at the whithers. This conspicuous marking is completely absent in the pale, yellow, and hornless females. At approximately one year, the young males still resemble the females in color. The color change is completed by about two to three years, but the white markings, which form at a later age, are not yet visible. The signficance of this conspicuous marking is not quite clear yet. Since these waterbucks also stand opposite each other with their heads lowered before a fight, this visible pattern may well be effective as a signal.

The mating ceremony of this species corresponds to the general "waterbuck-scheme." Only at the beginning of courtship does a rather strange, at least from the human observer's point of view, almost repellent action occur, which is found only in Mrs. Gray's waterbuck. The male lowers his head, as if he were going to dig the ground with his horns, and he urinates in a jet between his front legs into the extended hairs at his throat and cheeks. With his beard dripping, he walks over to the female and rubs his urine on her forehead and croup. This marking may have the same significance, although it is very different from the marking of females in gernuks and the dibatags (see Ch. 15).

All REEDBUCKS (genus *Redunca*) share a common mark, a patch below the base of the ear between the size of a nickle and an egg, but nothing is known about its significance. Only the ♂♂ have relatively small, crescent-shaped horns which are bent from the base backward and up, and then forward and up. There are three species: 1. REEDBUCK *(Redunca arundium)*, with an HRL of 120-160 cm, a TL of 18-30 cm, a BH of 65-105 cm, and a weight of 50-95 kg. The horns are evenly curved in a wide arc. The pseudo-claws are not fused. There are two subspecies. 2. BOHOR REEDBUCK *(Redunca redunca;* Color plate p. 402), with an HRL of 115-145 cm, a TL of 15-23 cm, a BH of 65-90 cm, and a weight of 35-65 kg. The upper part of the horns are strongly bent. The pseudo-claw bases are directed towards each other and they are completely or almost fused. There are seven subspecies. 3. MOUNTAIN REEDBUCK *(Redunca fulvorufula;* Color plate p. 402) with an HRL of 110-125, a TL of 17-26 cm, a BH of 60-80 cm, and a weight of 20-30 kg. This is the smallest species, with short horns which are curved only slightly in the upper part. The pseudo-claws are fused.

In contrast to the two other reedbuck species, the mountain reedbuck has a rather scattered area of distribution. It occurs only sporadically on isolated habitats in Cameroon, Northeast and Southeast

Reedbucks

Distinguishing characteristics

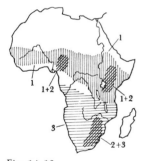

Fig. 14-13.
1. Bohor reedbuck *(Redunca redunca).*
2. Mountain reedbuck *(Redunca fulvorufula).*
3. Reedbuck *(Redunca arundium).*

Africa. While the two other species always live in the vicinity of water, the mountain reedbuck can go for a remarkably long time without water in his hilly to mountainous habitat. The males as well as females live singly, in pairs, and in small groups of five to 15 animals. All reedbuck species often lie down for long periods, and they even lie down when they are disturbed. They flee only in cases of imminent danger. Frequently they also take off in high jumps, always after uttering a piercing whistling sound to which other animal species in their habitat respond. In their mating ceremony, they are said to perform high jumps. As in many other horned ungulates, the courting male kicks with his front legs at the female. The gestation period is approximately seven months. Reedbucks are not very frequently seen in zoos.

The GREY RHEBUCKS have been separated from the water bucks and reedbucks and classified as a separate tribe, the Peleini. There is only one genus and species, the rhebucks (*Pelea capreolus;* Color plate p. 402). Their size is that of a roe deer. The HRL is 115–125 cm, the TL is 15–30 cm, the BH is 70–80 cm, and the weight is 20–30 kg. The horns are rather short, thin, almost smooth, straight, and vertical. They are of slender, graceful build, as the white-tailed deer, with a straight back and the croup slightly higher than the withers. The medium-large head is long and slender. The short, soft, dense, and woolly coat is like a rabbit's, and there is no beard or mane. Longevity in the wild is eight to ten years.

Here again is another misleading Boer name. Except for being approximately of the same size, this animal has nothing in common with the European roe deer. Like the mountain reedbuck, the grey rhebuck prefers mountainous and rocky areas with brush as a habitat. It lives in small groups of six to ten, or sometimes 20 to 30 animals. Although the grey rhebucks are rather shy and are becoming rarer as time goes on, they often stay close to human settlements. In flight they kick up their hindquarters with their tails erect. During the mating season, males have vehement fights with each other. After a gestation period of presemably six to seven months, two young are born in November and December. Details on the habits of these small antelopes are not known.

Fritz Walther

Grey rhebucks

Distinguishing characteristics

15   The Gazelles and Their Relatives

Subfamily:
Antilopinae, by
F. Walther

Distinguishing
characteristics

The dama gazelle

In the hunting and travel literature, gazelles and antelopes are usually treated as if they were different forms of animals. But according to an older classification, the gazelles belong with the antelopes, and in a more recent classification they are considered the only "true" antelopes.

The TRUE ANTELOPES (subfamily Antilopinae) are not quite as large as the white-tailed or fallow deer. The body and legs are slender and the back is straight or slightly higher at the croup. In some subspecies, only the ♂♂ have horns, while in others the ♀♀ have them too, but they are weaker. There are many skin glands on the body. The sexes usually have the same color (an exception is the blackbuck). They inhabit deserts, semi-deserts, short grass steppes, and steppes with trees and dense bush. Their diet consists of herbs, grass, leaves, buds, and shoots. There are six tribes (gazelles, blackbucks, gerenuks, dibatags, springbucks, and impalas) with 7 genera, 19 species, and 72 subspecies.

Most GAZELLES (Gazella) species are within the tribe Gazellini. The HRL is 85—170 cm, the TL is 15—30 cm, the BH is 50—110 cm, and the weight is 12—85 kg. Both sexes have horns which are medium-long to long, they are almost straight or curved like an "S". All horns have rings. The rather small head has a narrow muzzle. The eyes are relatively large, and the indented preorbital glands are well developed. There are three subgenera: Nanger, Gazella, and Trachelocele.

In the subgenus Nanger, the white hair patch on the anal region reaches upward over the base of the tail onto the back. We distinguish three species: 1. DAMA GAZELLE (Gazella dama; Color plate p. 422); 2. SOEMMERING'S GAZELLE (Gazella soemmeringi; Color plate p. 422); and 3. GRANT'S GAZELLE (Gazella granti; Color plate p. 422). The ranges of these three species are adjoining with small, overlapping zones (see fig. 15-1).

A genuine desert animal, the dama gazelle inhabits all of the Sahara from east to west, and the Sudan. However, of its five subspecies, two have already been exterminated, one of which is the MHORR-GAZELLE (*Gazella dama mhorr*) from Morocco. The rest, including the NUBIAN RED-NECKED GAZELLE (*Gazella dama ruficollis*), are greatly endangered.

Dama gazelles migrate according to the seasons. In the rainy season, they move north far into the Sahara, and in the dry season they wander back to the Sudan. They live in small groups of 10 to 30 animals, although during the rainy season they may form herds of 100 to 200 head. Since adult males are occasionally by themselves, this species may have seasonal territories. Unfortunately, no other details are known about the habits of these gradually disappearing animals.

In Ethiopia, Soemmering's gazelle has fared somewhat better. It lives in the bush-steppe region in small groups of five to 20 animals, or sometimes even in larger herds. Observations from captivity strongly suggest that the males become territorial at times. They are quite aggressive before mating and drive off any animal which is considered a potential rival. They defecate in a low squatting position at specific places, which in the wild could be a territorial marking. However, marking with the preorbital glands has never been observed. After initially chasing the female, the male approaches her with his neck and head stretched forward. He continues this advance until she tolerates his approach without fleeing. Then he drives her while walking with his head highly erect. Finally, he stands erect, at right angles to her, or in rarer cases behind her or reversed-parallel, and he mounts as soon as she starts walking. She moves her tail horizontally to the side and lowers her croup. Copulation takes place, as in all gazelles, while they are walking, with their heads carried high, and the male with his front legs angled towards the body.

We are somewhat better informed today about the East African Grant's gazelle. It lives in the open steppe as well as in the bush. Relative to their size, Grant's gazelles have extremely long and powerful horns. The tips are farthest apart in the Grant's gazelles of the Serengeti (*Gazella granti robertsi*). As food Grant's gazelles prefer herbs, leaves, and shoots. In the wide Serengeti steppe during the dry season, herds of up to 200 and even 400 animals, which consist of approximately equal numbers of males and females, may be found. In bush areas and during the rainy season, the herds, which break up into groups of males and females, are much smaller.

Single adult males become territorial, and groups of five to 20 females stay with them for several weeks or months. As "harem groups," they roam together with the male through his very large territory. Such Grant's gazelle territories have diameters of from 500 meters to almost two kilometers. The males have functioning preorbital glands, but they do not mark with them. Instead, they mark their territories by urinating, defecating, and constant beating of the grass and bush with their horns. Then they move their head to the right and

Soemmering's and Grant's gazelles

Fig. 15-1.
1. Dama gazelle (*Gazella dama*). 2. Soemmering's gazelle (*Gazella soemmeringi*). 3. Grant's gazelle (*Gazella granti*).
4. Springbuck (*Antidorcas marsupialis*).

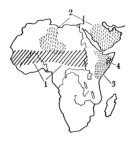

Fig. 15-2.
1. Red-fronted gazelle
(Gazella rufifrons).
2. Slender-horned gazelle
(Gazella leptoceros).
3. Gerenuk (Litocranius
walleri).
4. Dibatag (Ammodorcas
clarkei).

Fig. 15-3.
1. Dorcas gazelle (Gazella
dorcas).
2. Speke's gazelle (Gazella
spekei).

the left, to and fro, as if they were "weaving" (Fig. 11-24). The males in the bachelor herds occasionally do the same thing, especially in displays and before fighting.

Migrating herds march in single file. In the male groups, the youngest and lowest ranking animals often walk ahead, and the older males walk in the middle or at the rear. Before they start or when they meet another migrating herd, the males display and fight. The fight may last for almost two hours until the two groups separate again, and then an exchange of members may take place. Roaming bachelors will also join the female groups which are with territorial males. When threatening, male Grant's gazelles tilt their horns sideways and circle around each other. When the opponents do not "feel quite secure," they may perform displacement activities such as grooming, scratching, shaking, and grazing. In Grant's gazelles, grazing during the fighting ceremony is a clear indication of inferiority.

Male Grant's gazelles' fights consist predominantly of a ceremonial forehead pushing with the horns crossed over, although occasionally exchanges of blows may occur. The males display by walking towards each other. At a distance of approximtely ten meters, they turn the head out and to the side, pointing with their inclined horns towards each other. They approach to within one to three meters and, as soon as the front legs of one are on a level with the other's rear, both animals stand still and stretch their heads upward. Then with a simultaneous jerk, both turn their noses towards each other so that their white throats flash and the neck muscles are strained like the muscles of a person's upper arm when the arm is flexed. This is repeated several times. In about half of the cases, this display ends with one opponent walking off. Only in approximately 25 per cent of the cases will a fight follow this display. Thus the display often is a substitute for an actual fight. Females neither fight nor display as often as the males. Older males have fewer fights than young ones, but they display much more often.

During courtship, the male, with his nose held forward and upward and the tail stretched out horizontally, drives the female by walking behind her and uttering certain calls. After this mating march, the female stands still. At first the male also stands still; then he walks a few steps towards her and she moves for several meters and then stops. This is repeated for some time, and the male makes movements with one front leg which is like "foreleg-kick" in other species. When the female is ready for copulation, she displays by stretching her head upward and turning it to the side "into the empty space." After many mounting attempts he successfully copulates while they are walking.

To my knowledge, only one birth of a Grant's gazelle in the wild has ever been recorded to date. The mother alternately stood and reclined. All the difficult phases of birth, such as the expulsion of the head and shoulders, were performed lying down. The final expulsion was begun in a prone position and completed standing up. The birth lasted less than one hour. Twenty minutes later, the young was standing on its legs, and after two more minutes it nursed for the first time. The mother obviously was an experienced, older female, because she directed the young with soft butts of her nose straight to her udder. Forty-one minutes after birth, the young, led and guarded by the mother, left the place where it was born.

The young "stay put" in bushes and in high grass, and depending on the location the upper third of a hill. The mother keeps the hiding place under constant surveillance from a distance of 50 to 300 meters. She chases off jackals who may come close to the hiding place. When she sees a hyena from a long distance, she runs towards the young and passes it. Then the young will get up and follow her. She leads it around the top of the hill and returns alone after a while to watch the further progress of the hyena. Thus the mother behaves completely different with respect to these two predators.

The young is always called to nurse by the mother. When travelling, the young walks at first behind, then beside, and finally ahead of the mother. When the two walk over a great distance, they continually pass each other, although sometimes the young walks a larger part of the trip in front of the mother. Thus, even though there is in Grant's gazelles a slightly modified following reaction, the life of the young Grant's gazelles during the first days and weeks is determined by the stay put behavior which is characteristic of this species.

In the smaller gazelle of the subgenus *Gazella*, the white patches on the rear are not as large. There are eight species: 1. The MOUNTAIN GAZELLE (*Gazella gazella*; Color plate p. 426) has been largely exterminated. 2. The DORCAS GAZELLE (*Gazella dorcas*; Color plate p. 426) is endangered. 3. The SLENDER-HORNED GAZELLE (*Gazella leptoceros*; Color plate p. 426) is partially extinct. 4. The RED-FRONTED GAZELLE (*Gazella rufifrons*; Color plate p. 426). 5. THOMSON'S GAZELLE (*Gazella thomsoni*; Color plates p. 426, p. 435). 6. HEUGLIN'S GAZELLE (⚧ *Gazella tilonura*; Color plate p. 426) is endangered. 7. SPEKE'S GAZELLE (⚧ *Gazella spekei*; Color plate p. 426), and 8. PELZELN'S GAZELLE (⚧ *Gazella pelzelni*; Color plate p. 426) are both endangered.

Thus six of these gazelle species are seriously endangered. This applies especially to the species of North Africa, Arabia, and Asia Minor which, due to ruthless hunting, disappeared some time ago from many areas of their former distribution. The dorcas gazelle has been known since ancient times as *the* gazelle. At the end of the last century when it was still more abundant, Brehm and Chavanne described it in these beautiful words: "Repeatedly we noticed solidly packed, narrow tracks which led from one depression in the ground to another. By chance we found the explanation the next morning. In

▷

A male Thomson's gazelle *(Gazella thomsoni)* in the Serengeti.

▷▷

Impalas *(Aepyceros melampus)* are known for their elegant leaps. Simon Trevor photographed a female at the moment when she jumped over another female.

Gazelles in the narrow sense

The dorcas gazelle

1

2

3

♀

5

♂

F. Walth.

◄

Genus *Procapra*

1. Tibetan gazelle *(Procapra picticaudata)*
2. Mongolian gazelle *(Procapra gutturosa)*

Gazelles

3. Tibetan antelope *(Pantholops hodgsoni)*.
4. Goitered gazelle *(Gazella subgutturosa)*.
5. Saiga *(Saiga tatarica)*.

Fig. 15-4.
1. Mountain gazelle *(Gazella gazella)*.
2. Pelzeln's gazelle *(Gazella pelzelni)*.
3. Heuglin's gazelle *(Gazella tilonura)*.
4. Thomson's gazelle *(Gazella thomsoni)*.
5. Goitered gazelle *(Gazella subgutturosa)*.

The Speke's gazelle

the distance we saw a small group of slender gazelles who ran in single file along these paths. In the hammada, the waterless, rocky plateau of the Sahara, which is covered with sharp pebbles, they walk exclusively on such narrow trails, which they themselves made since centuries ago. This way they do not hurt their slender feet on the sharp pebbles. In the surrounding areas and in the mimosa-covered wadis of the steppe zones, in the sahara, and even on the hammada, man's path often crosses that of the gazelles. Man is their worst enemy because their meat is a much appreciated delicacy, and for a traveller in the desert who has endured several days of hardships, it means a delicious meal. The leopard, wild dog, lion, jackal, and the ever-present desert birds, the vultures, all prey upon them. The sight of a herd of gazelles in the desert is so enchanting and so graceful, that the poets of the Orient celebrated it with all the ardor of their souls. The stranger from the Occident who sees them in the wild is also deeply impressed by the beauty of their appearance, the gracefulness of their movements, and the charm of their tameness and tenderness. He is also caught up by the words of passionate praise for this animal in the fluent rhymes. The poet compares the eye which makes a Bedouin's heart glow to that of the gazelle's. The poet of the *Song of Songs* sings of the gazelle's neck and of the gazelle's twins, and the religious hermit finds in the graceful daughter of the desert a visible symbol which explains the heart's longing for the sublime."

In the Kronberg Zoo, I made friends with a rather aggressive male dorcas, which I called "Sputnik," by bringing my face close to his in the manner of gazelles, and opening and closing my lips in a rapid sequence. He immediately nibbled in the same manner and continued this eagerly and persistently long after I had stood up again. I was quite surprised to see that he not only moved his lips, but also the cheeks and, above all, the bottom of the lower jaw. Apparently, gazelles transport scent materials to their "Jacobson's organ," a duct in the upper part of the oral cavity which is lined with an olfactory mucus membrane, with these munching movements of the lips and the suckling-chewing movements of the lower interior of the mouth. This also aids in the recognition of an individual's scent. From then on Sputnik clearly distinguished me from all other people. Furthermore, nibbling not only promotes becoming acquainted, but, especially in combination with stretching the head forward, it is a gesture of appeasement and of friendship.

Obviously Speke's gazelle is rather closely related to the dorcas gazelle. Across the nose of the dorcas gazelle is a fold of skin, which males as well as females inflate and turn it into an air regulator or an amplifier. During the alarm call, it vibrates and sounds like the quacking of a large duck. This is much further developed in the Speke's gazelle. Both sexes have three folds across the nose. When they make an alarm call, it sounds like a pistol shot and these folds are inflated to the size of a tennis ball.

From zoos as well as from field research, we know most about the EAST AFRICAN Thomson gazelle. Like all gazelles and many other steppe antelopes, the "Tommies" walk in an amble but change into crosswalk as soon as they trot. Especially characteristic for them, as well as for all true antelopes, are the "stotting jumps." Thomson's gazelles groom and scratch themselves as do other horned ungulates, however, they do not roll or rub themselves against rocks or trees. Urination and defecation, which are linked in the males, are used to mark the territories. Excited gazelles regularly initiate urination and defecation with extensive ground pawing. Migrating males and females walk in mixed herds. In the Serengeti where Thomson's gazelle, with an estimated 500,000 head, are the most abundant game animal, they may often be seen by the hundreds and thousands.

Thomson's gazelle

Migrations of Thomson's gazelle

During the rainy season, they inhabit the open short grass-steppe where they apparently get enough moisture with the food they eat. In the dry season, they withdraw into the bush areas where they go rather regularly to water holes. They move into the bush only as far as it is necessary to find fresh greens and water. They are ready to return to the open steppe as soon as the first rains fall there. Although the Tommies live in herds, they are not really cohesive associations. The herds constantly split up into smaller groups, which then form into new groups, or they merge with others. When the animals do not migrate, these mixed associations split up into herds of females, males, and territorial males. Each territorial male has a territory of 100 to 200 meters in diameter, and these territories cover wide areas, one next to the other.

Territorial males

When the female herds come into the males' territories on their daily excursions, each male tries to keep a group of females there for a couple of hours. If one of the females is in the appropriate phase of estrous, he often separates her from the group. Since he has previously driven all the females toward the center of his territory, such a pair then frequently moves to its boundary. Females with very young offspring which still "stay put" separate from the others. They remain close to the young and do not join the moving groups and herds, but often stay in the "no-mans land" between the territories.

The territorial neighbors often fight with each other, but then the issue is usually the maintenance or alteration of boundaries and not chasing another from his territory. Bachelors, which are passing through, usually are driven off by the territory owners without fighting. In their fights, they may push with their foreheads against each other, wrestle from the side, or use still other methods, and exchange blows. The opponents jump towards each other, clash together, jump back, and repeat the process. Territorial males will then turn away

Fights with neighbors

from each other, begin to graze, and return into their own territory. When threatening, the male Tommy presents his horns with his head held high.

Reproductive behavior

When the male approaches the female in the mating ceremony, he first stretches his neck and head horizontally forward. With increased driving, he lifts his nose forward and up. These two actions may alternate. At the peak of the mating prelude the male starts drumming with his front legs; after that, driving becomes more relaxed. In the mating march, both animals walk closely together, one behind the other. When the female stands still, the male touches her from behind with his stretched-out forelegs. Mounting and copulation take place, like in all gazelles, while the pair is walking. If a female is not quite ready for copulation, the male may actually chase her. Then she usually dashes out of his territory, but she is immediately overtaken by the neighbor who continues to chase her. This may continue through several territories, until the female is exhausted. Interestingly enough, by then she has reached the peak of estrous which is indicated by the horizontally stretched-out tail. Then she is driven more slowly and copulation takes place. This guarantees that each female, in the first period of heat two to four weeks after giving birth, becomes pregnated again. Thus, with a gestation period of five to six months, Thomson's gazelles may have young twice a year. The territorial males play a crucial role here. Such a high birth rate is necessary to compensate for the many casualties due to predators.

The cheetah as a gazelle hunter

Of the many predators, the cheetah is probably the most specialized gazelle hunter. In most cases, it tries to approach to within 200 meters of the gazelles without being noticed by them. When the gazelles become aware of this and look in its direction, it chases after them. Even though gazelles may attain speeds of 60 kilometers per hour or more, the cheetah easily catches up with them. From the beginning the cheetah selects a specific animal and ignores the others. It may pass them, they may run beside it or cross in front of it, but it does not pay any attention to them. The animal it has selected zig-zags when it approaches it, and with some luck may escape by this method. In most cases, however, the cheetah in a rapid gallop strikes the gazelle with a blow of its forepaw, throws himself on it, and kills it by biting into the throat.

The goitered gazelle

Exclusively restricted to Asia are the GOITERED GAZELLES (subgenus *Trachocele*). There is only one species: the GOITERED GAZELLE (*Gazella subgutturosa*; Color plate p. 438). The ♂♂ have an enlargement of the throat and a thickening of the neck during the mating season which resembles a goiter. The ♀♀ are hornless or occasionally they have very small reduced horns. The dark pattern on the forehead and nose region occurs only when they are young. The mating season is from November until the beginning of January. Births, mostly of twins or more rarely three or four young, occur in April and May. Because of their colder habitat they give birth only once a year. For a distribution map, see fig. 15-4.

In captivity, the Eastern subspecies tolerates the Central European climate rather well, and they may stay outdoors all winter long. I vividly remember the terror I felt when I went to the pen of the goitered gazelles in the Kronberg Zoo after a night of heavy snowfall and found it empty. When I entered into the pen, the riddle was solved. The animals had been snowed over and only the tips of their noses stuck out of the snow. In recent years, the Persian goitered gazelle (*Gazella subgutturosa subgutturosa*) has been more frequently displayed in zoological gardens.

Two species of gazelles rather unknown in Europe which come from Central Asia are classified in the genus *Procapra*. The HRL is 95–148 cm, the TL is 2–12 cm, the BH is 54–84 cm, and the weight is 20–40 kg. Only the ♂♂ have medium-long, lyrate or S-shaped horns with ridges across. The coat is short and smooth, while in the winter it is dense and woolly. There are two species: 1. TIBETAN GAZELLE (*Procapra picticaudata*; Color plate p. 438) which is larger, ♂♂ without "goiter." 2. MONGOLIAN GAZELLE (*Procapra gutturosa*; Color plate p. 438), which is larger and ♂♂ with "goiter."

The mating and birth seasons of these two species correspond approximately to those of the goitered gazelle. While twins occur rarely in the Tibetan gazelle, they are the rule in the Mongolian gazelle. Except for the summer mating season, the males and the females form separate herds, while in the winter, they form mixed herds and migrate over great distances. For resting places, they paw small depressions into the ground. As Kleinschmidt recently found, the males of the Mongolian gazelle have, besides their well developed goiter, a large "goulla" as do camels. Nothing is known about its function nor about other details of their behavior.

The only member of the tribe Antilopini is the well known INDIAN BLACKBUCK or SASIN (*Antilope cervicapra*; Color plate p. 426). The HRL is 100–150 cm, the TL is 10–17 cm, the BH is 60–85 cm, and the weight is 25–45 kg. This animal is built like a gazelle. Only ♂♂ have the very long, straight horns which are loosely wound about the longitudinal axis with ridges across. There is great difference in coloration of the sexes.

Although the blackbuck's appearance differs somewhat from the gazelle's, its behavior shows that it is truly a gazelle. I know of no behavior which cannot be traced back to a common phylogenetic origin with the behavior of other gazelles. The blackbuck was the animal in which Hediger was able to demonstrate for the first time that ungulates mark with secretions from their preorbital glands. Prior to this, the zoologists had only observed that the excited male, especially while fighting and mating, would open the preorbital glands and

Tibetan and Mongolian gazelles

Fig. 15-5.
A male Thomson's gazelle is presenting his horns high up.

The blackbuck

Fig. 15-6.
The male Thomson's gazelle urinates in this position.

Fig. 15-7.
...and in this one he defecates.

Gerenuk and dibatag

Distinguishing
characteristics

The gerenuk

would even evert them. In the mating ceremony, the male approaches the female with his nose lifted foward and up. After an initial fast driving, a butting-driving follows where the male trips in place or takes a long step forward with one front leg. This same behavior is also found in Grant's gazelle. However, in the blackbuck, the steeply erect posture of the neck and head also occurs when the animal is displaying against an opponent. This is not as distinct in Grant's gazelle. The most frequent threat posture of the blackbuck is a flip with the horns, where he jerks his head downward, so that the tips of his horns point toward the opponent. Then he raises his head again. This, like other behaviors of the sasin, also occurs in gazelles, but not as frequently and perhaps not quite as emphasized.

Unfortunately, according to the latest information, the blackbuck's situation in its native habitat is disturbing. The populations, which have decreased due to severe hunting, have disappeared in some areas. Only in a few Indian provinces is the sasin still considered sacred to the moon, and therefore, it is protected. It is easy to imagine for how short a time this protection will remain effective in modern India because of old customs and religious ideas. Fortunately, the blackbucks reproduce quite well in captivity. However, not all zoos have realized the dangerous situation for this species. In some zoos, the sasin is considered "too ordinary," and because of a lack of space they even dispose of them in favor of other more unusual species of animals.

In the GERENUKS (tribe Litocraniini) and DIBATAGS (tribe Ammodorcatini), the neck and legs are strikingly thin and long. Only the ♂♂ have medium-long horns with bulges across them. There is only one species in each tribe: 1. GERENUK (*Litocranius walleri*; Color plates pp. 423-4-5-6), with an HRL of 140-160 cm, a TL of 25-35 cm, a BH of 90-105 cm, and a weight of 35-52 kg. The animals are dainty, with a straight back, and a small narrow head, long and flat. There is a narrow muzzle with very movable lips. The horns are curved from an S or a lyrate form. 2. DIBATAG (*Ammodorcas clarkei*; Color plate p. 426) with an HRL of 152-168 cm, a TL of 30-36 cm, a BH of 80-88 cm, and a weight of 22.5-31.5 kg. They are smaller than the gerenuk, and have crescent-shaped horns which are sickle-shaped and bent forward.

The gerenuk lives in small groups of usually no more than 12 animals exclusively in bush regions, where it feeds on leaves, buds, and shoots. While feeding, it often rises vertically on its hind legs (Color plate p. 423). It balances so well that it may stand without any support, although it usually leans with its front legs against a tree trunk or a branch.

Gerenuks are not kept very frequently in zoos. The Frankfurt Zoo has been able to keep a group of this species for more than ten years. They were bred successfully from only one initial pair. Most of our present knowledge on the habits and the behavior of the gerenuk comes from these Frankfurt animals. The gerenuks are quite different in some respects from other gazelles. For example, the Frankfurt gerenuks have never drunk water. Certainly, many species of gazelle in the wild may go for long periods without drinking, but they usually drink in captivity and do it regularly, if not in large quantities. Furthermore, the male gerenuk marks his female during the mating ceremony with the preorbital glands, especially at the shoulder region and the hindquarters. In this case, it seems to be true marking and not a "substitute action," as is occasionally found in captive red-fronted and Thomson's gazelles who deposit marking at the ear or horn tips. The shape of the shoulder region and hindquarters of the female are hence no substitute for twigs or stems, but they are the parts of the body which conspecifics sniff frequently in social contact. The gazelle's foreleg kick (Laufschlag) reaches extreme ritualization in the gerenuk. The male lifts his stretched front leg in an almost "solemn" manner almost to a horizontal level.

Fig. 15-8.
A courting blackbuck shows the "head-stretch-up" and the "long-foreleg-step."

The director of the Dresden Zoo, Wolfgang Ullrich, reports from the wild that the male's neck, probably during the mating season, becomes conspicuously thicker, especially in the upper third. This was not observed in the Frankfurt gerenuks. When gerenuks are disturbed they stand motionless at first with their necks erect. From my own experience, I know how difficult it is to detect the animals in the bush. When the Frankfurt pair arrived at the zoo, the female was less familiar with humans than the male. Therefore, during the first days, she stood in the pen almost constantly in this posture with her neck erect although there was no cover. The male adjusted well and walked about normally. Although they were only a few meters from the visitors, half of them saw only one animal in this pen. In the wild, a gerenuk may remain for a long time in this posture. Several times I approached solitary animals as close as 30 meters. Although they had been aware of my presence for a long time, they did not flee. When something dangerous approaches too closely, the gerenuk flees in a gallop or a trot. If he is not pressed too hard, he will soon stop behind some bushes. It is also said that males sometimes "sneak off" with the head and neck stretched forward.

Fig. 15-9.
1. Mongolian gazelle (Procapra gutturosa)
2. Tibetan gazelle (Procapra picticandata)
3. Blackbuck (Antilopa cervicapra)

Even rarer in captivity than the gerenuk is the dibatag. In Europe, only the Naples Zoo kept some dibatags for several years. Like the gerenuk, the dibatag is also a distinctive ambler, and it gallops only rarely. In flight, it usually carries the neck and tail erect, and moves off slowly at first and then changes to a trot. The long black tail is a very conspicuous flight signal when it is erect. The dibatag also rises upon its hind legs for feeding as does the gerenuk. Males mark the females in the same manner with secretions from the preorbital glands which they apply especially to the hindquarters.

The dibatag

Urination and defecation places are of great importance to male

Fig. 15-10.
A male gerenuk does a "foreleg kick" while courting (in this case from the side).

The springbuck

Distinguishing characteristics

What do their high jumps mean?

The migrations of the springbucks

dibatags. There he behaves in the same manner as the dik-dik (see Ch. 12). In the pair at the Naples Zoo, the lip-curl (flehmen) has evolved into a regular ceremony. At a specific place at a certain time of day, the male and female would alternatingly and in immediate succession allow the other's urine to run over their noses, and then they would show the lip-curl. The foreleg kick in the prelude of the mating ceremony is similar to the gerenuk's. The two animals at Naples would often nibble at each other's preorbital glands, as do many gazelle. Moreover, the dibatags of the Naples Zoo also never drank water in captivity.

The range of the dibatags is extremely small. According to Cuneo's observations, the animals occur mainly in places with red soil of a high aluminum content. The dibatag is not yet among the endangered species; nevertheless, it is necessary to protect them as soon as possible.

The oddest jumps in all true antelopes are performed by the SPRING-BUCKS (tribe Antidorcatini). There is only one species, the SPRINGBUCK (*Antidorcas marsupials;* Color plates pp. 391-2, p. 426), with an HRL of 120-150 cm, a TL of 20-32 cm, a BH of 68-90 cm, and weight of 18-45 kg. Both ♂♂ and ♀♀ have medium-long, lyrate horns with bulges across. The skin from the center of the back almost to the base of the tail is deeply folded, lined with many sebaceous glands, lined with white hair, and may be everted to form a conspicuous white dorsal crest. There are three subspecies, two of which have been almost completely exterminated. For a map of their distribution, see fig. 15-1.

The springbuck jumps by bringing his front and hind legs close together, similar to a gymnast who tries to touch the tips of his toes with his fingers while keeping his legs straight. It keeps its head low and arches its back. Then the back pouch opens and the white "mane" waves on the ridge of the arched back. In the sunshine, this is a blinking signal. When the predator is not too close, such jumps initiate the flight. They also may occur when the animal discovers something extraordinary or suspicious, for example, the tracks of car in deserted areas. These jumps, which always indicate that the animal is in a certain state of excitement, attract the conspecific's attention to the predator or object which had caused the leap. These jumps are "contagious," and "jumping excitement" often spreads like a wave throughout the herd.

Formerly, the Treckbokken (wandering bucks) of the Boers lived in South Africa in unbelievable numbers and migrated over great distances. Kretzschmar, a German physician, gave a vivid description in 1853 of such a herd of springbucks. He was on a farm in the Karroo when the approach of a large springbuck herd was announced. With a shooting party, he posted himself in a pass between the hills where the herd was expected to pass through. A hollow rumbling, which sounded like the surf of the distant ocean, preceded the animals. In order to avoid this flood of animals, the hunters hastily climbed rocks. Then the springbucks arrived—two, three, 20 of them, in groups increasing in size, and each bunch consisting of about 300 to 400 head; however there was still some space between them. Finally, a continuous solid white-brownish mass poured through the 800 yard-wide pass. Individual animals could no longer be distinguished as they became blurred in this living current of bodies. Without interruption, this flood of animals rolled along. There were no signs of the dogs which the springbucks had trampled in the first rush. An African was also seized in the whirlpool of the pushing animals. He tried in vain to escape, and for a moment he struggled like a drowning man in the flood before he disappeared in the turmoil of dust and bucks. Once again his head showed up, but it disappeared quickly as the crowded mass of bucks continued.

Kretzschmar's companions were Boers who knew how to count large herds of cattle. They estimated this one herd consisted of 25,000 head. Other travelers reported that even lions and leopards had been run over by the springbucks. The bucks, like a moving wall, had surrounded predators on all sides, and they were simply swept along until they were exhausted and were trampled to death. When springbucks came to gorges, hundreds of them often fell, pushed from behind, over the steep walls of rock. In the dry season, when drought and lack of food forced them to migrate, the springbucks gathered in huge masses and moved in masses over the arid steppes. Hyenas and vultures accompanied the fleeing, starving herd.

The days of such masses of springbucks are long passed. The subspecies of the Cape Province has been almost completely exterminated and only a few specimens live in protected areas. The subspecies of Angola has also largely disappeared. In Southwest Africa, springbucks still live in large numbers, but not in such masses as were formerly described. They still migrate in the rainy and the dry season. As long as they are young, springbucks still have the typical gazelle markings, a light brown band on the forehead and nose. It disappears completely in adult animals. The male's courtship gesture in the mating ceremony is the foreleg kick. The mating season peaks in May, and the gestation period is barely six months. At present, springbucks are kept in zoological gardens more frequently again, and they reproduce well.

The impala is classified by some zoologists into a subfamily of its own. Nevertheless, we shall consider it as a tribe (Aepycerini) with the gazelles. There is only one species, the IMPALA (*Aepyceros melampus;* Color plate p. 426), which is as large as fallow deer. The HRL is 130–180 cm, the TL is 25–42 cm, the BH is 75–100 cm, and the weight is 40–90 kg. These elegant animals have a slender build, with a croup slightly higher than the withers. The head is narrow and long. Only the ♂♂ have horns. There are no pseudo-claws. On the flanks, a naked spot

Fig. 15-11.
A male dibatag marks the female with his preorbital glands.

Fig. 15-12.
A female dibatag "nibbles" the male's preorbital region.

Fig. 15-13.
The urinating and "flehmen" ceremony of the dibatag.

The impala

Distinguishing characteristics

which is the size of a silver dollar has an unknown function. There is no difference in the color of the sexes. The females have four nipples. There are six subspecies.

The impala can jump as high as three meters and as far as ten meters. These jumps with the front and hind legs angled closely to the body are only slightly reminiscent of the gazelle's "stotting" jumps (Color plate pp. 436-7). Impalas form herds of six to 60 animals, although in well-populated areas, larger herds up to 200 head are not too rare. Occasionally the old males are solitary.

Typical in impalas are the "harem herds," where one male gathers many females and their young around him. Furthermore, there are bachelor herds with males of all ages. Often a bachelor herd joins a harem herd, and the harem owner is quite busy keeping the other males from his females. The closest distance two animals may approach each other without fighting (Hediger's individual distance) is very small in impalas. Their herds are often closely bunched up. They march in files as all other comparable species do. The impala always remains near water and comes to drink daily. It does not avoid meeting other game animals and may often be seen with various species of animals. Nevertheless, in most of these cases, it is not a true "association." According to my observation in East Africa, closer association seems to exist only with Grant's gazelles.

Schenkel has thoroughly studied the habits and behavior of the impala. According to his observations, the female herds are kept together mainly by the activity of a "herding" male, especially during the mating seasons which in the Nairobi National Park occur between February and April and between August and October. When they move from one place to another, usually an older female, the mother of a subadult young, leads the herd. Often the young of a herd stay closely together. Yearlings or mothers with very young offspring may form groups within the herd. The male circles the herd, and when single females separate from it, he brings them back to the others with a nodding "wink" with his horns. He chases young males 10 to 12 months old away from the herd. He does not directly attack full-grown opponents, but he demonstrates his superiority and his elevated "social" position by increased activity. When he separates his females from an adjoining herd of males, he alternatingly approaches one of the females and one of the bachelors with his head stretched forward and upward and his tail horizontally raised. He makes roaring sounds and chases them in a gallop, the females in one direction, the males in another.

During courtship the male will watch a specific female from a distance of approximately 20 meters. Suddenly, without roaring, but with tongue-licking "into empty space," he will chase this female. Soon the two will slow down, and the male will follow her at a distance. Finally,

Fig. 15-14.
Impala (*Aepyceros melampus*)

when she permits him to come nearer, he approaches her with his head stretched forward, and briefly licks her genital region. This is repeated several times, and the female may urinate in between, which causes the male to lip-curl (flehmen). Finally, he attempts the first mounting. In my observations, a male who is about to copulate rests his chest on the female's croup after a forward rush. His front legs are on both sides of the female's flanks, but he does not hold on with them. He carries his neck and head erect. From the momentum of the mounting jump, the female is pushed a few steps forward, but this is quite different from the steady walk during the gazelle's copulation.

After a gestation period of six and a half months, the mother leaves the herd and gives birth to only one young which "stays put" at first. The mothers of very small young often join groups of mothers, and later all return to the large herds. When two opponents stand opposite each other before a fight or during a break, they often turn the erect head aside to look at each other with one eye. This is the most frequent form of display. Threatening males present their horns very low, dropping them suddenly, turning them to both sides, and digging in the ground in front of their opponent. Before they clash they stand opposite each other with their heads deeply lowered. With forward pushes and jerks to the side, they lock their horns tightly together. The usual pushing against each other begins when they both use all their physical strength. If one of them flees, he always keeps his tail up and is often pursued vigorously.

According to Schenkel's observations the impala males are not territorial. Other informants testify to the contrary. It is possible that there are regional differences, similar to those in the brindled gnu. In zoological gardens the impalas are the pride and joy of their keepers, especially when they are kept in large outdoor pens. Some zoos keep large herds of them. The Hannover Zoo displays its handsome group of impala in an "African-enclosure," where by 1965, 15 young had been born.

A strange subfamily of the horned ungulates, which is unique because of the inflatable "bags" beside the nostrils, or a muscular, inflatable nose, is classified by some zoologists between the gazelles and the goats. They are the Saiga-like gazelles (subfamily Saiginae). We place them with the gazelles because they are closer to them than to any other horned ungulates.

Subfamily: Saiginae

The TIBETAN ANTELOPE (*Pantholops hodgsoni;* Color plate p. 438) belongs to those horned ungulates about which we actually know very little. They are the size of the white-tailed deer, with an HRL of 120–130 cm, a TL of 18–30 cm, a BH of 90–100 cm, and a weight of 25–35 kg. The head is not as graceful as a gazelle's, and both sides of the nose region can be inflated. Only ♂♂ have very long vertical horns. The bushy tail has hair on the underside. The hooves are narrow and long with bulg-

The Tibetan antelope, by F. Walther

ing, somewhat enlarged balls. The gestation period is about six months and one to two young are born.

In their Central Asiatic habitat, the Tibetan antelopes live in small groups or in herds. They make seasonal migrations to reach good grazing grounds. They eat in the morning and in the afternoon. At noon, they usually rest in deep depressions which they paw out of the ground as a windbreak; their young are probably born there. The Russian traveller and explorer Nikolaj Michajlowitsch Przewalski gave a vivid description of their mating in 1877:

"During the mating season, when the night temperatures may be as low as -30° centigrade, the male eats little and loses the fat he had stored in the summer. He is extremely excited and gathers a harem of 10 to 20 females around him. If he sees another male at a distance, he turns his horns toward him and roars with a hollow breaking sound. This means: 'Bucks will get a beating here.' Often there are hard fights with serious injuries. The harem is quite a burden for the male. As one of them—right before his eyes—walks off, he dashes after her, bleating, and tries to drive her back to the herd. We do not know whether the other females of the harem consider his efforts a personal affront or whether they had a 'secret agreement' right from the beginning, but in any case, while he tries to bring back this one female, some others run away. Now he chases after them, bleating, and four or five others go off into still another direction. Again bleating, he tries to catch this group—and now, to top it all, the rest of the harem scatters into all directions. This certainly is more amusing for the spectators than it is for the male who is still upset about his loss. He angrily paws the ground with his hooves, jerks his tail upward, shaped like a hook, and calling in all directions, he challenges the opponents to which his 'unfaithful' females have fled."

According to this description by Przewalski, it is very possible that Tibetan antelope males may also occupy territories in the mating season, while the females roam about. The pouches beside the nostrils may be inflated to the size of a dove's egg. They are like the structures which we already know in the gazelles, but there are no details about the function of this structure.

The saiga antelope, by B. Grzimek

In the SAIGA ANTELOPE (*Saiga tatarica*; Color plate p. 438), the nose is even larger. The HRL is 110-140 cm, the TL is 8-13 cm, the BH is 60-80 cm, and the weight is 23-40 kg. The head is large and plump, and the fleshy and inflatable nose has a longitudinal ridge on the front. Only the ♂♂ have medium long vertical horns. The tail is very short, and the underside is naked. The legs are slender as in the antelope, and the hooves are slightly broader to the rear. The summer coat is short and almost smooth, and in the winter, it is long and very dense with a thick fleece. There are two subspecies: 1. RUSSIAN SAIGA (*Saiga tatarica tatarica*). 2. MONGOLIAN SAIGA (*Saiga tatarica mongolica*).

During the Glacial Period when woolly rhinoceros and mammoth elephants roamed the earth, the saigas grazed on the steppes of Europe and Asia, from England through Germany and Russia as far as Siberia, Kamtschatka, and Alaska. Even as late as the 16th and 17th centuries, the lower hills of the Carpathian Mountains and the Bug River were still their western limit. The explorer and traveller Pallas writes in 1773 from the Kirghiz steppe that the saigas were so abundant that his Cossacks could kill as many as they pleased. Ewersmann saw them in 1850 swim across the lower Ural River in "unbelievable numbers." The steppe between the Volga River and the Ural was covered with them.

In the last century, the saigas were killed for the Chinese who believed that they could manufacture an aphrodisiac from the horns. The merchants of Buchara and Chibinsk alone sold no less than 344,747 pairs of the almost transparent, bright, wax-colored horns between 1840 and 1850. The males were shot from ambushes at the water holes, and the fleeing herds were driven by people on horseback toward long lines of shooters. The Kirghiz hunted them at the same time with greyhounds and trained steppe eagles. In the winter, they were chased from the bamboo and reed onto the ice of lakes where they could not move, and there they were clubbed to death. Another method was to build long fences which were five kilometers apart at the beginning, then gradually narrowed until only a small gap was left open at the end. There the animals were chased in by the thousands and ten thousands. In the bottleneck where the confused animals crowded in rapidly, there were pointed posts which tore their chests and stomachs open. In this manner thousands were cruelly captured, and even more escaped crippled.

Thus, the saigas which had survived thousands of years decreased rapidly in number. In the severe winter of 1828-29, they became extinct in the area between the Volga and the Ural Rivers. Since then, the "European saiga antelopes" west of the Volga River and the herds in Asia no longer came together. By the end of World War I, the saiga antelope was among the species that was expected to become extinct within a few years. There were barely a thousand of them left. At the last moment, in 1919, the Federation of the Soviet Republics issued a total prohibition against the hunting of saigas; the Republic of Kazachstan followed suit in 1923. Furthermore, Russian zoologists, especially Professor Bannikow in Moscow with his assistants, and Professor Sudski in Alma Ata, began to study the life of these animals that were on the verge of becoming extinct.

At first they tried to count the animals from cars and planes. Later they marked more than 10,000 newborn saigas. The scientists wanted to find out why and where the saigas migrated. Saigas, which are continuously on the move, are able to sense changes in the weather which will lead to drought or snow, and they immediately start to move.

They were almost destroyed

Fig. 15-15.
1. Saiga (Saiga tatarica)
2. Tibetan antelope
(Pantholops hodgsoni)

Research on their habits

While grazing, they do not move more than three to six kilometers per hour. But as soon as it becomes necessary, for example, when heavy snow begins to fall which might cut them off from any food or when severe frost comes, they may suddenly move 120 or even 200 kilometers farther to the south within two days. In the usually mild winters with periods of thawing, they move southward at freezing temperatures where the snow hardly ever is deeper than 10 centimeters or where it thaws more quickly on the salty soils around the Caspian Sea. When it thaws, they move back north where more food is available.

Saiga antelopes in hardship due to winter

In very cold weather and heavy snowfalls, they begin to run for their lives. The animals walk strictly in the direction in which the wind blows, they cross railroad tracks, they touch human settlements, ignoring all danger. They cross the Volga River only in winter. A snow cover of more than 20 centimeters, which stays for more than two weeks, means death for whole herds of saigas. Since during a snow storm all animals immediately move in the same direction, the herds grow like an avalanche. The distance between the saiga antelopes' grazing grounds in the summer and winter is between a total of 250 to 400 kilometers. Years of drought or winters with heavy snow force the saigas into extensive migrations. Even though in the severe winter 1953-54, 80,000 of the "European" saiga population of 180,000 animals starved and froze to death, the rest scattered over wide areas thus forcing the indefatigable animals to populate again vast parts of their former range.

The saiga antelope's diet

Of the many plant species which saiga antelopes eat, 13 per cent are not eaten by any other animals, especially domestic ones, because they contain poisonous substances and salt. These plants are the main diet of the steppe antelopes. As long as the saigas have moist plants to eat, they do not drink. The more dry the food becomes the further they have to move each day to find green plants. Only in times of severe drought do they go, single file, to water. Then their noses are tense, they move them slightly from one side to the other, and step into the water with their feet, and drink, turning their noses to one side. However, they will not stay in the water longer than seven or eight minutes.

Usually saigas walk in an amble. When they begin to flee they first make jumps into the air and then they gallop away. In flight they always head for flat, open areas, such as dried lakes, and they circle around obstacles instead of jumping over them. Their sense of hearing is poor, and olfaction is not important to them since the air currents in their native habitat usually rise swiftly. However, their vision is excellent. They are able to see danger at a distance of more than 1000 meters.

Reproductive behavior

The saiga males prefer older females over young ones. Their mating season begins at the end of November when the weather becomes cold and unpleasant. Then the males have three to four centimeters of fat, especially on their lower back. Their nose is more swollen than usual, and it hangs down like a trunk flapping to and fro. From a human perspective, this does not look very graceful. Large dense bunches of hair have grown below the eyes, and on both sides of the males' neck there are dirty-brown manes. During this time, they are easily recognized at a distance. From their preorbital glands a dark-brown viscous substance with a strong odor is secreted. One male has two to three, twenty or even fifty females which he keeps in a group and constantly herds together. During the mating season, the males eat only snow and have vigorous fights. The main mating season lasts only seven or eight days, but the animals stay together in a herd for forty-five additional days. In this period, the male copulates with the young females who were born in the spring of the same year and are only seven and a half to eight months old. Finally, the mating groups form large herds. The males, who had fought with each other so angrily a few days earlier, now graze peacefully next to each other, and they pay no more attention to each other.

They have every reason to behave in this way. While the females do not fast during the week of mating but actually gain weight, the exhausted males face the approaching severe winter in rather poor shape. By March many of them die, and in bad years, most of them are gone. Thus, the scarce food resources remain for the pregnant females, and they preserve the species. It is advantageous then that the young males do not become sexually mature in November of their first year as their female age-mates do. Therefore, more of them survive the winter and are ready in the following fall to mate and thus preserve the species. The young males are not driven from the harem by the leader of the mating group in their first year. Ninety-six per cent of the older, and eighty-six per cent of this year's females become pregnant. Three out of four saiga mothers have twins, but there never are triplets. The high number of young and their ability to survive in cold and poor weather is the secret that enables the saigas to survive in such arid regions and even to become fat.

They usually have twins

Even without man's ravaging, massive losses due to natural causes have always occurred among the saiga. After such widespread dying-off in 1949-50, half of the saiga population was dead. In the winter of 1953-54, forty to sixty centimeters of snow and temperatures to -40°C. killed forty per cent of the animals west of the Volga. The first to die were always the old males, then the young ones, and the last were the old and young females. However, already one year after this catastrophe, the saiga antelopes west of the Volga River had regained their former numbers. The birth rate is the same for males and females, but among adults the ratio is 1:2.6 of males to females. After very severe winters, sometimes only three to four per cent of the males survive.

At the end of March and the beginning of April, the saigas move out of the plains around the Caspian Sea and migrate north. They have their calving grounds, flat, slightly undulating country with short grass, where no predator can find cover or approach unseen. In the beginning of May, the steppe is alive as far as one can see. The air is filled with the bleating of the young and the mothers' replies. A large group may cover three hundred square kilometers, which is an area of fifteen by twenty kilometers. Counts from the air yielded one hundred-fifty to two hundred thousand animals. Ninety-five per cent of them are mothers with their young. The young saiga antelopes drink their mother's milk for the first time only six to eight hours after birth. During the first days, they lay pressed against the ground and nurse only for a few seconds or minutes at a time. The milk contains 6.7 per cent fat, which is more than in cow's milk, and 5.4 per cent protein. On the third or fourth day, the young saigas begin to eat plants in addition to milk, and by the age of two to two and a half months, they feed exclusively on greens. Only a few will then still occasionally nurse on their mothers.

Young saiga antelopes "stay put"

It is difficult for a person to catch a newborn saiga on its first day of life if it tries to get away. Usually, however, the young "stay put" so that they can be picked up easily. The mother may walk away from her young as far as five to six hundred meters before she returns. By four to five days, the young saigas usually run away. Like young gazelles, they drop to the ground from a full run and thus disappear from the pursuer's sight. When the mother comes back and bleats, often two to four other young also come running toward her and bleat. She sniffs them and leaves them alone. Her own young will continue to cry until the mother has reached it. Young saigas at five to six days are able to run at a speed of 30 to 35 kilometers per hour for 27 to 30 minutes. By ten days, they run as fast and as long as the adults. As early as eight or ten days after the beginning of the main calving season, large herds of mothers with young leave the "calving grounds" and, within a few days, they may be 200 to 250 kilometers away.

Their main enemies are men and wolves

Except for man, the saiga antelope's main predator is the wolf. Even though it is not able to catch single, healthy animals, the wolf may well catch the exhausted males after the winter mating season, females advanced in pregnancy, and large numbers of newborn young. They are also easily caught in high snow because saigas with their sharp hooves put a weight four times that of the wolf with his broad, hairy paws on the snow cover. Furthermore, the wolves hunt in packs. Like most of the other game animals, saigas flee in a circle and gradually return to the place where they started. There these wolves of the pack who did not run after them await them.

In the 1950's in Kazachstan, there was one wolf for every hundred saigas, and twenty to twenty-five per cent of the saigas fell victim to them each year. After that, 210,000 wolves were shot within ten years, which caused a sharp increase in the saiga population. In European Russia west of the Volga, there were hardly any wolves to begin with. It is hoped that no serious consequences will result as has happened so often when the natural predators of a species have been destroyed. These predators prevent explosive population growth and help to maintain a healthy prey animal by culling the weak and sick.

Young saiga antelopes are also endangered by steppe eagles, golden eagles, common ravens, and foxes. Every tenth young dies in the first days of its life. In poor weather, it may even be 20 or even 60 per cent. However, even in normal weather only two out of five newborn saigas are still alive at the end of the first year.

In former times saigas were also plagued by bot flies which would lay their eggs under the antelopes' skin. This caused abcesses around the larvae and was painful. Since the end of the 19th century, this horrible pest no longer plagues the migrating nose-antelopes. Probably at that time, the number of saigas had so decreased that not enough of them remained to sustain the reproductive cycle of this feared parasite. Thus, it became extinct sooner than the host animal. The humped-nosed animals, meanwhile, are almost back to their original numbers, but the bot flies have not returned. However, in Mongolia the saigas still suffer from the bot flies. Hoof and mouth disease sometimes spreads from domestic animals to the saiga herds. Then the upper part of the males' horns falls off, and a great many of the young die. However, a severe winter causes much greater losses. After the beginning of winter, the herds move south from the steppe to the desert and dig up, if necessary, mugwort plants which are rich in fat and protein from beneath the snow.

They suffered from the bot flies

As animals of the steppe, saigas are afraid of bushes in the winter. They stay at least a hundred meters away from them, and therefore, rarely damage new plantations. They react similarly to cultivated fields. In a rye field of one square kilometer, over which a herd of saigas had wandered, only fifteen to twenty ears torn off were found. The animals had carefully walked between the rows of the cultivated grain. Since they prefer plants of ten to fifteen centimeters high, they do not enter corn fields, not even at times of drought or in mid-summer. However, occasionally they do stamp down the edges of fields badly. Thus, the damage they do to agriculture is minimal.

The saiga's fate of being destroyed had been prevented at the very last moment. At first, the new hunting restriction had only the effect of keeping the last few hundred survivors from vanishing. In the European part west of the Volga, a slight increase was noted in the mid-20's, and in Kazachstan in the early 30's. During World War II,

How the saiga antelopes again increased in number

when there were few cattle and few people, their number increased more rapidly. In each of the larger villages, game wardens were appointed. In 1947 and 1948 in Kazakhstan there were as many saigas as a hundred years ago. In 1951 there were about nine hundred thousand, and in 1960 the number was up to 1.3 million. In 1954 the saigas had again reached the agricultural areas. Thus they had repopulated all of their presently available habitat, two million square kilometers in the Asiatic part of the Soviet Union, and a hundred-fifty thousand more in the European part, which together is almost one-fourth of the total of Europe. On these arid steppes and semi-deserts, two to seven hundred kilograms of plants per hectare grow at best. No more than about one saiga per square kilometer at best can exist.

Utilization of the saiga herds

Therefore, in the mid-50's there began the utilization of a part of the animals for human consumption. The "Astrachan Promchos" is a government association for game farming which was founded for this purpose. Every year the animals are counted by plane on the calving grounds, and this is repeated in the fall shortly before the hunting season. Saigas may also be hunted in the usual manner after the purchase of a hunting license, with a pointer, from October 1 until November 10 right before the beginning of the mating season. However, this alone does not yield the necessary quantities of meat, so government hunting brigades kill the animals on dark, windy nights in a less sportsmanlike but more humane manner by using floodlights. Private hunters are not permitted to do this. Such a brigade of five men drives over the steppe with a vehicle at a speed of fifteen to twenty kilometers per hour. With ordinary car lights, the saigas are easily recognized at a distance of two kilometers because their eyes reflect the light. After the team approaches to within one to two hundred meters of the animals, the cars stop and a powerful searchlight is switched on. The saigas stand still in the bright light or may even come closer slowly, as several other species of animals also do. Meanwhile, the shooters have left the car and kill the animals from a short distance of thirty to forty meters. In contrast to usual hunting, there are very few injured animals which can get away and thus die wretchedly. Furthermore, at this distance it is possible to distinguish clearly the age and sex. They usually kill the young males. In the beginning the old males were killed, which usually make up 10 to 20 per cent of the population. When they were reduced to one to two percent during an experiment in the following year, four to five times as many females were not pregnant. One brigade of shooters may kill 100 to 220 saigas within five to six hours. As long as there are only a few wolves and stray dogs in the area, up to 40 percent of the saigas may be killed annually, and with the natural losses that occur their overall numbers will be the same the following year. Presently, every year 250,000 to 300,000 saigas are shot in the European and Asiatic part of Russia. Even so, they have not yet reached their highest possible number; in Kazachstan, another increase of two or three more million head is expected. The meat, which tastes similar to lamb, and the hides of the animals are used. The horns are exported to China, where they are made into medicines of the "Pantokrin" type. A butchered male yields 25 kilograms of meat; a female, 16 to 17 kilograms. Thus, the Soviet Union produces 6,000 tons of meat per year and 20,000 square meters of leather from arid regions which would not be of any use if the saiga had been totally destroyed.

It is strange how difficult it is to reintroduce these hardy hump-nosed animals into an area once they have disappeared from it. Friedrich Falz-Fein tried it in the steppe of Askania Nova, but the animals vanished without a trace. A hundred years ago they had lived in immense herds on these very steppes in the southern Ukraine. Vain attempts were made during the last ten years to reintroduce them to the islands in Lake Asow and the Caspian Sea. In zoological gardens they rarely live long and reproduce only in exceptional cases. They prefer fresh greens over water, in captivity, and the water is better for them if some salt is added. On Barsa-Kelmes, a large island in Lake Aral, they were saved at the last moment. In 1929 there were only five females left. Eight more animals, including some males, were brought there, and this population had increased to 2000 head by 1961. This herd is not quite as shy as its continental conspecifics. When saigas need to be captured for experiments or for zoological gardens, they are usually taken from there. At first long nets were set up and the animals were driven towards them by beaters. Each such driving netted 10 or 20 animals which had entangled themselves in the nets. However, the saigas became smarter all the time. They turned around directly in front of the nets and dashed between the cars and motorcycles. Only newborn or animals a few days old were captured. It is useless to try to capture adults with a car. They give up only when they are so completely exhausted that they die of edema of the lungs. Within five or six days, thousands of young may be marked on their ears. If they are to be kept in captivity no more than 10 are raised together in pens, because in their excitement they step on each other so that they cut each other's skin with their sharp hooves.

The humped-nosed saigas once helped people to survive. When our ancestors in Europe and Asia dressed in fur and fled into caves during the glacial period, they lived almost exclusively on game animals. The saigas were probably the most readily available prey, and certainly they recovered quickly after years of scarcity. Their bones are found most frequently in places where people had lived for a long time. Who can foresee what catastrophes we and our descendents will face, what setbacks and famines will be caused by man in his nearsightedness.

Attempts to reintroduce them

Saiga antelope and man in the glacial period

The masses of man-made deserts increase steadily, and with every hour overpopulation becomes greater, although two thirds of mankind are undernourished.

The example of the saiga shows what an act of insanity it is to permit animals and things not made by man to vanish completely. A species of animals has evolved over millions of years, struggling for life, and they often have capacities of which we are not even aware. The saigas, in spite of severe winters, can turn salty deserts into meat and leather, but man, with his domestic animals, cannot do this. I shudder in retrospect at the thought that these miraculous, hump-nosed animals might have vanished from the earth by now if it had not been for a few active and sensible men 40 years ago.

Bernhard Grzimek
Fritz Walther

16   The Chamois-related

Those concerned with zoological classifications are again confronted by difficult questions with the chamois-related, takins, gorals, and serows. Formerly these animals had been classified with the antelopes, but lately they have been included with the sheep and goats in the Subfamily Caprinae. It is questionable to what degree these groups are related. At least, with respect to many characteristics, the takins are completely different from the chamois-like horned ungulates. We distinguish three tribes: 1. GORALS and SEROWS (Nemorhaedini); 2. TAKINS (Budorcatini); and 3. CHAMOIS (Rupicaprini).

The GORALS and SEROWS are about the size of goats. The horns are short and pointed in the ♂♂ and ♀♀. Longer hair appears as a brush or mane on the neck. The legs are sturdy. There are no differences in coloration between the sexes. Both have interdigital glands, but only the serow also has preorbital glands. The females have four nipples. There are two genera with one species each: 1. The GORAL (⊙Nemor-haedus goral; Color plate, p. 464) has an HRL of 95–130 cm. The TL is 10–45 cm, the BH is 55–75 cm, and the weight is 25–35 kg. The head is short and high, and the ears are medium-long with pointed tips. The dense shaggy coat has long fleece in winter. There are seven subspecies, three of which are endangered. 2. The SEROW (⊙Capricornis suma-traënsis: Color plate, p. 464) is larger. The HRL is 120–180 cm, the TL is 8–18 cm, the BH is 70–105 cm, and the weight is 55–140 kg. The head is rather flat and long, and the ears are usually longer than the horns. The rough, short to long coat is very dense, with a thick fleece in some subspecies, while in others it is sparse with thin fleece. The coat has hard bristles. There is a beard on the cheeks from the corner of the mouth to the base of the ear, and a mane on the neck crest from the horns to the whithers (only two subspecies have no mane). There are fourteen subspecies, seven of which are endangered or almost destroyed.

The goral lives in dry mountainous regions between one and four

Subfamily: Caprinae

Gorals and serows, by F. Walther

Distinguishing characteristics

The goral

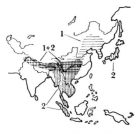

Fig. 16-1. 1. Goral (*Ne-morthaedus goral*). 2. Serow (*Capricornis sumatraensis*).

The serow

The takins, by
F. Walther

The takin

thousand meters. The animals form small groups, and older males are often loners. Excellent climbers, they feed mainly in the morning and the evening. When skies are cloudy, they are active all day long. They have lookouts on rocks or slanted tree trunks and rub their horns on bushes and trunks. Sanderson writes: "On precipitous slopes, these animals gallop along with breath-taking speed; it was possible to show on a movie film that their hooves had already left loose rocks by the time they began to roll under the goral's weight."

The gorals are sedentary. In the winter they will move to lower elevations, but not into the forests. Their mating season is in September and October. After a gestation period of approximately six months, usually only one young is born, but twins occur. The most eastern subspecies in the Soviet Union (*Nemorhaedus goral caudatus*), which lives near the lower course of the Amur River and in the mountains northeast of Vladivostok, are now protected. These gorals live on precipitous coastal rocks and steep wooded slopes. They feed mainly on branches and acorns which they do not scrape out of the snow, but can only pick up with their mouths. They spend most of the year in a very limited area. When they are grazing, they do not range farther away than two to three kilometers from the rocks and cliffs where they are safe from predators. An adult female was captured, marked, and then released at a distance of six kilometers from the place of capture. She soon was back on the familiar rocks. Because of their rather small size, the gorals avoid the loose snow in the winter. When the snow is thirty-five to forty centimeters deep, they become quickly exhausted. Therefore during heavy snowfalls the gorals limit their movements as much as possible and remain on the safe rocks until the snow becomes more packed. Many of the animals are starving during this time.

The larger and stronger serow is better adapted to humid air, and so it occurs in latitudes near the equator. It resembles the goral in its habits. These sedentary animals have regular paths and have distinct places for resting, defecating, and rubbing their horns. In some areas where gorals and serows live in the same habitat, the gorals inhabit the almost barren slopes while the serows stay mainly in the dense brush above the timber line. Female serows are pregnant for seven to eight months. In recent years, gorals from the Soviet Union were displayed in the zoos of Berlin-Friedrichsfelde and Halle.

Even less is known about the biology of the TAKINS (tribe Budorcatini). No one really seems to know how to classify or with whom to compare these animals. Some zoologists place them into a special subfamily with the Arctic musk ox. Others consider them to be the links between the antelopes and the Caprinae. Still others place them near the chamois. There is only one species, the TAKIN (*Budorcas taxicolor*; Color plate, p. 469), with an HRL of 170-220 cm. The TL is 15-20 cm, the BH is

100–130 cm, and the weight goes up to 350 kg. They have a build which is suggestive of an ox, with a large head and humped nose, and a broad muzzle. The horns in the ♂♂ and the ♀♀ rise flatly from the skull and curve into an angle. The base is bulky and bulged, but the ends are smooth and pointed. The legs are sturdy, and the front legs especially are very bulky. The hooves are broad and round, and the pseudo-claws are very long and strong. The dense and shaggy coat is longer in some places. The tail is short and bushy, with hair on the lower side. There are no skin glands, and the sexes do not differ greatly in color. The females have four nipples.

There are three subspecies: 1. The MISHMI TAKIN (◊Budorcas taxicolor taxicolor) is basically golden yellow to brownish-red, with black coloration in many places. 2. The SZECHWAN TAKIN (Budorcas taxicolor tibetana) is basically yellow, reddish-gray or gray-silver, with less black coloration. 3. The SHENSI TAKIN (◊ Budorcas taxicolor bedfordi) is basically white-yellow to gold, with only sparse black coloration.

The golden fleece which Jason once brought from Kolchis to Greece probably came from an animal which even now is almost as legendary as it was in classical times. The takin is among the least known horned ungulates. Like the musk ox, it is considered an archaic species of animal, a preserved relic of the past which, in the course of more recent geological ages, was forced into remote areas where it could survive until now. There are probably many primitive behavioral characteristics in takins which make this species especially interesting to zoologists and ethologists.

H. S. Wallace was able to observe "the animals with the golden fleece" in the Tsinglinschan Mountains: "We searched the area with our binoculars. After some moments of silence, we saw them: two large yellow objects moved between the rocks on the other side of the valley. I think we were most surprised at their color. It was the reincarnation of the golden fleece. Only the Tsinglinschan race is of this color. In the sunshine, they are a conspicuous golden yellow. The females are distinctly brighter and of a more silvery shade, while the males, which are clearly reddish in the neck, are not unlike the lion in color. Three males, three females, and two young were grazing on a rock plateau which dropped down vertically for four hundred meters.

"The head, which is carried low, the 'buffalo horns,' and the prominent 'Roman nose' are striking from the front. From behind, they look like huge teddy bears with their heavy body, short sturdy legs, and the tail which disappears into the long hair of the coat. When in a rage or in frantic flight, these 'teddy bears' may cover short distances with the speed of a rhinoceros. They are absolutely sedentary and hence they tolerate the presence and the noise of the Chinese lumbermen who work all over this area. When disturbed, they retreat into the bamboo jungle or other dense plant cover and remain there. The

Distinguishing characteristics

smartest old males press their bodies against the ground with their necks stretched out so that a person almost steps on them. Depending on the season, they move to lower elevations in the winter, to almost two thousand meters, and to higher altitudes of up to forty-five hundred meters in the summer. When they pass along in the light fog of the early morning on their way to other grazing grounds, a herd of one hundred of these unusual animals appears to be an incarnation of a legend. Along their well-trodden paths to the salt licks, commercial hunters often ambush them in hidden places."

So far only very few takins have been kept in captivity. In 1967 some were in the zoos of Rangoon (Burma), Peking, and one pair lived in the New York Bronx Zoo. For some time (1966/1967) the Hannover Zoo had a young female Mishmi takin who later went to New York. Since she had been captured at a very young age, she was completely tame and very attached to her keepers. Within five months, the young animal grew almost ten centimeters. On the heavy head, the horn plugs were still simple, straight, and slanted upward and back. They were ten centimeters long, and they were just about to show the first, rather weakly developed ornament circle. In touching the animal's coat, one's hand became oily and brown. This oily skin secretion had the strong scent of butyric acid and it tasted hot. Apparently it protects the takin from moisture. since its habitat is frequently and for a long period of time covered with fog. According to Lothar Dittrich's observations on orientation in a new environment, the sense of smell plays an important part. This is similar to what Hediger found in wild sheep.

Fig. 16-2. Takin (Budorcas taxicolor)

L. Dittrich reports: "She likes to play with the familiar keeper. She approaches him with her head stretched forward, presenting her horns, and she then pushes him away. Or she will hit forward or backward to the side with her head. The blows are short and painful. When she becomes more excited she comes dashing in a gallop, lifts her head high and hits downward. When she jumps away, she may stand on her hind legs, but we never could get her to strike from this posture as goats do. When she threatens the keeper, she presents her horns and makes a few jumps forward, accompanied by snorting. She can also lift the front legs slightly and hit with them."

Takins grow rather slowly. The New York animals' horns did not attain their final shape until the fourth year. Takins probably become sexually mature by two years of age. In 1964 in the Rangoon Zoo, a three and a half year old female became pregnant by a male two and a half years old, but unfortunately she had a miscarriage. The gestation period is probably seven to eight months. Hopefully, the zoos of New York, Rangoon, or Peking will successfully reproduce this "legendary" animal. Some of the behavior of these gregarious animals, which would come out clearly only in groups, cannot be studied in animals which are kept alone.

Even better adapted to life in rocky regions are the members of the Rupicaprini tribe. They are about the size of a goat. The horns are rather short in the ♂♂ and ♀♀. Glands occur only at the back of the head, and in some animals there are also interdigital glands. The hair is longer, especially at the neck and back. There are no differences in coloration between the sexes. There are two species: 1. CHAMOIS (*Rupicapra rupicapra*; Color plate, p. 463). The HRL is 110–130 cm, the TL is 10–15 cm, the BH is 70–85 cm, and the weight is 14–62 kg. The short head has a forehead which rises from the base of the horns. The eyes are rather large. The straight back appears to be uneven because of the crest of hair on the withers and croup. The summer coat is short and rather smooth, while in winter it is long and dense with a very soft fleece. It is much longer on the withers and croup. There are nine subspecies, at least three of which are endangered. 2. ROCKY MOUNTAIN GOAT (*Oreamnos americanus*; Color plate, p. 464 and p. 508). The HRL is 150–175 cm, the TL is 12–17 cm, the BH is 90–105 cm, and the weight is 80–135 kg. The head is rather flat and long, and the upper profile is even with a long nasal part. The eyes are right below the horns. The backline is higher at the withers and croup because of a mane, and therefore, it appears to be lower in the center. The hair of the coat is medium-long to long, very dense, and mane-like along the croup and from the neck to the withers, forming "cuffs" at the joints of the feet. It has a very delicate fleece. There are four subspecies.

The most important part of a mountain climber's equipment is his boots, and for the chamois (*Rupicapra rupicapra*) it is the hoof. Like all horned ungulates, the chamois has two toes on each foot. Behind them are the pseudo-claws which do not touch the ground when the surface is level. The hooves are very soft and have soles which are elastic like rubber, enabling them to cling to any unevenness of the rocks. In contrast, the outermost tips of the hooves are firmer. Therefore, they wear out less and protrude like rims. When a chamois slips, these rims are caught at the pointed surface of rocks. Furthermore, the digits are extremely moveable so that the chamois has firm support on eight points even on the roughest terrain. On steep grass slopes, with its head facing downhill, the toes of the front and the hind legs are pressed into the ground up to the heels, so that, in addition to the tips of the hooves, the pseudo-claws also dig tightly into the slope. Thus the chamois is moored by sixteen points in these difficult places. Even the individual joints of the legs are adjustable. Sometimes one can see the chamois whose legs seem quite "mixed up."

In the plains the chamois may appear to be rather plump, but it certainly does not in the mountains. Its downhill gallop is not elegant, but rather "bumpy" because of the length of the hind legs. However, the animals can dash uphill like an arrow, and they can climb a thousand meters of difficult terrain within a few minutes.

The Rupicaprini

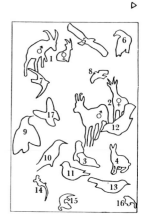

Animals of the Alps

1. Alpine ibex (*Capra ibex ibex*; see Ch. 17)
2. Chamois (*Rupicapra rupicapra*).
3. Alpine marmot (*Marmota marmota*)
4. Alpine snow rabbit (*Lepus timidus varronis*)
5. Golden eagle (*Aquila chrysaëtos*)
6. Bearded vulture (*Gypaëtus barbatus*)
7. Alpine swift (*Apus melba*)
8. Alpine jackdaw (*Pyrrhocorax graculus*)
9. Wall creeper (*Tichodroma muraria*)
10. Alpine crow (*Pyrrhocorax pyrrhocorax*)
11. Snow finch (*Montifringilla nivalis*)
12. Ring thrush (*Turdus torquatus*)
13. Water black bird (*Cinclus cinclus*)
14. Mountain lizard (*Lacerta vivipara*)
15. Alpine salamander (*Salamandra atra*)
16. Yellow-bellied toad (*Bombina variegata*)
17. *Parnassius delius*

The chamois, by F. Walther

Fig. 16-3. Chamois (*Rupi-capra rupicapra*)

The female and the young male chamois live in groups. Old males usually stay by themselves or tolerate one or two younger males in their presence. In May and June, when the young are born, the mothers separate from the groups. They chase off the yearlings which are still with them. When the newborn are a little older, they will often allow the yearlings to rejoin the group. Young chamois are strict "followers" who constantly "stick" to their mothers. Even though there are chamois who prefer the forest, the typical habitat of chamois is the high mountains. In new snow or when it thaws and avalanches threaten, the groups withdraw into the woods in the valleys, but they return to higher areas as soon as it is safe again.

The chamois' alarm call is a whistle. The males utter strange grunting sounds with their mouths open during the rutting season. Male chamois mark slender trees, branches, and high stems of grass with glands which are located behind the base of the horns. These marking glands are especially enlarged during the mating season, but they seem to function the whole year round. I observed male chamois perform marking movements at times other than the rut. They often do this when they see an opponent. They also mark between threat and display behavior. Sometimes a peculiar form of urination occurs. Younger males deeply flex their hind legs and urinate between their front legs with the penis extended, while the neck and head are erect.

Apparently males are temporarily territorial during the rut. The male chamois displays by showing his broadside to his opponent and circling him with stiff legs. When both display, they often stand in a reversed-parallel position. The back is arched and the mane on the back and the longer hair on the croup stand erect. According to my observations, young males keep the head stretched upward, while the old ones lower it to the ground. In any case, this keeping-the-head-low, which sometimes may change into grazing, plays an important role as a threatening gesture. The male also has a less serious form of display. He keeps his back slightly bent and the mane only moderately erect while he stretches his head upward from the shoulders.

As Burckhardt discovered, young males and females show the "submissive posture" when an older male threatens or displays. They stretch the neck vertically, flex the hind legs, and "crawl" in this posture past the higher ranking animal. Sometimes females will briefly sniff at the older male's penis. Males, when sexually aroused, sometimes approach a female in a similar posture.

The fight, which is as unritualized in chamois as it is in the mountain goat, is a serious attack. Pushing and tearing with crooked horns, the male chamois attacks his opponent and tries to strike him wherever he can. It is most conspicuous how the males chase each other during or after a severe fight. Suddenly one chases the other, and just as suddenly, the other turns around and dashes towards the pursuer, and

Chamois-related

1. Rocky Mountain goat (*Oreamnos americanus*)
2. Goral (*Nemorhaedus goral*) subspecies:
a. Cashmere goral (*Nemorhaedus goral goral*)
b. North Chinese goral (*Nemorhaedus goral caudatus*)
3. Serow (*Capricornis sumatraensis*)

Tahrs

4. Tahr (*Hemitragus jemlahicus*; see Ch. 17), subspecies: a. Nilgiri tahr (*Hemitragus jemlahicus hylocrius*) b. Himalayan tahr (*Hemitragus jemlahicus jemlahicus*)

now it is his turn to retreat. This change often takes place when the animal in flight has gained altitude on a slope and thus can attack the former pursuer from above. In the Kronberg Zoo, where chamois were kept with ibex, occasionally animals had to be captured for medical treatment or to have their hooves cut. In all these cases, the male chamois would take part in this chase. He went after the animal in question and vigorously attacked. According to the "chamois code," a fleeing opponent must be pursued inexorably. Incidentally, it was discovered that it is not advisable to keep chamois and ibex together in the same pen in a zoo. The unlike opponents fight with each other, something which probably never happens in the wild. The ibex, following his fighting method, rises on his hind legs, while the chamois' natural method of attack is level on the ground, so that he hits the ibex's belly with his sharp pointed crooks.

Chamois have many remarkable games. In 1853 the Swiss naturalist Johann Jakob von Tschudi described how the chamois drop down on their bellies with the hind legs flexed and the front legs stretched forward and slide down the hillside. This has been questioned more than once, but lately it was confirmed by such reputable observers as Burckhardt and Schloeth. Another conspicuous game, which I observed several times, is the "dance of the chamois." It is performed rather spontaneously, not only in rut but in the spring, and often when there is no partner to whom it might be directed. It can only be presumed that the male feels especially well and full of excess energy. In the beginning, he dashes around in narrow circles. Then he may throw his body into the air with various odd movements of the neck and legs. Finally he stands still with his hind legs on one spot, lowers the front part of the body to the ground, and then flings it upward so furiously that the front legs leave the ground. Then he circles around his own axis on his hind legs and raises up so high that it seems as if he will fall over backwards.

The rut is from mid October until January. After a gestation period of approximately six months, usually one young, or in rarer cases, twins, are born. The young and the adults, to a lesser extent, were formerly endangered by several mammalian predators and birds of prey. Since these enemies have been exterminated to a great extent, there is practically no further natural selection through predation on chamois. This was not advantageous to the population in many cases. Contagious diseases spread among them, especially the dreaded chamois rubbers, which in some areas destroyed large numbers of these animals. Now the hunter has taken the place of the predators. Sensible people are finally beginning to make efforts to protect the few remaining predators and birds of prey.

The attack of an eagle on a young chamois has been described by the zoologist Count Zedtwitz quite beautifully, although anthropo-

Attack from an eagle

morphically, but nevertheless correctly. "The old chamois was standing there, enjoying the warmth of the sunshine. She chewed a little, moved her jaws to and fro, bent her legs slightly, and indulged in the warmth of the sun. All this could not have lasted more than a few moments because the fog was still close. Then it seemed to her as if something dark and dreadful appeared. She stopped chewing, turned her head, and in the fraction of a heartbeat, she recognized the eagle. He had not been able to catch any prey this day because the fog had also restricted his view. When the clouds began to clear, he left his rock and let the wind carry him towards the rays of the sun which in gorgeous colors announced its rising. He broke through the top with great speed and deadly aim. The mother's alarm call sounded so piercingly that the young instantly came to her. It bounded up the rocks with rapid leaps.

"But the eagle had seen it and came whizzing along like an arrow. His cold eyes calculated exactly the place where he would meet up with the young. His claws shot out of his abdominal plumage, prepared for the strike. At the moment when he folded his wings so that he could strike, he was sure that he would kill the young right next to its mother. The little chamois wailed dolefully when it saw the huge bird swiftly approaching. It cowered between the herbs and the rocks. Paralyzed with terror, it resigned itself to its fate when at the last moment the mother threw herself between her young and the predator. The eagle could not change his course. With a jerk of his body, he opened his wings so that he could be carried aloft, but it was not enough. One of his wings hit the mother hard and almost knocked her over. But she only took one step backward to shield her young. Then she lowered her head and presented her weapons to the eagle.

"The huge bird could not believe that his game, this game which he never had lost before, was lost. He flew a curve so that the outer wing cut the air like a knife, and blind with rage, he attacked the older animal. He dropped from the sky like a projectile, but then he had to find a way to escape. The gentle mother had turned into a veritable fury. She thrust herself towards the eagle and struck her horns at him so that his feathers flew about. One of her blows had broken a primary feather and made him falter. He rose and circled around the two chamois. Their piercing whistles rose up to him. Then the world turned gray. Color and light vanished. The clouds had gathered again, forming a wall between the eagle and the chamois. The contours of the circling bird became more and more hazy and pale until they were absorbed by the gray fog. The young got up. It looked at its mother, still with fear. Her breath came in jerks from her nostrils, and her protruding eyes were rolling with fury. It was still young and foolish, and it did not quite understand that she had just saved its life. It went close to her and pushed her in the flank. Then the angry fire disappeared from her eyes and she let him have the food he had asked for."

Related to the chamois is the MOUNTAIN GOAT (*Oreamnos americanus;* Color plate, p. 464 and p. 508) of the North American Rocky Mountains. There are only 1200 mountain goats left in the United States (in two national parks of the northern states of Montana and Washington) and perhaps a few thousand in their native habitat, the Canadian Rockies. It is not too difficult for a traveller to see them now. The locations of the individual herds are well known and the animals do not roam in the mountains but stay for weeks, months, and often the whole year in the same small areas, mostly above the timber line. There they do not have any real enemies, especially not in winter when all other animals flee from this terrible glacial wilderness. Then the mountain goat's hooves and horns, which are whitish grey in the summer, change to a blue-black. Sometimes only their dark eyes, black lips, horns, and hooves are visible on the shining white plain. These bold climbing artists do not jump about the rocks, but they walk slowly, almost deliberately, and with the dignity of experienced mountaineers. Uphill a person may almost keep pace with them. A herd of mountain goats, when surprised and disturbed, never dashes off in flight like the antelopes on the plains. Instead they suddenly appear in places on steep slopes or rocky crags where one never would expect to find a creature with four feet and no wings.

A mountain goat does not jump up very high, rarely more than one or one and a half meters. But it does not hesitate to jump down a wall seven to eight meters to a small ridge on the rock even though it is covered with ice. Seymour, the Mayor of Seattle, once observed an almost unbelievably clever trick, which as a precaution, he had confirmed in writing by four men who accompanied him. "Last August we saw a mountain goat going up the almost smooth walls of Little Big Chief. When it had reached a spot which was obviously not traversable, I shouted: 'Don't go on, you idiot. You'll never make it!' But this is exactly what the goat did. In several places it jumped up the full length of its body to the next ridge. Finally, it attempted a leap which it could not make. The next ridge was more than a goat's length away. Its front legs could barely reach the edge of the ridge, but they could not reach over far enough for the animal to pull itself up. We expected the mountain goat to come crashing down. Instead, it thrust itself away with all four feet at the same time, turned over completely in the air, and landed safely on all four feet on the same ridge from which it had started."

However, Brewster witnessed a few decades ago in the Banff Park that even the unbelievable artistic climbing skills of these animals may fail them on the precipitous slopes of the Rocky Mountains. Five animals were pursued by hunters and became trapped on the ledge of an overhanging wall so that they could neither move on nor return.

The Rocky Mountain goat, by B. Grzimek

They climb cautiously and safely

Fig. 16–4. Rocky Mountain goat (*Oreamnos americanus*)

Takins:

1. Takin (*Budorcas taxicolor,* subspecies a) Shensi takin (*Budorcas taxicolor bedfordi*), b) Mishmi takin (*Budorcas taxicolor taxicolor*), c) Szechan takin (*Budorcas taxicolor tibetana*).

Sheep oxen:

2. Musk ox (*Ovibos moschatus*) in the faded winter coat shortly before shedding.

a

b

l

c

2   ♂

♀

Helmut Diller

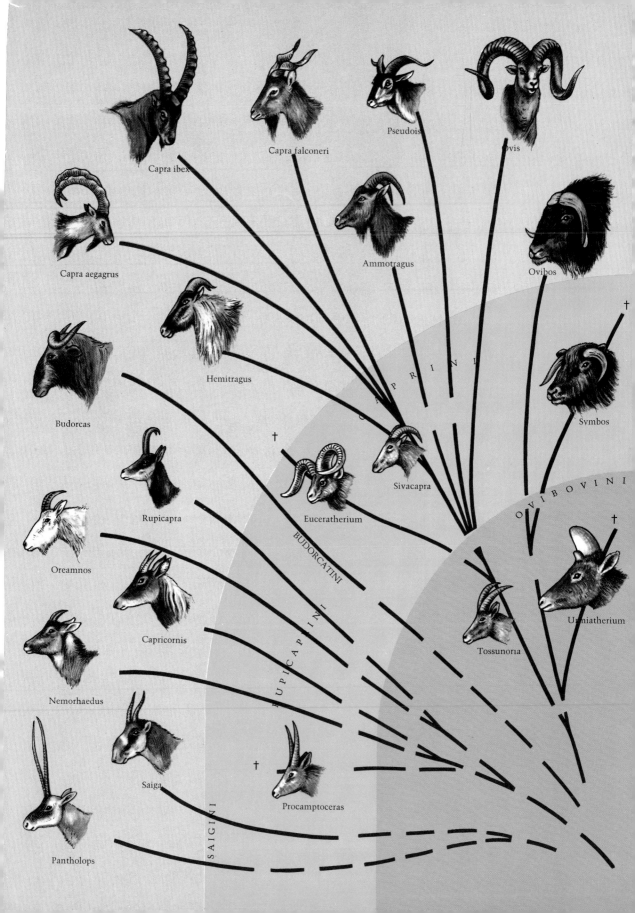

Capra ibex

Capra falconeri

Pseudois

Ovis

Capra aegagrus

Ammotragus

Ovibos

Hemitragus

Budorcas

Symbos

C A P R I N I

Rupicapra

Euceratherium †

Sivacapra

Oreamnos

O V I B O V I N I

BUDORCATINI

Capricornis

Urmiatherium †

Tossunoria

Nemorhaedus

R U P I C A P R I N I

Saiga

Procamptoceras †

Pantholops

S A I G I N I

The hunters had lost sight of the group, so they returned to their camp in the valley, from where they could see them again. The next morning the mountain goats were still standing on the same spot. During the following days, they fell, one after the other, usually at night, into the deadly abyss. The last of the animals fell after ten days.

At the end of the last century when the mountain goats in America were not protected as they are now, Mills described how hunters with their dogs cornered an old mountain goat buck at the end of a glacier in the area of the present Glacier National Park. The buck stood on the most precipitous part of the rocks at the edge of a steep wall. Waiting for his chance to escape, he made two attempts to flee. The dogs encircled him. He jumped toward one of them, and with a remarkably quick movement of his head, he pierced the dog with his sharp horns. Then, with a second movement he flung the animal into the gorge. In this manner he killed three of the dogs in rapid succession. The fourth was pushed alive over the edge. After that, the other dogs withdrew. Finally, he walked on over ledges and boulders so slowly and surely that it seemed as if the whole matter had not upset him in the least.

In alpine regions probably only the cougars are a threat to mountain goats. To our knowledge, only once has a young been killed by an eagle. Two catchers had separated a young mountain goat from its mother and it climbed away onto an almost smooth steep wall. At this moment, when it was without its mother's protection, it was seized by a bald eagle. Under normal circumstances, it certainly is not difficult for a mountain goat to ward off an eagle from her young. Probably the only case where mountain goats are seriously endangered is when they go down into the valleys or through woods. In some areas, they may be seen grazing regularly in the valley meadows. They have to cross valleys and woods in order to get from one mountain range to the other or to reach their salt licks. On these occasions, they may be attacked by grizzly bears, the smaller black bears, wolves, wolverines, and even the strong mountain coyotes. However, they are able to defend themselves, and a mountain goat often is not the only one to die. The bucks' horns, although only twenty-five centimeters long, are very sharp. Fenwick, a farmer in British Columbia, one day found a dead mountain goat which obviously had been killed by a grizzly bear. "I was surprised that the bear had not carried away and buried the goat as they usually do. Therefore, I began to look around and I found a huge grizzly bear dead and covered with blood. When I examined him, I found that the mountain goat had stabbed him twice right behind the heart. He had still been able to kill the goat, but then he had walked off and died."

Rain may be dangerous to mountain goats. In zoos, they need a shelter, or at least a roof; otherwise their woolly coat will be soaked with water and then the animals are likely to die of pneumonia. When

Warding off dogs

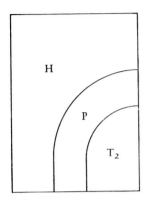

Phylogeny of the subfamily Caprinae. Thenius classifies them with the Caprinae as the tribe Saigini. He has also placed the Saiginae in a subfamily of its own.
1. Saigini: *Pantholops, Saiga.*
2. Rupicaprini, including gorals and serows: *Nemorhaedus* (Goral), *Capricornis* (Serow), *Oreamnos* (Rocky Mountain goat), *Rupicapra* (Chamois).
3. Budorcatini: *Budorcas* (Takin).
4. Caprini: *Tossunoria, Euceratherium, Sivacapra, Hemitragus* (Tahr), *Capra aegagrus* (Wild goat), *Capra ibex* (Ibex), *Capra falconeri* (Markhor), *Ammotragus* (Barbary sheep), *Pseudois* (Blue sheep), *Ovis* (Wild sheep).
5. Ovibovini: *Urmiatherium, Symbos, Ovibos,* (Musk ox). (T₂) Upper Tertiary, (P) Pleistocene ("Glacial Period"), (H) Holocene (present).

it rains in the wild, they find shelter under overhanging rocks or in caves. Their wool is more delicate than the famous cashmere wool. In some places, the clusters of their hair which have been shed may be found by the pound near bushes and protruding rocks. The Indians of the North American northwest coast formerly collected this wool and used it for spinning and weaving. This wool, the blankets, and their skins were the first things that the great discoverers saw of these alpine animals. Around 1860 it was fashionable to wear muffs and collars made of the black and white coat of the African guereza. Therefore, the price for these animals rose sharply, and the partly dyed skins of mountain goats were then used as imitations. This almost led to the extinction of these North American animals. Now these beautiful alpine animals are not only protected everywhere, they were also successfully introduced in 1923 on Baranow Island in Alaska and in the 1950's to Kodiak Island.

It is amazing that these alpine animals do rather well in zoos. A mountain goat buck lived in the New York Bronx Zoo from 1900 until 1909 and died at the age of ten years. When a young was born in May, 1963 in the zoo of Calgary (Canada), this was the first zoo birth of a mountain goat in thirty-five years. One of the catchers of the New York Zoo reported: "The little mountain goats did not present great problems for me. I fed them approximately every three hours, night and day. When they were hungry, they would either jump up at me or climb all over me, since I slept in the same compartment with them. Only when they were left alone did they become restless and try to get out. As soon as I had fed the poor little orphans, they all laid down, cuddled up closely against me, and went to sleep. When they became hungry again, they would kick me with their front legs." These little animals weigh about four kilograms at birth, and they are able to walk and jump half an hour later.

Lately, Geist has closely observed the mountain goats. According to his observations, the male mountain goats have a distinct inhibition against attacking females. They appear to be somewhat "hen-pecked." Usually the female, who stands higher in the rank order than the male, will occasionally soundly box him on the groin. In the beginning of mating, he approaches her in a crouching posture, his tongue stretched forward, and his legs flexed. This attitude, which looks very much like the "submissive posture" of the chamois, certainly is an emphasized "non-aggressive guarantee," or perhaps even an expression of inferiority. However, later on he treats her with the ritualized foreleg kicks (Laufschlag). While males and females usually stay together in small groups and the bucks often live singly, rutting males come to the female's herds and join them. In spite of the extensive similarity between the sexes, the males are easy to distinguish from the females at this time because they all have dark dirt spots on the rump and

Fig. 16-5. The courting Rocky Mountain goat buck approaches the female in a crouching posture.

thighs. These spots come from the bucks' habit of sitting down on the hind legs like dogs, and pawing with one front leg, throwing snow, sand, and soil against the bellies, flanks, and thighs. In this manner, they dig real "rutting dens." Between this pawing, or independently of it, they, like the chamois, mark branches or twigs with the large glands behind the horns.

According to Geist's observations, mountain goats are not territorial. The buck will occasionally defend the females, but not a place. Sometimes several courting males remain with the same herd of females and maintain a distinct distance between each other. If a quarrel breaks out between two bucks, the two stand in the reversed-parallel position beside each other. With stiff legs, they raise their back and erect the hair on the back. This behavior is in every detail the opposite of the humble "courting posture" in which the male approaches the female. If this display does not settle the argument, a fight will follow. The opponents circle each other, still standing reversed-parallel, and beat their horns sideways against the other's legs, flanks, shoulders, and thighs. Serious injuries and casualties result, as Geist proved from dissecting dead mountain goats.

Among all the horned ungulates, the mountain goat is one of the few species which rarely sustains great losses due to unfavorable climate or nature's catastrophes. They are exceptionally well adapted to their biotope, and because of the inaccessibility of their habitat, they are under very little "pressure" from natural predators, against which they can defend themselves quite well. Thus, of all horned ungulates, the mountain goat is the safest with respect to environment as well as enemies. Only in these animals do we find damaging fights between conspecifics. Is it too farfetched to see a parallel situation to that in humans?

Bernhard Grzimek
Fritz Walther

Fig. 16-6. Reversed-parallel stance at the beginning of an encounter between Rocky Mountain goat bucks.

17   Goats and Sheep

Everyone knows domestic goats and sheep and can easily tell them apart in spite of some similarities. Zoologists separate the GOATS (genus *Capra*) and the SHEEP (genus *Ovis*), especially the wild forms of the various species and within the genus, not only for conspicuous external characteristics, but also for differences in the bone structure, presence or absence of certain skin glands, characteristics in their behavior, and other aspects. Besides the sheep and the goats, three additional morphological groups which are clearly defined as genera of their own belong to the tribe Caprini.

The tribe Caprini includes five genera with nine species: 1. GOATS *(Capra)*, 2. BARBARY SHEEP *(Ammotragus)*, 3. TAHRS *(Hemitragus)*, 4. BLUE SHEEP *(Pseudois)*, and 5. SHEEP *(Ovis)*. The HRL is 100–200 cm, the TL is 3.5–25 cm, the BH is 60–125 cm, and the weight is 20–230 kg. The large and heavy horns have different shapes, and are weaker or absent in the ♀♀. The preorbital, postcornual, flank, and interdigital glands are absent in most of the genus, but the anal gland is usually present. There are one or (rarely) two pairs of nipples. The coat is often elongated into beards, manes, collars, crests, or cuffs.

The detailed phylogeny of the Caprini, as far as it is known, is illustrated on page 470. The genus *Tossunnoria* from the Early Pliocene of China (approximately ten million years ago) is considered the original form of all sheep and goats. Its skull structure combines characteristics of the tahrs and goats with those of the chamois-related species. It probably originated from rudimentary forms of the Rupicaprini. The actual wild sheep and wild goats are alpine animals from the high mountain regions. Therefore, their distribution is insular everywhere. Especially in the genera *Capra* and *Ovis*, the true sheep and goats, a specific form has developed on each isolated mountain region, separated from the others by valleys or plains through their overall range. Therefore, their classification into species is extremely difficult and the experts' opinions about the number of species and subspecies of these two genera differ greatly.

Keeping these alpine animals in zoological gardens usually does not present any problems as long as their special need for exercise is considered and they receive the appropriate food. However, some forms have become so rare in their native habitat that they are rarely seen in zoos. The pens should offer sufficient facilities for climbing, and an adequate ground covering should give sufficient abrasion for the hooves which, in all rock climbing animals, grow especially fast. Furthermore, it is necessary to have sufficient room so that the animals may get out of each other's way and sight, and can avoid having too many fights. It is not at all easy to construct the pen fences so that the wild goats, wild sheep and their relatives are not able to climb or to jump over them. Hediger reports on an ibex who jumped a fence of more than four meters high. Usually the animals do not attempt to leave the familiar, quiet pen and it is possible to get along with much lower fences.

The free ranging goats (genus *Capra*) are restricted to the mountain regions of Eurasia and North Africa. They are medium-large, compact, and have thick, strong legs. The parts of the skull behind the horns are rather long, and the apex region is arched. There are no lachrymal fossa. The horns of the ♂♂ are large, bulky, and mostly sabrelike, although in some species they have a different shape. The horns of the ♀♀ are small, straight, slightly bent backward, and flat on the sides. The coat is short but not shaggy. Males have a beard at the chin. The lower side of the tail is naked and in ♂♂ contains scent glands which secrete the typical strong "buck-scent." Following the suggestion of Haltenorth, we classify the wild goats into four species:

1. IBEX *(Capra ibex)*. 2. SPANISH IBEX *(Capra pyrenaica)*. 3. MARKHOR *(Capra falconeri)*. 4. WILD GOAT *(Capra aegagrus)*, the original form of the domestic goat.

Alpine, Nubian, Abyssinian, West Caucasian, East Caucasian, and Siberian ibex resemble one another, yet do not overlap in most of their range. Only in the ranges of the two Caucasian ibexes is there an area which is inhabited by a mixed or transitory form. Therefore, we

Ibexes

Distinguishing characteristics

include all these ibexes into one large species, the IBEX *(Capra ibex;* Color plates p. 463 and p. 481). The HRL is 115-170 cm, the TL is 10-20 cm, the BH is 65-105 cm, and the weight is 35-150 kg. The horns of the ♂♂ are different depending on the subspecies, but in no case do they have a sharp front edge or close spirals. The males' horns are 70-140 cm long, while those of the ♀♀ are 15-38 cm long, slender, and curved backwards. There are several subspecies, including: 1. The ALPINE IBEX *(Capra ibex ibex)*; 2. The WEST CAUCASIAN IBEX *(Capra ibex severtzovi)*; 3. The EAST CAUCASIAN IBEX *(Capra ibex cylindricornis)*; 4. The SIBERIAN IBEX *(Capra ibex sibirica)*; 5. The NUBIAN IBEX *(Capra ibex nubiana)*; and 6. The ABYSSINIAN IBEX *(Capra ibex walie)*.

Alpine ibex, by B. Grzimek and B. Nievergelt

In the European Alps the ibex lives at elevations up to approximately 3,500 meters. Emil Bächler found that "hot temperatures are very uncomfortable for ibexes, and they search for shade under overhanging rocks and rock walls, where they may stay for hours and ruminate. They also do this during their resting periods." In contrast, the ibex obviously can easily tolerate low, freezing temperatures. In the especially severe winter of 1962-63, the Swiss ibex colonies did not suffer any losses above average.

The ibex's habitat is above the timber line. They descend only in April and May to the upper tree zone. They almost never go into the dense forest. During the summer, the animals gradually migrate to higher regions. From the end of July until September or even October, they may be found in the highest regions of their habitat. The beginning of winter causes the ibexes to go to their winter range at lower elevations (at about 2300 meters in the Sapien Valley colony). These changes can probably be explained by the changing conditions of temperature and food. In summer the ibexes avoid the heat by moving to higher altitudes where they also find freshly sprouting forage plants. The ibexes need places either without or with little snow cover in winter. On the level valley floors the snow piles up, while it glides off the steep slopes which are more favorable. Furthermore, in winter the temperature decrease is lower depending on the altitude. On a beautiful day, it is often warmer on a sunny slope at a higher altitude than it is down in the valley. The highest parts of the region are also covered with deep snow, so that good winter quarters are usually at moderate altitudes. Only in the spring do ibexes migrate farther downhill where the fresh greens first begin to grow. The seasonal or daily marches usually follow along a slope, over a crest, or a summit. All the colonies I (Nievergelt) know avoid larger valleys below the timber line and the adjoining plain slopes. Only occasionally will single animals cross a valley. Thus, valleys and glaciers are effective but certainly not insurmountable boundaries for the distribution of the ibexes.

Formerly ibexes lived in most areas of the Alps, or at an earlier period even in other European mountains. The Romans imported them to Rome for their tournaments. The ibexes fell victim to medieval superstitions and popular remedies. People thought they had found in them a cure for all kinds of human maladies. Even in the last century, an ibex was, so to speak, a walking pharmacy. The blood was considered a remedy for callouses (because these animals live on rocks). Rings to be worn on the fingers were made from the horns and were believed to "protect" against many diseases. The Bezoar stones (round balls of hair, resin, pebbles, etc., which occasionally form in the ibex's stomach) were said to be effective against cancer. Even the feces were collected and eaten as a cure for tuberculosis and gout. The heart-shaped ossified tendons of the heart muscle were said to have miraculous curative powers. Therefore, whoever killed an ibex was a person whose fortune was assured. The bishops of Salzburg and

Fig. 17-1. Original distribution of the ibex (Capra ibex). Details on the following maps.

Fig. 17-2. 1. Nubian ibex (Capra ibex nubiana); triangles: present distribution; crosses: exterminated. 2. Only occurrence of the Abyssinian ibex (Capra ibex walie).

Fig. 17-3. In the black parts of the Alps in France, Italy, Switzerland, Austria, and Yugoslavia, Alpine ibexes (Capra ibex ibex) are living again. The state boundaries are striated and the rivers and the Mediterranean and Adriatic coast line are marked by an uninterrupted line.

Berchtesgaden maintained their own "ibex pharmacies" where these remedies were sold.

Therefore, it is not surprising that even the strictest laws for their protection could not save the ibexes. There were greater problems because the average person considered poaching more a daring, romantic sport than a crime like burglary, robbery, or theft. Therefore, the Bishop of Salzburg had no other choice than to dismiss, one by one, the six hunters which he had hired to protect the ibexes in the period from 1712 to 1720. There was nothing left for them to do.

The Swiss ibexes were no better off. In the Middle Ages, they were spread over wide areas of the Swiss Alps. Gradually they began to withdraw from the Central Alps and die out. The last ibex in Glarus was killed in 1558, and the last one in the Gotthard region died in 1583. In the Grisons in 1612, regulations for their protection were strengthened, but to no avail. The last ibex of the Grisons was seen in 1650. It was exterminated in the Waldstätten region in 1661. Ibexes vanished from the Bernese Alps between 1750 and 1800, and from their last refuge, the Valais, between 1800 and 1850.

How the ibex was saved

We would know about ibexes only from clumsy old wood cuts if it had not been for forester Josef Zumstein and natural historian Albert Girtanner who, in 1816, urged the administration of Piedmont to protect the last few dozen ibexes in the Gran Paradiso. Fortunately, the kings of Sardinia considered hunting of ibexes their exclusive privilege and they jealously saw to it that nobody interfered with it. In 1854 the Gran Paradiso became the personal property of King Victor Emmanuel II and he declared the ibex as royal game, so they were continually guarded by a company of foresters. All presently existing ibexes are derived from the Gran Paradiso.

At first people tried to preserve the last ibexes in zoos and game parks, but they were not experienced in breeding game species in captivity at that time. Therefore, a last remedy was to place domestic goats with the male ibexes because the female ibexes gradually died off. The offspring of such crossbreeds are fertile. Thus, in the Vienna Zoo Schoenbrunn and in the Salzburg Zoo Hellbrunn, which was dissolved in 1866, all the animals were exclusively crossbreeds of domestic goats and ibex. When their descendants later were returned to the wild in the Alps, they perished in a short time.

The Swiss increased their efforts to reintroduce this magnificent alpine animal. In 1892 near St. Gall, they founded the Peter and Paul Game Park. But, unfortunately, the only source for purebred ibexes were the Italians in the Gran Paradiso. For some time the Italians would not oblige. The ibex had become an Italian monopoly. At this time the Swiss seriously considered poaching and smuggling live ibexes from the Gran Paradiso, thus repeating in reverse that which had happened when the ibexes were exterminated. In 1906 the first purebred ibexes came to the Peter and Paul Game Park, and they had been smuggled. Since then they were bred purely, but were later augmented by legally purchased animals from Italy.

Breeding ibexes no longer presents a problem to modern zoologists. From one ibex pair which had been given to us by the game authorities of the Grisons in July of 1954, we had twenty-one offspring by 1967. One female had fourteen young.

Breeding in zoological gardens

In the beginning in the Gran Paradiso, no one knew how to capture free-ranging ibexes alive. This is much more difficult to accomplish than it is for the privileged hunter or poacher to kill "royal bucks." At first, they chose the easiest method. They killed the mothers and took the young. This method, which is also practiced with gorillas, orang-utans, and many other similar species, has the disadvantage of fertile, fully grown females being senselessly destroyed. Moreover, it was not possible in this manner to raise the young who were captured without proper care and the mother's milk. The people in charge of reintroducing ibexes experienced great frustration. This happened especially one time when they turned a mother with her young loose in the new environment. Immediately, a golden eagle came and killed the young not far from the escort team.

In Switzerland the first five ibexes were released in the wild on May 8, 1911, in the Grey Horns, a mountain range in the Canton of St. Gall. One of them refused to accept freedom and kept returning to the temporary enclosure in which they had been kept. He was finally brought back to the Peter and Paul Game Park. But the others became feral and reproduced. At present their protection in Switzerland is more effective than it was formerly. In 1953, when a man poached two ibexes, he was fined six thousand francs.

Meanwhile, ibexes have been reintroduced in many areas in the Alps of Italy, France, Switzerland, Austria, and Germany. In many of the places it was successful. In 1966 in Switzerland more than forty colonies of ibexes, totalling over thirty-five hundred animals, existed. Only in the German Alps did the plan fail. The ibexes, which were turned loose in 1922 in the Hagen Mountains south of the Koenigssee, succumbed to chamois rubbers and only ten of them survived. They moved over the border into Austria and joined the resident ibexes there. They migrate to German territory only occasionally. The total number of ibexes now living in the Alps is approximately eight thousand.

In the Swiss ibex colonies, Bernhard Nievergelt has thoroughly studied the social life of the alpine ibex from 1961 to 1964. From spring until fall, the males and the females with the young and subadult animals form groups of their own. During the rut in December and January, the males associate with the female herds. After rut abates, the bucks again begin to associate with others.

Social life of the ibex

Fig. 17-4. A fighting ibex tries, if possible, to get above his opponent, which he may accomplish by standing on a rock or by trying to circle upward around the other on a slope (F. Walther).

Fig. 17-5. In the mating behavior of the ibex, driving of the female is only weakly developed. The ibex in rut stretches his neck and head forward, sticks his tongue out, and flaps his tail on his back. Then he often vibrates his tongue, beating quickly up and down with the tip. Sometimes he makes a hesitant, perhaps "ritualized," pendulous movement of his front leg towards the goat. Walther compares this with the "Laufschlag" (fore-leg kick) of some of the antelopes (see fig. 11-9).

Certainly such an association of ibex does not constitute a permanent group. It has often been observed that the composition of the group varied. However, these groups are rather stable over a long period. Apparently, the members of a group know each other individually. One day a strange female ibex tried to associate with the ibexes of the Zurich game park, Langenberg. After she was missing for two days, a game warden found her dead, stabbed by one of the resident females.

The female herds are not formed at random. Usually female ibexes associate only with young or subadult animals. Mixed herds, where females live together with young and subadult animals, are rarer. The young bucks usually leave the female herds at the age of two to four years and then join the male groups. This transitionary age may vary in different colonies. There seems to be a correspondence between the time the young bucks join the male groups and the speed and strength with which they have matured. This occurs mainly in those ibex colonies which were reintroduced into an area and then reproduced especially well in the new habitat.

In summer the males frequently fight with each other, mainly in the morning and in the evening. But these fights are ritualized and usually are not meant to be very serious. Such playful fights occur almost exclusively between peers. Younger and weaker bucks avoid the superior ones. In peers, where the relative strength of the animals is not obvious, the rank order is established in a duel. During the winter rut, bucks of equal strength usually avoid each other, so the buck herds then consist of animals of different ages with a well established rank order and hardly any play fights. I observed during the rut in the Piz-Albris colony (in the Grisons) three fights which appeared to me to be without playful overtones. They gave the impression of serious fights. Such fights probably occur only when two rivals of equal strength from different herds accidentally meet.

Fighting ibexes rise on their hind legs (see fig. 11-5) and then beat their horns together from above. The bucks also cross their horns and push with their foreheads, or they wrestle sideways with their horns hooked while standing parallel to each other.

Rising on the hind legs, which is part of the threatening behavior of the alpine ibex, is the same in ritualized fights. But then the animals threaten from a distance, which does not invite an immediate fight. Furthermore, ibexes threaten by the mere presenting of their horns.

After a gestation period of 150 to 180 days, the female ibex gives birth to usually only one young at the end of May or in June. Twins are rare. As in all alpine game, close contact between the mothers and the young is conspicuous. Walther reports: "During the very first hours, the mother follows the young. Later the young follows the mother, usually in direct contact with her hind leg. If the young lose contact when their mother walks too rapidly, they gallop to catch up. At about fourteen days, the young reach a phase where they become noticeably more independent. They jump about on rocks from the first day, but only if the mother is with them or stands next to the rock. Now they play around the grazing mother in large circles and try their skill at climbing all by themselves. By four weeks, the young form groups with other juveniles, in which playful fights, where they rise on the hind legs, mount from the side, run and jump, or chase and pursue, are practiced, and where they probably have the first fights for rank order.

During this time, the mothers still watch their young carefully, but they do not interfere in their fights. Once, when a very lively and long-lasting fight, which was not altogether playful, broke out between two young six week old bucks, all the mothers approached in a trot, stood in a circle around the opponents and watched the fight. This is in striking contrast to fights between adult bucks, to which females pay no attention. However, none of them interfered."

Reproduction, growth, and longevity in ibexes are not dependent just on environmental conditions, but also on the density of the existing population. In captivity, female ibexes often have their first young as early as their second year. In the wild, they reproduce no earlier than three to six years of age. In the small number of recently settled colonies, the reproductive rate is distinctly higher and the sequence of generations is shorter than in heavily populated older stocks. Longevity seems to be a factor. In colonies with a high rate of increase and fast growth, the bucks die at a younger age. In the large ibex population of the Piz Albris colony, the animals now became pubescent at a later age than when it was founded forty years ago. Thus the growth of this ibex population has ceased. The males and probably also the females here reach an older age than animals in the more recently founded populations. The age of a male ibex can easily be determined by the annual growth rings of the horns (see fig. 17-17). This is much more difficult to do in female ibexes. In zoos ibexes may become quite old. Two females in the Vienna Zoo of Schoenbrunn had to be killed when they were seventeen years old because of old age symptoms.

Today the only predator of the ibex in Switzerland is the golden eagle, but it is usually dangerous only to very young animals. Wolves, lynx, bears, and the great bearded vulture, which are natural predators of the ibex, no longer exist in most of the alpine ibex's habitat.

The NUBIAN IBEX (Capra ibex nubiana; Color plate, p. 481) is small and delicate. The BH is hardly 80 cm. They have very large ears, and older ♂♂ have a long beard. The front edges of the horns are narrower than in the alpine ibex. The horns grow faster in the beginning so that the age of young ♂♂ may be easily overestimated. They live in rocky

Wild goats (only males are portrayed on the picture; the females have slightly bent horns which are much shorter):
1. Ibex (Capra ibex), subspecies a) Alpine ibex (Capra ibex ibex), b) Siberian ibex (Capra ibex sibirica), c) West caucasian ibex (Capra ibex severtzovi), d) East Caucasian ibex (Capra ibex cylindricornis), e) Nubian ibex (Capra ibex nubiana), f) Abyssinian ibex (Capra ibex walie, young male).
2. Spanish ibex (Capra pyrenaica)
3. Markhor (Capra falconeri), subspecies a) Astor markhor (Capra falconeri falconeri), b) Suleiman markhor (Capra falconeri jerdoni);
4. Wild goat (Capra aegagrus)

▷

The nubian ibex, by M. Schnitter

1♂ 2♂ 3 4♀ 5♂ 6♂

Helmut Diller

Races of the domestic goat *(Capra aegagrus hircus):*
1. Colored German goat
2. African pygmy goat
3. White angora goat
4. White German goat
5. Valais goat
6. Drooping-eared goat

deserts, since they are strictly dependent upon precipitous walls regardless of their geological nature or the elevation. They inhabit the granite Sinai Mountains up to 2,800 meters as well as the chalk cliffs near the Dead Sea, where they descend as low as 350 meters below sea level (see map, fig. 17-2).

Presently the distribution of the Nubian ibex is limited. In the Judaean and the Negev Deserts of Israel, where strict laws protect them, Schnitter estimated that there were approximately fifteen hundred head. Perhaps there are also ibexes on the Sinai Peninsula and in Arabia. In Egypt some live in protected areas, in mountains along the Red Sea, in the Karora Mountains in the Sudan, and on some other mountain ranges. The Nubian ibex is mainly endangered in areas where Arabs live. The Bedouins build blinds out of boulders like towers near the water holes and they kill the animals when they come daily to drink in the dry season.

In the rocky deserts of Israel, I observed the ibex in his habitat and recognized the unmistakable ibex characteristics. The buck likes to stand on a ridge, where he is outlined against the sky. In flight, in his social life, and in his mating behavior, he shows true ibex characteristics. In Israel the rut begins with the small winter rains in October and reaches its peak in November. The breeding season there lasts from the end of March into April. Twins or even triplets seem to occur frequently. In the Sudan the rut is said to begin with the July rains. There the main predator of the ibex, besides man, is the leopard.

The Abyssinian ibex, by B. Nievergelt

The ABYSSINIAN or WALIA IBEX (⚲ *Capra ibex walie;* Color plate, p. 481) is the southernmost form of ibex. Externally it more closely resembles the alpine ibex, since it is much larger and heavier than the dainty Nubian. It now lives only in the Simian Highlands of Ethiopia, a 4,550 meter high mountain northeast of Gondar. Its habitats are the heather forests (*Erica arborea*) and bands of meadows which section the northern precipices that are more than a thousand meters high. The Walia ibex is one of the most endangered species of animals; it would be a generous guess to say that there are still two hundred left. In two weeks spent in the ibex area, Nievergelt observed that among the fifty-six animals he saw, about one fifth were young, and one fourth were sub-adult animals. Thus the rate of reproduction is good. Unfortunately the animals are still severely limited. A national park, which would enclose the present ibex area in Ethiopia, is planned. Its establishment and effective enforcement of regulations would augur well for the preservation of this form of ibex (see map, fig. 17-2).

Asiatic ibexes, by A. G. Bannikow and W. G. Heptner

The WEST CAUCASIAN IBEX (*Capra ibex severtzovi;* Color plate, p. 481) and the EAST CAUCASIAN IBEX (*Capra ibex cylindricornis;* Color plate, p. 481) are considered by some zoologists to be different species, although the existence of a mixed or transitory population in the area between the boundaries of the two forms speaks in favor of their being one species. The oldest scientific name for the CAUCASIAN IBEXES, *Capra caucasica*, was given in 1783 by Güldenstaedt and Pallas to an animal of this mixed population.

The two forms of Caucasian ibex live at times in the mountain forests, where they feed mainly on trees, shrubs, tree bark, fir needles, and tree lichen during the winter. They also go to regions above the timber line where their main diet is grass and herbs. In contrast, the SIBERIAN IBEX (*Capra ibex sibirica*), which is similar to the alpine ibex, inhabits barren rocky slopes without trees. In the high mountains of Central Asia, it lives in a broad region between five hundred and five thousand meters elevation. In Pamir where the region of permanent snow is at a high altitude and human settlements are quite high, the ibexes live between three and five thousand meters, or even at higher elevations.

Fig. 17-6. Spanish ibex (*Capra pyrenaica*); triangles: still existing, but greatly endangered; crosses: recently destroyed completely.

In the Caucasus and in Central Asia, the ibex is not as endangered as it is in the Near East and in Europe, although the populations there have decreased in many areas. The Soviet Union has assured the existence of the subspecies within her borders in vast protected areas. The natural predators of ibexes in the Caucasus are the leopard, wolf, and lynx. In Central Asia the snow leopard takes the leopard's place. There the ibex and especially their young may also fall victim to the red wolf, the wolverine, or the bear. The golden eagle and the fox present additional dangers to the young ibex. In some areas, the ibex populations are threatened by the rubbers caused by the mite *Acarus siro* and from which masses of animals die miserably, emaciated, hairless, and blind.

The SPANISH IBEX (⊕*Capra pyrenaica*; Color plate, p. 481) is not as closely related to the true ibexes. The HRL is 130–140 cm in the ♂♂, and 100–110 cm in the ♀♀; the TL is 10–15 cm, the BH is 65–75 cm, and the weight is 35–80 kg. The horns of the ♂♂ are curved like the Caucasian ibex's. The inner side has edges in the front and rear and bulges across the front edge. The horns measure up to 100 cm. The ♀♀ horns, which reach 20 cm, are first curved backward and then upward again. The ♀♀ do not have a neck mane or black streak on their backs or flanks. The summer coat is light to red-brown. There are four subspecies, of which the PYRENEES IBEX (*Capra pyrenaica pyrenaica*) was exterminated approximately fifty years ago, and the NORTHWEST IBERIAN or SIERRA DOGOREZ IBEX (*Capra pyrenaica lusitanica*) died out in 1892.

The Spanish ibex lives in high rocky mountainous regions with scarce plant growth and feeds mainly on grass, herbs, and lichen. The Southeastern subspecies, the SIERRA NEVADA IBEX (*Capra pyrenaica hispanica*), occurs at lower elevations of one to two thousand meters in the dense brush of genista, rhododendron, and mastic. This subspecies is one of the most endangered large animals on the European continent; their total number in 1967 was estimated at about twenty animals.

The Spanish ibex

Distinguishing characteristics

The Markhor, by
A. G. Bannikow and
W. G. Heptner

The MARKHOR *(Capra falconeri)* at first glance seems to be completely isolated from the wild goats because of the different shape of its horns. At a closer look, however, it turns out that the conspicuous windings of the markhor's horns and the corresponding skull structure are present in the Spanish and the East Caucasian ibex to a lesser degree. The horn structure of the markhor, furthermore, shows a certain similarity to that of the wild goat.

Distinguishing
characteristics

The markhor is one of the largest species of this genus. The HRL is 161–168 cm in the ♂♂, and 140–150 cm in the ♀♀; the TL is 8–14 cm, the BH of the ♂♂ is 86–100 cm, and the weight is 80–110 kg in the ♂♂, and 32–40 kg in the ♀♀. The horns of the ♂♂, which are close to very close together, are V-shaped or at an acute angle which is directed backward-upward. The right horn turns left and the left horn turns right around its longitudinal axis. They are either loosely or tightly screwed, and the length of the horns along the windings are up to 160 cm. The horns of the ♀♀ are short (up to 25 cm), bent more backward, and screwed like the males'. The coat is short and smooth in summer and somewhat longer in winter. In the north its fleece is thicker, and in the south it is thinner or absent. The ♂♂ have a long chin beard, a long back mane, and long fringes of hair at the throat, chest, and shanks. The ♀♀ have a thin beard and no fringes. There are seven subspecies, among them the BUKHARA MARKHOR *(Capra falconeri heptneri)*, the ASTOR MARKHOR *(Capra falconeri falconeri;* Color plate, p. 481) and the SULEMAN MARKHOR *(Capra falconeri jerdoni;* Color plate, p. 481).

Habits of the
markhor

The markhors inhabit medium to high elevations in the Asiatic Mountains around and above the timber region. In the Soviet part of their distribution, the animals avoid the dense brush of the Central Asiatic juniper and may be found most frequently at the slopes of deep gorges where there are rocks and small meadows covered with grass, herb-like steppe plants, and isolated bushes. Their seasonal migrations are largely determined by environmental factors. In the Kugitangtau Mountains, the markhors move in winter only a few hundred meters deeper within the junipers. Only in winters with heavy snowfalls do some of them descend farther down to the steppe belt, near human settlements. In contrast, in the Kuljab region there is a difference of many hundreds of meters in altitude between their summer pastures and the winter quarters. The rut is, like in the ibex, in winter. After a gestation period of six months, the female gives birth to one or two young between the end of April and the beginning of June.

In their fighting behavior, markhors resemble the ibex. They rise less frequently on the hind legs to clash their horns together from above, and they turn their heads more to the side so that blows hit the opponent diagonally. Often they push straight against their rival's forehead with their horns; they lock the horns together and try to force the rival back. Sometimes they grapple with their horns while they

Fig. 17-7. Markhor *(Capra falconeri);* original distribution.

are walking parallel to each other. Threatening markhor bucks also rise on their hind legs much less frequently than do the ibexes. "The most frequent and conspicuous display posture of the adult markhor buck is the lateral display," reports Walther. "The markhor stands with his broadside to the opponent and pulls his chin toward his throat so that his horns point to the sky vertically or slightly forward. Thus the lower half of the face is hidden in the fluttering throat mane. The hair on his back stands erect, the horns point slightly to the side toward his opponent, and the usually erect tail wags to and fro. In this posture, the buck stands in front of the opponent or walks beside him."

The most important species of true goats for man is the WILD GOAT (*Capra aegagrus*; Color plate, p. 481), the original form of the domestic goat. The HRL is 120–160 cm, the TL is 15–20 cm, the BH is 70–100 cm, the weight is 25–40 kg. The horns of the ♂♂ are 80–130 cm (sometimes up to 150 cm) long, and laterally compressed with a narrow base. They are scimitar shaped; the lower and center parts are gently curved, while the upper part is more sharply curved. The windings are almost on one level, and the ends are bent slightly inward. The inner front edge is sharp, and in *Capra aegagrus blythi* it is almost smooth. Otherwise, there are six to twelve knobs with sharp edges, whose number does not necessarily depend upon the age. The rear edges of the horns are sharp or rounded, and the seams of annual growth are distinct. The horns of the ♀♀ are only 20–30 cm long, thin, and only slightly curved. The ♂♂ beard is dense and long with no fringes at the neck or chest. The ♀♀ do not have a beard. There are four (or five?) subspecies.

Formerly, the wild goat was widely distributed in the mountains of Asia Minor and on the Greek islands. At present their number has greatly decreased in large parts of their former distribution. In many areas they are completely extinct.

In the west of the Greek island of Crete are the Lefka Ori, the White Mountains, whose rocky crags of more than two thousand meters have snow until June, and whose bright chalky formations can be seen later in the summer. Here we find a subspecies of the wild goat, the CRETAN WILD GOAT or the AGRIMI, as the Cretans call it *(Capra aegagrus cretica)*. It is distinctly smaller and daintier, but at the same time it is prettier and more distinctly colored than the ASIA MINOR WILD GOAT *(Capra aegagrus aegagrus)*.

Wild goats were never evident, not even as fossils, on the European continent. Only one island in the Eastern Mediterranean, the small rocky island of Erimomilos near Milos in the Kyklades, has wild goats *(Capra aegagrus pictus)* as has Crete. But there they have been crossed with domestic goats and so are no longer pure. It is possible that the goats in the southwest of the Aegean island Samothrake have crossed with wild goats. All the other free-ranging goats on the Mediterranean islands are feral domestic goats and not true wild goats.

The wild goat, by Th. Schultze-Westrum

Distinguishing characteristics

Cretan wild goat or agrimi

Blue sheep *(Pseudois nayaur)*; the buck is above; below left is a goat with young; on the right is a sub-adult animal.

Helmut Diller

Fig. 17-8. Wild goat (*Capra aegagrus*), original distribution. Th. Schultze-Westrum reports on this page about the rather small habitats of the wild goat on the Mediterranean island of Crete.

The evidence of wild goats on Crete goes back as far as the Early Minoic Period. At this time, the animals were portrayed on seals, cameos, paintings, and on bronze and ceramic vases. At the beginning of our century, they occurred in all three large mountain ranges of the island (Lefka Ori, Ida, Lassithi). Then modern firearms were introduced. Now only a small number of them survive in the Lefka Ori, the most impassable mountains of the island. They outlived the guerilla warfare of World War II only because they lived in inaccessible habitats in the cliffs and the sheltering woods on the rocky slopes. Since then the Cretan wild goat has been threatened with extinction. Poachers in the mountain villages, who consider the goats their traditional game, have been armed since the war with ammunition and modern rifles. Furthermore, the agrimi's natural habitat is constantly decreasing due to excess felling of timber and the herds of domestic goats in the valleys and on the plateaus, which force the game animals to the small central regions.

Soon after 1945, Greeks in Athens and on Crete interceded for the preservation of the endangered agrimis. The government's game authorities in Athens, Cretan foresters, lawyers, associations of mountain climbers, hunters, and private persons organized campaigns to inform the people in the mountain villages. The agrimi was declared the national animal. These efforts were energetically supported by the Greek government and international associations for the conservation of nature. Presently a national park is being established in the wild goat areas in the White Mountains. There are still financial problems to be solved before the necessary expansion of this protected area may be guaranteed.

The number of free-ranging agrimi in the White Mountains in 1963 was estimated by the authorities to be about three hundred animals, although the exact figures are probably lower. Since the end of the war, only about five agrimis were seen by people other than natives. Fortunately, the population has remained pure in spite of many domestic goats in the area.

In 1930 one agrimi pair was set free on the small island of Theodoru off the coast of northern Crete. The animals have reproduced well and behave like true animals of the wild. They have a varied diet and good shelter in the cliffs. These animals constitute a reserve in the case of an emergency. They will be transferred to the highlands only if the population in the White Mountains decreases too greatly.

Other wild goats

In the small Soviet area of their distribution, according to A. A. Nasimowitsch, there are probably still several thousand wild goats, although they were much more abundant fifty to one hundred years ago.

Barbary sheep (*Ammotragus lervia*); below left is an old buck.

In all places where the wild goat is not effectively protected, it is hunted for its skin and horns. Unfortunately, an increasing number of these animals in the Near East fall victim to gun-toting European tourists, even in protected areas. In even greater demand than the skin and the horns of the wild goats were (and still are, to some extent) the "bezoars," which also caused the severe persecution of the alpine ibexes. These "stomach stones" are round, tightly matted balls of hair which they lick off. Since they are not digestible, they remain in the stomach and after a while turn as hard and as smooth as rocks. Popular superstition makes them into a panacea for all kinds of maladies. Due to these erroneous beliefs, many species of animals have been decimated and are still threatened with extinction; for example, the powdered horns of the rhinoceros are considered an aphrodisiac, or the planned killing of the South African horned ungulates occurred because of the misconception that Nagana and sleeping sickness could thus be eradicated.

In different parts of their distribution, the wild goats live in a variety of habitats. In the mountains of Dagestan and Groznyy, they frequently live at altitudes between 1,200 and 2,200 meters, on steep, rocky slopes, or on gentle slopes. They prefer the vicinity of forest or brush, while the ibexes of the same area are found much more frequently on slopes without trees. In the Armenian Mountains, which have been largely cleared of timber, they live at altitudes between 550 and 3,200 meters, and on Mount Ararat they even climb up to 4,200 meters. They also may be found on alpine steppes. In the mountains of Bolscoj Balchan, they were even found at elevations as low as one or two hundred meters above the Caspian Sea, where they inhabited the semi-desert at the foot of the mountains. Near Krasnowodsk they were seen in winter at sea level.

The rut of the wild goat also takes place in winter, but it does vary slightly in different areas. The females give birth, usually in April or May, to one or two young, or sometimes even to triplets. Their social life does not seem to be different from other wild goats.

The wild goat was domesticated by the seventh century B.C., as Charles Reed was able to prove by prehistoric fossils. This is even earlier than cattle. Ever since, the DOMESTIC GOAT (*Capra aegagrus hircus;* Color plate, p. 482) has served man as a source of meat, fine leather (Glace, Nappa, Saffian, Velour, and Chevreau), and mainly milk. Milk production is the main reason for keeping goats. In the 1930's, especially in Central Europe, the "poor man's cow" played an important part. Record yields of two thousand kilograms of milk annually were obtained. Goat's milk has approximately the same water and protein content as cow's milk, but more fat (4.3 per cent compared to 3.9 per cent) and less lactose (4.3 per cent compared to 4.9 per cent). Suddenly the bogy of "goat's milk anemia" appeared. This is an anemia allegedly caused by a heavy use of goat's milk. Doubtlessly, there were other causes which were primary or at least contributed to it. For ex-

The domestic goat, by H. Kraft

ample, the diet of the poorer people at this time was very low in vitamins. Goat's milk certainly is not harmful as long as the remainder of the diet is sufficiently varied.

As in wild goats, the udder has only two nipples. The goat gives birth to only one or two, or in very rare cases even three young. Births usually occur in April or May. As early as fall of the same year, the young goats may become pregnant, giving birth to their first offspring at the age of one year.

Keeping and care

Goats are clean and prefer tasty food which fills the rumen but is not necessarily very rich. Choosy animals, they test the food for scent and taste. They especially like leaves. When they are fed with leftovers from the kitchen and the barn, they give very little milk. Like their wild ancestors, the domestic goats prefer living on steep slopes. They appreciate their independence and it is hard for them to become accustomed to living in a barn. They do not like to be alone and prefer contact with conspecifics or other animals. The experienced goat keeper takes good care of the goat's grooming to keep it free from parasites and to prevent an unpleasant "goat smell" in the milk. The care of the hooves is important because the very strong hooves of these climbers do not wear off in a stall. When they are neglected, this may lead to serious and often painful damage to the legs.

Races of domestic goats

As in most domestic animals, a multitude of forms and colors also occurs in the domestic goat. Males and females may have horns or be hornless. Both sexes have two skin appendages at the throat, and the males have beards.

In earlier centuries, the center of goat breeding was in the Alpine regions of Switzerland. In 1796, 7,676 goats were counted in the Canton of Glarus alone. There are no longer that many today. The most famous Swiss goat breed is the hornless white Saanen's goat. Because of its ample milk production, it is widely distributed in the Alps. They are eighty to one hundred centimeters high, and well fed and castrated males may weigh eighty kilograms. The white Appenzell's goat is smaller and more compact, as is the light brown, short or long haired Toggenburg goat. The Wallis goat (Color plate, p. 482), which is black at the front part of the body and white at the rear, has been known for centuries. While this long haired breed is sturdy and easily satisfied, it does not produce much milk. Different breeds of the chamois-colored Alp goat are also distributed in the countries around Switzerland, but they are no longer of great importance.

In Central Europe and especially in Germany, there are only a few breeds of goats. The best known are the WHITE GERMAN GOAT (Edelziege) (Color plate, p. 482) and the COLORED or LIGHT BROWN GERMAN GOAT (Edelziege) (Color plate, p. 482). They are two hornless breeds with good milk production, but their numbers are greatly reduced at present.

In Asia Minor and South Asia there are, among other breeds, the fine-haired CASHMERE GOAT and the especially long-haired ANGORA GOAT (Color plate, p. 482), from whose hair mohair is made. In Africa the DROOPING-EARED GOATS (Color plate, p. 482) and the PYGMY GOATS (Color plate, p. 482) are widely distributed, and they are kept by the Africans for their milk and meat. There the herds of goats are often driven over long distances because of rather scarce pasture. Single goats there may become the prey of leopards.

The appearance of the BARBARY SHEEP (*Ammotragus lervia;* Color plate, p. 488) is more like a goat than a sheep. It has a goat-like short tail with a naked underside and glands. They do not have preorbital, flank, and interdigital glands, and they have some bone characteristics which resemble goats rather than sheep. Therefore, it has been suggested several times that this species be called the barbary goat instead of the barbary sheep. However, J. Schmitt was able to show in the Frankfurt Zoo by electrophoresis that the protein in the barbary sheep's serum is more like the sheep's than the goat's, so that the barbary sheep is closer to the sheep after all. Grzimek found the best compromise for the special situation in the genus *Ammotragus* when he applied the German word "Mähnen-springer" (maned jumper) to them.

The HRL is 155-165 cm in the ♂ and 130-140 cm in the ♀. The TL is 15-25 cm, the BH is 90-100 cm in the ♂ and 75-90 in the ♀, and the weight is 100-140 kg in the ♂ and 40-55 in the ♀. The long head has a straight facial profile. The eyes are large, the ears are small and pointed, and the nostrils are oblique. The naked nose region consists of a narrow field around the nostrils which extends from there in a small center line to the edge of the upper lip. The horns, which are similar to the ibex's, are triangular when cut across, and relatively strong in the ♀♀. The length of horns extends up to 80 cm in the ♂ and to 40 cm in the ♀. They have no beard, but they do have a short mane on the crest of the neck, and a long mane at the throat, chest, and fore-limbs. In strong ♂♂, the manes almost reach to the ground, while in ♀♀ they are considerably weaker and shorter. The ♀♀ have two nipples. There are six subspecies.

Barbary sheep inhabit the rocky, barely accessible desert mountains of North Africa. They live singly or in small groups and are excellent, surefooted jumpers and climbers. From the prolific breeding group of the Frankfurt Zoo, some barbary sheep were to be captured to be transported to another zoo. When several keepers tried to catch the chosen animals with nets, all members of the group easily jumped over the fence of the enclosure, which was more than two meters high. When the keepers withdrew, the animals, one after another, jumped back over the fence into the familiar pen. In their desert habitat, the animals rest during the day time, especially at noon, and they roam about at night. In the morning and in the evening, they search for food.

Barbary sheep, tahr, and blue sheep, by D. Heinemann

Barbary sheep

Distinguishing characteristics

Habits

Fig. 17-9. Barbary sheep (*Ammotragus lervia*); approximate present distribution of the greatly endangered subspecies: 1. Mauritanian barbary sheep (*Ammotragus lervia lervia*), 2. Tripolis barbary sheep (*Ammotragus lervia fassini*), 3. Egyptian barbary sheep (*Ammotragus lervia ornatus*, destroyed), 4. Kordofan barbary sheep (*Ammotragus lervia blainei*), 5. Air barbary sheep (*Ammotragus lervia angusi*), 6. Sahara barbary sheep (*Ammotragus lervia sahariensis*).

Fig. 17-10. Two barbary sheep bucks try their strength by "grappling" with one another.

The tahr

In the summer they eat the nutritious herbs that grow in the mountains, but during the winter they have to settle for dry grass and lichen. They seem to appreciate water and like to wallow in moist places and muddy pools, but they can get along well with very little water. E. P. Walker reports: "In their habitat, accumulations of water are rare and located great distances apart. But these desert animals seem to be able to obtain sufficient moisture from the green plants and from the dew which condenses on the leaves during the cold desert nights."

During the rut in November, the bucks, who until then have been living singly, join the small female groups. Then they often have vehement fights. "The bucks often stand opposite each other at a distance of ten to fifteen meters," reports Walker. "Then they walk towards each other with increasing speed, starting to run shortly before they clash together. Immediately before the clash, they lower their heads. However, a buck will not attack when the other one has lost his balance or is not prepared." Nevertheless, deaths may occur in such fights, especially in zoo pens where the inferior animal has no chance to escape. Besides the forehead to forehead clash, the barbary sheep have two additional forms of fighting which are closely related to the shape of their horns, which are curved to the sides. Standing reverse-parallel, the rivals reach with one horn over the opponent's back and try to force him down. Or they stand beside each other like a team or horses, hook their horns together, and test their strength by pulling and tearing, somewhat like "Indian wrestling."

The females give birth to one or two young, or sometimes to triplets, 150 to 165 days after copulation. The young, which jump about vigorously a few hours after birth, are soon able to climb as skillfully as the adults. They are nursed for half a year, but by one week of age they begin to supplement their diet with grass.

Barbary sheep become pubescent at one and a half years. In a zoo they may reach an age of fifteen or sixteen years, but generally they become only ten to twelve years old. From crosses between barbary sheep and domestic goats, viable offspring have been obtained. Hans Petzsch of the Halle Zoo even managed to successfully cross such a hybrid with an ibex. But this fact alone is not sufficient to permit the conclusion that there is an especially close relationship between the two genera. Once in a while, crossbreeding between sheep and goats has occurred.

In North America, free ranging barbary sheep have lived for a number of years. In 1950 they were successfully introduced into the canyons of the Canadian River in New Mexico and in San Louis Obispo County in California. In their African native habitat, they are ruthlessly pursued by the natives, and their complete extermination is merely a matter of time.

Like barbary sheep, the TAHR (*Hemitragus jemlahicus;* Color plate, p.

464) is the only species of a genus which is equally related to goats and sheep. The scientific name *Hemitragus* (semi-goat) underscores this problem rather well. The HRL is 130-170 cm, the TL is 10-20 cm, the BH is 60-100 cm, and the weight reaches 105 kg. The ♀♀ are about one fourth smaller than the ♂♂. The head is short to medium long, and the facial profile is straight. The animals have rather large eyes, ears which are small and pointed, and a small naked nose region. The horns of the ♂♂ extend up to 43 cm, and those of the ♀♀ reach 35 cm. The horns are closely set together at the base, are curved, and have a sharp front edge. The direction and shape of the horns is somewhat different in the various subspecies. The tail is short and goat-like, and the underside is naked and has scent glands. There are no preorbital and flank glands, and the interdigital glands, which are missing on the front legs, may be rudimentary on the hind legs. There are carpal joints, and the chest has callous pads. Depending on the subspecies, the females have two or four nipples. The dense coat with soft fleece varies in length and softness depending on the subspecies. The ♂♂ of the two northern subspecies have a strong collared mane on the neck, shoulders, and chest. The southern subspecies have only a short mane on the crest of the neck. The Arabian tahrs have, in addition to the cheek beards, longer hair on the chest and cuffs. There are four subspecies: 1. HIMALAYAN TAHR (*Hemitragus jemlahicus jemlahicus*; Color plate, p. 464); 2. SIKKIM TAHR (*Hemitragus jemlahicus schaeferi*); 3. NILGIRI TAHR (◊*Hemitragus jemlahicus hylocrius*; Color plate, p. 464); and 4. ARABIAN TAHR (◊*Hemitragus jemlahicus jayakiri*).

The Arabian tahr lives in small groups in the higher elevations of the desert mountains where there is a little more vegetation. The three other subspecies live on rocky, wooded mountains and often join large herds of thirty to forty animals. Except during the rut, the males form groups of their own. The nilgiri tahr prefers the grass-covered slopes above the woods, while the tahr bucks in the Himalayas remain in the dense forest and only the females move up to the mountain pastures in the summer. Occasionally, they may be found with markhors and monal pheasants.

The northern subspecies mate in December, and the southern forms probably do not have a seasonal rut. After a gestation period of 180 to 242 days, the female tahr gives birth to only one young, although Nilgiri tahrs frequently have twins. The young are nursed for half a year and are fully grown by one and a half years. They probably live no longer than ten to fourteen years, but some of them have lived up to twenty years in zoos. Tahrs are seen in zoos less often than barbary sheep. In the beginning of this century, Himalayan tahrs were introduced into New Zealand where they have reproduced as well as European, Asiatic, and North American deer, which were introduced there, unfortunately for the characteristic native fauna. The Nilgiri tahr is

Fig. 17-11. In this manner one tries to force the other down.

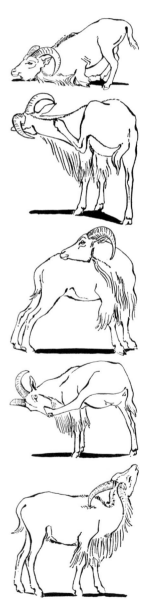

greatly endangered; according to Gustav Kirk, only about four hundred remain in several protected areas.

The blue sheep

Distinguishing characteristics

The BLUE SHEEP (*Pseudois nayaur*; Color plate, p. 487) is neither sheep nor goat but represents a genus of its own, the *Pseudois*, which means "fake sheep." The HRL is 115-165 cm, the TL is 10-20 cm, the BH is 75-90 cm, and the weight is 25-80 kg. The ♀♀ are one fourth to one third weaker than the ♂♂. The short to medium-long head has a straight facial profile. The eyes are rather large, the ears are small and pointed, and the nostrils are oblique. There is a small, naked seam on the nose region. The horns, which are close together at the base, are up to 80 cm long in the ♂♂. They are V-shaped, diverge outward, and curve in an arch or S-shape to the sides and back. There is a keel-shaped edge along the inner backside. The horn surfaces are densely ringed with narrow bulges. The horns of the ♀♀ are up to 20 cm long and more straight up; they have no keel or ridges. The tail, which is naked only at the base of the underside, has glands. The ♂♂ have no scent. There are no preorbital, flank, or interdigital glands. The females have two nipples. The short, dense coat has less fleece in the summer and far more during the winter. A blue coloration occurs only in the second juvenile coat, which is the first winter coat. There are three subspecies: 1. HIMALAYAN BLUE SHEEP (*Pseudois nayaur nayaur*); 2. SZECHWAN BLUE SHEEP (*Pseudois nayaur szechuanensis*); and 3. PYGMY BLUE SHEEP (*Pseudois nayaur schaeferi*), which is much smaller and short-legged, with a horn length in the ♂♂ of only up to 45 cm.

◁

Fig. 17-12. These grooming actions were observed by Gerhard Hass in the Frankfurt Zoo: wallowing, scratching with the foot, nibbling, and scratching with the tip of the horn.

Fig. 17-13. Original distribution of the tahr (*Hemitragus jemlahicus*).
1. Himalaya and Sikkim tahr (*Hemitragus jemlahicus jemlahicus* and *Hemitragus jemlahicus schaeferi*).
2. Arabian tahr (*Hemitragus jemlahicus jayakiri*).
3. Nilgiri tahr (*Hemitragus jemlahicus hylocrius*)

Fig. 17-14. Original distribution of the blue sheep (*Pseudois nayaur*).

Blue sheep are true mountain animals, inhabiting open slopes, canyons, and plateaus between altitudes of three thousand and five thousand, five hundred meters. They avoid areas with wood and brush cover. Only in winter when the food becomes scarce in the open areas do they go to the edges of the brush land. The habitat of the pygmy blue sheep in the canyons of the upper Yangtze-Kiang is separated from the habitat of the large subspecies of the same region by the forest and brush belt. The pygmy blue sheep live in the open areas below the forest at elevations of two thousand, five hundred and three thousand, five hundred meters. Ernst Schaefer discovered this new form of animal in 1934. They live only in small groups, while the larger blue sheep, which occur in large herds, join associations of several hundred animals in the winter. In the summer old bucks usually form groups of their own. Blue sheep are diurnal animals which feed in the morning and afternoon. However, the pygmy blue sheep are active mainly at dawn. The animals eat predominantly grass, herbs, and lichen. In their habitat, which is poor in cover, blue sheep, when sensing danger, often remain motionless on the spot, and the rocky gray color of their coats prevents them from being seen. The same has been reported of the sand-yellow barbary sheep.

The rut lasts from October into November, and the lambs are born at the end of May or in the beginning of June. The exact period of gestation is not yet known. Each mother gives birth to only one young which is nursed for half a year. The young become pubescent by the age of one and a half years. Blue sheep may reach an age of twelve to fifteen years, or occasionally, even eighteen years.

The SHEEP (genus *Ovis*) have the widest distribution of all horned ungulates in the wild. In the large area extending from Corsica to Asia Minor and Central Asia to western North America and into Mexico, thirty-seven different forms, classified into two species, the wild sheep and the bighorn sheep, have been described.

The WILD SHEEP (*Ovis ammon*; Color plate, p. 497) varies greatly in size. The HRL is 110-200 cm, the TL is 3.5-13 cm, the BH is 65-125 cm, and the weight is 20-230 kg. The ♀♀ weigh one fourth to one third less than the ♂♂. The profile of the nose is straight or humped (Roman nose), and the forehead bulges. The eyes are rather large, the ears are short and pointed, and the nostrils are oblique. The bare nose region is limited to a narrow seam above the nostrils and a small center line leading to the edge of the upper lips. The horns of the ♂♂ are strong and broad-fronted, with bulges or edges along the sides. They curve in spirals which are bent backwards. The length along the curvature is 50-190 cm. The horns of the ♀♀ are short (10-30 cm) and scimitar-shaped; they may be absent. The base of the tail has a very small patch of glands on the underside. They have preorbital, flank, and interdigital glands. The ♀♀ have two nipples. Most species have no beard; in some subspecies there is a neck mane which is either barely indicated or very pronounced. The coat color varies with the subspecies, age, sex, and season.

The MOUFLON (*Ovis ammon musimon*; Color plate, p. 497) is the smallest subspecies of the wild sheep. The BH is 70-90 cm in the ♂♂ and 65-75 cm in the ♀♀. The weight reaches 50 kg in the ♂♂ and 35 kg. in the ♀♀. The figure shows the ♂ in the summer coat. The winter coat is darker with a conspicuous whitish saddle patch. The ♀♀ and the lambs are rather gray or darker in winter, but they have no saddle patch.

During the beginning of the Upper Stone Age, the mouflon was distributed from Hungary, Moravia, and Southern Germany to the Mediterranean area. However, there are few fossils. Presently, its actual distribution is limited to the islands of Corsica and Sardinia, where only small populations live in a few, insufficiently protected reserves. On the island of Corsica in the Bavella Preserve, there are, according to Pfeffer (1967), about one hundred and fifty head, and perhaps fifty more outside the preserve. The Sardinian population consisted, according to an estimate by Mottl in 1955, of approximately seven hundred animals. The latest figures speak of three or four hundred animals at the most.

Sheep

Sheep (only rams portrayed): 1. Bighorn sheep (*Ovis canadensis*), subspecies a) Kamtschatka sheep (*Ovis canadensis nivicola*), b) Rocky Mountain bighorn sheep (*Ovis canadensis canadensis*), c) Stone's sheep (*Ovis canadensis stonei*), d) Dall's sheep (*Ovis canadensis dalli*). 2. Wild sheep (*Ovis ammon*), subspecies a) Pundjab urial (*Ovis ammon cycloceros*), b) Marco Polo's sheep (*Ovis ammon polii*); c) Elbur's urial (*Ovis ammon orientalis*); d) European mouflon (*Ovis ammon musimom*), summer coat.

Mouflon, by D. Müller-Using

1♂ 2♂ 3♂ 4○ 5♂ 6♂

Helmut Diller

Fig. 17-15. Original distribution of the wild sheep (Ovis ammon). The distribution of the European subspecies is presented in more detail on the following map.

Fig. 17-16. European mouflon (Ovis ammon musimon):
1. Original distribution on the islands of Corsica and Sardinia. 2. Areas of introduction on the European continent.

Races of domestic sheep (Ovis ammon aries):
1. North German moorland sheep
2. Merino Country sheep
3. Black Zackel sheep
4. Black-headed sheep (Fat buttocks sheep from Kenya)
5. Cameroon sheep
6. Syrian fat-tailed sheep

Therefore the situation of this wild sheep would have been rather desperate if it had not been for animal lovers and hunters who have taken care of it since the eighteenth century. Prince Eugen of Savoy kept Corsican mouflons in his game park near Vienna, and from there, some of the animals came to the Imperial hunting grounds at Lainz near Vienna. In 1840 in Lainz, nineteen mouflons were turned loose which had come directly from Corsica and Sardinia. In the following decades, mouflons were brought into several game parks, mainly in the area of present-day Czechoslovakia and Silesia, where they were set free in the wild.

After 1900 a Hamburg merchant, O. L. Tesdorpf, made efforts to introduce mouflons to the Lüneburger Heide, an environment not suitable for wild sheep, and to the Harz Mountains, both in Germany. The same was done in the Solling, the Taunus Mountains, and in several parts of Austria-Hungary. This new species of game was particularly appreciated because the Corsican mouflon does not eat bark from trees.

When the additional imports from Corsica and Sardinia stopped, and some of the hunters preferred a somewhat bigger game with larger horns as trophies, it unfortunately happened that crosses with domestic sheep, rather than purebred island mouflons, were set free. Especially the crosses with "Zackel" sheep produce rams that are heavier and make "better" trophies. But these mongrels eat the bark of various kinds of trees, and so discredit the mouflons. However, mouflons may eat bark when too many of them live in a small area.

During World Wars I and II, mouflons decreased in number everywhere, since the fauna usually suffered from the human unrest. In 1954 Motti estimated the number of mouflons to be approximately forty-five hundred. Meanwhile, the populations have recovered nicely, due to careful preservation. Presently about seven thousand head live in the Federal Republic of Germany, and two thousand more live in Austria, Hungary, and Czechoslovakia. Mouflons have also been introduced to most other South, East, and Central European countries, so that the total presently may be around twenty thousand.

The reason for the wide distribution of the mouflon on the continent, besides its beautiful shape and color, is undoubtedly the handsome head decoration which the hunters desire as a trophy. The old rams' spiral horns keep growing throughout the animal's life, reaching a length of seventy-five centimeters, and the circumference at the base measures almost twenty-five centimeters. The females have only very short horns, and are often hornless. Some of the continental forms do not have the saddle patch, which may be due to crossbreeding with domestic sheep.

Details on the social life of free-ranging sheep are known from Val Geist's studies on the American bighorn sheep (see p. 514). The voice of the mouflon is similar to the domestic sheep's bleating. The rams are almost mute, but in the females Pfeffer distinguishes between a bleating alarm call and the loud, anxious bleating of a mother searching for her lamb, who will usually answer with the familiar "bah" sound. Occasionally, one may hear a dog-like barking, which indicates an uneasiness not necessarily caused by threatening danger. Besides these vocal expressions, the animals also make a hissing warning sound which is probably uttered through the nose. This sound may also be heard in the rut when the ram pursues females. Sometimes a grunting sound is heard from the old rams. The warning whistle is often accompanied by a loud stamping with the front leg. A rutting ram bangs his horns resoundingly against rocks or tree trunks, and this may be heard over a distance of more than one kilometer. It occurs at intervals of about one second, and like the beating of horns in elk, it seems to be an audible marking of territory. It may also include olfactory marking, because the secretion from the large preorbital glands may be sprayed about.

Vision is their most important sense. "A hair lost by a hunter is heard by deer, smelled by the boar, and seen by the mouflon ram," goes a Corsican hunter's saying. Being an alpine animal which occasionally may be threatened by the golden eagle, the mouflon may see a hunter from his platform blind, a behavior most unlikely in European deer. In addition to this keen sense of sight, they have an excellent ability of recognition. Even a camouflaged hunter will be recognized from a distance of hundreds of meters. A hidden camera with a disguised lens was seen from a distance of two hundred meters. In Pfeffer's opinion, they cannot see very well at dusk or dawn. I would like to question this opinion, since no game species considers man to be dangerous in the darkness of night. The flight distance then is always very short. Incidentally, in areas where they are not disturbed, mouflons are diurnal animals.

The mouflon's sense of smell, like in all ruminants, is excellent, particularly at a close range. Its effectiveness at a greater distance naturally depends upon the direction of the wind. Once the animals have smelled an enemy, which even with moderate wind velocity may be at a distance of approximately three hundred meters, they will take flight immediately. However, when they become aware of danger through their sense of sight, they are able to judge the critical distance rather accurately and flight may not necessarily occur at three hundred meters. The two senses thus complement each other. Hearing does not seem to play as important a part as the just discussed senses. However, the clicking sound from the releaser of a camera may be heard at distances of more than fifty meters. Little is known about their senses of touch and taste.

The mouflons are not at all choosy about food. They may even eat

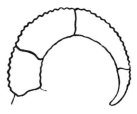

Fig. 17-17. Horn of a four year old mouflon ram. The annual growth rings are usually clearly visible in the horned ungulates of the temperate zone. The horn section at the tip, which developed during the first year, originally had been the longest but is slightly worn off. With increasing age the annual growth decreases.

Fig. 17-18. Two year old, six years old, and twelve year old bighorn rams (*Ovis canadensis*).

Fig. 17-19. Fighting behavior in Marco Polo's sheep *(Ovis ammon polii)*. Above: the rams run toward each other on two legs...and...Below:... clash their horns together.

Fig. 17-20. Difference in size of wild sheep. From above to below: Argali *(Ovis ammon ammon)*, Pundjab urial *(Ovis ammon cycloceros)*, Armenian usial *(Ovis ammon gmelini)*.

plants which are poisonous to other species of animals, like spurge and the deadly nightshade *(Atropa bella donna)*, without being harmed. In hours of gregarious grazing, they eat all the herbs and grass they can get. Furthermore, they like the buds and shoots of bushes and young trees. In artificially settled habitats, it is possible in the winter to uncover food with a snowplow and at other times to establish meadows and fields, since mouflons are sedentary and do not roam far even in times of need.

Mouflons reach sexual maturity at one and a half years. The rams begin to produce sperm around this time, but under natural conditions they do not copulate before three and a half years of age because the very presence of older rams greatly inhibits the sexual activity of the young rams. We find the same situation in many other ungulates, particularly red deer. Türcke observed that under very favorable conditions strong young females may be impregnated as early as seven months of age, and they have their lambs after five months like the older females. The rut begins in October and often lasts until December. Usually the one and a half year old females are the first to come into heat. The rams, who return from their solitary way of life, try to separate a female in heat from the herd and drive it into cover. The young rams stay in groups of their own or in herds with females who are not yet in heat.

As in most ungulates, some preliminaries precede copulation. It begins with the ram's flehmen (lip curl), where he exposes his teeth and "greedily" sniffs the female's scent. He walks behind her and often touches her gently on her hind leg with his stretched out front leg. He licks her, rests his head on her croup, and he may occasionally push her with his nose or nibble at the hair on her flanks and back. Real copulation, which lasts only three or four seconds, may be repeated at intervals from half an hour to an hour.

Fights between the males occur only when a searching old ram meets a serious rival. The rivalries are settled with enormous power and force, but they never lead to serious injuries, skull fractures, or concussions of the brain. Reinhard Schober describes such a fight: "In the midst of the sheep, two older rams kept attacking each other with such force that the clashing of the horns sounded like two heavy tree trunks hitting each other. The two rivals dashed against each other, crashed their horns together resoundingly, walked beside each other apparently peacefully, and then suddenly turned aside and again rushed towards each other" (fig. 17-19). The only accident which may happen at such tournaments is that the opponents will lock their horns together; they then are not able to separate and so come to a dreadful end.

Shortly before giving birth, the sheep chase off their previous year's offspring; the newborn lamb usually does not weigh more than two kilograms. Twins may occur in purebred strains, but they are rare. Where they are more frequent, it usually indicates a cross with domestic sheep. In one recorded case, the birth of a young lasted approximately one hour. The mother licked her lamb "dry" several times but, contrary to the rules, did not eat the placenta. The observer, Chief Forester Dr. Türcke, reported: "After twenty-three minutes, the lamb stood up, stretched its body, and sat down again. After several more attempts to get up and walk, it followed the mother, sometimes clumsily and sometimes with little jumps. After one and three quarters hours, the lamb began to try to suckle and, after several attempts, it was successful." Lambs of mouflon double their weight at birth in about three weeks. After three months, they weighed four times as much; after one year, they are ten times as heavy as they were at birth. The record longevity for a healthy ram was sixteen years.

Fig. 17-21. Flipping the front leg in the punjab urial ram is part of the fighting and mating behavior.

The CYPRUS URIAL OR RED SHEEP (⊘*Ovis ammon ophion*), which is close to extinction, also lives in light forests with brush in sub-alpine mountains. The other subspecies are found predominantly in open, rough terrain at medium or high altitudes, where they inhabit rocky hill country, lowland and highland steppes, and rocky semi-deserts, as well as grass covered slopes and alpine meadows. The alpine forms spend the summer at the highest elevations, up to six thousand meters, right below the permanent snow. In winter they move to lower elevations and may come into the valleys. They live gregariously in small or larger herds, and in the summer the older males live singly or in separate groups. The rutting seasons vary depending on the climate. Gestation is 147 to 188 days, and the females give birth to one or two lambs between March and June. The young are nursed for half a year, but at two to four weeks they begin to eat additional solid food. The smaller subspecies become pubescent at one and a half years, and in two and a half years for the larger ones. The rams of the largest forms reach their sexual maturity at three and a half years. They live twelve to eighteen years.

Other wild sheep, by A. G. Bannikow and W. G. Heptner

The color plate on page 497 shows the various subspecies. The type of mouflon described on page 496 has evolved in Asia Minor into the Transcaucasian-Asia Minor type. The basic coloration of the ram's winter coat becomes lighter, but the whitish saddle patch is still expanded and very distinct. This type of mouflon with its yellow coloration became adapted to the desert. To the wild sheep of this color belong the above mentioned cyprus urial or red sheep, the ARMENIAN URIAL (*Ovis ammon gmelini*), and the ELBURS URIAL (*Ovis ammon orientalis*; Color plate, p. 497). Farther to the east and northeast the basic color continues to become lighter, the saddle patch in the winter coat disappears, and the rams have well developed hair fringes. At the same time, the horn shapes change as shown in Fig. 17-22 and as first indicated in the elburs urial. This light type which lacks the saddle patch includes the PUNJAB

Fig. 17-22. Shapes of horns in wild sheep From above to below: Armenian urial (*Ovis ammon gmelini*), Elburs urial (*Ovis ammon orientalis*), Punjab urial (*Ovis ammon cycloceros*), Marco Polo's sheep (*Ovis ammon polii*).

URIAL (*Ovis ammon cycloceros;* Color plate, p. 497).

Larger than all the wild sheep described so far are the subspecies of the Pamir and the adjoining Central Asiatic mountain ranges. The largest form is the ALTAI ARGALI *(Ovis ammon ammon)* whose rams reach a body height of 125 centimeters. The Altai argali and the closely related subspecies may be of various colors, but they have only little developed hair fringes on the neck. Their horns are particularly large and strangely curved. The KARA TAU ARGALI *(Ovis ammon nigrimontana;* Color plate, p. 507) and the PAMIR ARGALI (Color plate, p. 497), with the scientific name of *Ovis ammon polii* after the Venetian East Asia explorer Marco Polo (1254–1324), also belong to this group.

Domestic sheep, by H. Kraft

The DOMESTIC SHEEP (*Ovis ammon aries:* Color plate, p. 498) originated from the wild sheep. There are large and small breeds of sheep, with and without horns, and with or without wool. The Tibetan unicorn sheep has only one horn, while northern island forms may have four or more horns. The ears may be enlarged, short, or even absent. The different races may be distinguished by the differences in their coats.

A. HAIRY SHEEP. The short coat has coarse, pithed bristles and fine, fluffy hairs without piths. The bucks usually have a neck mane. Regular shedding occurs in the spring and fall. They yield milk and meat. These breeds include the Fessan sheep, the narrow-tailed, horned, or hornless Senegal, the Guinea, and the Cameroon sheep (Color plate, p. 498), the Southwest African fat-tailed sheep, the Abyssinian short-eared sheep, and the hornless, short-tailed Persian, Masai, and Somali sheep (Color plate, p. 498), and others.

B. WOOLLY SHEEP. The coat has fluffier hair with a few or no bristles and no seasonal shedding. In one group of sheep the wool is fleecy and mats easily. These mainly northern breeds include the small, short-tailed, horned North German Moorland sheep (Color plate, p. 498) of the Lüneburger Heide which yields wool, meat, and milk; the Central and East Asiatic fat-rumped sheep, the fat-tailed sheep from Western Asia, the Balkans, and Africa, including the Karakul sheep,

Domestic American sheep, by Leslie Laidlaw

and the East European sheep (Zackelschafe) (Color plate, p. 498). Another group of sheep has fluffy hair and pithless bristles, and includes sheep with drooping ears of the European Alps and the Moorland sheep from the North Sea. Merino wool sheep form a large group with fleece of fluffy hair. Among this group are the Merino sheep (Color plate, p. 498), with a world-wide distribution of various breeds, the Anglo-Merinos (crossbreeds from Merinos with English longhaired sheep), and the improved country sheep of South Germany, France, and Southern Europe with their delicate wool. They are mainly kept for their wool.

The Merino (fine-wool breed)

The first successful importation of the MERINO from Spain to the United States was in 1801. The Delaine Merino was developed in Pennsylvania and Ohio. It is about the only type popular in the United

States today. Rams fitted for show weigh 91 kg or more. The Delaine Merino is free of neck folds and body wrinkles. It is open-faced (no wool around the eyes to block vision). The quantity and quality of the wool produced is outstanding. It has a distinct crimp and grades fine with a spinning count of sixty-eight to seventy. Rams shear 7 to 9 kg per year, and ewes shear 4.5 to 5.5 kg of fleece. There is a forty to forty-five percent shrinkage of wool produced under range conditions. Merinos have medium-short heads. The rams have well-developed horns. There is also a polled (naturally hornless) strain. The hair on the face, ears, and legs is white and of fine quality. Wool is present on the fore and hind legs down to the tops of the hoofs. The Merino is well adapted to range conditions, especially semi-arid and rough ranges. The breed is able to get along with little feed, water, and shelter in comparison with more strictly mutton-type sheep. They are not as early maturing as some breeds, and they produce fair carcasses. Lambing percentages seldom exceed 105% on the range, but approach 125% in farm flocks. The Delaine Merino is decreasing in popularity, since the purebred breeder has not kept up with the desires of commercial sheepmen. It is mainly of use in crossbreeding programs and for production of out-of-season lambs.

The RAMBOUILLET was developed in France and was originally known as the French Merino. It was first imported into the United States in 1840. The present day Rambouillet does not have heavy folds of skin around the neck, but may exhibit extra skin in the brisket area. Today's breeders prefer a sheep with little or no wool on the face below the eyes. The rams show strong, wide heads with well balanced, outward-curving horns. Ewes are generally hornless, and hornless rams are becoming more popular. Rams in show condition weigh 110 to 135 kg. Fitted ewes weigh 70 to 100 kg with fleece making up 9 to 11 kg of the weight. Ewes produce a clean wool yield averaging 2.14 kg before cleaning. The spinning count is usually around sixty-one. Rambouillet show straighter tops, deeper bodies, and thicker rear quarters than the Merino. However, many are crooked in their front legs and have too much angle at the hook. The hair on the face and legs is white, but lacks the quality of the hair found on the Merino. The fleece is not of as high quality as that of the Merino, but usually is of a fine wool grade. The breed is large and rugged, and does well under range conditions. Carcasses are not equal in muscling to those of the mutton breeds, but are satisfactory. Strong flocking instinct is present, making the breed an easy one to herd. Rambouillet live about two years longer than the mutton breeds. The ewes are good milkers and excellent mothers. Farm flocks average 125% lamb crops, and ewes will readily breed out of season. Occasionally there is too much variation in the fineness of the wool on different areas of the body.

The SOUTHDOWN was imported from England in 1823. There is some

The Rambouillet (fine-wool breed)

Table 17–1. In wool terminology for domestic American sheep, the clean fleece yield is the percentage of clean wool remaining after all foreign material (grease, dirt, sand, burrs, etc.) have been removed by washing and scouring. Grease wool is wool that has not been washed or scoured. Fleece is classified according to fiber diameter, with braid being the coarsest. Spinning counts are based on the amount of yarn that can be spun from one pound of clean wool.

Blood Grade	Spinning Count
Fine	64–80
½ Blood	60–62
⅜ Blood	56–58
¼ Blood	50–54
Low ¼ Blood	46–48
Common	
Braid	

The Southdown (medium-wool breed)

wool on the head and ears, but the face is clear below the eyes. The face and legs are between dark gray and light brown. Both rams and ewes are naturally hornless. The ears are carried farther down on the head than in most other breeds. Fitted rams weigh from 84 to 90 kg and fitted ewes weigh from 61 to 70 kg. Southdowns shear from 2.7 to 3.6 kg of wool per year. The wool is of short length, and grades from ½ to ⅜ blood. The breed is short-legged and wide-chested, with straight tops and bulging rear quarters. They are unsurpassed in natural muscling along the top. Southdowns are good, efficient feeders and are unsurpassed in their ability to take on flesh. Lamb crops average 135 to 140 percent. The ewes are good mothers and fair milkers. Southdowns lack the size and ruggedness desired for range conditions. The fleece is lighter than that of the other mutton breeds. Southdowns have an ambling walk, and may turn under on the outside hoof wall. Rams are especially useful for upgrading commercial herds.

The Hampshire (medium-wool breed)

The HAMPSHIRE was imported from England prior to 1840, but there were no registrations until 1889. The predominant, well-shaped ears are carried well down on the head and may be held out horizontally. The face, ears, and legs vary in color from dark brown to black. Rams and ewes are naturally hornless. Rams in show condition weigh over 125 kg. Ewes weigh between 82 to 90 kg. Fleece weight averages about 3.5 kg. The fleece is dense, but doesn't exceed 7.6 cm in length. The wool is coarse and grades from ¼ to ⅜ blood. Undesirable black fibers are frequently scattered throughout the fleece. Lamb crops average 125 to 140 percent. Lambs are rapid-growing and reach market weight up to six weeks earlier than Southdown lambs, but require good feeding to attain top production. Hampshires are rugged, and do well under range conditions. They do not have the flocking instinct of the fine wool breeds, and are thus harder to herd. They can withstand more hot weather than some of the other mutton breeds. Hampshires have well-muscled mutton conformation, but may commonly be crooked in the front or hind legs, and may show too great an angle at the hocks. Although the Hampshire is discriminated against because of the light weight fleece with black fibers, it has still become one of the dominant breeds in the United States.

The Shropshire (medium-wool breed)

The SHROPSHIRE is a dual-purpose breed, first imported from England in 1860. There is little wool around the eyes. The ears are not carried in an erect manner. The face, ears, and legs are dark brown or black. White hairs are common on the tip of the nose. Rams in breeding condition weigh around 100 kg. Ewes in breeding condition weigh around 75 kg, and the present tendency is to increase the size and robustness. The fleece averages ⅜ blood and may reach ⅜ to ½ blood grade. The fleece is dense and uniform in grade over different parts of the body. The average length of the fiber is over 6.35 cm. Most flocks shear an average fleece of 4.5 kg per sheep per year. The wool is unsurpassed by any mutton breeds of equal size. Ewes lamb easily and nurse well. They are prolific, frequently averaging lamb crops over 150 percent. The Shropshire is low set with well placed legs and a deep body. The top is strong, and the leg is full and bulging. Many are too small or lack the ruggedness needed for range conditions. Shropshires frequently are too straight in their hind legs, which interferes with long time usefulness. The carcasses are not equal to those of the Southdown, but are very acceptable. Shropshire rams are less active and less fertile in hot weather than rams of other breeds.

The OXFORD originated in England and was imported into the United States in 1846. It is a large, rugged breed with rams fitted for show weighing three hundred pounds or more, and ewes in show condition weighing more than two hundred pounds. There is a trend toward slightly smaller, earlier-maturing animals. The breed is characterized by a topknot. There is also wool on the face and ear, but not enough to block vision. The face color varies from light gray to almost black, and white hairs may be present at the tip of the nose. The Oxford has a mild temperament. Ewes usually shear from 4.5 to 5.5 kg of wool that will grade about ¼ blood. Rams shear 5.5 to 7 kg, but the wool is slightly coarser. The length of the wool fiber is between 9 and 13 cm. Occasionally, black fibers are present in the wool, but this is discriminated against by breeders. Ewes are prolific and lamb crops average around 150%. The ewes produce enough milk to enable them to nurse twins with little effort. The Oxford is a good feeder and can consume large quantities of pasture and hay. It does not do as well when it must forage widely to get its feed. Oxfords usually have straight tops, well-sprung ribs, and wide loins. Lambs frequently lack the fleshing and fat covering along the top that is seen in the other breeds. There is also a smaller proportion of fat within the muscle. The meat has a coarser grain than that of the smaller mutton breeds. The breed is popular in the northern farming states and in Canada.

The Oxford (medium-wool breed)

The DORSET breed was developed in southern England. It was first imported in the United States in 1885. Ewes and rams are horned. The horns of rams are very large and angular. They curve downward and forward, making a large spiral. Ewes have smaller, flatter horns that curve down and forward, but do not make a spiral. Since 1956 polled (naturally hornless) animals have been registered and are rapidly increasing in popularity. Dorsets have wool on the cheeks and down to the eyes, but seldom show any wool below the eyes. Many Dorsets have very little wool on the underline. The face is covered with short white hair. Flocks usually shear an average of 3.6 kg per sheep per year. The wool grades from ⅜ to ¼ blood, and usually exceeds 7.6 cm in length. Rams fitted for show weigh about 100 kg, and ewes in show condition weigh around 80 kg. The breed is low set. Ewes have smooth straight tops, but rams are high over their shoulders and low

The Dorset (medium-wool breed)

A ram of the kara tau argali *(Ovis ammon nigri-montana).* This subspecies belongs to the largest wild sheep.

behind. The ewes have good mutton-type carcasses, and castrated males are similar to the ewes in conformation. Dorset ewes lamb easily and are good mothers. Lamb crops exceeding 150 percent are common. Triplets are more frequently seen in this breed than in any other. Dorsets will breed early to produce out-of-season lambs. The rams are active in hot weather. Breeders are striving to improve the quality and fleshing of the carcass.

The Cheviot (medium-wool breed)

The CHEVIOT was developed in northern England and in Scotland; it first entered the United States in 1838. The Cheviot has a distinctive appearance created by the high carriage of the head and the erect, forward-pointing ears. There is no wool on the head or face, and no wool below the knees and hocks. Very fine white hair covers the un-wooled areas. The breed is naturally hornless. Well-fitted rams weigh around 80 kg, and ewes fitted for show weigh 64 kg or more. Ewes shear about 2.7 to 3.2 kg per year, and rams shear about .45 to .90 kg heavier. The fleece usually grades about ¼ blood and is 10 to 13 cm in length. The Cheviot is exceptionally hardy, and does well with little shelter. The lamb crop is about average. The ewes are good mothers despite their nervous temperament. They are fair milkers. The Cheviot produces a high quality carcass with a high proportion of fat within the muscle. The carcass is similar to that of the southdown in conformation. The breed is faulted for lack of size and lack of fleece weight. The sheep are very active grazers. Lambs mature slowly. The breed is popular around country homes because of its distinctive, stylish appearance.

The Suffolk (medium-wool breed)

SUFFOLK SHEEP were developed in southeastern England and introduced into the United States in 1888. The Suffolk breed has black hair on the face, ears, and legs. A small amount of white wool may be present on the forehead. The ears are medium to long and are held horizontally or droop forward. The breed is naturally hornless. Rams fitted for show weigh around 137 kg, and ewes fitted for show weigh about 115 kg. The Suffolk is low set and carries good muscling on the back. They are more heavily muscled in the rear quarter than most of the other breeds. They are a rugged, easy keeping breed. The Suffolk shears 2.7 to 3.2 kg of wool which is 5 to 6.3 cm in length, and grades about ⅜ blood. Black fibers are sometimes prevalent in the fleece, especially in the line between hair and wool. The breed is popular under range conditions, especially in warm areas. The lack of wooling on the face and legs decreases the amount of trouble caused by grass types in this region, and the breed can withstand heat. Ewes lamb easily and are good mothers and excellent milkers. Lamb crops exceeding 150% are prevalent. The lambs mature as rapidly as those of any other breed. Suffolks produce well-muscled, high-quality carcasses. Suffolks have excellent conformation, but frequently lack spring of forerib, and are light in the foreflank. Suffolks are difficult

Rocky Mountain goat (*Oreamnos americanus*), a relative of the European chamois.

to herd since they spread out to make the most of available foods. The primary criticism of the breed is the small amount and low quality of the wool produced.

The CORRIEDALE was developed in New Zealand from Romney-Merino crosses, and was first imported to the United States in 1914. The face and ears of the Corriedale are covered with white hair. Wool is present on both front and rear legs and on the forehead down to, but not below, the eyes. Rams in show condition weigh from 84 to 90 kg. Ewes fitted for show weigh from 57 to 84 kg. Wool production is the strongest feature of the breed. Rams shear between 7 to 11 kg of wool per year, and it is generally coarser than the ⅜ blood grade sheared from the ewe. The ewe produces 5.5 to 9 kg per year. The length of the fiber usually exceeds ten cm. The wool is uniform and of high commercial value. Corriedales are lacking in width, depth, and fullness of leg, and the front legs are too close together, reflecting the Merino ancestry. Corriedales do well under range conditions and are easy to herd. The ewes lamb easily and average lamb crops of 125 to 140 percent. The lambs are hardy at birth and grow at a reasonable rate. Corriedales are a long-lived breed, and their use in farm flocks has recently been increasing. Their carcass is less desirable than those of the strictly mutton breeds but is more acceptable than the carcasses of the fine-wool breeds.

The COLUMBIA was developed from crosses of the Rambouillet and Lincoln breeds. The breed association was established in 1941. The Columbia is naturally hornless. Rams in show condition weigh around 137 kg, and ewes fitted for show weigh about 90 kg. The face and ears are covered with fine white hair. There is little wool below the knees and hocks. The Columbia ewe shears about 5.5 kg of fleece per year. The fleece is about 9 cm long and grades ¼ to ⅜ blood; ½ blood grades are also known. The fleece is fairly constant with not more than two grades in an individual fleece. The fleeces stay together in storms and repel moisture. The ewes lamb easily and are good milkers. Lamb crops average 125 to 140 percent. The lambs are hardy and active at birth, in keeping with the ruggedness of the breed. The Columbias have strong, well-placed legs. They are wide-chested and deep-bodied. They sometimes lack depth and fullness of leg. The Columbia lacks uniformity in fleece and carcass conformation. It is frequently used in crossbreeding. The breed is popular in the western states and has recently been increasing in popularity in the Midwest of the United States.

The DEBOUILLET originated in New Mexico from crosses between Rambouillet and Delaine Merinos. The breed association was formed in 1954. The rams may be horned or hornless. Rams in breeding condition weigh about 90 kg, and ewes average about 57 kg. Ewes produce about 2.7 kg of clean wool per year, and rams produce approximately

The Corriedale (medium-wool breed)

The Columbia (medium-wool breed)

The Debouillet (medium-wool breed)

4 kg. The fleece grades 64 or finer and is approximately 7.6 cm in length. There is no wooling below the eyes. The Debouillet does quite well under poor range conditions.

The Montadale (medium-wool breed)

The MONTADALE was developed in Missouri. It consists of sixty percent Columbia and forty percent Cheviot breeding, although some breeders have been emphasizing the Cheviot characteristics. The breed association was organized in 1945. The Montadale is naturally hornless. There is no wool on the head or on the legs below the knees and hocks. The head and legs are covered with white hair. Ewes weigh 55 kg at one year of age and shear 4 to 5.5 kg of wool per year. The fleece is of ⅜ to ½ blood quality. Rams are heavier and shear more wool. The ewes are good mothers. The Montadale is free of skin folds, and the conformation is mutton type.

The Targhee (medium-wool breed)

The TARGHEE was developed in Idaho from Rambouillet, Lincoln, and Corriedale crosses. The breed association was formed in 1951. The breed is naturally hornless. The hair on the face is white. No wool is present below the eyes. There are no skin folds on the neck. The fleece is about 7.6 cm long and grades about ½ blood. Ewes shear an average of 5 kg of wool per year. The breed is exceptionally rugged and does well under adverse conditions on the range. Rams in breeding condition weigh around 90 kg. Ewes weigh around 60 kg. Conformation and carcass quality are acceptable.

The Panama (medium-wool breed)

The PANAMA breed was developed in Idaho, and the breed association was formed in 1951. Rambouillet rams were bred to Lincoln ewes to form the breed. It closely resembles the Columbia, in which Lincoln rams were bred to Rambouillet ewes.

The Lincoln (long-wool breed)

The LINCOLN breed originated in England and was first imported after 1780. The breed is naturally hornless. The ears are held horizontally and are inclined forward. There is no wool present below the eyes or below the knees. The hair on the face is white. The fleece is carried in heavy locks which are 25 to 37 cm in length. There is a tendency for the wool to part along the back, thus offering little protection in cold rains. Ewes shear 5.5 to 7.3 kg of fleece per year, and rams shear in excess of 9 kg. The wool is coarse and grades as braid. It has a distinctive luster. Rams in show condition weigh around 150 kg. Ewes fitted for show weigh at least 100 kg. The Lincoln is a wide, deep-sided sheep, but is lacking in fullness of leg. Ewes are average in prolificacy and milking qualities. Carcasses carry excess external fat and lack the fine texture of the mutton breed carcasses. The Lincoln is of limited use where feed is not abundant or where weather conditions are severe. It is not found in large numbers in the United States, but has been extensively used in the past. It is used today to increase size and wool production in commercial crossbred herds.

The Cotswald (long-wool breed)

The COTSWALD originated in England and was introduced into the United States in 1832. The Cotswald is hornless, but small horny growths known as scurs may be present on some rams. The ears are fairly long and are held slightly erect. There is a pronounced foretop; this wool may hang in front of the eyes and in some cases reaches to the nose tip. The hair on the face is white. Cotswalds shear 4.5 to 7 kg of wool per year. The fiber is 25 to 30 cm in length and grades as braid. There is a part down the back; thus the fleece offers little protection in hard, cold rains. The breed is straight-topped, deep-sided, and shows good width over the top. However, it lacks substance of bone. The Cotswald is slightly above the Lincoln in prolificacy and milking ability. Carcasses show good mutton conformation, but the meat has a coarse texture. The breed is more hardy than the Lincoln and can do without abundant feed. The Cotswald is primarily used for crossbreeding.

The ROMNEY was developed in southeastern England. It was first imported in 1904. Romneys are naturally hornless. There is a tuft of short wool on the forehead, but the rest of the face, except for the lower part of the cheeks, is free of wool. There is little wool below the knees or hocks. The hair on the face and legs is white. Rams in show condition weigh between 100 to 115 kg. Ewes in show condition weigh between 73 to 80 kg. Romneys shear to 5.5 kg of wool per year. The fleece has a fiber length of 18 cm and grades to ¼ blood. There is no part down the back, so the fleece offers more protection than is found in the other long-wool breeds. It is wide-chested and deep-bodied with a full, thick rear quarter. Strength of back and fullness in back of the shoulders are sometimes lacking. Ewes are fairly prolific and are very good milkers. The breed is more hardy on lowland pastures than others, but is not particularly well adapted to hot, dry climates. The Romney presents a sluggish appearance, but they are active grazers. The carcass is more acceptable in muscling and quality than the carcasses of the other long-wool breeds. The Romney crosses well with the fine-wool and medium-wool breeds. The lambs require abundant feed to give full development.

The Romney (long-wool breed)

The KARAKUL originated in Bokhara. The breed was first imported into the United States in 1909. Mature rams in breeding condition weigh from 77 to 90 kg. Ewes weigh from 59 to 73 kg. Rams may have horns, but ewes are hornless. The face and legs are brown or black. The Karakul has poor mutton conformation. The body is narrow and uneven, the neck is long, and the loin is high. The leg lacks fullness. The rump is steep with an overhanging mass of fat, which interferes with copulation if ewes are in fat condition. Karakul wool is low in quality. It is coarse and wiry, and of a brown or black color. There are two classes of fibers in the fleece. The outer coat is from 15 to 25 cm in length. The undercoat is shorter, finer, and darker in color. The wool grades as carpet wool and sells thirty to fifty percent lower than other wools. The fleece production averages 3 kg per year. Lamb pelts

The Karakul (fur-bearing breed)

Fig. 17-23. Shapes of horns in bighorn sheep Above: Kamtschatka sheep *(Ovis canadensis nivicola).* Below: Stone's sheep *(Ovis canadensis stonei).*

Fig. 17-24. Original distribution of the bighorn sheep *(Ovis canadensis)* in North America and East Asia.

are the chief feature of the breed. The pelts are classified as Broadtail, Persian Lamb, Krimmer, and Karakul. Broadtail pelts have the highest value and are obtained from lambs born prematurely or killed shortly after birth. These pelts have a "watery" design, short fibers with a pronounced pattern. Persian Lamb has a longer fiber with a short, tight curl. Like the Broadtail, they are usually black in color. Krimmer is similar to Persian Lamb, but is gray. The Karakul classification indicates a pelt with an open, lustrous fur of coarser fibers which is usually gray. Prices vary subject to fashion change and foreign competition. The Karakul is a very hardy breed of sheep. It is one to two years longer-lived than the mutton-type breeds. Prolificacy averages 105 percent. The ewes are good milk producers. The breed lacks uniformity of fur types and quality. The Karakul is not found in large numbers in the United States due to low income from the mutton and wool, the uncertainty of the fur market, and the competition from imported pelts.

By the Lower Stone Age, approximately 6000 B.C., sheep had been domesticated. Since then they have supplied man with meat, milk, and wool. Among the early breeds was the Moorland sheep which was similar to the mouflon and which matured at a later age. Man probably domesticated sheep in different places on earth, but the most important region of sheep domestication seems to have been Asia Minor. Long-tailed domestic sheep were developed from steppe sheep, among them the fat-tailed sheep, which are well adapted to the dry steppe with the fat that they store in their tails. The short-tailed hairy sheep, which has little wool, probably came to Southern Europe and North Africa from Central Asia.

In the fourteenth century in Spain, the intensive wool production from sheep began. There were herds of ten thousand or more Merinos in the pastures. For a long time, the export of living sheep from Spain was prohibited. Only around 1800 did the first Merinos come to Australia and South Africa, countries which presently rank first in the world's wool trade. At the same time, breeding of Merinos was begun in Germany, particularly in Saxony.

The sheep increased in numbers, and around 1900 there were approximately six hundred million of them. However, after World War I sheep raising decreased in many countries, and cattle grazed on the same pastures. Later the flocks of sheep again increased and were improved with crossbreeding, particularly overseas. Thus, presently more than eight hundred million sheep exist on earth. In Europe, sheep raising has decreased; in West Germany (in the area of the present Federal Republic) there were 1.9 million sheep in 1935, and in 1957 there were 1.12 million. The greatest number of sheep is in Australia where there are 149 million, followed by the Soviet Union with 120 million. The annual wool production in Australia alone is 389,000 tons.

Since herds of sheep may easily be kept together by sheep dogs, the expenses in breeding sheep are low. The animals require very little. Steppe and bush areas, especially on plateaus, are excellent grazing grounds for sheep.

After a gestation period of five months, an ewe gives birth to from one to four lambs. Weak animals fall victim to predators or are killed by the shepherd, because extensive care for a longer period is not possible on the pasture; hence the present breeds of sheep are in good condition.

Slightly smaller than the larger subspecies of wild sheep but more heavily built and with thick, relatively short horns, is the BIGHORN SHEEP (*Ovis canadensis;* Color plate p. 497). Some zoologists include it with the wild sheep in only one species. The North American bighorn sheep also belong in this group, such as the ROCKY MOUNTAIN BIGHORN (*Ovis canadensis canadensis;* Color plate p. 497), the WHITE or ALASKAN BIGHORN SHEEP or DALL'S SHEEP (*Ovis canadensis dalli;* Color plate p. 497), the BLACK BIGHORN SHEEP or STONE'S SHEEP (*Ovis canadensis stonei;* Color plate p. 497). In addition, there are the Siberian snow sheep; among them are the KAMCHATKAN SNOW SHEEP (*Ovis canadensis nivicola;* Color plate p. 497), which is endangered in many areas, and AUDUBON'S BIGHORN SHEEP (*Ovis canadensis auduboni*), which has been exterminated.

Bighorn sheep

Distinguishing characteristics

The social life of bighorn sheep has recently been studied by Val Geist in the Canadian Rocky Mountains. According to his findings, the horn size is significant for rank order among the rams. Only rams with horns of approximately equal size fight with each other (comp. fig. 17-19). Sometimes the fights last up to twenty hours. Lower ranking as well as defeated rams behave like females, and they are treated accordingly by the higher ranking males. They will drive and mount the lower ranking animals regularly and ejaculate. This "homosexual" behavior allows younger or weaker males to remain within the herds without being chased off. The highest-ranking rams are interested in the females only during the short rutting season, while other rams are interested in them throughout the year. The dominant rams with big horns have a shorter life span than the other males.

Homosexual behavior of the rams

Spencer estimates the gestation period to be approximately 180 days. In European zoos, the bighorns are rarely seen, even though they are not any more difficult to keep than the Eurasian wild sheep.

Bernhard Grzimek
W. G. Heptner
and contributors

18   The Musk Ox

Formerly the musk oxen, those large, cattle-like ruminants of the far north, were classified with the wild oxen. But according to recent tests by Moody, the protein composition of the musk ox's blood serum is closer to that of the sheep, goat, and mountain goat, than to that of cattle or bison. According to the studies of Mrs. Gijzen, the musk ox's external characteristics are more like those of sheep and goat than of cattle.

Subfamily: Ovibovini

Distinguishing characteristics

Therefore we now classify the Ovibovini as a tribe within the subfamily Caprinae. There is only one genus (*Ovibos*) with only one species, the MUSK OX (*Ovibos moschatus;* Color plate p. 469); the body is long, low, and plump. The HRL is 180-245 cm, the TL is 7-10 cm, the BH is 110-145 cm, and the weight is 200-300 kg. The ♀♀ are approximately one fourth lighter than the ♂♂. They have a large head with a straight profile on the upper side. The eyes are small and right below the horns; the ears are medium long, pointed, and partly hidden in the coat; and the broad muzzle has a small naked nose region between the nostrils and on their upper edge. The base of the horns, which is enlarged in the shape of a shield in old ♂♂, covers the crest like a helmet. Otherwise the crosscut through the horns is round. The horns curve downward close to the sides of the head, then curve upward again, and have pointed ends. The horns of the ♀♀ are smaller. The short, sturdy legs have short, broad hooves and large pseudo-claws which are almost completely hidden in the coat. The shaggy coat is unusually long, and even the scrotum and udder are covered with hairs. The calves are evenly covered with a woolly coat. The summer coat of the adults (only from the end of June to the end of July) is dark brown; the winter coat, with longer hair, is a dark to black brown, and in spring it bleaches to brown. The back of old ♂♂ becomes yellow brown (see Color plate p. 470). The short tail has very long hair and no glands on the lower side. They have preorbital glands. The females have four nipples. There are three subspecies: 1. The ALASKAN MUSK ox (*Ovibos moschatus moschatus*) has almost been exterminated except for a few animals. Recently it has been increasing in number due to effective laws for its protection. 2. The WAGER MUSK OX (*Ovibos moschatus niphoecus*), also largely decimated, has lately been increasing in numbers. 3. The GREENLAND MUSK OX (*Ovibos maschatus wardi*).

"With a musk ox coat, a person will not freeze to death even at the north pole." These are the words of the Arctic explorer Alwin Pedersen, who should know since he has spent several winters in the Arctic. One day he hung the freshly skinned coat of an old bull over a post at -27°C. He placed a thermometer between the hair. The sun was slightly above the horizon. Ten minutes later, the thermometer showed two degrees centigrade above freezing. Therefore, musk oxen probably never suffer from cold, and they may have trouble with summer warmth. The mosquitoes which torment reindeer and moose during the Arctic spring and summer may only reach the musk ox's eyes.

Musk oxen probably have the longest hair of all wild animals. The back hair is sixteen centimeters long, but this reaches sixty and occasionally ninety cemtimeters on the neck, the chest, and the hindquarters. When the hair is shed in the spring, it hangs in clusters at the edges of rocks, in brush, or it hangs in masses to the ground. Usually musk oxen look moth-eaten in zoological gardens. Therefore I did not recognize them when I first saw them in Canada. They were deep brown to black, magnificently smooth animals, which seemed to me to be much larger than those in the Copenhagen Zoo. A bull musk ox is only 130 centimeters tall at the back, but with his thick coat he appears to be much larger and thicker than he actually is. So why not domesticate these large Arctic relatives of the sheep and use their wool for spinning? Unfortunately it is not easy to get the wool. When the animals are sheared, they often die afterwards from pneumonia. Furthermore, there are many hard, long bristles between the warm wool which are hard to remove.

During the Glacial Period, musk oxen lived in Northern Germany, Mongolia, and even in the Northern parts of the United States. They withdrew with the ice so far to the north that they were only discovered in 1869. At the same time, very difficult times set in for them which almost led to their total extinction. The Hudson Bay Company alone purchased 5,408 musk oxen skins from 1888-1891. When we read the thrilling reports on the great North Pole expeditions, we certainly are not aware that musk oxen were massacred. They are the easiest of all Arctic animals to kill. Not only did the explorers and their companions live on the meat, but the many sleigh dogs which had to pull the heavy equipment and the provisions for the Europeans across the ice did so as well.

In the end musk oxen were saved by the invention of cars, and especially airplanes. Modern North Pole expeditions no longer employ dog sleds, and supplies are flown into the camps by airplanes. Inci-

Why doesn't the musk ox freeze to death?

Fig. 18-1. Musk ox (*Ovibos moschatus*):
1. Original distribution in North America, Greenland, and on the Arctic Archipelago.
2. Introduction to the Nunivak Island, on Spitsbergen, and in Norway.

Fig. 18-2. In the Glacial Period musk oxen lived in Europe. In the black places their fossils were found.
1-1': Southern boundary of the northern glaciation;
2-2': Northern boundary of the Alpine glaciation.

dentally, not even the "musky odor" saved the musk oxen from being eaten. The bulls have a rather strong scent in the rut. The aromatic musk secretion, which is a component of all good perfumes, comes from the musk deer (see Ch. 8).

Musk oxen are courageous animals which do not run away from anything, neither wolves nor bears. When an attacker appears, a small herd of musk oxen forms a real fortress. The fully grown bulls and cows stand with their heads towards the outer edge, the calves remain in the center, and this bastion of defiance does not waver. If a dog or a wolf should come really close, it is thrown into the air with a lightening quick blow from the sharp horns, and then crushed with the hard hooves. The bulls, which make completely unexpected attacks, will dash from the group towards the attacker. The long fleece may well protect them against bites in the center of their bodies. But they quickly return and back into their places in the defensive perimeter.

It is easy to make musk oxen stand still with barking dogs, and then to approach them to within shooting range without being endangered in any way. However, it is necessary to kill off all of the herd in order to get the prey. Such a "hunter" has all the time he needs to aim, which is necessary. The Arctic traveller Vitalis Pantenburg, once shot an 9.3 mm armour-piercing shell from a distance of only thirty meters into the forehead of a bull without even getting any reaction. The two horns are joined above the eyes by a plate. This forehead plate, which is elastic and as hard as steel, is about ten centimeters thick and so had easily resisted this modern projectile.

To capture the calves it is necessary to kill all members of the herd. The seal and whale slaughterers of the turn of the century certainly did not mind this as long as they could earn some extra money from some musk oxen calves. In this manner, zoological gardens in the first quarter of our century doubtlessly unwittingly contributed to the destruction of the musk oxen of Greenland. Between 1900 and 1925, at least 250 musk oxen calves went to zoos, and it is a good guess that for each of the captured calves, five to six adult animals had been shot. The New York Zoo alone purchased twenty-six musk oxen calves from 1902 until 1939. Generally these animals did not live for a long time. Since they come from a region which is rather free from bacteria, they hardly have any resistance against the diseases of our sheep and cattle. Once acclimatized, they may live for quite some time in a zoo. The Boston Zoo obtained a pair of acclimatized musk oxen in 1925; the bull died after eleven and one half years, but the cow was still there after fifteen years. They live much longer in the wild. One marked cow was twenty-three years old. After the zoo directors had found out what methods were used to capture the musk oxen, they decided not to buy any more. This rule was generally followed during the last quarter of the century, and this ended the seal slaughterers' extra profits.

Because they are hardy and yield good meat and wool, it was natural to try to introduce them into other places. However, all of those which were freed on Iceland soon died. The first six calves to be turned loose in Sweden also died of pneumonia after a short while. In South Norway on the Douvre Mountain, thirty-eight calves have been released since 1932. Five of them died from an avalanche, and not many of the others remained alive either. The young bulls left their female calves, disappeared, and reappeared in Sundalen among the herds of domestic cattle. It became clear that domestic cows, people, and musk oxen easily adjust to each other. After a while, the musk oxen even went with the cows into the barns at night. Outdoors they always behaved peacefully; bulls of domestic cattle often are more dangerous. Only in rare instances have free-ranging musk oxen attacked people. These animals in the Douvre Mountains later came to a completely remote valley of Stolsdalen where they now live. However, it is only a small herd. In past attemps to resettle them, the young calves were probably left to fend for themselves too soon, and, furthermore, musk oxen should not be brought into a habitat which is too far south.

The American zoologist Hornaday approached this problem more cautiously and took everything into account. He wanted to introduce musk oxen to Alaska to provide the needy Eskimos with a source of food. Therefore, Congress granted him $40,000. In 1930 a total of thirty-four musk oxen were captured in Greenland for this project (fifteen bulls and nineteen cows) and were brought by boat via Oslo to New York. After four weeks of quarantine, they went to Seattle by train, then by boat as far as Seward, Alaska, and by train to Fairbanks in Central Alaska.

Resettlement of musk oxen

These animals survived the transport on this 23,000 kilometer trip in good condition. It certainly does make a difference whether they are captured and tended on the transport by specialists or by seal catchers. Furthermore, they were not left on their own in Alaska, but were kept in pens at first. During the following year, six musk oxen were killed by bears and three more died from other causes. After that there were no more casualties. Meanwhile, reindeer had been imported to Alaska and they were much better suited as semi-domestic animals for the Eskimos. Therefore the musk oxen, which had reproduced, were again captured five years later and brought to the Nunivak Island in the Bering Sea off the coast of Alaska. In 1943 there were more than one hundred of them, and a recent airplane survey showed that they continued to increase. Thus the project has been successful.

In east and northeast Greenland, the number of musk oxen has greatly decreased, probably because of winters which are too wet or too severe. For example, in 1953-54 the snow cover was three meters high instead of the usual half meter, and it had an ice cover. This made it extraordinarily difficult for the animals to dig up the Arctic pastures which are their main winter diet. The Danish zoologist Chr. Vibe feared that a series of severe winters might destroy all musk oxen. Therefore, he suggested that some of them be transferred to the southwestern part of Greenland where severe weather conditions do not occur simultaneously with those in the east. In 1961 drugs were used in capture guns to obtain fourteen calves. Again in 1964, seven bulls and eleven female calves and yearlings were captured this way. The animals were kept through the winter in the Copenhagen Zoo, and in the following July, they were brought to southwest Greenland where they were released. Meanwhile, they have reproduced there. As far as we know, musk oxen never occurred in the southwest of Greenland. This successful translocation is another proof of the important part zoological gardens may play in the conservation of nature.

There was a similar success in Spitsbergen. Of the seventeen musk oxen which were brought there, one fell from a cliff and died, but the others stayed alive in the Advent Bay area. By 1942 their number had increased to seventy. Musk oxen reproduce very slowly. A herd of twenty animals rarely has more than three or four calves. They are born in April, when the weather is very cold and the nights are longer than the days. They often freeze to death right after birth, before they have a chance to dry. During World War II, most of the musk oxen of Spitsbergen were shot, but by 1960 their number had increased again to 150. In 1964 the Canadian government consented to the export of a herd of musk oxen to the Soviet Union. They were set free on the Wrangel Island in the Arctic Ocean offshore Siberia.

Musk oxen in zoological gardens

Today when some zoos keep musk oxen, it is in cooperation with the authorities for the Conservation of Nature. The animals are carefully captured by modern methods without unnecessarily shooting other members of the same herd. Al Oeming of the Alberta Game Farm near Edmonton, Canada, has been particularly successful in breeding musk oxen. There the animals begin to reproduce at two years of age, and the cows have calves every year. In the far North, they have young only every second year. Six musk oxen calves, three males and three females, were captured in the Thelon Protected Area of the Northwest Territory of Canada, which is the southernmost natural distribution of musk oxen. These animals, which appeared to me to be in excellent health and condition when I visited the Alberta Game Farm, had increased to nineteen head by 1967. Then a breeding group could be given to the Whipsnade Zoo in England. In the United States in 1966, twenty-one musk oxen lived in zoos, and three of them had been born there that year.

At Al Oeming's game park, musk oxen eat not only willows and other bushes and tree branches, but also pasture grass. They like to go into the brush forest. In the winter they are fed hay. Furthermore, they receive a grain mixture of twenty percent wheat, twenty percent barley, and sixty percent crushed oats with supplement vitamins and pellets of deer feed, containing fish oil, linseed oil, and minerals, the whole year round. Rutting is in September when the large bulls are very aggressive even against people. When they have their first calf, the mothers are two years old. The young are born during the second half of May and the first half of June.

There are not too many musk oxen left in the world. Alwin Pedersen estimated that there were ten thousand. The Canadians are somewhat more optimistic. They have established very severe laws to protect these courageous, black-brown animals. The possession of a musk oxen hide is severely punished. They estimate the total number of musk oxen in Canada, the continent, and the islands in the Northern Arctic Sea, to be thirteen thousand. There they exclusively inhabit the Barren Grounds, except for the southernmost population in the Thelon Protected Area. On the north and the east coasts of Greenland, another eleven thousand musk oxen are said to exist. However, no one knows details about their progress during the last few years. Weather stations and military bases in the Arctic, as mentioned before, have garrisons whose men are bored and like to hunt in the area. Nevertheless, the chances for survival have increased for these robust and brave northern animals.

There is a clear lesson to be learned from the example of this hardy "sheep ox." It is possible to prevent endangered species of animals from vanishing completely. They may reproduce and increase in number when a government, with the necessary authority, issues laws for their protection and then enforces them.

We hope that our encyclopedia on the animal kingdom may help to prevent more species from disappearing from earth. To help these animals, it is necessary to know what they need to exist and to survive. It is only possible to help when one knows what is needed. Millions of people who are condemned to live in cities and large towns may be moved by the multitude and wonder of the animal kingdom. It may inspire them to know the real, live animals of the wild and to ask their political leaders to establish national parks and protected areas for them. Hopefully, our efforts may contribute to the rescue and preservation of many other species of animals, of which the courageous musk oxen have been an example.

Knowing the animals is the basic prerequisite for their conservation

Bernhard Grzimek

Systematic Classification

Fossil species have not been included. The numbers refer to the pages of the text; when a species and subspecies is not mentioned in the text, references to color plates and pages are cited.

Order Odd-toed Ungulates (Perissodactyla)

Continued from Volume XII

Suborder Rhinoceros related (Ceratomorpha)

Order Even-toed Ungulates (Artiodactyla)

Suborder Nonruminants (Nonruminantia)

On the Zoological Classification and Names

For many years, zoologists and botanists have tried to classify animals and plants into a system which would be a survey of the abundance of forms in fauna and flora. Such a system, of course, may be established under very different aspects. Since Charles Darwin, his predecessors, and his successors have found that all creatures have evolved out of common ancestors, species of animals and plants have been classified according to their natural relationships. Our knowledge about the phylogeny, and thus the relationship of each living being to the other, is augmented every year by new discoveries and insights. Old ideas are replaced with more recent and more appropriate ones. Therefore, the natural classification of the animal kingdom (and the plant

kingdom) is subject to changes. Furthermore, the opinions of zoologists, who are working on the classification of animals into the various groups, are anything but uniform. These differences and changes are usually insignificant. The classification of vertebrates into the classes of fish, amphibians, reptiles, birds, and mammals has been fixed for many decades. Only the Cyclostomata were recently separated from the fish and all other classes of vertebrates as the "jawless" Agnatha (comp. Vol. 4).

The animal kingdom has been split into several subkingdoms and these were again divided into further sections, subsections, and so on. The scale of the most important systematic categories follows in a descending rank order:

Kingdom
Subkingdom
Phylum
Subphylum
Class
Subclass
Superorder
Order
Suborder
Infraorder
Family
Subfamily
Tribe
Genus
Subgenus
Species
Subspecies

The scientific names of the animals and their spelling follow the international rules for the zoological nomenclature as agreed upon by the XV International Congress for Zoology and are obligatory for all zoological publications. The name of the genus, which is a Latin or Latinized noun, is singular and capitalized. After the name of the genus follows the name of the species and of the subspecies. The names of the species and subspecies may be nouns or adjectives, and they are spelled in the lower case. The name of a subgenus, which is formed in the same manner as a genus, may be added in brackets following the name of the genus. The names of the tribes, subfamilies, families, and superfamilies are plural capitalized nouns. They are formed from the name of a given genus by adding to the principal

word the endings -ini for the tribe, -inae for the subfamily, -idae for the family, and -oidea for the superfamily. The names of the authors who were the first to describe and to name a species, subspecies, or group of animals should be cited with the year of this naming at least once in each scientific publication. The name of the author and year are not enclosed in brackets when the species or subspecies is classified as belonging to the same genus with which the author had originally classified it. They are in brackets when another genus name is used in the present publication. The scientific names of the genus, subgenus, species, and subspecies are supposed to be printed with different letters, usually italics.

ANIMAL DICTIONARY

I. English—German—French—Russian

For scientific names of species see the German-English-French-Russian section of this dictionary or the index.

In most cases names of subspecies are formed by putting an adjective or geographical specification before the name of species. These English names of subspecies will, as a rule, not appear in this part of the zoological dictionary.

ENGLISH NAME	GERMAN NAME	FRENCH NAME	RUSSIAN NAME
Abbot's Duiker	Abbotducker	Cèphalophe d'Abbot	Дукер Аббота
Addax	Mendesantilope	Addax au Nez Tacheté	Антилопа мендес
African Buffalo	Kaffernbüffel	Buffle d'Afrique	Кафрский буйвол
Alfred's Sambar	Prinz-Alfreds-Hirsch	Sambar d'Alfred	Олень принца Альфреда
Alpaca	Alpaka	Alpaga	Альпака
Alpine Ibex	Alpensteinbock	Bouquetin des Alpes	Альпийский горный козел
American and Roe Deer	Trughirsche	Odocoiléinés	Телеметакарпальные олени
– Bison	Bison	Bison Américain	Бизон
– Deer	Amerikahirsche	Cerfs Américains	Американские олени
Anoa	Anoa	Anoa	Целебесский карликовый буйвол
Antelopes	Antilopen	Antilopes	Антилопы
Argali	Argali	Argali	Алтайский аргали
Arna	Arni	Buffle de l'Inde	Буйвол
Asiatic Two-horned Rhinos	Halbpanzernashörner	Rhinocéros Bicornes d'Asie	Азиатские двурогие носороги
Aurochs	Ur	Aurochs	Первобытный бык
Axis Deer	Axishirsch	Axis	Аксис
Babirusa	Hirscheber	Babiroussa	Бабируссы
Bactrian Camel	Hauskamel	Chameau domestique	Домашний двугорбый верблюд
Banded Duiker	Zebraducker	Céphalophe Rayé	Дукер-зебра
Banteng	Banteng	Banting	Бантенг
Barasingha	Barasingha	Barasingha	Барасинга
Barbary Sheep	Mähnenspringer	Aoudad	Североафриканский гривистый баран
Bate's Dwarf Antelope	Batesböckchen	Antilope de Bates	Антилопа Батеса
Bay Duiker	Schwarzrückenducker	Céphalophe à Bande Dorsale Noire	Черноспинный дукер
Bearded Pig	Bartschwein	Sanglier à Moustache	Бородатая свинья
Beira	Beira	Dorcatrague	Бейра
Bighorn Sheep	Dickhornschaf	Mouflon du Canada	Толсторог
Blackbuck	Hirschziegenantilope	Antilope cervicapre	Винторогая антилопа сасси
Black Duiker	Schwarzducker	Céphalophe Noir	Черный дукер
Black-fronted Duiker	Schwarzstirnducker	– à Front Noir	Чернолобый дукер
Black Rhinoceros	Spitzmaulnashorn	Rhinocéros Noir	Африканский острорылый носорог
Black-tailed Deer	Schwarzwedelhirsch	Cerf de Mulet	Длинноухий олень Скалистых гор
Blesbok	Bleßbock	Blesbok	
Blue Duiker	Blauducker	Céphalophe Bleu	Голубой Дукер
– Sheep	Blauschaf	Bouc Bleu	Нахур
– Wildebeest	Blaues Gnu	Gnou Bleu	Южный полосатый гну
Bohor Reedbuck	Riedbock	Nagor de Buffon	Болотный козел нагор
Bongo	Bongo	Bongo	Бонго
Bontebok	Buntbock	Bontebok	
Brindled Gnu	Streifengnu	Gnou Bleu	Белобородый гну
Brockets	Mazamas	Daguets	Мазамы
Buffalos	Büffel	Buffles	Буйволы
Buffon's Kob	Moor-Antilope	Cobe Buffon	Эквитун
Bushbuck	Buschbock	Guib Harnachée	Антилопа гуиб
Bush Duiker	Kronenducker	Céphalophe de Grimm	Антилопа-дукер
– Pig	Buschschwein	Potamochère de l'Afrique	Кустарная свиня
Camels	Kamele	Chameaux	Верблюды
Caribu	Karibu	Renne d'Amérique	Западноканадский карибу
Cattle	Rinder	Bœufs	Быки
Central American Tapir	Mittelamerikanischer Tapir	Tapir de Baird	Тапир Бэрда
Chamois	Gemse	Chamois	Серна
Chevrotains	Hirschferkel	Tragulidés	Оленьки
Chilenian Huemul	Südandenhirsch	Hippocamelus	Южноандский гуэмал
Chinese Water Deer	Wasserreh	Hydropote	Водяной олень
Collared Peccary	Halsbandpekari	Pecari à Collier	Ошейниковый пекари

ENGLISH NAME	GERMAN NAME	FRENCH NAME	RUSSIAN NAME
Cretan Wild Goat	Kretische Wildziege	Chèvre de la Crète	Критский бородатый козел
Dama Gazelle	Damagazelle	Gazelle Dama	Сахарская газель
Deer	Hirsche	Cerfs	Плоторогие
Defassa-Waterbuck	Defassa-Wasserbock	Cobe Defassa	Синг-синг
Dibatag	Stelzengazelle	Dibatag	Дибатаг
Dik-Diks	Windspielantilopen	Dik-Diks	Дик-дики
Domestic Buffalo	Hausbüffel	Buffle Domestique	Домашний буйвол
– Pig	Hausschwein	Cochon Domestique	Домашняя свинья
– Sheep	Hausschaf	Mouton Domestique	Домашняя овца
Dorcas Gazelle	Dorkasgazelle	Dorcas	Обыкновенная газель
Dromedary	Dromedar	Dromadaire	Одногорбый верблюд
Duikers	Ducker	Céphalophes	Хохлатые антилопы
Dwarf Antelopes	Böckchen	Neotraguinés	Карликовые антилопы
Dybowski's Deer	Dybowskihirsch	Cerf de Dybowski	Пятнистый олень Дыбовского
Eland	Elenantilope	Eland	Антилопа канна
Eld's Deer	Leierhirsch	Cerf d'Eld	Таменг
Elk (American)	Wapiti	Wapiti	
– (English)	Elch	Elan	Лось
European Mouflon	Mufflon	Mouflon d'Europe	Европейский муфлон
European Bison	Wisent	Bison d'Europe	Зубр
Even-toed Ungulates	Paarhufer	Artiodactyles	Парнокопытные
Fallow Deer	Damhirsch	Daim	Лань
Four-horned Antelope	Vierhornantilope	Tetracère	Четырехрогая антилопа
Gaboon Duiker	Weißbauchducker	Céphalophe à Ventre Blanc	Белобрюхий дукер
Gaur	Gaur	Gaur	Гаур
Gayal	Gayal	Gayal	Гаял
Gazelles	Gazellen	Gazelles	Газели
Gerenuk	Giraffengazelle	Gazelle Girafe	Газель-жирафа
Giant Duiker	Gelbrückenducker	Céphalophe Géant	Желтоспинный дукер
– Forest Pig	Riesenwaldschwein	Hylochère	Исполинская лесная свинья
Giraffe	Giraffe	Girafe	Обыкновенная жирафа
Gnus	Gnus	Gnous	Антилопы гну
Goat	Ziege	Chèvre	Коза
Goitred Gazelle	Kropfgazelle	Gazelle à Goitre	Джейран
Goral	Goral	Goral	Гималайский горал
Grant's Gazelle	Grantgazelle	Gazelle de Grant	Газель Гранта
Greater Kudu	Großer Kudu	Grand Koudou	Антилопа куду
Great Indian Rhinoceros	Panzernashorn	Rhinocéros Unicorne des Indes	Большой однорогий носорог
Grey Brocket	Graumazama	Daguet Gris	Серый мазам
– Duiker	Kronenducker	Céphalophe de Grimm	Антилопа-дукер
– Dwarf Brocket	Zwergmazama	Daguet Nain Gris	Карликовый мазам
– Rhebuck	Rehantilope	Rhebouk	Антилопа-косуля
Grysbok	Greisböckchen	Raphicère du Cap	Антилопа грис
Guanaco	Guanako	Guanaco	Гуанако
Guemals	Andenhirsche	Guemals	Андские олени
Guenther's Dik-Dik	Güntherdikdik	Dik-Dik de Guenther	Дик-дик Гюнтера
Hartebeest	Kuhantilope	Bubale	Коровья антилопа
Harvey's Duiker	Harveyducker	Céphalophe d'Harvey	Дукер Харвея
Heuglin's Gazelle	Heuglingazelle	Gazelle d'Heuglin	
Hippopotamus	Flußpferd	Hippopotame Amphibie	Обыкновенный бегемот
Hog Deer	Schweinshirsch	Cerf Cochon	Свиной олень
Horned Ungulates	Hornträger	Bovidés	Пологие
Ibex	Steinbock	Bouquetin	Гоный козел
Impala	Schwarzfersenantilope	Impalla	Чернопятая антилопа
Indian Sambar	Indischer Sambar	Sambar de l'Inde	Самбар
Javan Pig	Pustelschwein	Sanglier Pustule	Бородавчатая свинья
– Rhinoceros	Javanashorn	Rhinocéros de la Sonde	Малый однорогий носорог
Jentink's Duiker	Jentinkducker	Céphalophe de Jentink	Либерийский лесной дукер
Kaama	Kaama	Caama	Антилопа каама
Kirk's Dik-Dik	Kirkdikdik	Dik-Dik de Kirk	Дик-дик Кэрка
Klipspringer	Klippspringer	Oreotrague	Антилопа-прыгун
Kongoni	Kongoni	Kongoni	Коровья антилопа Кука
Korrigum	Korrigum	Korrigum	Антилопа корригум
Kouprey	Kouprey	Kouprey	Индокитайский серый бык
Kudu	Kudu	Koudou	Антилопа куду
Larger Malay Mouse Deer	Großkantschil	Chevrotain Malais	Оленек напу
Lechwe Waterbuck	Litschi-Wasserbock	Lechwe	Водяной козел лихи
Lelwel	Lelwel-Hartebeest	Lelwel	Коровья антилопа Лельveля
Lesser Kudu	Kleiner Kudu	Petit Koudou	Малый куду
– Malay Mouse Deer	Kleinkantschil	Chevrotain Malais	Яванский канчили
Llama	Lama	Lama	Лама
Lowland Tapir	Flachlandtapir	Tapir Terrestre	Обыкновенный тапир

ENGLISH NAME	GERMAN NAME	FRENCH NAME	RUSSIAN NAME
Malayan Tapir	Schabrackentapir	– de l'Inde	Индийский тапир
Markhor	Schraubenziege	Markhor	Винторогий козел
Marsh Deer	Sumpfhirsch	Cerf Marécageux	Болотный олень
Maxwell's Duiker	Maxwellducker	Céphalophe de Maxwell	Дукер Максвелла
Mongolian Gazelle	Mongoleigazelle	Gazelle de la Mongolie	Дзерен
Moose	Elch	Elan	Лось
Mountain Gazelle	Edmigazelle	Edmi	
– Nyala	Bergnyala	Nyala de Montagne	Горная ниала
– Reedbuck	Bergriedbock	Redunca de Montagne	Горный болотный козел
– Tapir	Bergtapir	Tapir des Andes	Горный тапир
Mouse Deer	Kantschile	Chevrotains d'Asie	Азиатские оленьки
Mrs. Gray's Lechwe	Weißnacken-Moorantilope	Cobe de Mme. Gray	Белошейная антилопа
Mule Deer	Großohrhirsch	Cerf de Mulet	Длинноухий олень
Musk Deer	Moschustier	Porte-Musc	Мускусная кабарга
– Ox	Moschusochse	Bœuf Musqué	Мускусный овцебык
Muntjac	Muntjak	Muntjac	Мунтжак
Natal Duiker	Rotducker	Céphalophe de Natal	Красный дукер
Nile-Lechwe	Weißnacken-Moorantilope	Lechwe du Nil	Белошейная антилопа
Nilgai	Nilgauantilope	Nilgaut	Антилопа нильгау
Northern Pudu	Nordpudu	Pudu du Nord	Северный пуду
Nyala	Nyala	Nyala	Антилопа ниала
Ogilby's Duiker	Fernando-Po-Ducker	Céphalophe d'Ogilby	Дукер о. Фернандо-По
Odd-toed Ungulates	Unpaarhufer	Périssodactyles	Непарнокопытные
Okapi	Okapi	Okapi	Окапи Джонстона
Old World Pigs	Altweltliche Schweine	Suidés	Свиньи Старого света
Oribi	Bleichböckchen	Ourebie	Бледный ориби
Oryx	Spießbock	Oryx	Сернобык
Ox	Hausrind	Bœuf	Домашний быт
Pampas Deer	Pampashirsch	Cerf de Pampas	Пампасский олень
Peccaries	Pekaris	Pecaris	Пекари
Pelzeln's Gazelle	Pelzelngazelle	Gazelle de Pelzeln	
Père David's Deer	Davidshirsch	Cerf du Père David	Давидов олень
Peruvian Guemal	Nordandenhirsch	Guemal	Североандский олень
Philippine Sambar	Philippinensambar	Sambar des Philippines	Филиппинский самбар
Pig-like Mammals	Schweineartige	Suoidea	Свиные
Pigmy Hippopotamus	Zwergflußpferd	Hippopotame Nain	Либерийский карликовый бегемот
– Hog	Zwergwildschwein	Sanglier Nain	Карликовая свинья
Pronghorn	Gabelbock	Antilope-Chèvre Américaine	Американский вилорог
Pudus	Pudus	Pudus	Пуду
Red Belly Dik-Dik	Rotbauchdikdik	Dik-Dik de Phillips	Краснобрюхий дик-дик
– Brocket	Großmazama	Daguet Rouge	Большой мазам
– Deer	Rothirsch	Cerf Rouge	Благородный олень
– Duiker	Rotducker	Céphalophe Rouge	Красный дукер
Red-flanked Duiker	Rotflankenducker	– aux Flancs Roux	Рыжебокий дукер
Red-fronted Gazelle	Rotstirngazelle	Gazelle à Front Roux	Рыжелобая газель
Reedbuck	Großer Riedbock	Redunca Grand	Большой болотный козел
Reindeer	Ren	Renne	Северный олень
Rhinoceroses	Nashörner	Rhinocérotidés	Носороги
Roan Antelope	Pferdeantilope	Antilope Rouanne	Чалая лошадиная антилопа
Rocky Mountain Goa	Schneeziege	Chèvre des Montagnes Rocheuses	Снежная коза
Roe Deer	Reh	Chevreuil	Косуля
Royal Antelope	Kleinstböckchen	Antilope Royale	Антилопа-карлик
Ruminants	Wiederkäuer	Ruminants	Жвачные
Sable Antelope	Rappenantilope	Hippotrague Noir	Черная антилопа
Saiga Antelope	Saiga	Saiga	Сайга
Salt's Dik-Dik	Eritreadikdik	Dik-Dik de Salt	Эритрейский дик-дик
Sambars	Sambarhirsche	Sambars	Самбары
Sassaby	Sassaby	Damalisque	Антилопа сассаби
Schomburgk's Deer	Schomburgkhirsch	Cerf de Schomburgk	Сиамский олень
Scimitar-horned Oryx	Säbelantilope	Oryx algazelle	Нубийский саблерогий сернобык
Serow	Serau	Serow	Суматранский серао
Sheep	Schaf	Mouton	Горные бараны
Sika Deer	Sikahirsch	Sika	Пятнистый олень
Sitatunga	Sitatunga	Sitatunga	Антилопа наконг
Slender-horned Gazelle	Dünengazelle	Gazelle à Cornes Grêles	
Soemmering's Gazelle	Sömmeringgazelle	Gazelle de Soemmering	Абиссинская газель
Southern Pudu	Südpudu	Pudu du Sud	Южный пуду
Spanish Ibex	Iberiensteinbock	Bouquetin d'Espagne	Иберийский тур
Speke's Gazelle	Spekegazelle	Gazelle de Speke	
Spiral-horned Antelopes	Drehhörner	Tragelaphini	Винторогие антилопы
Spotted Mouse Deer	Fleckenkantschil	Chevrotain Tacheté	Маминна
Springbuck	Springbock	Antidorcas	Антилопа-прыгун
Steinbok	Steinböckchen	Raphicère Champêtre	Степной штейнбок
Suiformes	Nichtwiederkäuer	Suiformes	Нежвачные

ENGLISH NAME	GERMAN NAME	FRENCH NAME	RUSSIAN NAME
Sumatran Rhinoceros	Sumatranashorn	Rhinocéros de Sumatra	Суматранский двурогий носорог
Sunda Sambar	Mähnenhirsch	Sambar de la Sonde	Гривистый олень
Suni	Suniböckchen	Suni	Суни
Swayne's Dik-Dik	Kleindikdik	Dik-Dik de Swayne	Малый дик-дик
Tahr	Tahr	Tahr	Тар
Takin	Takin	Takin	Такин
Tamarou	Tamarau	Tamarou	Филиппинский буйвол
Tapirs	Tapire	Tapirs	Тапиры
Thamin	Thamin	Thameng	Бирманский таменг
Thomson's Gazelle	Thomsongazelle	Gazelle de Thomson	Газель Томсона
Thorold's Deer	Weißlippenhirsch	Cerf de Thorold	Тибетский горный олень
Tibetan Antelope	Tschiru	Antilope du Thibet	Чиру
– Gazelle	Tibetgazelle	Gazelle du Thibet	Тибетский дзерен
Topi	Topi	Topi	Антилопа топи
Tora	Tora	Tora	Тора
True Antelopes	Gazellenartige	Antilopinés	Газели и карликовые антилопы
Tufted Deer	Schopfhirsch	Cerf Touffe	Хохлатый олень
Two-humped Camel	Zweihöckriges Kamel	Chameau à Deux Bosses	Двугорбый верблюд
Tylopodes	Schwielensohler	Tylopodes	Мозоленогие
Vicugna	Vikunja	Vigogne	Викунья
Wapiti	Wapiti	Wapiti	
Wart Hog	Warzenschwein	Phacochère	Абиссинский бородавочник
Waterbuck	Wasserbock	Cobe	Обыкновенный водяной козел
Water Buffalo	Wasserbüffel	Buffle de l'Inde	Буйвол
– Chevrotain	Afrikanisches Hirschferkel	Chevrotain Africain	Африканский оленек
– Deer	Wasserhirsche	Hydropotes	Водяные олени
White-lipped Peccary	Weißbartpekari	Pecari à Barbe Blanche	Белогубый пекари
White-tailed Deer	Weißwedelhirsch	Cerf de Virginie	Белохвостый олень
– – Gnu	Weißschwanzgnu	Gnou à Queue Blanche	Белохвостый гну
Wide-mouthed Rhinoceros	Breitmaulnashorn	Rhinocéros Blanc	Африканский широкорылый носорог
Wild Boar	Wildschwein	Sanglier	Кабан
– Camel	Wildkamel	Chameau Sauvage	Дикий двугорбый верблюд
– Cattle	Wildrinder	Bœufs Sauvages	Дикие быки
– Goat	Bezoarziege	Chèvre véritable	Безоаровый козел
– Goats	Wildziegen	Chèvres Sauvages	Горные козлы
– Sheep	Wildschaf	Mouflon	Горный баран
Yak	Yak	Yack	Як
Zebu	Zebu	Zébu	Зебу

II. German–English–French–Russian

Unterartnamen werden meist aus den Artnamen durch Voranstellen von Eigenschaftswörtern oder geographischen Bezeichnungen gebildet. In diesem Teil des Tierwörterbuchs sind so gebildete deutsche Unterartnamen sowie die wissenschaftlichen Unterartnamen in der Regel nicht aufgeführt.

GERMAN NAME	ENGLISH NAME	FRENCH NAME	RUSSIAN NAME
Abbotducker	Abbot's Duiker	Céphalophe d'Abbot	Дукер Аббота
Addax nasomaculatus	Addax	Addax au Nez Tacheté	Антилопа мендес
Adenota kob	Buffon's Kob	Le Cobe Buffon	Эквитун
Aepyceros melampus	Impala	Impalla	Чернопятая антилопа
Afrikanischer Büffel	African Buffalo	Buffle d'Afrique	Африканский буйвол
Afrikanisches Hirschferkel	Water Chevrotain	Chevrotain Africain	Африканский оленек
Agrimi	Cretan Wild Goat	Chèvre de la Crète	Критский бородатый козел
Alcelaphus buselaphus	Hartebeest	Bubale	Коровья антилопа
Alces alces	Elk (englisch), Moose (amerikanisch)	Elan	Лось
Alpaka	Alpaca	Alpaga	Альпака
Alpensteinbock	Alpine Ibex	Bouquetin des Alpes	Альпийский горный козел
Ammodorcas clarkei	Dibatag	Dibatag	Дибатаг
Ammotragus lervia	Barbary Sheep	Aoudad	Североафриканский гривистый баран
Andenhirsche	Guemals	Guemals	Андские олени
Andentapir	Mountain Tapir	Tapir des Andes	Горный тапир
Anoa	Anoa	Anoa	Целебесский карликовый буйвол

GERMAN NAME	ENGLISH NAME	FRENCH NAME	RUSSIAN NAME
Antidorcas marsupialis	Springbuck	Antidorcas	Антилопа-прыгун
Antilocapra americana	Pronghorn	Antilope-Chèvre Américaine	Американский вилорог
Antilope cervicapra	Blackbuck	Antilope cervicapre	Винторогая антилопа сасси
Antilopinae	True Antelopes	Antilopinés	Газели и карликовые антилопы
Argali	Argali	Argali	Алтайский аргали
Arni	Arna	Buffle de l'Inde	Буйвол
Artiodactyla	Even-toed Ungulates	Artiodactyles	Парнокопытные
Asiatische Büffel	Asiatic Buffalos	Buffles d'Asie	Азиатские буйволы
Auerochse	Aurochs	Aurochs	Первобытный бык
Axis axis	Axis Deer	Axis	Аксис
– *porcinus*	Hog Deer	Cerf Cochon	Свиной олень
Axishirsch	Axis Deer	Axis	Аксис
Babirusa	Babirusa	Babiroussa	Бабируссы
Babyrousa babyrussa	Babirusa	Babiroussa	Бабирусса
Bairds Tapir	Central American Tapir	Tapir de Baird	Тапир Бэрда
Banteng	Banteng	Banting	Бантенг
Barasingha	Barasingha	Barasingha	Барасинга
Bartschwein	Bearded Pig	Sanglier à Moustache	Бородатая свинья
Batesböckchen	Bate's Dwarf Antelope	Antilope de Bates	Антилопа Батеса
Beira	Beira	Beira	Бейра
Bergnyala	Mountain Nyala	Nyala de Montagne	Горная ниала
Bergriedbock	Mountain Reedbuck	Redunca de Montagne	Горный болотный козел
Bergtapir	Mountain Tapir	Tapir des Andes	Горный тапир
Bezoarziege	Wild Goat	Chèvre véritable	Безоаровый козел
Bisamschwein	White-lipped-Peccary	Pecari à Barbe Blanche	Белогубый пекари
Bison	American Bison	Bison Américain	Бизон
Bison bison	American Bison	Bison Américain	Бизон
– *bonasus*	European Bison	Bison d'Europe	Зубр
Blauböckchen	Blue Duiker	Céphalophe Bleu	Голубой дукер
Blauducker	Blue Duiker	Céphalophe Bleu	Голубой Дукер
Blaues Gnu	Blue Wildebeest	Gnou Bleu	Южный полосатый гну
Blaurückenducker	Red-flanked Duiker	Céphalophe aux Flancs Roux	Рыжебокий дукер
Blauschaf	Blue Sheep	Bouc Bleu	Нахур
Bleichböckchen	Oribi	Ourebie	Бледный ориби
Bleßbock	Blesbok	Blesbok	
Böckchen	Dwarf Antelopes	Neotraguinés	Карликовые антилопы
Bongo	Bongo	Bongo	Бонго
Bos gaurus	Gaur	Gaur	Гаур
– – *frontalis*	Gayal	Gayal	Гаял
– *javanicus*	Banteng	Banting	Бантенг
– *mutus*	Yak	Yack	Як
– *primigenius*	Aurochs	Aurochs	Первобытный бык
– – *taurus*	Ox	Bœuf	Домашний бык
– *sauveli*	Kouprey	Kouprey	Индокитайский серый бык
Boselaphus tragocamelus	Nilgai	Nilgaut	Антилопа нильгау
Bovidae	Horned Ungulates	Bovidés	Полорогие
Bovinae	Cattle	Bœufs	Быки
Breitmaulnashorn	Wide-mouthed Rhinoceros	Rhinocéros Blanc	Африканский широкорылый носорог
Bubalus	Asiatic Buffalos	Buffles d'Asie	Азиатские буйволы
– *arnee*	Water Buffalo	Buffle de l'Inde	Буйвол
– – *bubalis*	Domestic Buffalo	Buffle Domestique	Домашний буйвол
– – *mindorensis*	Tamarou	Tamarou	Филиппинский буйвол
– *depressicornis*	Anoa	Anoa	Целебесский карликовый буйвол
Buckelrind	Zebu	Zébu	Зебу
Budorcas taxicolor	Takin	Takin	Такин
Budorcatini	Takins	Takins	Такины
Büffel	Buffalos	Buffles	Буйволы
Buntbock	Bontebok	Bontebok	
Buschbock	Bushbuck	Guib Harnaché	Антилопа гуиб
Buschducker	Grey Duiker	Céphalophe de Grimm	Антилопа-дукер
Buschschwein	Bush Pig	Potamochère de l'Afrique	Кустарная свинья
Camelidae	Camels	Chameaux	Верблюды
Camelus dromedarius	Dromedary	Dromadaire	Одногорбый верблюд
– *ferus*	Two-humped Camel	Chameau à Deux Bosses	Двугорбый верблюд
– – *bactrianus*	Bactrian Camel	– Domestique	Домашний двугорбый верблюд
– – *ferus*	Wild Camel	– Sauvage	Дикий двугорбый верблюд
Capra aegagrus	Wild Goat	Chèvre véritable	Безоаровый козел
– – *cretica*	Cretan Wild Goat	– de la Crète	Критский бородатый козел
– – *hircus*	Domestic Goat	– Domestique	Домашняя коза
– *falconeri*	Markhor	Markhor	Винторогий козел
– *ibex*	Ibex	Bouquetin	Горный козел

GERMAN NAME	ENGLISH NAME	FRENCH NAME	RUSSIAN NAME
— — ibex	Alpine Ibex	Bouquetin des Alpes	Альпийский горный козел
— pyrenaica	Spanish Ibex	— d'Espagne	Иберийский тур
Capreolus capreolus	Roe Deer	Chevreuil	Косуля
Capricornis sumatrensis	Serow	Serow	Суматранский серао
Cephalophinae	Duikers	Céphalophinés	Хохлатые антилопы
Cephalophus dorsalis	Bay Duiker	Céphalophe à Bande Dorsale Noire	Черноспинный дукер
— harveyi	Harvey's Duiker	— d'Harvey	Дукер Харвея
— jentinki	Jentink's Duiker	— de Jentink	Либерийский лесной дукер
— leucogaster	Gaboon Duiker	— à Ventre Blanc	Белобрюхий дукер
— maxwelli	Maxwell's Duiker	— de Maxwell	Дукер Максвелла
— monticola	Blue Duiker	— Bleu	Голубой дукер
— natalensis	Red Duiker	— Rouge	Красный дукер
— niger	Black Duiker	— Noir	Черный дукер
— nigrifrons	Black-fronted Duiker	— à Front Noir	Чернолобый дукер
— ogilbyi	Ogilby's Duiker	— d'Ogilby	Дукер о. Фернандо-По
— rufilatus	Red-flanked Duiker	— aux Flancs Roux	Рыжебокий дукер
— spadix	Abbot's Duiker	— d'Abbot	Дукер Аббота
— sylvicultor	Giant Duiker	— Géant	Желтоспинный дукер
— zebra	Banded Duiker	— Rayé	Дукер-зебра
Ceratotherium simum	Wide-mouthed Rhinoceros	Rhinocéros Blanc	Африканский широкорылый носорог
Cervidae	Deer	Cerfs	Плоторогие
Cervinae	Eurasian Deer	— véritables	Плезиметакарпальные олени
Cervus albirostris	Thorold's Deer	Cerf de Thorold	Тибетский горный олень
— duvauceli	Barasingha	Barasingha	Барасинга
— — schomburgki	Schomburgk's Deer	Cerf de Schomburgk	Сиамский олень
— elaphus	Red Deer	— Rouge	Благородный олень
— — canadensis	Wapiti (englisch), Elk (amerikanisch)	Wapiti	Восточный вапити
— — nelsoni	Wapiti (englisch), Elk (amerikanisch)	Wapiti	Благородный олень
— eldi	Eld's Deer	Cerf d'Eld	Таменг
— — thamin	Thamin	Thameng	Бирманский таменг
— mariannus	Philippine Sambar	Sambar des Philippines	Филиппинский самбар
— — alfredi	Alfred's Sambar	— d'Alfred	Олень принца Альфреда
— nippon	Sika Deer	Sika	Пятнистый олень
— — dybowskii	Dybowski's Deer	Cerf de Dybowski	Пятнистый олень Дыбовского
— timorensis	Sunda Sambar	Sambar de la Sonde	Гривистый олень
— unicolor	Indian Sambar	— de l'Inde	Самбар
Choeropsis liberiensis	Pigmy Hippopotamus	Hippopotame Nain	Либерийский карликовый бегемот
Connochaetes gnou	White-tailed Gnu	Gnou à Queue Blanche	Белохвостый гну
— taurinus	Brindled Gnu	— Bleu	Белобородый гну
Dama dama	Fallow Deer	Daim	Лань
Damagazelle	Dama Gazelle	Gazelle Dama	Сахарская газель
Damaliscus dorcas dorcas	Bontebok	Bontebok	
Damaliscus dorcas philippsi	Blesbok	Blesbok	
— lunatus corrigum	Korrigum	Korrigum	Антилопа корригум
— — lunatus	Sassaby	Damalisque	Антилопа сассаби
— — topi	Topi	Topi	Антилопа топи
Damhirsch	Fallow Deer	Daim	Лань
Davidshirsch	Père David's Deer	Cerf du Père David	Давидов олень
Defassa-Wasserbock	Defassa-Waterbuck	Cobe Defassa	Синг-синг
Dibatag	Dibatag	Dibatag	Дибатаг
Dicerorhininae	Asiatic Two-horned Rhinos	Rhinocéros Bicornes d'Asie	Азиатские двурогие носороги
Dicerorhinus sumatrensis	Sumatran Rhinoceros	— de Sumatra	Суматранский двурогий носорог
Diceros bicornis	Black Rhinoceros	— Noir	Африканский острорылый носорог
Dickhornschaf	Bighorn Sheep	Mouflon du Canada	Толсторог
Dikdiks	Dik-Diks	Dik-Diks	Антилопы-левретки
Dorcatragini	Beira-Antelopes	Beiras	Антилопы бейра
Dorcatragus megalotis	Beira	Dorcatrague	Бейра
Dorkasgazelle	Dorcas Gazelle	Dorcas	Обыкновенная газель
Drehhörner	Spiral-horned Antelopes	Tragelaphini	Винторогие антилопы
Dromedar	Dromedary	Dromadaire	Одногорбый верблюд
Ducker	Duikers	Céphalophes	Хохлатые антилопы
Dünengazelle	Slender-horned Gazelle	Gazelle à Cornes Grêles	
Dybowskihirsch	Dybowski's Deer	Cerf de Dybowski	Пятнистый олень Дыбовского
Echthirsche	Eurasian Deer	— véritables	Млезиметакарпальные олени
Edelhirsch	Red Deer	— Rouge	

GERMAN NAME	ENGLISH NAME	FRENCH NAME	RUSSIAN NAME
			Благородный олень
Edmigazelle	Mountain Gazelle	Edmi	
Elaphodus cephalophus	Tufted Deer	Cerf Touffe	Хохлатый олень
Elaphurus davidianus	Père David's Deer	Cerf du Père David	Давидов олень
Elch	Elk (englisch), Moose (amerikanisch)	Elan	Лось
Elchhirsche	Elk deer	Elans	Лоси
Elenantilope	Eland	Eland	Антилопа канна
Eritreadikdik	Salt's Dik-Dik	Dik-Dik de Salt	Эритрейский дик-дик
Fernando-Po-Ducker	Ogilby's Duiker	Céphalophe d'Ogilby	Дукер о. Фернандо-По
Flachlandtapir	Lowland Tapir	Tapir Terrestre	Обыкновенный тапир
Fleckenhirsche	Axis Deer	Cerfs Axis	Южноазиатские пятнистые олени
Fleckenkantschil	Spotted Mouse Deer	Chevrotain Tacheté	Мавинна
Flußpferd	Hippopotamus	Hippopotame Amphibie	Обыкновенный бегемот
Flußschwein	Bush Pig	Potamochère de l'Afrique	Кустарная свинья
Frau Grays Wasserbock	Nile-Lechwe	Lechwe du Nil	Белошейная антилопа
Gabelbock	Pronghorn	Antilope-Chèvre Américaine	Американский вилорог
Gabelhorntiere	Pronghorns	Antilopes-Chèvres Américaines	Вилороги
Gabunducker	Gaboon Duiker	Céphalophe à Ventre Blanc	Белобрюхий дукер
Gaur	Gaur	Gaur	Гаур
Gayal	Gayal	Gayal	Гаял
Gazella	Gazelles	Gazelles	Газели
– *dama*	Dama Gazelle	Gazelle Dama	Сахарская газель
– *dorcas*	Dorcas Gazelle	Dorcas	Обыкновенная газель
– *gazella*	Mountain Gazelle	Edmi	
– *granti*	Grant's Gazelle	Gazelle de Grant	Газель Гранта
– *leptoceros*	Slender-horned Gazelle	– à Cornes Grêles	
– *pelzelni*	Pelzeln's Gazelle	– de Pelzeln	
– *rufifrons*	Red-fronted Gazelle	– à Front Roux	Рыжелобая газель
– *soemmeringi*	Soemmering's Gazelle	– de Soemmering	Абиссинская газель
– *spekei*	Speke's Gazelle	– de Speke	
– *subgutturosa*	Goitred Gazelle	– à Goitre	Джейран
– *thomsoni*	Thomson's Gazelle	– de Thomson	Газель Томсона
– *tilonura*	Heuglin's Gazelle	– d'Heuglin	
Gazellen	Gazelles	Gazelles	Газели
Gazellenartige	True Antelopes	Antilopinés	Газели и карликовые антилопы
Gelbrückenducker	Giant Duiker	Céphalophe Géant	Желтоспинный дукер
Gemsbüffel	Anoa	Anoa	Целебесский карликовый буйвол
Gemse	Chamois	Chamois	Серна
Gerenuk	Gerenuk	Gazelle Girafe	Газель-жирафа
Giraffa camelopardalis	Giraffe	Girafe	Обыкновенная жирафа
Giraffe	Giraffe	Girafe	Обыкновенная жирафа
Giraffengazelle	Gerenuk	Gazelle Girafe	Газель-жирафа
Gnus	Gnus	Gnous	Антилопы гну
Goral	Goral	Goral	Гималайский горал
Grantgazelle	Grant's Gazelle	Gazelle de Grant	Газель Гранта
Grauer Spießhirsch	Grey Brocket	Daguet Gris	Серый мазам
Graumazama	Grey Brocket	Daguet Gris	Серый мазам
Greisböckchen	Grysbok	Raphicère du Cap	Антилопа грис
Großer Kudu	Greater Kudu	Grand Koudou	Антилопа куду
Großer Riedbock	Reedbuck	Redunca Grand	Большой болотный козел
Großkantschil	Larger Malay Mouse Deer	Chevrotain Malais	Оленек напу
Großmazama	Red Brocket	Daguet Rouge	Большой мазам
Großohrhirsch	Mule Deer	Cerf de Mulet	Длинноухий олень
Guanako	Guanaco	Guanaco	Гуанако
Güntherdikdik	Guenther's Dik-Dik	Dik-Dik de Guenther	Дик-дик Гюнтера
Halbmondantilope	Sassaby	Damalisque	Лиророгая антилопа
Halbpanzernashörner	Asiatic Two-horned Rhinos	Rhinocéros Bicornes d'Asie	Азиатские двурогие носороги
Halsbandpekari	Collared Peccary	Pecari à Collier	Ошейниковый пекари
Hartebeest	Hartebeest	Bubale	Коровья антилопа
Harveyducker	Harvey's Duiker	Céphalophe d'Harvey	Дукер Харвея
Hausbüffel	Domestic Buffalo	Buffle Domestique	Домашний буйвол
Hauskamel	Bactrian Camel	Chameau Domestique à Deux Bosses	Домашний двугорбый верблюд
Hausrind	Ox	Bœuf	Домашний бык
Hausschaf	Domestic Sheep	Mouton Domestique	Домашняя овца
Hausschwein	– Pig	Cochon Domestique	Домашняя свинья
Hausziege	– Goat	Chèvre Domestique	Домашняя коза
Hemitragus jemlahicus	Tahr	Tahr	Тар
Heuglingazelle	Heuglin's Gazelle	Gazelle d'Heuglin	
Hippocamelus antisiensis	Peruvian Guemal	Guemal	Североандский олень
– *bisculus*	Chilenian Huemul	Hippocamelus	Южноандский гуэмал
Hippopotamidae	Hippopotamuses	Hippopotamidés	Бегемоты

GERMAN NAME	ENGLISH NAME	FRENCH NAME	RUSSIAN NAME
Hippopotamus amphibius	Hippopotamus	Hippopotame Amphibie	Обыкновенный бегемот
Hippotraginae	Roan and Sable Antelopes	Hippotraginés	Лошадиные антилопы
Hippotragus equinus	Roan Antelope	Antilope Rouanne	Чалая лошадиная антилопа
– *niger*	Sable Antelope	Hippotrague Noir	Черная антилопа
Hirschantilope	Waterbuck	Cobe	Обыкновенный водяной козел
Hirsche	Deer	Cerfs	Плоторогие
Hirscheber	Babirusa	Babiroussa	Бабируссы
Hirschferkel	Chevrotains	Tragulidés	Оленьки
Hirschziegenantilope	Blackbuck	Antilope cervicapre	Винторогая антилопа сасси
Hornträger	Horned Ungulates	Bovidés	Полорогие
Huemul	Chilenian Huemul	Hippocamelus	Южноандский гуэмал
Hydropotes inermis	Chinese Water Deer	Hydropote	Водяной олень
Hydropotinae	Water Deer	Hydropotes	Водяные олени
Hydrotragus leche	Lechwe Waterbuck	Lechwe	Водяной козел лихи
Hyelaphus	Hog Deer	Cerf Cochon	Свиной олень
Hyemoschus aquaticus	Water Chevrotain	Chevrotain Africain	Африканский оленек
Hylochoerus meinertzhageni	Giant Forest Pig	Hylochère	Исполинская лесная свинья
Iberiensteinbock	Spanish Ibex	Bouquetin d'Espagne	Иберийский тур
Impala	Impala	Impalla	Чернопятая антилопа
Indischer Sambar	Indian Sambar	Sambar de l'Inde	Самбар
Isabellantilope	Bohor Reedbuck	Nagor de Buffon	Болотный козел нагор
Javanashorn	Javan Rhinoceros	Rhinocéros de la Sonde	Малый однорогий носорог
Jentinkducker	Jentink's Duiker	Céphalophe de Jentink	Либерийский лесной дукер
Kaama	Kaama	Caama	Антилопа каама
Kaffernbüffel	African Buffalo	Buffle d'Afrique	Кафрский буйвол
Kamel, Einhöckriges	Dromedary	Dromadaire	Одногорбый верблюд
Kamel, Zweihöckriges	Two-humped Camel	Chameau à Deux Bosses	Двугорбый верблюд
Kamphirsch	Pampas Deer	Cerf de Pampas	Пампасский олень
Kantschile	Mouse Deer	Chevrotains d'Asie	Азиатские оленьки
Karibu	Caribou	Renne d'Amérique	Западноканадский карибу
Kerabau	Domestic Buffalo	Buffle Domestique	Домашний буйвол
Kirkdikdik	Kirk's Dik-Dik	Dik-Dik de Kirk	Дик-дик Кэрка
Kleindikdik	Swayne's Dik-Dik	– de Swayne	Малый дик-дик
Kleiner Kudu	Lesser Kudu	Petit Koudou	Малый куду
Kleinkamele	Llamas	Lamas	Ламы
Kleinkantschil	Lesser Malay Mouse Deer	Chevrotain Malais	Яванский канчили
Kleinstböckchen	Royal Antelope	Antilope Royale	Антилопа-карлик
Klippspringer	Klipspringer	Oreotrague	Антилопа-прыгун
Kob	Buffon's Kob	Cobe Buffon	Эквитун
Kobus ellipsiprymnus	Waterbuck	Cobe Buffon	Обыкновенный водяной козел
Kongoni	Kongoni	Kongoni	Коровья антилопа Кука
Korrigum	Korrigum	Korrigum	Антилопа корригум
Kouprey	Kouprey	Kouprey	Индокитайский серый бык
Kretische Wildziege	Cretan Wild Goat	Chèvre de la Crète	Критский бородатый козел
Kronenducker	Grey Duiker	Céphalophe de Grimm	Антилопа-дукер
Kropfgazelle	Goitred Gazelle	Gazelle à Goitre	Джейран
Kudu	Kudu	Koudou	Антилопа куду
Kuhantilope	Hartebeest	Bubale	Коровья антилопа
Lama	Llama	Lama	Лама
Lama	Llamas	Lamas	Лама
– *guanicoë*	Guanaco	Guanaco	Гуанако
– – *glama*	Llama	Lama	Лама
– – *pacos*	Alpaca	Alpaga	Альпака
– *vicugna*	Vicugna	Vigogne	Викунья
Lamagazelle	Dibatag	Dibatag	Дибатаг
Leierantilope	Sassaby	Damalisque	Лиророгая антилопа
Leierhirsch	Eld's Deer	Cerf d'Eld	Таменг
Lelwel-Hartebeest	Lelwel	Lelwel	Коровья антилопа Лельвеля
Litocranius walleri	Gerenuk	Gazelle Girafe	Газель-жирафа
Litschi-Moorantilope	Lechwe Waterbuck	Lechwe	Водяной козел лихи
Litschi-Wasserbock	Lechwe Waterbuck	Lechwe	Водяной козел лихи
Madoqua phillipsi	Red Belly Dik-Dik	Dik-Dik de Phillips	Краснобрюхий дик-дик
– *saltiana*	Salt's Dik-Dik	– de Salt	Эритрейский дик-дик
– *swaynei*	Swayne's Dik-Dik	– de Swayne	Малый дик-дик
Madoquini	Dik-Diks	Dik-Diks	Антилопы-левретки
Mähnenhirsch	Sunda Sambar	Sambar de la Sonde	Гривистый олень
Mähnenschaf	Barbary Sheep	Aoudad	Североафриканский гривистый баран
Mähnenspringer	Barbary Sheep	Aoudad	Североафриканский гривистый баран
Marchur	Markhor	Markhor	Винторогий козел
Maultierhirsch	Mule Deer	Cerf de Mulet	Длинноухий олень

GERMAN NAME	ENGLISH NAME	FRENCH NAME	RUSSIAN NAME
Maxwellducker	Maxwell's Duiker	Céphalophe de Maxwell	Дукер Максвелла
Mazama americana	Red Brocket	Daguet Rouge	Большой мазам
– *bricenii*	Grey Dwarf Brocket	Daguet Nain Gris	Карликовый мазам
– *gouazoubira*	Grey Brocket	Daguet Gris	Серый мазам
Mazamas	Brockets	Daguets	Мазамы
Mendesantilope	Addax	Addax au Nez Tacheté	Антилопа мендес
Milu	Père David's Deer	Cerf du Père David	Давидов олень
Mindorobüffel	Tamarou	Tamarou	Филиппинский буйвол
Mittelamerikanischer Tapir	Central American Tapir	Tapir de Baird	Тапир Бэрда
Mongoleigazelle	Mongolian Gazelle	Gazelle de la Mongolie	Дзерен
Moor-Antilope	Buffon's Kob	Cobe Buffon	Эквитун
Moschinae	Musk Deer	Porte-Muscs	Кабарги
Moschus moschiferus	Musk Deer	Porte-Musc	Мускусная кабарга
Moschusböckchen	Suni	Suni	Суни
Moschushirsche	Musk Deer	Porte-Muscs	Кабарги
Moschusochse	– Ox	Bœuf Musqué	Мусусный овцебык
Moschustier	– Deer	Porte-Musc	Мускусная кабарга
Mufflon	European Mouflon	Mouflon d'Europe	Европейский муфлон
Muntiacus muntjak	Muntjac	Muntjac	Мунтжак
Muntjak	Muntjac	Muntjac	Мунтжак
Muntjakhirsche	Muntjacs	Muntjacs	Мунтжаки
Nabelschweine	Peccaries	Pecaris	Пекари
Nashörner	Rhinoceroses	Rhinocérotidés	Носороги
Natalducker	Red Duiker	Céphalophe Rouge	Красный дукер
Nemorhaedini	Gorals and Serows	Gorals et Serows	
Nemorhaedus goral	Goral	Goral	Гималайский горал
Neotraginae	Dwarf Antelopes	Neotraguinés	Карликовые антилопы
Neotragus pygmaeus	Royal Antelope	Antilope Royale	Антилопа карлик
Nesotragus batesi	Bate's Dwarf Antelope	– de Bates	Антилопа Батеса
– *moschatus*	Suni	Suni	Суни
Nichtwiederkäuer	Suiformes	Suiformes	Нежвачные
Nilgauantilope	Nilgai	Nilgaut	Антилопа нильгау
Nilpferd	Hippopotamus	Hippopotame Amphibie	Обыкновенный бегемот
Nonruminantia	Suiformes	Suiformes	Нежвачные
Nordandenhirsch	Peruvian Guemal	Guemal	Североандский олень
Nordpudu	Northern Pudu	Pudu du Nord	Северный пуду
Nototragus melanotis	Grysbok	Raphicère du Cap	Антилопа грис
Nyala	Nyala	Nyala	Антилора ниала
Odocoileinae	American and Roe Deer	Odocoiléinés	Телеметакарпальные олени
Odocoileini	American Deer	Cerfs Américains	Американские олени
Odocoileus bezoarticus	Pampas Deer	Cerf de Pampas	Пампасовые олени
– *dichotomus*	Marsh Deer	– Marécageux	Болотный олень
– *hemionus*	Mule Deer	– de Mulet	Длинноухий олень
– *virginianus*	White-tailed Deer	– de Virginie	Белохвостый олень
Ogilby-Ducker	Ogilby's Duiker	Céphalophe d'Ogilby	Дукер о. Фернандо-По
Okapi	Okapi	Okapi	Окапи Джонстона
Okapia johnstoni	Okapi	Okapi	Окапи Джонстона
Onototragus megaceros	Nile-Lechwe	Lechwe du Nil	Белошейная антилопа
Oreamnus americanus	Rocky Mountain Goat	Chèvre des Montagnes Rocheuses	Снежная коза
Oreotragus oreotragus	Klipspringer	Oreotrague	Антилопа-прыгун
Oribi	Oribi	Ourebie	Бледный ориби
Oryx	Oryx	Oryx	Сернобык
Oryx gazella	Oryx	Oryx	Сернобык
Ourebia ourebi	Oribi	Ourebie	Бледный ориби
Ovibos moschatus	Musk Ox	Bœuf Musqué	Мускусный овцебык
Ovibovini	– Oxen	Bœufs Musqués	Овцебыки
Ovis	**Sheep**	Moutons	Горные бараны
– *ammon*	Wild Sheep	Mouflon	Горный баран
– – *aries*	Domestic Sheep	Mouton Domestique	Домашняя овца
– – *musimon*	European Mouflon	Mouflon d'Europe	Европейский муфлон
– *canadensis*	Bighorn Sheep	– du Canada	Толсторог
Paarhufer	Even-toed Ungulates	Artiodactyles	Парнокопытные
Pampashirsch	Pampas Deer	Cerf de Pampas	Пампасский олень
Pantholops hodgsoni	Tibetan Antelope	Antilope du Thibet	Чиру
Panzernashorn	Great Indian Rhinoceros	Rhinocéros Unicorne des Indes	Большой однорогий носорог
Passan	South African Oryx	Oryx d'Afrique du Sud	Южноафриканский сернобык
Pecora	Pecora	Pécores	Рогатые
Pekaris	Peccaries	Pecaris	Пекари
Pelea capreolus	Grey Rhebuck	Rhebouk	Антилопа-косуля
Pelzelngazelle	Pelzeln's Gazelle	Gazelle de Pelzeln	
Perissodactyla	Odd-toed Ungulates	Périssodactyles	Непарнокопытные
Peruanischer Gabelhirsch	Peruvian Guemal	Guemal	Североандский олень
Pferdeantilope	Roan Antelope	Antilope Rouanne	Чалая лошадиная антилопа

GERMAN NAME	ENGLISH NAME	FRENCH NAME	RUSSIAN NAME
Pferdeböcke	Roan and Sable Antelopes	Hippotraginés	Лошадиные антилопы
Pferdehirsch	Indian Sambar	Sambar de l'Inde	Самбар
Phacochoerus aethiopicus	Wart Hog	Phacochère	Абиссинский бородавочник
Philippinenhirsch	Philippine Sambar	Sambar des Philippines	Филиппинский самбар
Philippinensambar	Philippine Sambar	Sambar des Philippines	Филиппинский самбар
Potamochoerus porcus	Bush Pig	Potamochère de l'Afrique	Кустарная свинья
Prinz-Alfreds-Hirsch	Alfred's Sambar	Sambar d'Alfred	Олень принца Альфреда
Procapra gutturosa	Mongolian Gazelle	Gazelle de la Mongolie	Дзерен
– *picticauda*	Tibetan Gazelle	– du Thibet	Тибетский дзерен
Pseudois nayaur	Blue Sheep	Bouc Bleu	Нахур
Pudu mephistopheles	Northern Pudu	Pudu du Nord	Северный пуду
– *pudu*	Southern Pudu	– du Sud	Южный пуду
Pudus	Pudus	Pudus	Пуду
Pustelschwein	Javan Pig	Sanglier Pustule	Бородавчатая свинья
Pyrenäensteinbock	Spanish Ibex	Bouquetin d'Espagne	Пиренейский тур
Rangiferinae	Reindeer	Rennes	Северные олени
Rangifer tarandus	Reindeer	Renne	Северный олень
– – (Amerikanische Unterarten)	Caribu	– d'Amérique	Западноканадский карибу
Raphicerus campestris	Steinbok	Raphicère Champêtre	Степной штейнбок
Rappenantilope	Sable Antelope	Hippotrague Noir	Черная антилопа
Redunca arundium	Reedbuck	Redunca Grand	Большой болотный козел
– *fulvorufula*	Mountain Reedbuck	– de Montagne	Горный болотный козел
– *redunca*	Bohor Reedbuck	Nagor de Buffon	Болотный козел нагор
Reh	Roe Deer	Chevreuil	Косуля
Rehantilope	Grey Rhebuck	Rhebouk	Антилопа-косуля
Rehböckchen	Grey Rhebuck	Rhebouk	Антилопы-косули
Ren	Reindeer	Renne	Северный олень
Renhirsche	Reindeer	Rennes	Северные олени
Rentier	Reindeer	Renne	Северный олень
Rhinoceros sondaicus	Javan Rhinoceros	Rhinocéros de la Sonde	Малый однорогий носорог
– *unicornis*	Great Indian Rhinoceros	– Unicorne des Indes	Большой однорогий носорог
Rhinocerotidae	Rhinoceroses	Rhinocérotidés	Носороги
Rhynchotragus guentheri	Guenther's Dik-Dik	Dik-Dik de Guenther	Дик-дик Гюнтера
– *kirki*	Kirk's Dik-Dik	– de Kirk	Дик-дик Кэрка
Riedbock	Bohor Reedbuck	Nagor de Buffon	Болотный козел нагор
Riesenducker	Giant Duiker	Céphalophe Géant	Желтоспинный дукер
Riesenwaldschwein	Giant Forest Pig	Hylochère	Исполинская лесная свинья
Rinder	Cattle	Bœufs	Быки
Rindergemsen	Takins	Takins	Такины
Roan	Roan Antelope	Antilope Rouanne	Чалая лошадиная антилопа
Rotbauchdikdik	Red Belly Dik-Dik	Dik-Dik de Phillips	Краснобрюхий дик-дик
Rotducker	Red Duiker	Céphalophe Rouge	Красный дукер
Roter Spießhirsch	– Brocket	Daguet Rouge	Большой мазам
Rotflankenducker	Red-flanked Duiker	Céphalophe aux Flancs Roux	Рыжебокий дукер
Rothirsch	Red Deer	Cerf Rouge	Благородный олень
Rotstirngazelle	Red-fronted Gazelle	Gazelle à Front Roux	Рыжелобая газель
Ruminantia	Ruminants	Ruminants	Жвачные
Rupicapra rupicapra	Chamois	Chamois	Серна
Rusa	Sambars	Sambars	Самбары
Säbelantilope	Scimitar-horned Oryx	Oryx algazelle	Нубийский саблерогий сернобык
Saiga	Saiga Antelope	Saiga	Сайга
Saiga tatarica	Saiga Antelope	Saiga	Сайга
Sambarhirsche	Sambars	Sambars.	Самбары
Sasin	Blackbuck	Antilope cervicapre	Винторогая антилопа сасси
Sassaby	Sassaby	Damalisque	Антилопа сассаби
Schabrackentapir	Malayan Tapir	Tapir de l'Inde	Индийский тапир
Schafe	Sheep	Moutons	Горные бараны
Schafkamele	Llamas	Lamas	Ламы
Schafochsen	Musk Oxen	Bœufs Musqués	Овцебыки
Schneeziege	Rocky Mountain Goat	Chèvre des Montagnes Rocheuses	Снежная коза
Schomburgkhirsch	Schomburgk's Deer	Cerf de Schomburgk	Сиамский олень
Schopfducker	Duikers	Céphalophes	Лесные дукеры
Schopfhirsch	Tufted Deer	Cerf Touffe	Хохлатый олень
Schraubenziege	Markhor	Markhor	Винторогий козел
Schuppennashorn	Javan Rhinoceros	Rhinocéros de la Sonde	Малый однорогий носорог
Schwarzbüffel	African Buffalo	Buffle d'Afrique	Кафрский буйвол
Schwarzducker	Black Duiker	Céphalophe Noir	Черный дукер
Schwarzes Nashorn	– Rhinoceros	Rhinocéros Noir	Африканский острорылый носорог
Schwarzfersenantilope	Impala	Impalla	Чернопятая антилопа
Schwarzrückenducker	Bay Duiker	Céphalophe à Bande Dorsale Noire	Черноспинный дукер

GERMAN NAME	ENGLISH NAME	FRENCH NAME	RUSSIAN NAME
Schwarzstirnducker	Black-fronted Duiker	− à Front Noir	Чернолобый дукер
Schwarzwedelhirsch	Black-tailed Deer	Cerf de Mulet	Длинноухий олень Скалистых гор
Schwarzwild	Wild Boar	Sanglier	Кабан
Schweine, Altweltliche	Old World Pigs	Suidés	Свиньи Старого света
Schweineartige	Pig-like Mammals	Suoidea	Свиные
Schweinshirsch	Hog Deer	Cerf Cochon	Свиной олень
Schwielensohler	Tylopodes	Tylopodes	Мозоленогие
Serau	Serow	Serow	Суматранский серао
Sikahirsch	Sika Deer	Sika	Пятнистый олень
Sitatunga	Sitatunga	Sitatunga	Антилопа наконг
Sömmeringgazelle	Soemmering's Gazelle	Gazelle de Soemmering	Абиссинская газель
Spanischer Steinbock	Spanish Ibex	Bouquetin d'Espagne	Иберийский тур
Spekegazelle	Speke's Gazelle	Gazelle de Speke	
Spießbock	Oryx	Oryx	Сернобык
Spießhirsche	Brockets	Daguets	Мазамы
Spitzmaulnashorn	Black Rhinoceros	Rhinocéros Noir	Африканский острорылый носорог
Springantilopen	True Antelopes	Antilopinés	Газели и карликовые антилопы
Springbock	Springbuck	Antidorcas	Антилопа-прыгун
Steinbock	Ibex	Bouquetin	Гоный козел
Steinböckchen	Steinbok	Raphicère Champêtre	Степной штейнбок
Stelzengazelle	Dibatag	Dibatag	Дибатаг
Stirnwaffenträger	Pecora	Pécores	Рогатые
Streifengnu	Brindled Gnu	Gnou Bleu	Белобородный гну
Südandenhirsch	Chilenian Huemul	Hippocamelus	Южноандский гуэмал
Südpudu	Southern Pudu, Pudu	Pudu du Sud	Южный пуду
Suidae	Old World Pigs	Suidés	Свиньи Старого света
Sumatranashorn	Sumatran Rhinoceros	Rhinocéros de Sumatra	Суматранский двурогий носорог
Sumpfantilope	Sitatunga	Sitatunga	Антилопа наконг
Sumpfhirsch	Marsh Deer	Cerf Marécageux	Болотный олень
Suniböckchen	Suni	Suni	Суни
Suoidea	Pig-like Mammals	Suoidea	Свиные
Sus barbatus	Bearded Pig	Sanglier à Moustache	Бородатая свинья
− salvanius	Pigmy Hog	Sanglier Nain	Карликовая свинья
− scrofa	Wild Boar	Sanglier	Кабан
− − domesticus	Domestic Pig	Cochon Domestique	Домашняя свинья
− verrucosus	Javan Pig	Sanglier Pustule	Бородавчатая свинья
Sylvicapra grimmia	Grey Duiker	Céphalophe de Grimm	Антилопа-дукер
Syncerus caffer	African Buffalo	Buffle d'Afrique	Кафрский буйвол
Tahr	Tahr	Tahr	Тар
Takin	Takin	Takin	Такин
Tamarau	Tamarou	Tamarou	Филиппинский буйвол
Tapire	Tapirs	Tapirs	Тапиры
Tapiridae	Tapirs	Tapirs	Тапиры
Tapirus bairdi	Central American Tapir	Tapir de Baird	Тапир Бэрда
− indicus	Malayan Tapir	− de l'Inde	Индийский тапир
− pinchaque	Mountain Tapir	− des Andes	Горный тапир
− terrestris	Lowland Tapir	− Terrestre	Обыкновенный тапир
Taurotragus euryceros	Bongo	Bongo	Бонго
− oryx	Eland	Eland	Антилопа канна
Tayassu albirostris	White-lipped Peccary	Pecari à Barbe Blanche	Белогубый пекари
− tajacu	Collared Peccary	− à Collier	Ошейниковый пекари
Tayassuidae	Peccaries	Pecaris	Пекари
Tetracerus quadricornis	Four-horned Antelope	Tetracère	Четырехрогая антилопа
Thamin	Thamin	Thameng	Бирманский таменг
Thomsongazelle	Thomson's Gazelle	Gazelle de Thomson	Газель Томсона
Tibetgazelle	Tibetan Gazelle	− du Thibet	Тибетский дзерен
Tieflandnyala	Nyala	Nyala	Антилопа ниала
Timorhirsch	Sunda Sambar	Sambar de la Sonde	Гривистый олень
Topi	Topi	Topi	Антилопа топи
Tora	Tora	Tora	Тора
Tragelaphini	Spiral-horned Antelopes	Tragelaphini	Винторогие антилопы
Tragelaphus angasi	Nyala	Nyala	Антилопа ниала
− buxtoni	Mountain Nyala	− de Montagne	Горная ниала
− imberbis	Lesser Kudu	Petit Koudou	Малый куду
− scriptus	Bushbuck	Guib Harnaché	Антилопа гуиб
− spekei	Sitatunga	Sitatunga	Антилопа наконг
− strepsiceros	Greater Kudu	Grand Koudou	Антилопа куду
Tragulidae	Chevrotains	Tragulidés	Оленьки
Tragulus	Mouse Deer	Chevrotains d'Asie	Азиатские оленьки
− javanicus	Lesser Malay Mouse Deer	Chevrotain Malais	Яванский канчили
− meminna	Spotted Mouse Deer	− Tacheté	Маминна

GERMAN NAME	ENGLISH NAME	FRENCH NAME	RUSSIAN NAME
– *napu*	Larger Malay Mouse Deer	– Malais	Оленек напу
Trampeltier	Two-humped Camel	Chameau à Deux Bosses	Двугорбый верблюд
Trughirsche	American and Roe Deer	Odocoiléinés	Телеметакарпальные олени
Tschiru	Tibetan Antelope	Antilope du Thibet	Чиру
Tur	Ibex	Bouquetin	Гоный козел
Tylopoda	Tylopodes	Tylopodes	Мозоленогие
Unpaarhufer	Odd-toed Ungulates	Périssodactyles	Непарнокопытные
Ur	Aurochs	Aurochs	Первобытный бык
Vierhornantilope	Four-horned Antelope	Tetracère	Четырехрогая антилопа
Vikunja	Vicugna	Vigogne	Викунья
Virginiahirsch	White-tailed Deer	Cerf de Virginie	Белохвостый олень
Waldziegenantilopen	Gorals and Serows	Gorals et Serows	
Wapiti	Wapiti (englisch), Elk (amerikanisch)	Wapiti	
Warzenschwein	Wart Hog	Phacochère	Абиссинский бородавочник
Wasserbock	Waterbuck	Cobe	Обыкновенный водяной козел
Wasserbüffel	Water Buffalo	Buffle de l'Inde	Буйвол
Wasserhirsche	Water Deer	Hydropotes	Водяные олени
Wassermoschustier	Water Chevrotain	Chevrotain Africain	Африканский оленек
Wasserreh	Chinese Water Deer	Hydropote	Водяной олень
Weißbartgnu	Brindled Gnu	Gnou Bleu	Белобородный гну
Weißbartpekari	White-lipped Peccary	Pecari à Barbe Blanche	Белогубый пекари
Weißbartschwein	White-lipped Peccary	Pecari à Barbe Blanche	Белогубый пекари
Weißbauchducker	Gaboon Duiker	Céphalophe à Ventre Blanc	Белобрюхий дукер
Weißes Nashorn	Wide-mouthed Rhinoceros	Rhinocéros Blanc	Африканский широкорылый носорог
Weißlippenhirsch	Thorold's Deer	Cerf de Thorold	Тибетский горный олень
Weißnacken-Moorantilope	Nile-Lechwe	Lechwe du Nil	Белошейная антилопа
Weißschwanzgnu	White-tailed Gnu	Gnou à Queue Blanche	Белохвостый гну
Weißwedelhirsch	White-tailed Deer	Cerf de Virginie	Белохвостый олень
Wiederkäuer	Ruminants	Ruminants	Жвачные
Wildkamel	Wild Camel	Chameau Sauvage	Дикий двугорбый верблюд
Wildrinder	– Cattle	Bœufs Sauvages	Дикие быки
Wildschaf	– Sheep	Mouflon	Горный баран
Wildschwein	– Boar	Sanglier	Кабан
Wildziegen	– Goats	Chèvres Sauvages	Горные козлы
Windspielantilopen	Dik-Diks	Dik-Diks	Дик-дики
Windungshörner	Spiral-horned Antelopes	Tragelaphini	Винторогие антилопы
Wisent	European Bison	Bison d'Europe	Зубр
Yak	Yak	Yack	Як
Zackenhirsch	Barasingha	Barasingha	Барасинга
Zebraducker	Banded Duiker	Céphalophe Rayé	Дукер-зебра
Zebu	Zebu	Zébu	Зебу
Zwergflußpferd	Pigmy Hippopotamus	Hippopotame Nain	Либерийский карликовый бегемот
Zwergmazama	Grey Dwarf Brocket	Daguet Nain Gris	Карликовый мазам
Zwergwildschwein	Pigmy Hog	Sanglier Nain	Карликовая свинья

III. French—German—English—Russian

Les noms des subspecies se forment pour l'ordinaire des noms des species qui sont succédés par des adjectifs ou des notations géographiques. En cette partie du dictionnaire des animaux les noms français des subspecies formés de cette maniére ne sont cités qu'en cas exceptionnels.

FRENCH NAME	GERMAN NAME	ENGLISH NAME	RUSSIAN NAME
Addax au Nez Tacheté	Mendesantilope	Addax	Антилопа мендес
Alcélaphes	Eigentliche Kuhantilopen	Alcelaphini	Коровьи антилопы
Alpaga	Alpaka	Alpaca	Альпака
Anoa	Anoa	Anoa	Целебесский карликовый буйвол
Antidorcas	Springbock	Springbuck	Антилопа-прыгун
Antilope cervicapre	Hirschziegenantilope	Blackbuck	Винторогая антилопа сасси
Antilope-Chèvre Américaine	Gabelbock	Pronghorn	Американский вилорог
Antilope de Bates	Batesböckchen	Bate's Dwarf Antelope	Антилопа Батеса
– du Thibet	Tschiru	Tibetan Antelope	Чиру
– Rouanne	Pferdeantilope	Roan Antelope	Чалая лошадиная антилопа

FRENCH NAME	GERMAN NAME	ENGLISH NAME	RUSSIAN NAME
– Royale	Kleinstböckchen	Royal Antelope	Антилопа-карлик
Antilopes	Antilopen	Antelopes	Антилопы
Antilopinés	Gazellenartige	True Antelopes	Газели и карликовые антилопы
Aoudad	Mähnenspringer	Barbary Sheep	Североафриканский гривистый баран
Artiodactyles	Paarhufer	Even-toed Ungulates	Парнокопытные
Aurochs	Ur	Aurochs	Первобытный бык
Axis	Axishirsch	Axis Deer	Аксис
Babiroussa	Hirscheber	Babirusa	Бабируссы
Banting	Banteng	Banteng	Бантенг
Barasingha	Barasingha	Barasingha	Барасинга
Beiras	Beira-Antilopen	Beira-Antelopes	Антилопы бейра
Bison Américain	Bison	American Bison	Бизон
Bison d'Europe	Wisent	European Bison	Зубр
Bœuf	Hausrind	Ox	Домашний бык
– Musqué	Moschusochse	Musk Ox	Мускусный овцебык
Bœufs	Rinder	Cattle	Быки
– Sauvages	Wildrinder	Wild Cattle	Дикие быки
Bongo	Bongo	Bongo	Бонго
Bouc Bleu	Blauschaf	Blue Sheep	Нахур
Bouquetin	Steinbock	Ibex	Гоный козел
– des Alpes	Alpensteinbock	Alpine Ibex	Альпийский горный козел
– d'Espagne	Iberiensteinbock	Spanish Ibex	Иберийский тур
Bovidés	Hornträger	Horned Ungulates	Полорогие
Bubale	Kuhantilope	Hartebeest	Коровья антилопа
Buffle d'Afrique	Kaffernbüffel	African Buffalo	Кафрский буйвол
– de l'Inde	Wasserbüffel	Water Buffalo	Буйвол
– Domestique	Hausbüffel	Domestic Buffalo	Домашний буйвол
Buffles	Büffel	Buffalos	Буйволы
Caama	Kaama	Kaama	Антилопа каама
Céphalophe à Bande Dorsale Noire	Schwarzrückenducker	Bay Duiker	Черноспинный дукер
– à Front Noir	Schwarzstirnducker	Black-fronted Duiker	Чернолобый дукер
– aux Flancs Roux	Rotflankenducker	Red-flanked Duiker	Рыжебокий дукер
– à Ventre Blanc	Weißbauchducker	Gaboon Duiker	Белобрюхий дукер
– Bleu	Blauducker	Blue Duiker	Голубой Дукер
Céphalophe d'Abbot	Abbotducker	Abbot's Duiker	Дукер Аббота
– de Grimm	Kronenducker	Grey Duiker	Антилопа-дукер
– de Jentink	Jentinkducker	Jentink's Duiker	Либерийский лесной дукер
– de Maxwell	Maxwellducker	Maxwell's Duiker	Дукер Максвелла
– d'Harvey	Harveyducker	Harvey's Duiker	Дукер Харвея
– d'Ogilby	Fernando-Po-Ducker	Ogilby's Duiker	Дукер о. Фернандо-По
– Géant	Gelbrückenducker	Giant Duiker	Желтоспинный дукер
– Noir	Schwarzducker	Black Duiker	Черный дукер
– Rayé	Zebraducker	Banded Duiker	Дукер-зебра
– Rouge	Natalducker	Red Duiker	Красный дукер
Céphalophes	Schopfducker	Duikers	Лесные дукеры
Céphalophinés	Ducker	Duikers	Хохлатые антилопы
Cerf Cochon	Schweinshirsch	Hog Deer	Свиной олень
– de Dybowski	Dybowskihirsch	Dybowski's Deer	Пятнистый олонь Дыбовского
– d'Eld	Leierhirsch	Eld's Deer	Таменг
– de Mulet	Großohrhirsch	Mule Deer	Длинноухий олень
– de Pampas	Pampashirsch	Pampas Deer	Пампасский олень
– de Schomburgk	Schomburgkhirsch	Schomburgk's Deer	Сиамский олень
– de Thorold	Weißlippenhirsch	Thorold's Deer	Тибетский горный олень
– de Virginie	Weißwedelhirsch	White-tailed Deer	Белохвостый олень
– du Père David	Davidshirsch	Père David's Deer	Давидов олень
– Marécageux	Sumpfhirsch	Marsh Deer	Болотный олень
– Rouge	Rothirsch	Red Deer	Благородный олень
– Touffe	Schopfhirsch	Tufted Deer	Хохлатый олень
Cerfs	Hirsche	Deer	Плотнорогие
Chameau à Deux Bosses	Zweihöckriges Kamel	Two-humped Camel	Двугорбый верблюд
– Sauvage	Wildkamel	Wild Camel	Дикий двугорбый верблюд
Chameaux	Kamele	Camels	Верблюды
Chamois	Gemse	Chamois	Серна
Chèvre	Ziege	Goat	Коза
– de la Crète	Kretische Wildziege	Cretan Wild Goat	Критский бородатый козел
– des Montagnes Rocheuses	Schneeziege	Rocky Mountain Goat	Снежная коза
Chèvres Sauvages	Wildziegen	Wild Goats	Горные козлы
Chèvre véritable	Bezoarziege	Wild Goat	Безоаровый козел
Chevreuil	Reh	Roe Deer	Косуля
Chevrotain Africain	Afrikanisches Hirschferkel	Water Chevrotain	Африканский оленек
– Tacheté	Fleckenkantschil	Spotted Mouse Deer	Маминна

FRENCH NAME	GERMAN NAME	ENGLISH NAME	RUSSIAN NAME
Chevrotains d'Asie	Kantschile	Mouse Deer	Азиатские оленьки
Cobe	Wasserbock	Waterbuck	Обыкновенный водяной козел
– Buffon	Moor-Antilope	Buffon's Kob	Эквитун
– Defassa	Defassa-Wasserbock	Defassa-Waterbuck	Синг-синг
– de Mme. Gray	Frau Grays Wasserbock	Mrs. Gray's Lechwe	Белошейная антилопа
Daguet Gris	Graumazama	Grey Brocket	Серый мазам
– Nain Gris	Zwergmazama	– Dwarf Brocket	Карликовый мазам
– Rouge	Großmazama	Red Brocket	Большой мазам
Daguets	Mazamas	Brockets	Мазамы
Daim	Damhirsch	Fallow Deer	Лань
Damalisque	Leierantilope	Sassaby	Лиророгая антилопа
Dibatag	Stelzengazelle	Dibatag	Дибатаг
Dik-Dik de Guenther	Güntherdikdik	Guenther's Dik-Dik	Дик-дик Гюнтера
– de Kirk	Kirkdikdik	Damara Dik-Dik, Kirk's Dik-Dik	Дик-дик Кэрка
– de Phillips	Rotbauchdikdik	Red Belly Dik-Dik	Краснобрюхий дик-дик
– de Salt	Eritreadikdik	Salt's Dik-Dik	Эритрейский дик-дик
– de Swayne	Kleindikdik	Swayne's Dik-Dik	Малый дик-дик
Dik-Diks	Windspielantilopen	Dik-Diks	Дик-дики
Dorcas	Dorkasgazelle	Dorcas Gazelle	Обыкновенная газель
Dorcatrague	Beira	Beira	Бейра
Dromadaire	Dromedar	Dromedary	Одногорбый верблюд
Elan	Elch	Elk (Anglais), Moose (Américain)	Лось
Eland	Elenantilope	Eland	Антилопа канна
Gaur	Gaur	Gaur	Гаур
Gayal	Gayal	Gayal	Гаял
Gazelle à Cornes Grêles	Dünengazelle	Slender-horned Gazelle	
– à Front Roux	Rotstirngazelle	Red-fronted Gazelle	Рыжелобая газель
– à Goitre	Kropfgazelle	Goitred Gazelle	Джейран
– Dama	Damagazelle	Dama Gazelle	Сахарская газель
– de Grant	Grantgazelle	Grant's Gazelle	
– de Pelzeln	Pelzelngazelle	Pelzeln's Gazelle	
– de Soemmering	Sömmeringgazelle	Soemmering's Gazelle	Абиссинская газель
– de Speke	Spekegazelle	Speke's Gazelle	
– de Thomson	Thomsongazelle	Thomson's Gazelle	Газель Томсона
– de la Mongolie	Mongoleigazelle	Mongolian Gazelle	Дзерен
– d'Heuglin	Heuglingazelle	Heuglin's Gazelle	
– du Thibet	Tibetgazelle	Tibetan Gazelle	Тибетский дзерен
– Girafe	Giraffengazelle	Gerenuk	Газель-жирафа
Gazelles	Gazellen	Gazelles	Газели
Girafe	Giraffe	Giraffe	Обыкновенная жирафа
Gnou à Queue Blanche	Weißschwanzgnu	White-tailed Gnu	Белохвостый гну
– Bleu	Streifengnu	Brindled Gnu	Белобородый гну
Gnous	Gnus	Gnus	Антилопы гну
Goral	Goral	Goral	Гималайский горал
Grand Koudou	Großer Kudu	Greater Kudu	Антилопа куду
Guanaco	Guanako	Guanaco	Гуанако
Guemal	Nordandenhirsch	Peruvian Guemal	Североандский олень
Guib Harnaché	Buschbock	Bushbuck	Антилопа гуиб
Hippocamelus	Südandenhirsch	Chilenian Huemul	Южноандский гуэмал
Hippopotame Amphibie	Flußpferd	Hippopotamus	Обыкновенный бегемот
– Nain	Zwergflußpferd	Pigmy Hippopotamus	Либерийский карликовый бегемот
Hippopotamidés	Flußpferde	Hippopotamuses	Бегемоты
Hippotraginés	Pferdeböcke	Roan and Sable Antelopes	Лошадиные антилопы
Hippotrague Noir	Rappenantilope	Sable Antelope	Черная антилопа
Hydropote	Wasserreh	Chinese Water Deer	Водяной олень
Hydropotes	Wasserhirsche	Water Deer	Водяные олени
Hylochère	Riesenwaldschwein	Giant Forest Pig	Исполинская лесная свинья
Impalla	Schwarzfersenantilope	Impala	Чернопятая антилопа
Kongoni	Kongoni	Kongoni	Коровья антилопа Кука
Korrigum	Korrigum	Korrigum	Антилопа корригум
Koudou	Kudu	Kudu	Антилопа куду
Kouprey	Kouprey	Kouprey	Индокитайский серый бык
Lama	Lama	Llama	Лама
Lamas	Kleinkamele	Llamas	
Lechwe	Litschi-Wasserbock	Lechwe Waterbuck	Водяной козел лихи
– du Nil	Weißnacken-Moorantilope	Nile-Lechwe	Белошейная антилопа
Lelwel	Lelwel-Hartebeest	Lelwel	Коровья антилопа Лельвеля
Markhor	Schraubenziege	Markhor	Винторогий козел
Mouflon	Wildschaf	Wild Sheep	Горный баран
– d'Europe	Mufflon	European Mouflon	Европейский муфлон

FRENCH NAME	GERMAN NAME	ENGLISH NAME	RUSSIAN NAME
Mouton Domestique	Hausschaf	Domestic Sheep	Домашняя овца
Moutons	Schafe	Sheep	Горные бараны
Muntjac	Muntjak	Muntjac	Мунтжак
Muntjacs	Muntjakhirsche	Muntjacs	Мунтжаки
Nagor de Buffon	Riedbock	Bohor Reedbuck	Болотный козел нагор
Néotraguinés	Böckchen	Dwarf Antelopes	Карликовые антилопы
Nilgaut	Nilgauantilope	Nilgai	Антилопа нильгау
Nyala	Nyala	Nyala	Антилопа ниала
Okapi	Okapi	Okapi	Окапи Джонстона
Oreotrague	Klippspringer	Klipspringer	Антилопа-прыгун
Oryx	Spießbock	Oryx	Сернобык
Ourebie	Bleichböckchen	Oribi	Бледный ориби
Pecari à Barbe Blanche	Weißbartpekari	White-lipped Peccary	Белогубый пекари
– à Collier	Halsbandpekari	Collared Peccary	Ошейниковый пекари
Pecaris	Pekaris	Peccaries	Пекари
Pécores	Stirnwaffenträger	Pecora	Рогатые
Périssodactyles	Unpaarhufer	Odd-toed Ungulates	Непарнокопытные
Petit Koudou	Kleiner Kudu	Lesser Kudu	Малый куду
Porte-Musc	Moschustier	Musk Deer	Мускусная кабарга
Potamochère de l'Afrique	Buschschwein	Bush Pig	Кустарная свинья
Pudu du Nord	Nordpudu	Northern Pudu	Северный пуду
– du Sud	Südpudu	Southern Pudu, Pudu	Южный пуду
Pudus	Pudus	Pudus	Пуду
Raphicère Champêtre	Steinböckchen	Steinbok	Степной штейнбок
– du Cap	Greisböckchen	Grysbok	Антилопа грис
Redunca de Montagne	Bergriedbock	Mountain Reedbuck	Горный болотный козел
– Grand	Großer Riedbock	Reedbuck	Большой болотный козел
Renne	Ren	Reindeer	Северный олень
Rhebouk	Rehantilope	Grey Rhebuck	Антилопа-косуля
Rhinocéros Blanc	Breitmaulnashorn	Wide-mouthed Rhinoceros	Африканский широкоры-лый носорог
– de la Sonde	Javanashorn	Javan Rhinoceros	Малый однорогий носорог
– de Sumatra	Sumatranashorn	Sumatran Rhinoceros	Суматранский двурогий носорог
– Noir	Spitzmaulnashorn	Black Rhinoceros	Африканский острорылый носорог
– Unicorne des Indes	Panzernashorn	Great Indian Rhinoceros	Большой однорогий носорог
Rhinocérotidés	Nashörner	Rhinoceroses	Носороги
Ruminants	Wiederkäuer	Ruminants	Жвачные
Saiga	Saiga	Saiga Antelope	Сайга
Sambar d'Alfred	Prinz-Alfreds-Hirsch	Alfred's Sambar	Олень принца Альфреда
– de la Sonde	Mähnenhirsch	Sunda Sambar	Гривистый олень
– de l'Inde	Indischer Sambar	Indian Sambar	Самбар
– des Philippines	Philippinensambar	Philippine Sambar	Филиппинский самбар
Sambars	Sambarhirsche	Sambars	Самбары
Sanglier	Wildschwein	Wild Boar	Кабан
– à Moustache	Bartschwein	Bearded Pig	Бородатая свинья
– Nain	Zwergwildschwein	Pigmy Hog	Карликовая свинья
Sanglier Pustule	Pustelschwein	Javan Pig	
Serow	Serau	Serow	Суматранский серао
Sika	Sikahirsch	Sika Deer	Пятнистый олень
Sitatunga	Sitatunga	Sitatunga	Антилопа наконг
Suidés	Altweltliche Schweine	Old World Pigs	Свиньи Старого света
Suiformes	Nichtwiederkäuer	Suiformes	Нежвачные
Suni	Suniböckchen	Suni	Суни
Tahr	Tahr	Tahr	Тар
Takin	Takin	Takin	Такин
Tamarou	Tamarau	Tamarou	Филиппинский буйвол
Tapir de Baird	Mittelamerikanischer Tapir	Central American Tapir	Тапир Бэрда
– de l'Inde	Schabrackentapir	Malayan Tapir	Индийский тапир
– des Andes	Bergtapir	Mountain Tapir	Горный тапир
– Terrestre	Flachlandtapir	Lowland Tapir	Обыкновенный тапир
Tapirs	Tapire	Tapirs	Тапиры
Tetracère	Vierhornantilope	Four-horned Antelope	Четырехрогая антилопа,
Thameng	Thamin	Thamin	Бирманский таменг
Topi	Topi	Topi	Антилопа топи
Tora	Tora	Tora	Тора
Tragulidés	Hirschferkel	Chevrotains	Оленьки
Tylopodes	Schwielensohler	Tylopodes	Мозоленогие
Vigogne	Vikunja	Vicugna	Викунья
Wapiti	Wapiti	Wapiti (Anglais), Elk (Américain)	
Yack	Yak	Yak	Як
Zébu	Zebu	Zebu	Зебу

IV. Russian—German—English—French

Названия подвидов отличаются от видовых чаще всего лишь дополнительным прилагательным, главным образом географического характера. Такие русские названия подвидов как правило не включены в данную часть зоологического словаря.

RUSSIAN NAME	GERMAN NAME	ENGLISH NAME	FRENCH NAME
Абиссинская газель	Sömmeringgazelle	Soemmering's Gazelle	Gazelle de Soemmering
Абиссинский бородавочник	Warzenschwein	Wart Hog	Phacochère
Азиатские буйволы	Asiatische Büffel	Asiatic Buffalos	Buffles d'Asie
Азиатские двурогие носороги	Halbpanzernashörner	Asiatic Two-horned Rhinos	Rhinocéros Bicornes d'Asie
Азиатские оленьки	Kantschile	Mouse Deer	Chevrotains d'Asie
Аксис	Axishirsch	Axis Deer	Axis
Альпака	Alpaka	Alpaca	Alpaga
Альпийский горный козел	Alpensteinbock	Alpine Ibex	Bouquetin des Alpes
Алтайский аргали	Argali	Argali	Argali
Американские олени	Amerikahirsche	American Deer	Cerfs Américains
Американский вилорог	Gabelbock	Pronghorn	Antilope-Chèvre-Américaine
Андские олени	Andenhirsche	Guemals	Guemals
Антилопа Батеса	Batesböckchen	Bate's Dwarf Antelope	Antilope de Bates
Антилопа грис	Greisböckchen	Grysbok	Raphicère du Cap
Антилопа гуиб	Buschbock	Bushbuck	Guib Harnaché
Антилопа-дукер	Kronenducker	Grey Duiker	Céphalophe de Grimm
Антилопа каама	Kaama	Kaama	Caama
Антилопа канна	Elenantilope	Eland	Eland
Антилопа-карлик	Kleinstböckchen	Royal Antelope	Antilope Royale
Антилопа корригум	Korrigum	Korrigum	Korrigum
Антилопа-косуля	Rehantilope	Grey Rhebuck	Rhebouk
Антилопа куду	Großer Kudu	Greater Kudu	Grand Koudou
Антилопа мендес	Mendesantilope	Addax	Addax au Nez Tacheté
Антилопа наконг	Sitatunga	Sitatunga	Sitatunga
Антилопа ниала	Nyala	Nyala	Nyala
Антилопа нильгау	Nilgauantilope	Nilgai	Nilgaut
Антилопа-прыгун	Springbock	Springbuck	Antidorcas
Антилопа сассаби	Sassaby	Sassaby	Damalisque
Антилопа топи	Topi	Topi	Topi
Антилопы бейра	Beira-Antilopen	Beira-Antelopes	Beiras
Антилопы гну	Gnus	Gnus	Gnous
Антилопы-леаретки	Dikdiks	Dik-Diks	Dik-Diks
Африканский буйвол	Afrikanischer Büffel	African Buffalo	Buffle d'Afrique
Африканский оленек	Afrikanisches Hirschferkel	Water Chevrotain	Chevrotain Africain
Африканский острорылый носорог	Spitzmaulnashorn	Black Rhinoceros	Rhinocéros Noir
Африканский широкорылый носорог	Breitmaulnashorn	Wide-mouthed Rhinoceros	Rhinocéros Blanc
Бабируссы	Hirscheber	Babirusa	Babiroussa
Бантенг	Banteng	Banteng	Banting
Барасинга	Barasingha	Barasingha	Barasingha
Бегемоты	Flußpferde	Hippopotamuses	Hippopotamidés
Безоаровый козел	Bezoarziege	Wild Goat	Chèvre véritable
Бейра	Beira	Beira	Beira
Белобородый гну	Streifengnu	Brindled Gnu	Gnou Bleu
Белобрюхий дукер	Weißbauchducker	Gaboon Duiker	Céphalophe à Ventre Blanc
Белогубый пекари	Weißbartpekari	White-lipped Peccary	Pecari à Barbe Blanche
Белохвостый гну	Weißschwanzgnu	White-tailed Gnu	Gnou à Queue Blanche
Белохвостый олень	Weißwedelhirsch	White-tailed Deer	Cerf de Virginie
Белошейная антилопа	Weißnacken-Moorantilope	Nile-Lechwe	Lechwe du Nil
Верблюды	Kamele	Camels	Chameaux
Бизон	Bison	American Bison	Bison Américain
Бирманский таменг	Thamin	Thamin	Thameng
Благородный олень	Rothirsch	Red Deer	Cerf Rouge
Бледный ориби	Bleichböckchen	Oribi	Ourebie
Болотный козел нагор	Riedbock	Bohor Reedbuck	Nagor de Buffon
Болотный олень	Sumpfhirsch	Marsh Deer	Cerf Marécageux
Большой болотный козел	Großer Riedbock	Reedbuck	Redunca Grand
Большой мазам	Großmazama	Red Brocket	Daguet Rouge
Большой однорогий носорог	Panzernashorn	Great Indian Rhinoceros	Rhinocéros Unicorne des Indes
Бонго	Bongo	Bongo	Bongo
Бородавчатая свинья	Pustelschwein	Javan Pig	Sanglier Pustule
Бородатая свинья	Bartschwein	Bearded Pig	Sanglier à Moustache
Буйвол	Wasserbüffel	Water Buffalo	Buffle de l'Inde
Буйволы	Büffel	Buffalos	Buffles

RUSSIAN NAME	GERMAN NAME	ENGLISH NAME	FRENCH NAME
Быки	Rinder	Cattle	Bœufs
Викунья	Vikunja	Vicugna	Vigogne
Вилороги	Gabelhorntiere	Pronghorns	Antilopes-Chèvres Américaines
Винторогая антилопа сасси	Hirschziegenantilope	Blackbuck	Antilope cervicapre
Винторогие антилопы	Drehhörner	Spiral-horned Antelope	Tragelaphini
Винторогий козел	Schraubenziege	Markhor	Markhor
Водяной козел лихи	Litschi-Wasserbock	Lechwe Waterbuck	Lechwe
Водяной олень	Wasserreh	Chinese Water Deer	Hydropote
Водяные олени	Wasserhirsche	Water Deer	Hydropotes
Восточный вапити	Wapiti	Wapiti (englisch), Elk (amerikanisch)	Wapiti
Газели	Gazellen	Gazelles	Gazelles
Газели и карликовые антилопы	Gazellenartige	True Antelopes	Antilopinés
Газель Гранта	Grantgazelle	Grant's Gazelle	Gazelle de Grant
Газель-жирафа	Giraffengazelle	Gerenuk	Gazelle Girafe
Газель Томсона	Thomsongazelle	Thomson's Gazelle	Gazelle de Thomson
Гаур	Gaur	Gaur	Gaur
Гаял	Gayal	Gayal	Gayal
Гималайский горал	Goral	Goral	Goral
Голубой дукер	Blauducker	Blue Duiker	Céphalophe Bleu
Горная ниала	Bergnyala	Mountain Nyala	Nyala de Montagne
Горные бараны	Schafe	Sheep	Moutons
Горные козлы	Wildziegen	Wild Goats	Chèvres Sauvages
Горный баран	Wildschaf	Wild Sheep	Mouflon
Горный болотный козел	Bergriedbock	Mountain Reedbuck	Redunca de Montagne
Гоный козел	Steinbock	Ibex	Bouquetin
Горный тапир	Bergtapir	Mountain Tapir	Tapir des Andes
Гривистый олень	Mähnenhirsch	Sunda Sambar	Sambar de la Sonde
Гуанако	Guanako	Guanaco	Guanaco
Давидов олень	Davidshirsch	Père David's Deer	Cerf du Père David
Двугорбый верблюд	Zweihöckriges Kamel	Two-humped Camel	Chameau à Deux Bosses
Джейран	Kropfgazelle	Goitred Gazelle	Gazelle à Goitre
Дзерен	Mongoleigazelle	Mongolian Gazelle	Gazelle de la Mongolie
Дибатаг	Stelzengazelle	Dibatag	Dibatag
Дик-дик Гюнтера	Güntherdikdik	Guenther's Dik-Dik	Dik-Dik de Guenther
Дик-дики	Windspielantilopen	Dik-Diks	Dik-Diks
Дик-дик Кэрка	Kirkdikdik	Kirk's Dik-Dik	Dik-Dik de Kirk
Дикие быки	Wildrinder	Wild Cattle	Bœufs Sauvages
Дикий двугорбый верблюд	Wildkamel	Wild Camel	Chameau Sauvage
Длинноухий олень	Großohrhirsch	Mule Deer	Cerf de Mulet
Длинноухий олень Скалистых гор	Schwarzwedelhirsch	Black-tailed Deer	Cerf de Mulet
Домашний бык	Hausrind	Ox	Bœuf
Домашний буйвол	Hausbüffel	Domestic Buffalo	Buffle Domestique
Домашний двугорбый верблюд	Hauskamel	Bactrian Camel	Chameau Domestique à Deux Bosses
Домашняя коза	Hausziege	Domestique Goat	Chèvre Domestique
Домашняя овца	Hausschaf	Domestic Sheep	Mouton Domestique
Домашняя свинья	Hausschwein	Domestic Pig	Cochon Domestique
Дукер о. Фернандо-По	Fernando-Po-Ducker	Ogilby's Duiker	Céphalophe d'Ogilby
Дукер Аббота	Abbotducker	Abbot's Duiker	Céphalophe d'Abbot
Дукер-зебра	Zebraducker	Banded Duiker	Céphalophe Rayé
Дукер Максвелла	Maxwellducker	Maxwell's Duiker	Céphalophe de Maxwell
Дукер Харвея	Harveyducker	Harvey's Duiker	Céphalophe d'Harvey
Европейский муфлон	Muffelwild	European Moufflon	Mouflon d'Europe
Жвачные	Wiederkäuer	Ruminants	Ruminants
Желтоспинный дукер	Gelbrückenducker	Giant Duiker	Céphalophe Géant
Западноканадский карибу	Karibu	Caribou	Renne d'Amérique
Зебу	Zebu	Zebu	Zébu
Зубр	Wisent	European Bison	Bison d'Europe
Иберийский тур	Iberiensteinbock	Spanish Ibex	Bouquetin d'Espagne
Индийский тапир	Schabrackentapir	Malayan Tapir	Tapir de l'Inde
Индокитайский серый бык	Kouprey	Kouprey	Kouprey
Исполинская лесная свинья	Riesenwaldschwein	Giant Forest Pig	Hylochère
Кабан	Wildschwein	Wild Boar	Sanglier
Кабарги	Moschushirsche	Musk Deer	Porte-Muscs
Карликовая свинья	Zwergwildschwein	Pigmy Hog	Sanglier Nain
Карликовые антилопы	Böckchen	Dwarf Antelopes	Neotraguinés
Карликовый мазам	Zwergmazama	Grey Dwarf Brocket	Daguet Nain Gris
Кафрский буйвол	Kaffernbüffel	African Buffalo	Buffle d'Afrique
Коровья антилопа	Kuhantilope	Hartebeest	Bubale
Коровья антилопа Кука	Kongoni	Kongoni	Kongoni
Коровья антилопа Лельвеля	Lelwel-Hartebeest	Lelwel	Lelwel

RUSSIAN NAME	GERMAN NAME	ENGLISH NAME	FRENCH NAME
Косуля	Reh	Roe Deer	Chevreuil
Краснобрюхий дик-дик	Rotbauchdikdik	Red Belly Dik-Dik	Dik-Dik de Phillips
Красный дукер	Natalducker	Red Duiker	Céphalophe Rouge
Критский бородатый козел	Kretische Wildziege	Cretan Wild Goat	Chèvre de la Crète
Кустарная свинья	Buschschwein	Bush Pig	Potamochère de l'Afrique
Лама	Lama	Llama	Lama
Ламы	Kleinkamele	Llamas	Lamas
Лань	Damhirsch	Fallow Deer	Daim
Лесные дукеры	Schopfducker	Duikers	Céphalophes
Либерийский карликовый бегемот	Zwergflußpferd	Pigmy Hippopotamus	Hippopotame Nain
Либерийский лесной дукер	Jentinkducker	Jentink's Duiker	Céphalophe de Jentink
Лиророгая антилопа	Leierantilope	Sassaby	Damalisque
Лось	Elch	Elk (englisch), Moose (amerikanisch)	Elan
Лоси	Elchhirsche	Elk deer	Elans
Лошадиные антилопы	Pferdeböcke	Roan and Sable Antelopes	Hippotraginés
Мазамы	Mazamas	Brockets	Daguets
Малый дик-дик	Kleindikdik	Swayne's Dik-Dik	Dik-Dik de Swayne
Малый куду	Kleiner Kudu	Lesser Kudu	Petit Koudou
Малый однорогий носорог	Javanashorn	Javan Rhinoceros	Rhinocéros de la Sonde
Млезиметакарпальные олени	Echthirsche	Eurasian Deer	Cerf véritables
Мозоленогие	Schwielensohler	Tylopodes	Tylopodes
Мунтжаки	Muntjakhirsche	Muntjacs	Muntjacs
Мускусная кабарга	Moschustier	Musk Deer	Porte-Musc
Мускусный овцебык	Moschusochse	Musk Ox	Bœuf Musqué
Нахур	Blauschaf	Blue Sheep	Bouc Bleu
Нежвачные	Nichtwiederkäuer	Suiformes	Suiformes
Непарнокопытные	Unpaarhufer	Odd-toed Ungulates	Périssodactyles
Носороги	Nashörner	Rhinoceroses	Rhinocérotidés
Нубийский саблерогий сернобык	Säbelantilope	Scimitar-horned Oryx	Oryx algazelle
Окапи Джонстона	Okapi	Okapi	Okapi
Обыкновенная газель	Dorkasgazelle	Dorcas Gazelle	Dorcas
Обыкновенная жирафа	Giraffe	Giraffe	Girafe
Обыкновенный бегемот	Flußpferd	Hippopotamus	Hippopotame Amphibie
Обыкновенный водяной козел	Wasserbock	Waterbuck	Cobe
Обыкновенный тапир	Flachlandtapir	Lowland Tapir	Tapir Terrestre
Овцебыки	Schafochsen	Musk Oxen	Bœufs Musqués
Одногорбый верблюд	Dromedar	Dromedary	Dromadaire
Оленек напу	Großkantschil	Larger Malay Mouse Deer	Chevrotain Malais
Оленьки	Hirschferkel	Chevrotains	Tragulidés
Олень принца Альфреда	Prinz-Alfreds-Hirsch	Alfred's Sambar	Sambar d'Alfred
Ошейниковый пекари	Halsbandpekari	Collared Peccary	Pecari à Collier
Пампасский олень	Pampashirsch	Pampas Deer	Cerf de Pampas
Парнокопытные	Paarhufer	Even-toed Ungulates	Artiodactyles
Пекари	Pekaris	Peccaries	Pecaris
Первобытный бык	Ur	Aurochs	Aurochs
Пиренейский тур	Pyrenäensteinbock	Spanish Ibex	Bouquetin d'Espagne
Плотнорогие	Hirsche	Deer	Cerfs
Полорогие	Hornträger	Horned Ungulates	Bovidés
Пуду	Pudus	Pudus	Pudus
Пятнистый олень	Sikahirsch	Sika Deer	Sika
Пятнистый олень Дыбовского	Dybowskihirsch	Dybowski's Deer	Cerf de Dybowski
Рогатые	Stirnwaffenträger	Pecora	Pécores
Рыжебокий дукер	Rotflankenducker	Red-flanked Duiker	Céphalophe aux Flancs Roux
Рыжелобая газель	Rotstirngazelle	Red-fronted Gazelle	Gazelle à Front Roux
Сайга	Saiga	Saiga Antelope	Saiga
Самбар	Indischer Sambar	Indian Sambar	Sambar de l'Inde
Самбары	Sambarhirsche	Sambars	Sambars
Сахарская газель	Damagazelle	Dama Gazelle	Gazelle Dama
Свиной олень	Schweinshirsch	Hog Deer	Cerf Cochon
Свиные	Schweineartige	Pig-like Mammals	Suoidea
Свиньи Старого света	Schweine, Altweltliche	Old World Pigs	Suidés
Северные олени	Renhirsche	Reindeer	Rennes
Северный олень	Ren	Reindeer	Renne
Северный пуду	Nordpudu	Northern Pudu	Pudu du Nord
Североандский олень	Nordandenhirsch	Peruvian Guemal	Guemal
Североафриканский гривистый баран	Mähnenspringer	Barbary Sheep	Aoudad
Серна	Gemse	Chamois	Chamois
Сернобык	Spießbock	Oryx	Oryx
Серый мазам	Graumazama	Grey Brocket	Daguet Gris

RUSSIAN NAME	GERMAN NAME	ENGLISH NAME	FRENCH NAME
Сиамский олень	Schomburgkhirsch	Schomburgk's Deer	Cerf de Schomburgk
Синг-синг	Defassa-Wasserbock	Defassa-Waterbuck	Cobe Defassa
Снежная коза	Schneeziege	Rocky Mountain Goat	Chèvre des Montagnes Rocheuses
Степной штейнбок	Steinböckchen	Steinbok	Raphicère Champètre
Суматранский двурогий носорог	Sumatranashorn	Sumatran Rhinoceros	Rhinocéros de Sumatra
Суматранский серао	Serau	Serow	Serow
Суни	Suniböckchen	Suni	Suni
Такин	Takin	Takin	Takin
Такины	Rindergemsen	Takins	Takins
Таменг	Leierhirsch	Eld's Deer	Cerf d'Eld
Тапир Бэрда	Mittelamerikanischer Tapir	Central American Tapir	Tapir de Baird
Тапиры	Tapire	Tapirs	Tapirs
Тар	Tahr	Tahr	Tahr
Телеметакарпальные олени	Trughirsche	American and Roe Deer	Odocoiléinés
Тибетский горный олень	Weißlippenhirsch	Thorold's Deer	Cerf de Thorold
Тибетский дзерен	Tibetgazelle	Tibetan Gazelle	Gazelle du Thibet
Толсторог	Dickhornschaf	Bighorn Sheep	Mouflon du Canada
Тора	Tora	Tora	Tora
Филиппинский буйвол	Tamarau	Tamarou	Tamarou
Филиппинский самбар	Philippinensambar	Philippine Sambar	Sambar des Philippines
Хохлатые антилопы	Ducker	Duikers	Céphalophinés, Céphalophes
Хохлатый олень	Schopfhirsch	Tufted Deer	Cerf Touffe
Целебесский карликовый буйвол	Anoa	Anoa	Anoa
Черная антилопа	Rappenantilope	Sable Antelope	Hippotrague Noir
Чернолобый дукер	Schwarzstirnducker	Black-fronted Duiker	Céphalophe à Front Noir
Чернопятая антилопа	Schwarzfersenantilope	Impala	Impala
Черноспинный дукер	Schwarzrückenducker	Bay Duiker	Céphalophe à Bande Dorsale Noire
Черный дукер	Schwarzducker	Black Duiker	Céphalophe Noir
Четырехрогая антилопа	Vierhornantilope	Four-horned Antelope	Tetracère
Чиру	Tschiru	Tibetan Antelope	Antilope du Thibet
Зквитун	Moor-Antilope	Buffon's Kob	Cobe Buffon
Эритрейский дик-дик	Eritreadikdik	Salt's Dik-Dik	Dik-Dik de Salt
Южноазиатские пятнистые олени	Fleckenhirsche	Axis Deer	Cerfs Axis
Южноандский гуэмал	Südandenhirsch	Chilenian Huemul	Hippocamelus
Южноафриканский сернобык	Passan	South African Oryx	Oryx d'Afrique du Sud
Южный полосатый гну	Blaues Gnu	Blue Wildebeest	Gnou Bleu
Южный пуду	Südpudu	Southern Pudu, Pudu	Pudu du Sud
Яванский канчили	Kleinkantschil	Lesser Malay Mouse Deer	Chevrotain Malais
Як	Yak	Yak	Yack

Conversion Tables of Metric to U.S. and British Systems

U.S. Customary to Metric   Metric to U.S. Customary

—— Length ——

To convert	Multiply by	To convert	Multiply by
in. to mm.	25.4	mm. to in.	0.039
in. to cm.	2.54	cm. to in.	0.394
ft. to m.	0.305	m. to ft.	3.281
yd. to m.	0.914	m. to yd.	1.094
mi. to km.	1.609	km. to mi.	0.621

—— Area ——

sq. in. to sq. cm.	6.452	sq. cm. to sq. in.	0.155
sq. ft. to sq. mi.	0.093	sq. m. to sq. ft.	10.764
sq. yd. to sq. m.	0.836	sq. m. to sq. yd.	1.196
sq. mi. to ha.	258.999	ha. to sq. mi.	0.004

—— Volume ——

cu. in. to cc.	16.387	cc. to cu. in.	0.061
cu. ft. to cu. m.	0.028	cu. m. to cu. ft.	35.315
cu. yd. to cu. m.	0.765	cu. m. to cu. yd.	1.308

—— Capacity (liquid) ——

fl. oz. to liter	0.03	liter to fl. oz.	33.815
qt. to liter	0.946	liter to qt.	1.057
gal. to liter	3.785	liter to gal.	0.264

—— Mass (weight) ——

oz. avdp. to g.	28.35	g. to oz. avdp.	0.035
lb. avdp. to kg.	0.454	kg. to lb. avdp.	2.205
ton to t.	0.907	t. to ton	1.102
l. t. to t.	1.016	t. to l. t.	0.984

Abbreviations

U.S. Customary   Metric

U.S. Customary	Metric
avdp.—avoirdupois	cc.—cubic centimeter(s)
ft.—foot, feet	cm.—centimeter(s)
gal.—gallon(s)	cu.—cubic
in.—inch(es)	g.—gram(s)
lb.—pound(s)	ha.—hectare(s)
l. t.—long ton(s)	kg.—kilogram(s)
mi.—mile(s)	m.—meter(s)
oz.—ounce(s)	mm.—millimeter(s)
qt.—quart(s)	t.—metric ton(s)
sq.—square	
yd.—yard(s)	

By kind permission of Walker: Mammals of the World
©1968 Johns Hopkins Press, Baltimore, Md., U.S.A.

TEMPERATURE

AREA

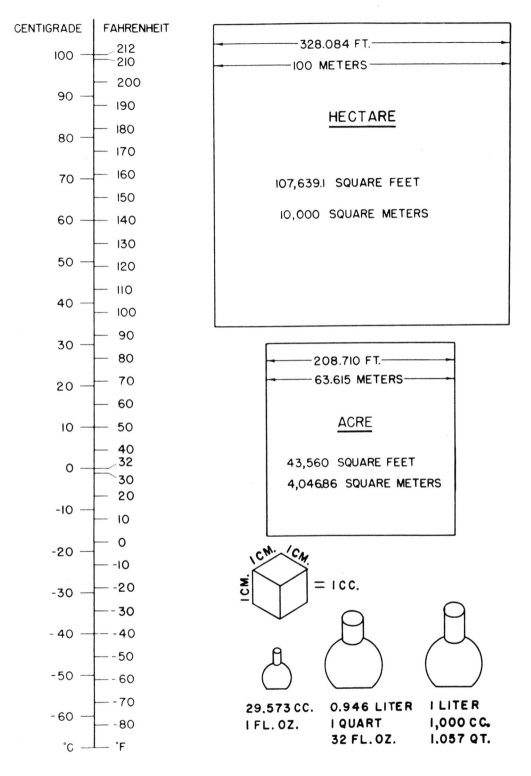

CENTIGRADE	FAHRENHEIT
100	212 / 210
90	200 / 190
80	180 / 170
70	160 / 150
60	140 / 130
50	120 / 110
40	100 / 90
30	80
20	70 / 60
10	50 / 40
0	32 / 30 / 20
-10	10 / 0
-20	-10
-30	-20 / -30
-40	-40 / -50
-50	-60 / -70
-60	-80
°C	°F

328.084 FT.

100 METERS

HECTARE

107,639.1 SQUARE FEET

10,000 SQUARE METERS

208.710 FT.

63.615 METERS

ACRE

43,560 SQUARE FEET

4,046.86 SQUARE METERS

I CM. I CM. I CM. I CM. = I CC.

29.573 CC.
I FL. OZ.

0.946 LITER
I QUART
32 FL. OZ.

I LITER
1,000 CC.
1.057 QT.

WEIGHT

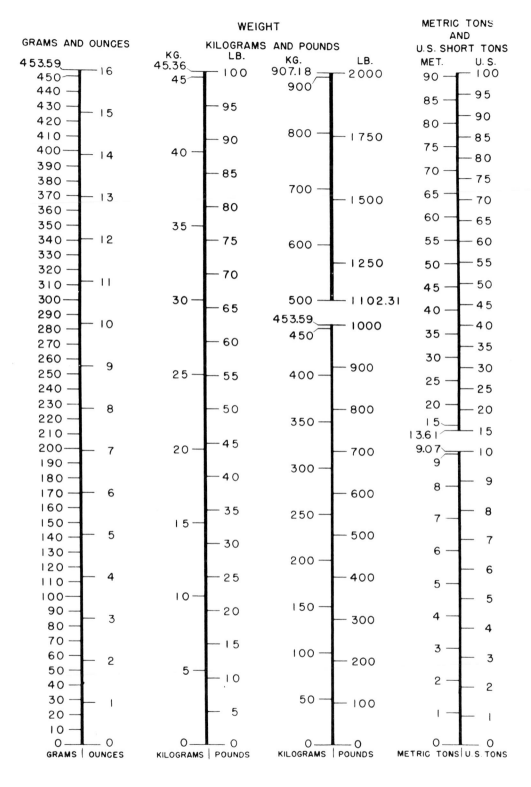

GRAMS AND OUNCES

KILOGRAMS AND POUNDS

METRIC TONS AND U.S. SHORT TONS

LENGTH: MILLIMETERS AND INCHES

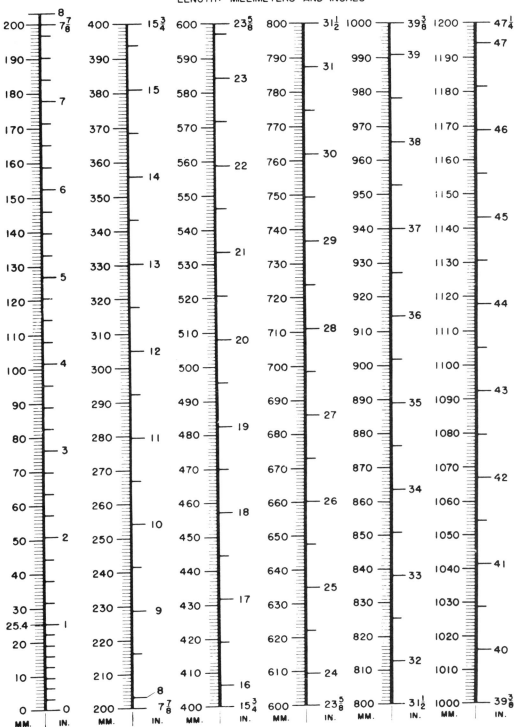

LENGTH

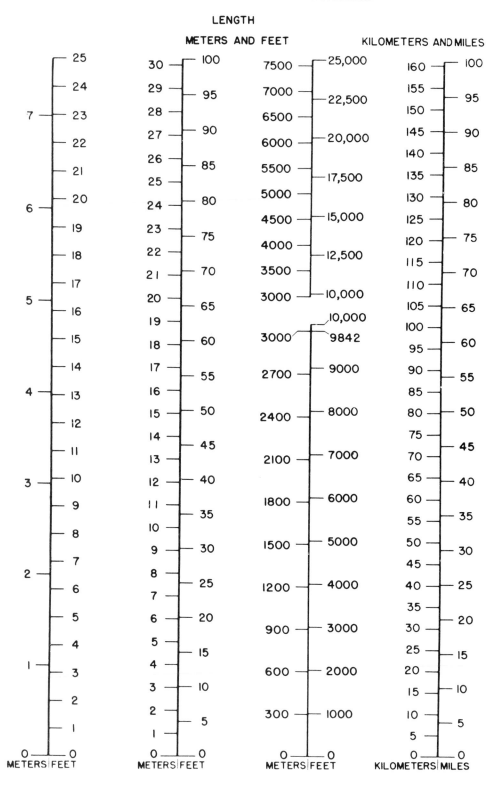

METERS AND FEET

KILOMETERS AND MILES

Supplementary Readings

These references of books and articles published in scientific journals deal with animals and topics that are covered in this volume. Some of these were the original sources on which the content of this book is based. These titles are intended as an aid to readers who are interested in additional information and more detailed coverage of the subjects contained in this book.*

BOOKS:

Aitken, R. B. 1969. *Great Game Animals of the World.* The Macmillan Company, New York.

Annison, E. F. and D. Lewis. 1959. *Metabolism in the Rumen.* Menthuen and Co., Ltd., London and John Wiley and Sons, Inc., New York.

Bere, R. 1970. *Antelopes.* Arthur Baker, London.

Clark, J. L. 1964. *The Great Arc of the Wild Sheep.* University of Oklahoma Press, Norman.

Cloudsley-Thompson, J. L. 1969. *The Zoology of Tropical Africa.* W. W. Norton, New York.

Crowe, P. K. 1970. *World Wildlife: The Last Stand.* Charles Scribner's Sons, New York.

Cullen, E. 1957. *The Migratory Springbucks of South Africa.* Fisher Unwin, London.

Darling, F. F. 1937. *A Herd of Red Deer.* Oxford University Press, London.

Dorst, J. and P. Dandelot. 1970. *A Field Guide to the Larger Mammals of Africa.* Houghton Mifflin, Boston.

Drimmer, Frederick, ed. 1954. *The Animal Kingdom,* Vol. II. Greystone Press, New York.

Egorov, O. V. 1967. *Wild Ungulates of Yakutia.* Clearing House of Scientific and Technical Information, Springfield, Virginia.

Einarsen, A. S. 1947. *The Pronghorn Antelope and its Management.* Wildlife Management Institute.

Fischer, Ralph A., 1957. *The Guide to Javelina.* Naylor Co., San Antonio.

Fitter, R. 1968. *Vanishing Wild Animals of the World.* Franklin Watts, New York.

Fraser, A. F. 1968. *Reproductive Behavior in Ungulates.* Academic Press, London.

French, M. H. *European Breeds of Cattle,* Vols. I & II. F.A.O. Agricultural Studies No. 67, Columbia University Press International Documents Service, New York.

Geist, V. 1971. *The Mountain Sheep: A Study of Behavior and Evolution.* University of Chicago Press, Chicago.

Grzimek, Bernard. *Rhinos Belong to Everybody.* Hill and Wang, New York.

Guggisberg, C. A. W. 1966. *S.O.S. Rhino.* Andre Deutsch, London.

Hafez, E. S. E. 1969. *The Behavior of Domestic Animals,* 2nd ed. Baillière, Tindall and Cassell, London.

Harris, R. A. and K. R. Duff. 1970. *Wild Deer in Britain.* Taplinger, New York.

Hediger, H. 1968. *The Psychology and Behavior of Animals in Zoos and Circuses.* Dover Publications, Inc., New York.

Heller, Edmund. 1913. *The White Rhinoceros.* Smithsonian Institution, Washington.

Hey, D. 1966. *Wildlife Heritage of South Africa.* Oxford University Press, New York.

Hoover, H. 1966. *The Gift of the Deer.* Alfred A. Knopf, New York.

Hungate, R. E. 1966. *The Rumen and its Microbes.* Academic Press, New York.

Kingdon, Jonathan, 1971. *East African Mammals. An Atlas of Evolution in Africa,* Vol. I. Academic Press, New York.

Linsdale, J. and P. Tomich, 1953. *A Herd of Mule Deer.* Berkeley, California.

Lydekker, R. 1893. *Horns and Hoofs.* London.

—. 1898. *The Deer of All Lands.* London.

—. 1913-16. *Catalogue of Ungulate Mammals in the British Museum.* 5 vols. London.

—. 1924. *The Game Animals of India, Burma, Malaya, and Tibet.* London.

McCullough, Dale R. 1969. *The Tule Elk: Its History, Behavior and Ecology.* University of California Press, Berkeley and Los Angeles.

Mochi, Ugo and T. D. Carter. 1953. *Hoofed Mammals of the World.* Charles Scribner's Sons, New York.

Murie, O. 1951. *The Elk of North America.* Washington, D.C.

— and M. Murie. 1966. *Wapiti Wilderness.* Alfred A. Knopf, Inc., New York.

Newsom, W. M. 1926. *White-tailed Deer.* Charles Scribner's Sons, New York.

Olaniyan, C. I. O. 1968. *An Introduction to West African Animal Ecology.* Heinemann Educational Books, London.

Orr, R. T. 1970. *Animals in Migration.* The Macmillan Company, New York.

Owen, D. F. 1966. *Animal Ecology in Tropical Africa.* W. H. Freeman, San Francisco.

Park, E. 1969. *The World of the Bison.* J. B. Lippincott, Co., Philadelphia and New York.

Prior, R. 1968. *The Roe Deer of Cranborne Chase.* Oxford University Press, London.

Roe, Frank G. 1951. *The North American Buffalo.* University of Toronto Press, Toronto.

Roedelberger, Frank A. and Vera I. Groschoff. 1965. *African Wildlife.* (English version by Nieter O'leary and Pamela Paulet). The Viking Press, New York.

Rue, Leonard Lee. 1969. *The World of the White-tailed Deer.* J. B. Lippincott, Philadelphia and New York.

Sadlier, R. M. F. S. 1969. *The Ecology of Reproduction in Wild and Domestic Mammals.* Barnes and Noble, New York.

Schaller, George B. 1967. *The Deer and the Tiger.* University of Chicago Press, Chicago.

Sclater, P. L. and O. Thomas. 1894-1900. *The Book of Antelopes.* 4 vols. R. H. Porter, London

Sidney, J. 1965. *The Past and Present Distribution of some African Ungulates.* Zoological Society of London, London

Smithers, R. H. N. 1966. *The Mammals of Rhodesia, Zambia and Malawi.* Collins, London.

Stümpke, H. *The Snouters: Form and Life of the Rhinogrades.* Doubleday, New York.

Taylor, W. P. 1956. *The Deer of North America: Their History and Management.* Stackpole Co., Harrisburg, Pennsylvania, and Wildlife Management Institute, Washington, D.C.

Van Wormer, J. 1969. *The World of the Pronghorn.* J. B. Lippincott Co., Philadelphia and New York.

Walker, E. P. 1968. *Mammals of the World,* Vol. II. John Hopkins Press, Baltimore.

Wood, F. and D. Wood. 1968. *Animals in Danger: The Story of Vanishing American Wildlife.* Dodd, Mead, New York.

SCIENTIFIC JOURNALS:

Altmann, M. 1960. The Role of Juvenile Elk and Moose in the Social Dynamics of their Species. *Zoologica* 45(4):35-39.

Anthony, H. 1929. Horns and Antlers: Their Occurrence, Development and Function in the Mammalia. *New York Zoological Society Bulletin* 32:2-33.

Bramley, P. S. 1970. Territoriality and Reproductive Behavior of Roe Deer. *Journal of Reproductive Fertility* 11:43*.

Bromley, Peter T. 1969. Territoriality in Pronghorn Bucks on the National Bison Range, Moiese, Montana. *Journal of Mammalogy* 50(1):81-89.

Brooks, A. C. 1961. A Study of Thomson's Gazelle (*Gazella thomsoni* Günther) in Tanganyika. *Colonial Research Publication,* No. 25, Her Majesty's Stationary Office, London.

Buechner, H. K. 1960. The Bighorn Sheep in the United States. *Wildlife Monographs,* No. 4. The Wildlife Society, Washington, D.C.

—. 1961. Territorial Behavior in Uganda Kob. *Science* 133:698-99.

—. 1963. Territoriality as a Behavioral Adaptation to Environment in Uganda Kob. *Proceedings of the 16th International Congress of Zoology* 3:59-63.

Coe, M. J. 1967. "Necking" Behavior in the Giraffe. *Journal of Zoology, London* 151:313-321.

Dagg, A. I. 1968. External Features of Giraffe. *Mammalia* 32:657-669.

—. Tactile Encounters in a Herd of Captive Giraffes. *Journal of Mammalogy* 51(2):279-287.

Dasmann, R. F. and A. S. Mossman. 1962. Population Studies of Impala in Southern Rhodesia. *Journal of Mammalogy* 43:375-395.

— and—. 1962. Reproduction in some Ungulates in Southern Rhodesia. *Journal of Mammalogy* 43(4):533-37.

De Vos, A. 1965. Territorial Behavior Among Puku in Zambia. *Science* 148:1752-53.

Espmark, Y. 1964. Rutting Behavior in Reindeer (*Rangifer tarandus* L.). *Animal Behavior* 12(1):159-63.

Estes, R. D. 1966. Behavior and Life History of the Wildebeest. *Nature* 212:999-1000.

—. 1967. The Comparative Behavior of Grant's and Thompson's Gazelles. *Journal of Mammalogy* 48:189-209.

—. 1969. Territorial Behavior of the Wildebeest (*Connochaetes taurinus* Burchell, 1823). *Zeitschrift für Tierpsychologie* 26:284-370.

Etkin, W. 1954. Social Behavior of the Male Blackbuck under Zoo Conditions. *Anatomical Record* 120(3): Abstract No. 81.

Foster, J. B. 1966. The Giraffe of Nairobi National Park: Home Range, Sex Ratios, the Herd, and Food. *East African Wildlife Journal* 4:139-148.

Gee. E. 1953. The Life History of the Great One-horned Rhinoceros (*Rhinoceros unicornis* Linn.). *Journal of the Bombay Natural History Society* 51(2):341-48.

Geist, V. 1963. On the Behavior of the North American Moose (*Alces alces andersoni* Peterson, 1950). *Behaviour* 20(3-4):377-416.

—. 1964. On the Rutting Behavior of the Mountain Goat. *Journal of Mammalogy* 45(4):551-68.

—. 1964. Social Mechanisms involved in Habitat Retention of Stone's Sheep (*Ovis dalli stonei*). *American Zoologist* 4(3).

Goddard, J. 1966. Mating and Courtship of the Black Rhinoceros (*Diceros bicornis* L.). *East African Wildlife Journal* 4:69-75.

Graf, W. 1956. Territorialism in Deer. *Journal of Mammalogy* 37(2):165-70.

Heppes, J. B. The White Rhinoceros in Uganda. *African Wildlife* 12:273-280.

Hershkovitz, Philip. 1956. Mammals of Northern Columbia, Preliminary Report No. 7: Tapirs. *Proceedings of the United States National Museum* (Vol. 103) 1956.

Hosley, N. W. 1949. The Moose and its Ecology. *U.S. Fish and Wildlife Service,* Leaflet 312.

Hubback, T. 1937. The Malayan Gaur or Seladang. *Journal of Mammalogy* 18(3):267-79.

Hunsaker, Don F. and Thomas C. Hahn. 1965. Vocalizations of the South American Tapir, *Tapirus terrestris. Animal Behavior* 13(1):69-73.

Innis, A. C. 1958. The Behavior of the Giraffe, *Giraffa camelopardalis,* in the Eastern Transvaal. *Proceedings of the Zoological Society of London* 131:245-278.

Kelsall. 1968. The Migratory Barren-Ground Caribou of Canada. *Canadian Wildlife Service Monograph* 3. The Queen's Printer, Ottawa, Canada.

Kiley-Worthington, M. 1965. The Waterbuck (*Kobus defassa* Ruppel, 1835 and *Kobus ellipsiprymnus* Ogilby, 1933) in East Africa. Spatial Distribution. A Study of Sexual Behavior. *Mammalia* 29:177-204.

Koford, C. 1957. The Vicuna and the Puna. *Ecology Monographs* 27:153-219.

Laws, R. M. and G. Clough. 1966. Observations on Reproduction in the Hippopotamus (*Hippopotamus amphibius* Linn.). *Symposia of the Zoological Society of London,* No. 15:117-140.

Lent, P. 1965. Rutting Behavior in a Barren-Ground Caribou Population. *Animal Behavior* 13(2-3):259-64.

Leuthold, W. 1966. Variations in Territorial Behavior of Uganda Kob, *Adenota kob thomasi* (Newman, 1896). *Behavior* 28:215-258.

—. 1970. Observations on the Social Organization of Impala (*Aepyceros melampus*). *Zeitschrift für Tierpsychologie* 27(6):693-721.

McCabe, R. and A. Leopold. 1951. Breeding Season of the Sonora White-tailed Deer. *Journal of Wildlife* Management 15(4):433-34.

McHugh, T. 1958. Social Behavior of the American Buffalo (*Bison bison bison*). *Zoologica* 43(1):1-40.

Ogilvie, C. 1953. The Behavior of the Seladang (*Bibos gaurus*). *Oryx* 2:167-69.

Packard, R. L. 1955. Release, Dispersal and Reproduction of Fallow Deer in Nebraska. *Journal of Mammalogy* 36:471-73.

Powell, A. 1964. The Gaur or Indian Bison. *Journal of the Bengal Natural History Society* 32(2):73-80.

Prakash, I. 1960. Shikar in Rajasthan. *The Cheetal* 2(2):68-72.

Prater, S. 1945. Breeding Habits of Swamp Deer (*Rucervus duvaucelli*) in Assam. *Journal of the Bombay Natural History Society* 45(3):415-16.

Pruitt, W. O. 1954. Rutting Behavior of the White-tailed Deer (*Odocoileus virginianus*). *Journal of Mammalogy* 35:129-130.

—. 1960. Behavior of the Barren-Ground Caribou. *Biology Papers of the University of Alaska,* No. 3, College, Alaska.

Pucek, Zdzislaw, ed. 1967. The European Bison, Current State of Knowledge and Need for Further Studies. Proceedings of the 2nd Symposium, Mammal Section, The Polish Zoological Society. Reprinted from *Acta Theriologica* 12:323-501.

Schenkel, R. 1966. On Sociology and Behavior in Impala (*Aepyceros melampus suara* Matschi). *Zeitschrift für Säugetierkunde* 31:177-205.

— and Schenkel-Hulliger. 1969. Ecology and Behavior of the Black Rhinoceros (*Diceros bicornis* L.): A Field Study. *Mammalia Depicta.* Paul Parey, Hamburg.

Talbot, L. and M. Talbot. 1963. The Wildebeest in Western Masailand, East Africa. *East African Wildlife Monographs,* No. 12.

Walther, Fritz R. 1969. Flight Behavior and Avoidance of Predators in Thomson's Gazelle (*Gazella thomsoni* Günther, 1884). *Behavior* 34:184-221.

Wharton, C. 1957. An Ecological Study of the Kouprey, *Novibus sauveli* (Urbain). *Monographs of the Institute of Science and Technology* (Manila), No. 5.

*Supplementary Readings prepared by John B. Brown.

Picture Credits

Animal painters: H. Diller (pp. 346, 469, 482, 488, 498). W. Eigner (pp. 61/62, 93, 94, 263, 285/286, 342, 343, 344, 345, 391/392, 407, 463, 464, 481, 487, 497). E. Hudecek-Neubauer (pp. 31, 32, 264, 270). F. Reimann (pp. 163/164). F. Walther (pp. 302, 401, 402, 408, 426, 438). W. Webber (pp. 111, 132, 133, 134, 209-224, 262, frontispiece). R. Zieger (pp. 22, 37, 92, 341). Scientific advisors of the animal painters: Prof. H. Dathe (Zieger, Reimann), Dr. Th. Haltenorth (H. Diller), Prof. Heck (Weber, pp. 134, lower 209-224), Dr. D. Heinemann (Eigener, Weber, pp. 111, 132, 133), Prof. E. Thenius (Hudecek-Neubauer, Weber, p. 134 upper). Color photos: Bavaria (pp. 351, 352/353). Grzimek (pp. 21, 40, 114, 190, 192, 299, 354/355, 356, 413, 416, 508). Klages (p. 507). Miller/Collignon (pp. 188/189). E. Müller (pp. 424/425). Myers/PIP (pp. 38/39, 112/113). Paysan (p. 131). Quraishy/PIP (p. 261). Root/Grzimek (pp. 300/301). Six/PIP (p. 91 upper). Tönges (p. 91 lower, 191). Tönges/PIP (pp. 185, 186/187, protective cover). Trevor/PIP (pp. 414/415, 423, 436/437). Line drawings: J. Kühn (maps of distributions); Bannikow, Heptner, and Nasimowitsch (Fig. 13-10, 13-11, 13-12); J. Boessneck (Fig. 13-5); R. Geigy (Fig. 4-10); G. Haas (Fig. 17-11, 17-12); by Haltenorth/Trense (Fig. 8-1); E. M. Lang (Fig. 2-2). E. Thompson Seton (Fig. 10-2); F. Walther (Fig. 6-7, 6-8; 11-1 through 11-26; 12-15 through 12-21, 12-24; 14-3, 14-7; 15-5, 15-7 through 15-11; 16-5, 16-6; 17-5, 17-19, 17-21, 17-22); E. Diller (all others; most from sketches of the authors).

Index

Abbreviations and Symbols

C, °C . . . . Celsius, degrees centigrade

C.S.I.R.O. . . . Commonwealth Scientific and Industrial Res. Org. (Australia)

f . . . . following (page)

ff . . . . following (pages)

L . . . . total length (from tip of nose [bill] to end of tail)

I.R.S.A.C. . . . Institute for Scientific Res. in Central Africa, Congo

I.U.C.N. . . . . Intern. Union for Conserv. of Nature and Natural Resources

BH . . . . body height

HRL . . . . . . head-rump length (from nose to base of tail or end of body)

N, N- . . . . . . North, Northern, North-

NE, NE- . . . . Northeast, Northeastern, Northeast-

E, E- . . . . East, Eastern, East-

S, S- . . . . South, Southern, South-

TL . . . . tail length

SE, SE- . . . . . Southeast, Southeastern, Southeast-

SW, SW- . . . Southwest, Southwestern, Southwest-

W, W- . . . . . West, Western, West-

♂ . . . . male

♂♂ . . . . males

♀ . . . . female

♀♀ . . . . females

♂♀ . . . . pair

✝ . . . . extinct

$\frac{2 \cdot 1 \cdot 2 \cdot 3}{2 \cdot 1 \cdot 2 \cdot 3}$ . . . tooth formula, explanation in Volume X

▷ . . . . following (opposite page) color plate

▷▷ . . . . Color plate or double color plate on the page following the next

▷▷▷ . . . . Third color plate or double color plate (etc.)

) . . . . Endangered species and subspecies